e-portals Ltd
Nov 01

MONTE CARLO

Methodologies and Applications for Pricing and Risk Management

MONTE CARLO

Methodologies and Applications for Pricing and Risk Management

RISK BOOKS

Published by Risk Books, a specialist division of Risk Publications.

Haymarket House
28–29 Haymarket
London SW1Y 4RX
Tel: +44 (0)171 484 9700
Fax: +44 (0)171 930 2238

ISBN 1 899 322 86 3 (case bound)
ISBN 1 899 332 91 X (softback)

British Library Cataloguing in Publication Data
A catalogue record for this book is available from the British Library

Risk Books Commissioning Editor: Laurie Donaldson
Desk Editor: Ben Mullane
Copy Editor: Romilly Hambling
Typesetter: Marie Doherty

Printed and bound in Great Britain by Bookcraft.
Covers printed by Bookcraft.

The introduction to each section and Chapters 3, 4, 12, 13, 17 and 22 are published here for the first time. The remaining chapters are slightly revised versions of articles previously published in the following journals:

Chapter 1 *Journal of Financial Economics* 4 (1977); Chapter 2 *Journal of Economic Dynamics and Control* 21 (1997); Chapter 5 *Risk* (February, 1992); Chapter 6 *Risk* (November, 1992); Chapter 7 *Risk* (July, 1993); Chapter 8 *Risk* (May, 1996); Chapter 9 *Risk* (March, 1994); Chapter 10 *Risk* (May, 1994); Chapter 11 *Risk* (October, 1996); Chapter 14 *Transactions of the Society of Actuaries* XLV (1993); Chapter 15 *Journal of Financial and Quantitative Analysis* (1995); Chapter 16 *Journal of Derivatives* (Fall, 1997); Chapter 18 *Journal of Fixed Income* 5 (1995); Chapter 19 *Journal of Fixed Income* 2 (1992); Chapter 20 *Risk Management for Financial Institutions*, Risk Books (1997); Chapter 21 *VAR: Understanding and Applying Value-at-Risk*, Risk Books (1997); Chapter 23 *VAR: Understanding and Applying Value-at-Risk*, Risk Books (1997); Chapter 24 *Management Science* 42 (1996); Chapter 25 *Mathematics of Derivative Securities*, Cambridge University Press (1997); Chapter 26 *Journal of Computational Finance* 1 (1997); Chapter 27 *Journal of Mathematical Computer Modelling* 23 (1996); Chapter 28 *Risk* (June, 1996); Chapter 29 *Risk* (December, 1994); Chapter 30 *Risk* (December, 1995); Chapter 31 *Risk* (October, 1992).

CONTENTS

AMERICAN-STYLE

FIXED INCOME

VAR

DETERMINISTIC METHODS

AUTHORS

Leif Andersen is a vice president in General Re Financial Products (GRFP), a global dealer in fixed income, foreign exchange, equity, and credit derivatives. He has been with GRFP since 1993 and is responsible for developing pricing and risk management models for all areas of their business. Leif holds MS degrees in electrical and mechanical engineering from the Technical University of Denmark, an MBA from the University of California at Berkeley, and a PhD in finance from the Aarhus School of Business.

Jérôme Barraquand is a vice president in the Arbitrage Trading Group at Salomon Smith Barney. He devises model-based trading strategies for European fixed income and equity derivative products. Jerome was previously head of European fixed income derivatives research at Salomon Brothers. Prior to joining Salomon in 1994, he was head of the Computational Finance Group at Digital's Paris Research Laboratory. The group developed computational methods for complex derivative pricing and dynamic portfolio allocation in multi-factor economies. Jerome also worked as a research associate at Stanford University in the computer science department from 1988 until 1990. He holds a PhD in computer science from the University of Nice-Sophia-Antipolis, with his undergraduate degree being from Ecole Polytechnique, both in France.

John W. Barrett is reader in numerical analysis at the Department of Mathematics of Imperial College, London. His main research interest is non-linear partial differential equations and their numerical approximation by finite difference and finite element methods. John has published over 50 research papers in this area, and is an associate editor of the IMA *Journal of Numerical Analysis*.

Professor Phelim P. Boyle moved to the University of British Columbia in Vancouver after gaining his degree from Queen's University and Trinity College, and subsequently joined the University of Waterloo. He currently holds the J. Page R. Wadsworth Chair of Finance in the School of Accountancy and is the director of the Centre for Advanced Studies in Finance. Professor Boyle has been an invited professor at leading universities in North America, Europe, Australia and Japan. He is an active researcher in the valuation of derivative securities, specialising in numerical valuation methods and computational finance, and his work has appeared in all leading finance journals. He is the finance editor of *Management Science*, and has been instrumental in establishing an advanced masters program in quantitative finance at the University of Waterloo.

Mark Broadie is a professor at the Graduate School of Business at Columbia University. He received a BS from Cornell University and his PhD from Stanford University. Professor Broadie's research focuses on problems in risk management, the pricing of derivative securities, and portfolio optimisation, much of it looking at the effective use of Monte Carlo methods for pricing and risk management. Professor Broadie is an associate editor of *Management Science*, *Mathematical Finance*, and a founding co-editor of the *Journal of Computational Finance*. He has done extensive consulting for financial firms and previously was a vice president at Lehman Brothers in their fixed income research group.

Rupert Brotherton-Ratcliffe is director of research at General Re Financial Products based in New York, where he has worked since the company was formed in 1990. His responsibilities include the development of quantitative techniques for pricing interest rate, foreign exchange and equity derivatives, and credit and market risk measurement. Prior to GRFP, Rupert worked for Manufacturers Hanover Trust in New York and London. He earned a PhD in fluid dynamics from London University in 1986 and an MA in mathematics from University of Cambridge.

Russel Caflisch is a professor of mathematics at UCLA and a research scientist with NumeriX LLC, a technology company offering accelerated computational tools for pricing and derivative risk management. He has been on the faculty of Stanford University and New York University, and was a Sloan Research Fellow. Russel's research fields include epitaxial crystal growth, fluid dynamics and partial differential equations, as well as computational finance. Much of his work has focused on Monte Carlo methods, especially quasi-Monte Carlo methods, and their application to finance. Russel has been a principal speaker at many conferences and professional courses on Monte Carlo methods and computational finance.

Andrew Carverhill is a member of the Department of Finance, University of Science and Technology, Hong Kong. Prior to this, he spent four years in the Financial Options Research Centre, University of Warwick, England. Andrew has also worked for a short time at the Bank of England, and spent a year at Indiana University at Bloomington. Before entering finance in 1987, he was a research mathematician for a number of years. Andrew's current research includes interest rate dynamics, efficient algorithms for computational evaluation of derivatives and dynamic equilibrium in financial markets.

Oren Cheyette is senior manager of fixed income research at BARRA, a California-based provider of analytical models, software, consulting and money management services. His responsibilities include design and development of valuation and risk models for interest rate and currency products. Oren has been at BARRA since July 1994. Prior to joining BARRA, he was senior vice president of quantitative research at Capital Management Sciences. His experience includes development of trading and portfolio management software at BARRA and CMS, bond portfolio management for home insurance, and fixed income R&D at County NatWest Government Securities and Merrill Lynch. Oren has written several technical and non-technical articles on interest rate models, derivatives valuation and mortgage-backed securities. His academic background is in theoretical physics. Oren received his PhD in 1987 from the University of California at Berkeley and his AB from Princeton in 1981.

Dr Les Clewlow is an associate research fellow with the Financial Options Research Centre, University of Warwick, the Centre for Financial Mathematics, Australian National University and the School of Finance and Economics, University of Technology Sydney. He is a principal of Lacima Consultants, implementing derivatives models for financial institutions and software companies. Les has a first class honours degree in physics with astrophysics (Birmingham University), and a PhD in computer science (University of Warwick). He has recent publications in *Risk* magazine, *Journal of Fixed Income*, *Journal of Derivatives*, *European Journal of Finance* and *Journal of Economic Dynamics and Control*. Les has presented papers at leading academic and practitioner conferences and taught derivatives courses in Europe, North and South America, Asia and Australia.

Michael Curran has worked as a quant for a variety of banks and investment houses since 1990. His work has focused on numerical methods, much of which has been published in academic journals, industry magazines and books. Prior to his working career, Mike obtained a bachelor degree in applied mathematics from the University of Virginia, a masters degree in operations research from Clemson University and conducted doctoral studies at Columbia Business School. He now works for RiskCare, Europe's largest derivatives-based consultancy.

Raphaël Douady is head of the mathematical finance division of the Centre for Applied Mathematics, Ecole Normale Supérieure, Cachan, France, and a consultant for both CIBC and Murex. He was previously co-head of research on fixed income and foreign exchange market activities at Société Générale, Paris. A former student of Ecole Normale Supérieure, Raphaël holds a PhD in mathematics and has 15 years research experience in pure mathematics, including chaos theory and dynamical systems. He has taught dynamical systems in many universities across the world, including St Petersburg, Berkeley and Rio de Janeiro. Between 1995 and 1997 he was visiting professor at New York University, where he created the Courant Institute Seminar of Mathematical Finance. Raphaël's professional experience spans 10 years, during which he has been a consultant for the space industry, several trading rooms and software companies, including NatWest (London) and KPMG (Paris). He has also been invited to speak at various mathematical finance conferences (Princeton, New York) on interest rate models on the calibration of complex models.

Dr Bruno Dupire heads the quantitative research team at Nikko Financial Products. Prior to the last 10 years in finance, he studied mathematics at the Ecole Normale Suprieure and obtained his masters degree in artificial intelligence and his PhD in numerical analysis. Dr Dupire was later involved in biology modelling and neural networks, which he applied to finance as early as 1988. After heading the derivatives research teams at Socit Gnrale (two years in Paris) and at Paribas (more than four years in London), he joined Nikko Financial Products in London in September 1997. Dr Dupire is best known for his research into stochastic volatility, the pricing of exotic options in the presence of volatility smiles and improvements to Monte Carlo simulations.

Paul Glasserman is a professor in the Columbia University Graduate School of Business. He earned an AB in mathematics from Princeton University in 1984 and a PhD in applied mathematics from Harvard University in 1988. From 1988 to 1991, Paul was with AT&T Bell Laboratories, and in 1991 he joined the Columbia faculty. His research addresses modelling and computational issues in option pricing and risk management. Paul's most recent projects are in the areas of term structure models and the development of efficient Monte Carlo techniques. He has published several articles in academic journals and two books, including one on derivative estimation by simulation. Professor Glasserman is a recipient of a National Young Investigator Award from the US National Science Foundation, of the Erlang Prize in Applied Probability, the TIMS Outstanding Simulation Publication Award, and a University Partnership Award from IBM. He currently serves on the editorial board of the *Journal of Computational Finance*.

Corwin Joy is a principal at Positron Energy Consulting, where he heads a group that specialises in quantitative tools and software for the deregulating energy industry. Prior to founding Positron in 1996, he was a senior researcher at Enron under the direction of Vince Kaminski, where he worked on sophisticated VAR tools, credit risk engines, and derivative pricing models for Enron's commodity trading division. Corwin received an MS in applied mathematics from Carnegie-Mellon University in 1993, and a BA in mathematics and statistics from Rice University in 1991.

Edmond Levy is an assistant director in the Specialised Derivatives Group at HSBC MIDLAND, where he is responsible for the development and risk management of foreign exchange exotic options and currency-related structured products. The group covers a wide range of structured equity and capital market products. Prior to this he was with BZW-Barclays Bank and with Nomura Bank International. Edmond previously lectured in finance at Liverpool University and econometrics at Southampton University, where he received his PhD.

David Loebell is a vice president at Chase Manhattan and head of a software development team focused on the risk management of exotic products. He has been with the bank since 1995, in a role which bridges the gap between research and technology. Software development in this area has been centred around performing risk calculations in a distributed computing environment and building data models to store this risk for future analysis. David gained his BSc (hons) in computer systems engineering from the University of Warwick in 1989.

Steve Marshall is vice president, trading structured interest rate derivatives at Chase Manhattan Bank, where he first joined as head of the Capital Markets Derivatives Technology Group before the merger with Chemical Bank. Prior to this, he worked on technology and research projects within the swaps and interest rate options groups at JP Morgan in London and New York. Steve graduated from London University in 1992, where he gained a BSc in mathematics and computing.

Didier Martineau is a strategist at Long-Term Capital Management (UK) in London. He has been with the fund since 1994, and his responsibilities include market risk management and market strategy analyses. Didier received a masters degree in engineering from the Ecole Polytechnique while living in his native France.

Gerald Moore is reader in numerical analysis at the Department of Mathematics of Imperial College, London, having previously worked at Brunel, Dublin City, Reading, and Bath universities. His research interests include the construction and analysis of efficient algorithms for many types of computational problem, but especially for those arising from dynamical systems, non-linear equations and bifurcation theory, stochastic differential equations, optimisation and linear algebra.

William Morokoff is an associate at Goldman, Sachs & Co, in the firmwide risk department, where his responsibilities include developing covariance estimation methods and modelling emerging market risks. His work in finance has been focused on developing numerical methods, including quasi-Monte Carlo methods for option pricing, error analysis for stochastic simulations, and missing data algorithms for likelihood estimation. Prior to joining Goldman, William held a National Science Foundation postdoctoral fellowship in mathematics, during which he worked on kinetic theory and simulation of many particle systems with applications to microelectronics and space flight. He holds a PhD in mathematics from the Courant Institute at New York University.

Bradley Moskowitz is a senior mathematician at Bettis Laboratory, a government research facility located near Pittsburgh, Pennsylvania, where he is involved in Monte Carlo methods development work. Prior to working at Bettis he attended the University of California, Los Angeles, as a graduate student in the applied program of the department of mathematics. Bradley received his PhD from the university in 1993, completing his thesis on the application of quasirandom sequences to Monte Carlo methods. His research has focused on identifying ways to improve the performance of practical Monte Carlo methods by applying quasirandom (low discrepancy) sequences.

Art Owen is a professor of statistics at Stanford University. His recent research has centred on statistical approaches to high-dimensional numerical problems, including integration and approximation. Art is the inventor of scrambled nets and other methods for numerical integration over high-dimensional spaces (scrambled nets are at least as accurate as low-discrepancy methods, sometimes much more accurate, and allow sample-based estimates of accuracy). He is an editor of *Statistical Science* and an associate editor of the *Journal of Computational Finance*. Art teaches doctoral applied statistics and consults on financial and industrial problems.

Kin Pang is working in fixed income research at Morgan Stanley Dean Witter & Co in London. His primary responsibilities include the pricing and modelling of interest rate and credit derivatives. Kin studied for his PhD at the Financial Options Research Centre, University of Warwick. He also has a MSc in economics and finance from University of Warwick and an MA in engineering and management studies from the University of Cambridge.

Anargyros Papageorgiou is a research scientist in the Department of Computer Science at Columbia University. He has been studying the use of quasi-Monte Carlo methods for problems of mathematical finance since 1995. Among his interests is the use of quasi-Monte Carlo for Value-at-Risk calculations. His responsibilities include the Finder project, which develops software using low-discrepancy sequences for the efficient solution of problems in finance. He received his PhD from Columbia University in 1990.

Spassimir Paskov is a vice president in the Global Market Risk Management Department at Barclays Capital, New York. His responsibilities encompass model validation for derivative products and development of market risk management techniques, methodologies and systems. Prior to joining Barclays Capital, Spassimir designed the methodology and implementation of the Value-at-Risk Global Risk Management System at UBS. He has published many papers in leading scientific and professional journals in the area of numerical analysis. Spassimir research interests and expertise include the application of the low-discrepancy sequences for pricing and risk management of complex financial instruments. He holds PhD and MS degrees in computer science from Columbia University, and received his BS and MS degrees in mathematics from the University of Sofia, Bulgaria.

Dr Evan Picoult is a managing director and head of the Global Markets Risk Analytics unit of Citibank. The unit's responsibilities include validating models, managing Citibank's Global Market Risk system and developing and implementing methods for measuring the market and counterparty risks of trading. He is a frequent lecturer on risk topics. Dr Picoult joined Citibank in 1980 and has worked in internal risk management since 1988. He is currently co-head of ISDA's North American Risk Management Committee. Evan has a PhD in experimental particle physics and an MBA in finance, both from Columbia University.

Gerald Quinlan is a financial engineer at TrueRisk Inc in Toronto, where he works on simulation algorithms for measuring market risk and credit risk, and tree-based algorithms for pricing convertible bonds. Prior to joining TrueRisk in 1997, he held several academic positions, where he did research in theoretical astrophysics. He has a BS degree from McGill University and a PhD from Cornell University, both in physics.

Antoine Savine is a vice president at General Re Financial Products in London, where he has been responsible for quantitative research on fixed income, equity and credit derivatives since August 1998. Prior to this position, he created and ran the derivatives research activity in Nikko, with Bruno Dupire and Guillaume Blacher. Antoine is a former student from ESSEC and holds a DEA (French diploma equivalent to a PhD) in financial mathematics and probability theory. He started his career at Paribas, where he worked for three years in the late Swaps and Options Research Team, on the fixed income side.

Julian Shaw, until recently the director for derivatives risk at BZW, is head of risk policy at NatWest Markets in London. Prior to that, he was vice president for new products in the capital markets division of CIBC Wood Gundy in New York and Toronto and, before that, the quantitative analyst for Gordon Capital in Toronto.

Craig Stetson is manager, analytics and structured products trading, at Arizona Public Service (APS), the operator of the Palo Verde Nuclear Station, the delivery point of the first Nymex electricity future. His responsibilities include the valuation and hedging of non-standard electricity products, risk management of real assets, and analytic support to gas, coal, and allowance trading. Prior to joining APS, Craig was employed at Chase Manhattan Bank in a variety of quantitative research roles, including senior derivatives research analyst, Europe, and global manager of derivatives research IT. His education includes doctoral fellowships at Duke University and the University of Florida, a master's degree from Harvard University in public policy (quantitative analysis), and an undergraduate degree in mathematics from Haverford College. In addition to his research in Monte Carlo methods, Craig has also contributed a paper to *Risk* on Markov Chain models for multi-asset Bermudan options.

Chris Strickland holds associate research positions at the Financial Options Research Centre, University of Warwick, the University of Technology Sydney, the Centre for Financial

Mathematics at the Australian National University, and the Instituto de Estudios Superiores de Administration in Caracas, Venezuela. He is also a director of Lacima Consultants, as a financial consultant implementing pricing and hedging models and numerical algorithms for financial institutions and software companies. Chris has a first class honours degree in pure mathematics from Liverpool University, an MSc in mathematics from the University of Warwick, where he is currently completing his PhD in finance. He has recent papers in the *Review of Futures Markets*, the *Journal of Fixed Income*, and the *European Journal of Finance*, as well as writing many chapters in derivatives books, and co-authoring and editing other books. Chris' research interests include advanced models for the term structure of interest rates, including stochastic volatility, pricing energy derivatives, pricing and hedging exotic options, introducing derivatives into developing markets and applying numerical algorithms to finance.

Ken Seng Tan is an assistant professor in the mathematics faculty at the University of Waterloo. He holds a PhD from the University of Waterloo. With Professor Phelim Boyle, Dr Tan provides consulting services to financial institutions and other organisations on the valuation of complex derivatives.

Domingo Tavella is president and CEO of Align Risk Analysis Inc, a company of consultants specialising in corporate and institutional risk management. Previously, he was with Bankers Trust Securities, where he pioneered the use of analytical techniques in the prediction of mergers and acquisitions. Dr Tavella is the founder and chief editor of the *Journal of Computational Finance*. Dr Tavella holds a PhD in aeronautical engineering from Stanford University and an MBA in finance from UC Berkeley.

James A. Tilley is the managing director, CIO of Information Technology at Morgan Stanley. In March of 1997, prior to being named CIO, he was the head of securities, asset gathering and finance area in the Application Products Group (AGP). In September 1995, Jim became head of the Insurance Capital Markets Group, a joint venture among Morgan Stanley's equity, fixed income, and investment banking divisions. Prior to this, he was head of fixed income research worldwide. Mr Tilley is a graduate of McGill University (BSc honours in physics) and Harvard University (PhD in physics). He joined Morgan Stanley in 1983, having previously been employed by Sun Life of Canada, John Hancock, and Equitable Life Assurance Society. Jim is a fellow of the Society of Actuaries, and served on its board of governors from 1984 to 1986. He is recognised internationally as an expert in asset/liability management and has published prize-winning papers on strategic asset allocation, the pricing of interest-contingent claims, and the valuation of American options. He has been involved with the insurance industry for more than 20 years, and has created several innovative instruments and transactions involving the aggregate issuance of securities in excess of $5 billion.

Joseph F. Traub is the Edwin Howard Armstrong Professor of Computer Science at Columbia University and external professor at the Santa Fe Institute. Starting in 1992, his former PhD student, Dr Spassimir Paskov, showed that contrary to conventional wisdom quasi-Monte Carlo methods were superior to Monte Carlo methods for the valuation of financial instruments. Professor Traub's research group is continuing to conduct research and develop software for financial computations. He was executive head of the Computer Science Department at Carnegie-Mellon University between 1971 and 1979, and founding chairman of the Computer Science Department at Columbia University between 1979 and 1989. Professor Traub was founding chair of the Computer Science and Telecommunications Board of the National Academy of Sciences, 1986–92.

Stuart M. Turnbull is Bank of Montreal Professor of Banking and Finance, Queen's University (Canada), and a research fellow, Institute for Policy Analysis (Toronto). He is a graduate of the Imperial College of Science and Technology (London) and the University of British Columbia. Stuart is the author of *Option Valuation* and, with Robert Jarrow, *Derivative Securities*. He has published over 30 articles in major finance and economic journals, and in law and economics journals, as well as many articles in practitioner journals. Professor Turnbull's current research interests include the pricing of credit derivatives, exotic options, risk management and asset pricing theory. He is an associate editor of *Mathematical Finance*, *NetExposure*, the *International Journal of Theoretical and Applied Finance*, and has served as an associate editor of the *Journal of Finance*.

Paul Wilmott is a Royal Society University research fellow at the University of Oxford and Imperial College, London. He is the author of three books on derivatives and over 60 research papers on mathematical modelling and finance. Paul is the leader of the Mathematical Finance Group at Oxford and the academic director of their part-time diploma in mathematical finance. He is a founding editor-in-chief of *Applied Mathematical Finance* and an associate editor of other journals. Paul is a director of two software companies, partner in a publishing firm and in a training firm. He consults to many banks through Wilmott Associates. He lives in Surrey, London and Buenos Aires.

Introduction

Bruno Dupire
Nikko Financial Products

People do not usually like randomness. Randomness is associated with uncertainty, chance (this, interestingly, is translated in French as "hasard" – which is close to hazard), or risk (which is potential pain). However, it is commonly said that any risk can be turned into an opportunity, and in fact randomness, often seen as an evil to be eliminated, can be turned into an operational technique. In more scientific terms, by estimating an expectation as an average over many random realisations, it can be used to compute an integral (as in the central limit theorem) or the solution to a partial differential equation (the Feynman–Kacs formula). Numerous problems are beyond the reach of analysis but can be solved by drawing many random samples to reveal a general rule. This technique has been termed *Monte Carlo simulation* after the Monaco casinos, where chance reigns.

A brief history of Monte Carlo simulation

The first documented account of Monte Carlo simulation seems to date back to Buffon's needle (1768), which is a simulation technique used to estimate the value of π by throwing needles on a striped floor and counting the proportion of needles that intersect a line. Indeed, many people had made use of Monte Carlo simulations before, without knowing it, as in the estimation of the probability of the face of a biased die by rolling it hundreds of times. In the current century, the world of physics has been the main user of Monte Carlo simulation, from problems of neutron diffusion in nuclear reactors to quantum transport or Ising spins. In the 1940s, von Neumann laid the foundations for pseudo-random generators, inverse cumulative distribution functions, and even for the digital computers to run the simulations. The Monte Carlo approach then pervaded a variety of fields, including engineering, biology, seismology, experimental psychology, astronomy, economics and demography. The application to finance is due to Phelim Boyle in 1977.

A brief history of option pricing

The first attempts to value options made use of probability theory (and even contributed to it). The first scientific approach dates back to Louis Bachelier who, in 1900, created a mathematical object – the Brownian motion (five years before Einstein discovered it in physics) – to model the evolution of stock prices. In the 1960s, modelling stock prices as drifted geometric Brownian motions became the norm, and option pricing formulas were derived by taking discounted expectation. The early 1970s witnessed the breakthrough of the hedging argument, which renders pricing independent of preferences and anticipations and leads to a partial differential equation (PDE). The solution to the equation is the Black–Scholes–Merton formula, the main success story of modern finance. The late 1970s saw the advent of the (so-called) martingale approach, which establishes that asset prices relative to a numeraire are fair game under a probability measure associated with this numeraire. The martingale approach prices contingent claims as a properly discounted expectation of their terminal value, which paves the way for Monte Carlo simulation as the natural method to estimate the expectation. In simple cases, the two methods are linked by the Feynman–Kacs formula, which states that the solution of a PDE is an expectation and, conversely, that an expectation as a function of the starting point satisfies a PDE. In the 1980s and 1990s, the PDE and martingale approach have competed with and complemented each other, with a rough division of the pricing world into PDE for American-style and martingales for path-dependent, multi-asset and complex dynamics.

Problems in finance

PRICING

Financial instruments are becoming increasingly complex. Exotic options give a final payout which may depend on the full path of an asset price or may depend on several asset prices. Their value depends greatly on model assumptions, which tend to be increasingly complex. Finance theory tells us that their fair price can be written as a discounted expectation – in other words, it is an expectation under a complex probability measure of a complex function. Indeed, this makes Monte Carlo simulation a method of choice (actually, sometimes the only one). Along with the fair price, the sensitivity coefficients (or "Greeks") must be computed to establish the proper hedge. This can be done within a Monte Carlo framework, taking due care to control numerical instabilities.

VALUE-AT-RISK

Along with this requirement for accurate exotic option pricing, there is also a need to measure risk. Adverse conditions, whether it be market movements or credit events, may result in major losses – therefore, risk needs to be assessed. The common measure for market exposure, value-at-risk (VAR), is defined as the maximum loss over a certain time horizon, excluding a few percentage points. It is also used to compute capital requirements. Although less precision is needed than for pricing, the size of the portfolios and the complexity of the assumptions make speed issues particularly acute.

Thus, different tasks require the computation of different statistics: pricing requires estimating an expectation; and VAR computation requires estimating a quantile.

Central limit theorem

The core principle of Monte Carlo method is the central limit theorem (CLT), which establishes how the empirical average of random samples converge to the true expectation. The CLT states that the error is asymptotically Gaussian, with a standard deviation of the order of one over the square root of the number of samples, which is the crucial point in the convergence analysis. The Gaussian shape comes from the fact that the law of a sum is obtained by convolution from the laws of the components. In the case of the average, taking the Fourier transform and performing a Taylor expansion before exponentiating gives a Gaussian characteristic function, hence a Gaussian law. The magnitude of the standard deviation comes from the fact that the variance of independent random variables adds up (which is similar to Pythagoras' theorem, which tells us that the squares of the lengths of orthogonal sides add up). Thus, the variance of the average of n samples is the individual variance divided by n, and the standard deviation is divided by the square root of n.

Variance reduction

The irritating slowness of the Monte Carlo method (100 times as much effort to gain a mere decimal place in accuracy) has motivated the development of many ingenious improvements. Most of these consist of reducing the variance which, in turn, determines the accuracy of the computation. I will now look at the most common variance reduction techniques.

ANTITHETIC SAMPLING

When the samples are drawn from a symmetric law, it is natural to draw symmetric samples. This means that each sample will see its symmetric, or antithetic, being drawn. A typical example is when a Gaussian sample or a Brownian path is drawn – as its antithetic is also generated. By doing so, the empirical mean of the samples is 0, as it should be. If the value function is linear (or, more generally, anti-symmetric plus a constant) in those Gaussian values, the expectation would be perfectly estimated (and, by the way, would not require a simulation). In general, the value function is not linear, and antithetic sampling will estimate perfectly the anti-symmetric part of the value function but not its symmetric part, for which the antithetic samples are wasted as they purely duplicate their original. In summary, antithetic sampling usually accelerates the convergence – at worst it will double the computation time.

MOMENT MATCHING/CALIBRATION

Antithetic sampling aims at getting the right empirical mean, which can be achieved in other ways, such as shifting all samples by the difference between the theoretical mean and the empirical mean. Higher moments can be matched (for instance, covariance) by applying a proper linear transform to the original samples. Imposing the fact that the empirical distribution has the same moments as the theoretical one can be interpreted as requiring a certain set of functions (monomials) to be properly valued by the empirical samples. It is equally possible to adjust the empirical samples to price correctly a certain set of instruments whose market value is known. A good choice is to take the hedging instruments of the product to be priced, as the latter will in some way be extrapolated from its hedge.

CONTROL VARIATE

This method stems from the natural idea that, to value something unknown (the *target*), we can approach it through something known (the *control variate*) and try to value the difference – which should be small. To find a good control variate is an art more than a science; it is easier to give examples than rules. If the product to be priced is close to another (the value of which is known), the latter will be used as a control variate for the former. The difference will be priced by Monte Carlo with a higher degree of accuracy as the variance will be lower. An example is an arithmetic average option that is approached by a geometric option in the Black–Scholes model. If the product has to be priced in a complex model and can easily be priced in a simpler model, the latter will be used as a control variate for the former. For instance, to price a barrier option in a model which fits the initial market smile, pricing it in the Black–Scholes model provides a useful control variate. Another possibility is to use the hedge of the product in a simple model as a control variate. Finally, antithetic sampling can to some extent be interpreted as a special case of the control variate method, in that the result of the antithetic samples is subtracted as a control variate.

IMPORTANCE SAMPLING

A classical method in numerical analysis is to adapt the scheme to fit the problem, focusing on singularities or critical regions. For instance, it is common in fluid dynamics to refine a finite element mesh around singular points – the same idea can be applied in finance, where it is called importance sampling. It consists of generating more points where needed by changing the probability measure. Indeed, this oversampling requires a renormalisation of the value function, which has to be multiplied by the Radon–Nikodym derivative (a typical example is given by a deep out-of-the-money call option). Only a few Monte Carlo samples will end in the money, making the valuation inaccurate. It is possible to add a drift to the process to make around half of the points in-the-money to sample the region which gives its value to the option better. The payout then has to be reduced appropriately (computed by Girsanov's theorem) to compensate for the oversampling. Changing the drift can be applied fruitfully in other situations, such as a double knock-out barrier, where the two barriers are quite close. A standard Monte Carlo simulation would generate very few surviving paths. Applying a mean reversion (ie a drift which is positive for low values and negative for high values) substantially increases the number of contributors to the option value. However, it is recommended not to apply too big a drift as it would overly distort the payout.

STRATIFICATION/LATIN HYPERCUBE

The requirement that no important region be ignored can be met by a sampling technique known as *stratification*. This consists of partitioning the event space into subsets, and then sampling each of the subsets. Generally, it is not easy to draw samples from within a complex region, and this method is usually applied in simple cases. A very simple method, in dimension 1, is to draw samples from different intervals. A joint stratification in high dimension is difficult to achieve due to the "curse of dimensionality" (exponential explosion). An alternative is to require each marginal law to be well sampled, as the Latin hypercube method does. Indeed, it cannot be fully satisfactory as it favours the projections on to the initial axes – which fits some problem structures, but not all. As a simple rule of thumb, it works better for several assets over one time step (as multi-underlying instruments usually have a strong dependence on each of the underlyings) than for path-dependent instruments (as good sampling of the increment does not guarantee the proper sampling of the asset value or of its average/minimum/maximum).

Error estimates

The central limit theorem states that the Monte Carlo estimate of the expectation of a random variable is approximately normally distributed. The mean is the proper expectation and the variance is the variance of the random variable divided by the number of samples. A Monte Carlo simulation provides a number whose significance relies on confidence intervals. To obtain useful confidence intervals, it is necessary to estimate the variance itself, which can be performed by … Monte Carlo simulation. The expectation is estimated as the empirical average of the draws, while the variance is estimated as the expectation of the square minus the square of the expectation, which merely requires one to take the square of the samples.

Another possibility is to partition the set of draws into p equal subsets. The standard deviation of the p mean is the square root of p multiplied by the standard deviation of the Monte Carlo estimate.

Discretisation

A major application of Monte Carlo simulation is stochastic process path generation. The perfect simulation would generate every possible path and each path would be defined for all possible dates. Unfortunately, machines are confined to finiteness and two discretisations need to be performed. The first of these is a discretisation in time, where the continuous time line is replaced by a finite set of points on which the process is simulated. This is the time approximation error. The second is a discretisation in the space of all possible events, where an expectation over all possible paths is estimated on a finite set of paths. This is defined as sampling error.

DISCRETISATION IN TIME

Discretisation in time stems from the impossibility of computing a process path for every possible value of the time variable. As a result, each path is defined on a finite number of dates, usually regularly spaced. For instance, this makes stochastic differential equation (SDE) path generation difficult. Such a SDE path is obtained from a Brownian path by twisting it – the drift coefficient shifts the path and the volatility coefficient dilates its variations. Perfect knowledge of the continuous-time Brownian path entails perfect knowledge of the continuous SDE path. However, this is no longer the case when the Brownian path is only known at a finite number of dates. This is the problem of the pathwise approximation of the SDE, as opposed to the approximation in law. The classical construction of an SDE path is performed by a first-order scheme (called the Euler scheme), which assumes that the coefficients of the SDE remain constant over the time step. More precise schemes exist – such as the second-order Milstein scheme, which anticipates the spot dependency of the coefficients.

DISCRETISATION IN SPACE

The discretisation in space of events, or sampling, is the essence of the Monte Carlo method. Improvements on the standard Monte Carlo method are numerous, with some of them being reviewed earlier in this introduction.

American Monte Carlo

Two main methods dominate the universe of pricing. The first is the partial differential equation (PDE) approach, which can be implemented in a variety of ways (finite difference, finite element, trees) and can be adapted to take into account early exercise. The second is Monte Carlo simulation, which easily handles path-dependence and high dimensionality.

The PDE approach cannot deal with path-dependence in its most general form, although in many applications path-dependence is captured by a single additional state variable, such as the average or the maximum. The cost of path-dependence is that it increases the dimension of the PDE, which cannot realistically be more than three or four, preventing its use for multi-asset instrument pricing.

Monte Carlo methods have no problem in dealing with high dimension, but they are not designed to handle early exercise. A standard Monte Carlo simulation generates final payouts independently and combines them in a linear fashion, while early exercise requires knowledge of the option value at intermediate dates and performs non-linear operations along the way. However, recent years have seen the increasing application of twists in Monte Carlo simulation in order to endow them with this capability.

The core of American option pricing is dynamic programming, which requires the ability to compute the live value of the option at any time in order to decide whether it is preferable to exercise it or to hold it. The live value is the discounted expectation one step ahead.

In a standard Monte Carlo simulation, each path is generated independently and the option value is known at only two dates: at the end, conditional on the path that has been followed, and at the beginning, by taking the discounted average of the final values. To compute the value at an intermediate date requires structuring the simulation in a manner which makes it a sort of hybrid between a Monte Carlo approach and a tree. It involves creating links between different paths, and various path-bundling schemes have been investigated.

Low discrepancy

The superiority of the Monte Carlo method over classical numerical integration schemes comes from the difficulty of evenly populating a high-dimensional space. High dimensionality is not only a challenge for the intuition, it also poses intrinsically different problems. For instance, in high dimension most of the volume of a ball is close to the edge, and the volume of a hypercube of length $\frac{1}{2}$ vanishes to 0 when the dimension increases, which suggests that the unit ball is compact in finite dimension but not in infinite dimension. This may appear to be a mere topological curiosity, but in infinite dimension you can lodge an infinite number of non-intersecting hypercubes of length $\frac{1}{2}$ in a hypercube of length 1, and any finite sequence of points will leave an infinite number of small hypercubes untouched. This explains why, on the way to infinite dimension, it becomes increasingly difficult to fill volumes. Most common numerical schemes are based on Taylor expansions and require that the diameter of the area controlled by a point is small, which in turn requires an exponential number of integration points. Monte Carlo methods are immune to this "curse of dimensionality" as their order of convergence does not depend on the dimension, although the random points may gather in some regions and ignore some others (which explains the poor order of convergence).

Recently, remarkable advances have been accomplished in improving deterministic integration through the development of low-discrepancy sequences. The discrepancy of a sequence measures its maximum departure from uniformity. Many low-discrepancy sequences have been developed, most of them based on p-adic digit manipulation. It is possible to conduct an error analysis – thanks to a result known as the Koksma–Hlawka inequality, which states that the integration error on a function is less than the discrepancy multiplied by the variation of the function. The discrepancy measures the maximum integration error on products of interval-characteristic functions, and the variation of a function measure its approximatability by a combination of such functions. Low-discrepancy sequences perform impressively in low dimensions, but their effectiveness in high dimension is highly debatable. First, it is virtually impossible to compute the variation of a function in high dimension; and it may be very big (if the function if not smooth enough, its variation may be infinite). Second, although the theoretical convergence order of the good low-discrepancy sequences is almost 1, it seems it is attained only after a very long cycle has been completed. In some ways, even low-discrepancy sequences are eventually caught up by the curse of dimensionality.

Adaptive methods

When Monte Carlo samples are generated, even if we do not know them in advance because of their random nature, at the end of the day they are still just a given set of scenarios. It is natural to wonder whether this set could have been "improved" in some sense. We can give a more formal meaning, for instance gauging the quality of a set of points by their ability to price correctly a class of functions. This is essentially what low-discrepancy sequences do, as the discrepancy is the maximum error on multidimensional digital functions. Other choices can be made, such as the requirement that European options have to be correctly priced, or to demand – in the case of path generation – that important asset value statistics such as final, average, maximum and minimum values are well sampled. In these cases, the point positions can be optimised to fit the chosen criterion.

Another approach is to make use of the knowledge of the function to be priced. For example, singularities and high-gradient (or high-convexity) zones should be more finely sampled. As we have seen with importance sampling, oversampling needs to be compensated by renormalising the function; or, equivalently, by attaching smaller-probability weights to the samples of the overrepresented regions.

Another method which makes use of non-uniform weights is numerical integration, which estimates an integral as a linear combination of point values. It is tempting to take advantage of the freedom of choice of both the point positions and of the weights; however, it is quite impractical in high dimension. As an example, it seems to be a good idea to assign to each sample a weight equal to the volume of the region it controls, defined as the set of points closer to this sample than to any other sample. In dimension 1, this weighted Monte Carlo scheme has an improved order of convergence of one. However, it does not extend to higher dimension, and the computation of the region (known as Delaunay tesselation) volumes is very involved.

Malliavin calculus

A path-dependent option gives, by definition, a payout which depends on the realised path. If this path has been generated by discretising an SDE, it is characterised by n Gaussian increments and the option price is an expectation under a Gaussian probability in dimension n. It can be written as an n-dimensional integral of the payout multiplied by the Gaussian density, which can be integrated by parts – which leads to uncommon interpretations. Two interesting examples can be given to illustrate this approach, termed Malliavin calculus (which was developed in the late 1970s and applied to finance in the late 1990s). The first of these is Greeks computation, in which calculations are made along each original path without having to shift them. For instance, the delta of an option can be computed by multiplying the final payout by a certain quantity, as opposed to differentiating it, which gives smoother estimates for the Greeks. The second example is the computation of conditional expectations, where the integration by part transforms a Dirac function into a Heaviside function. This allows conditioning by a given asset price to be transformed into conditioning by being above that asset price.

Conclusion

Monte Carlo simulation does not necessarily have the best reputation. It does not give the same result twice in a row and it does not provide a worst-case upper bound on the error. It is sometimes considered as a last recourse when both knowledge and imagination have failed, somewhat like resorting to black magic when there is no hope left. It seems to be considered as an admission of defeat, like tossing a coin when unable to decide. However, it is used daily in all major financial institutions – to price complex products and to manage the risk of huge positions.

The Monte Carlo method has come a long way since its birth. It has gained the status of a well-established discipline with firm scientific foundations, combining a rich fund of techniques. It is now a powerful and intriguing tool that no-one can afford to ignore.

The book

This book is a blend of reference papers, short *Risk* magazine articles and specially commissioned papers. I have attempted to structure the chapters in a useful way for the reader: the first section ("Generalities") establishes the subject with the paper which introduced Monte Carlo simulation to finance, a global overview and a short summary of useful basic algorithms. In the second section ("Pricing") we present various speed-up techniques and applications to specific options. The third section ("American-Style"), is devoted to the handling of early exercise by Monte Carlo methods, a task long believed to be out of the reach of a simulation approach. In the next section ("Fixed Income"), application of the simulation approach to implement yield curve models is tackled. This is followed by a section on value-at-risk ("VAR"), which investigates different approaches to VAR computations. The book concludes with a look at "Deterministic Methods", which analyses low-discrepancy sequences, a popular contender to simulation, and lattice methods. A glossary of the main technical terms then follows.

Acknowledgments

I would like to thank my ex-colleagues from ***SORT*** at Paribas and colleagues from Quant at Nikko Financial Products, especially Guillaume Blacher, Edwin Kwok and Antoine Savine, for their technical advice and the nice working environment they provide.

GENERALITIES

Introduction

Bruno Dupire

This introductory section gathers an "historical" paper, a recent overview, and a brief memento of practical algorithms. It paves the way for the following sections, which focus more on precise improvements or on applications to specialised problems.

Indeed, it is impossible not to start this section with the ubiquitously quoted paper by Phelim Boyle, "Options: A Monte Carlo Approach", which introduced Monte Carlo methods into finance in 1977. It is the foundation for what followed, and is the paper most referred to. It precedes the famous Cox–Ross–Rubinstein paper, which introduced binomial tree pricing, by two years. This is quite surprising, as for many years practitioners have extensively used and twisted binomial trees, virtually without using Monte Carlo simulations. It has to be acknowledged that the recent interest in Monte Carlo simulations is mainly due to the fact that the surge of exotic options, which proved to be their main consumer, occurred at the turn of the 90s. Chapter 1 is a very clear introduction to Monte Carlo simulations. One should keep in mind that, at the time, not that many people in the financial world were aware of this obscure method used by physicists. The chapter presents antithetic sampling and control variate methods, with an option on a dividend-paying stock as an illustration. Due to the structure of option markets at the time of its writing, Boyle pays more attention to the dynamics of the spot than to the option payout.

The Boyle, Broadie and Glasserman paper, "Monte Carlo Methods for Security Pricing" (Chapter 2), is a much more recent analysis, which provides an excellent modern overview of the field. First, they review variance reduction methods such as antithetic sampling, control variates, moment-matching methods, stratified and Latin hypercube sampling, importance sampling, and conditional Monte Carlo, before shifting their attention to low-discrepancy sequences, studying their performance in various dimensions. The problem of computing sensitivities, or Greeks, is then addressed. They compare the merits of finite difference approximations and of direct methods. Finite difference approximations consist, as in the case of computing the delta of an option, of shifting the initial point while keeping the same Brownian paths. In direct methods, several computations are performed along each path to constitute a Monte Carlo estimate of the Greeks. The last section of the chapter is a review of American option pricing by Monte Carlo simulations. This paper provides a nice journey in simulationland, and if I had to recommend a single paper in the whole book to a newcomer, it would certainly be this one.

Chapter 3 is a brief but pragmatic summary of what the practitioner needs to get started. It is a collection of basic algorithms, which should help to face simple situations. It explains how to draw Gaussian samples, Brownian motion paths, stochastic differential equation paths, and n-dimensional paths, detailing how to implement various variance reduction methods. It is a simple account of standard techniques, but I hope it will prove to be of interest to practitioners.

1

Options: A Monte Carlo Approach

Phelim P. Boyle
School of Accountancy, University of Waterloo

This paper develops a Monte Carlo simulation method for solving option valuation problems. The method simulates the process generating the returns on the underlying asset and invokes the risk-neutrality assumption to derive the value of the option. Techniques for improving the efficiency of the method are introduced. Some numerical examples are given to illustrate the procedure and additional applications are suggested.

Option valuation models are very important in the theory of finance since many corporate liabilities can be expressed in terms of options or combinations of options. Smith (1976) has recently summarised the major advances in the theory of option pricing under conditions of general equilibrium. The seminal paper by Black and Scholes (1973) yields the option values that would obtain in conditions of market equilibrium. Black and Scholes assume that the stock price follows a lognormal distribution and show how a hedge position can be formed with a portfolio of one unit (long) of the stock and a short position of a number of options. Arbitrage arguments lead to a second-order linear partial differential equation governing the value of the option. A simple closed-form solution of this equation exists in the case of a non-dividend-paying stock and also in the case of a stock which pays a continuous dividend proportional to the stock price. For other dividend policies numerical methods must be used to solve the differential equation. Schwartz (1977) has recently developed a method which involves replacing the differential equation by a series of difference equations.

Cox and Ross (1976) have analysed the structure of option valuation models and presented an alternative approach to their solution. Essentially, they show that as long as a hedge position can be constructed the value of a European call option can be obtained by discounting the expected maturity value of the option at the risk-free rate. The distribution of the maturity value of the option can be obtained from the distribution of the terminal stock value. Thus, if the distribution of the terminal stock value is known the value of the option can be obtained by integration. In general the integrals involved will not have analytic solutions and they must be evaluated by numerical methods. Chen (1969) used a numerical integration method in his thesis on warrant valuation and more recently Parkinson (1976) has used this approach to obtain values of American put options.

The purpose of the present paper is to show that Monte Carlo simulation provides a third method of obtaining numerical solutions to option valuation problems The technique proposed is simple and flexible in the sense that it can easily be modified to accommodate different processes governing the underlying stock returns. This method should provide a useful supplement to the two approaches mentioned above. Furthermore, it has distinct advantages in some specialised situations – eg when the underlying stock returns involve jump processes. Essentially, the method uses the fact that the distribution of terminal stock prices is determined by the process generating future stock price movements. This process can be simulated on a computer, thus generating a series of stock price trajectories. This series determines a set of terminal stock values which can be used to obtain an estimate of the option value. Furthermore, the standard deviation of the estimate can be obtained at the same time so that the accuracy of the results can be established.

The second section of the paper provides a brief description of the Monte Carlo method. One potential drawback of the method arises from the fact that the standard error of the estimate is inversely proportional to the square root of the number of simulation trials. Although any desired degree of accuracy can be obtained by performing enough simulation trials, there are usually more efficient ways of reducing the error. One such method, known as the control variate approach, has proved effective in dealing with some option valuation problems and it is described in the second section. Another technique for increasing the precision of the estimates uses so-called antithetic variates, and the second section also includes a short account of this approach.

In the third section the simulation method is used to obtain numerical estimates of a European call option on a stock which pays discrete dividends. In this case it is assumed that the returns on the underlying stock follow a lognormal distribution. The introduction of an appropriate control variate considerably improves the accuracy of the estimates. The true answers are obtained using numerical integration and it is verified that they lie as predicted within the calculated confidence limits.

The fourth section discusses further possible applications of the method and suggests when the Monte Carlo method should be used either in preference to or in conjunction with the alternative approaches available.

The Monte Carlo method

In this section the Monte Carlo method is described and two techniques for improving the efficiency of the method are discussed.[1] An excellent exposition of the Monte Carlo method is given by Hammersley and Handscomb (1964). Shreider (1966), Fishman (1973) and Meyer (1956) provide additional useful references.

It is convenient to couch the discussion in terms of the evaluation of a definite integral, although Monte Carlo methods have a much wider range of applicability. Consider the integral

$$\int_A g(y)f(y)dy = \bar{g} \qquad (1)$$

where $g(y)$ is an arbitrary function and $f(y)$ is a probability density function with

$$\int_A f(y)dy = 1$$

A denotes the range of integration and is omitted for convenience in the sequel.

To obtain an estimate of $\bar{g}$ a number (n) of sample values (y_i) is picked (at random) from the probability density function $f(y)$. The estimate of $\bar{g}$ is given by

$$\hat{g} = \frac{1}{n}\sum_{i=1}^{n} g(y_i) \qquad (2)$$

The standard deviation of the estimate is given by $\hat{s}$ where

$$\hat{s}^2 = \frac{1}{(n-1)}\sum_{i=1}^{n}\left(g(y_i) - \hat{g}\right)^2 \qquad (3)$$

For large n the error involved in replacing $(n - 1)$ by n in this formula is of little consequence. The distribution

$$\frac{\hat{g} - \bar{g}}{\sqrt{\dfrac{\hat{s}^2}{n}}} \qquad (4)$$

tends to a standardised normal distribution with increasing values of n. For the values of n considered in this paper ($n > 1000$), the distribution can be regarded as normal and confidence limits on the estimate of $\hat{g}$ can be obtained on this basis.

Since the standard deviation of $\hat{g}$ is equal to $\hat{s}/\sqrt{n}$, the confidence limits can be reduced by increasing n. To reduce the standard deviation by a factor of ten the number of simulations trials has to be increased one hundredfold. An alternative approach is to concentrate on reducing the size of $\hat{s}$. Such techniques are known as variance reduction techniques and a number of them are described in Chapter 5 of Hammersley and Handscomb (1964). The thrust of these methods is to modify or distort the original problem in such a way as to improve the accuracy of the results obtained by crude Monte Carlo methods. One such approach is known as the control variate method.

The basic idea underlying this method is to replace the problem under consideration by a similar but simpler problem which has an analytic solution. The solution of the simpler problem is used to increase the accuracy of the solution to the more complex problem. Suppose that the integral

$$\int g(y)h(y)dy = G \qquad (5)$$

can be evaluated analytically, where h is a probability density function. From equations (1) and (5) it is clear that

$$\bar{g} = G + \int g(y)[f(y) - h(y)]dy \qquad (6)$$

A revised estimate of $\bar{g}$, g^*, can be obtained by evaluating the integral on the right-hand side of (6) by crude Monte Carlo methods. The function h is called the control variate. The gain in efficiency will be measured by the reduction in the variance of g^* as compared to the variance of $\hat{g}$ The increase in efficiency will depend on the degree to which h mimics the behaviour of f. Thus, in selecting an appropriate control variate there are two (usually conflicting) requirements. First, h must give rise to an integral that is easy to evaluate. Second, h must model the behaviour of f.

In evaluating the integral in equation (6) by crude Monte Carlo the same random number is used in the ith simulation trial to generate a value y_i from $g(y)$ and a value Y_i from $h(y)$. For a given value of n let

$$\hat{g} = \int g(y)f(y)dy \quad \text{and} \quad \hat{G} = \int g(y)h(y)dy$$

be the estimates obtained by crude Monte Carlo under these conditions. Then g^* is given by

$$g^* = G + (\hat{g} - \hat{G}) \tag{7}$$

This is an unbiased estimate and its variance is

$$\text{var}(\hat{g}) + \text{var}(\hat{G}) - 2\,\text{cov}(\hat{g}, \hat{G})$$

This will be less than the variance of $\hat{g}$ as long as

$$\text{cov}(\hat{g}, \hat{G}) > \frac{\text{var}(\hat{G})}{2}$$

or

$$\text{corr}(\hat{g}, \hat{G}) > \frac{1}{2}\sqrt{\frac{\text{var}(\hat{G})}{\text{var}(\hat{g})}} \tag{8}$$

This confirms the earlier observation that the efficiency gain is a function of the relationship between f and h. Furthermore, equation (8) furnishes a prescription for testing if a particular control variate will result in an efficiency gain.

Whereas the control variate approach used a second estimate of the integral with a high positive correlation with the estimate of interest, the antithetic variate approach exploits the existence of negative correlation between two estimates. In a given problem there may be different methods of introducing an antithetic variate. One possible method is the following. Suppose that the series $y_1, y_2, \ldots, y_n$ has been generated using the random number sequence $u_1, u_2, \ldots, u_n$, where the us are selected at random from the interval (0,1). The sequence $(1 - u_1), (1 - u_2), \ldots, (1 - u_n)$ is used to generate a second set of variates from the distribution $f(y)$. Suppose the estimate of (1) using the first set is $g(u)$ and the estimate of (1) using the second set is $g(1 - u)$. Then

$$\tfrac{1}{2}[g(u) + g(1 - u)]$$

will be an unbiased estimate of (1) with variance equal to

$$\tfrac{1}{4}[\text{var}(g(u)) + \text{var}(g(1 - u))] + \tfrac{1}{2}[\text{cov}(g(u), g(1 - u))] \tag{9}$$

If the covariance between $g(u)$ and $g(1 - u)$ is negative, this will yield a smaller estimate of the variance than an independent estimate. A more detailed discussion of this point is given by Fishman (1973).

Valuation of European call options on dividend-paying stocks

In this section the Monte Carlo method is used to obtain estimates of option values on dividend-paying stocks in the case of European call options. It is demonstrated that the introduction of an appropriate control variate significantly increases the accuracy of the results. The accuracy and reliability of the Monte Carlo method has been verified for a range of parameter values using the accurate results obtained by numerical integration and a representative sample of the results is given here. The effect of using antithetic variates is also examined and a modest gain in efficiency is observed.

Let

S_t = the current stock price at time t,
r = the risk-free rate per quarter compounded continuously,
σ^2 = the (assumed constant) variance rate per quarter on the underlying stock,
D_t = the dividend payable at time t; dividends payable quarterly,
E = the exercise price of the option,
T = the expiration date of the option.

Cox and Ross (1976) have shown that the assumption of risk neutrality can be used to obtain solutions of option valuation problems.[2] This assumption implies that the equilibrium rate of return on all assets, and in particular the common stock, is equal to the risk-free rate. Hence the expected return on the stock is given by

$$E\left(\frac{S_T}{S_t}\right) = \exp(r[T - t])$$

In addition, it is assumed that the ratio of successive stock values follows a lognormal distribution and so the ratio S_{t+1}/S_t has a lognormal distribution with mean equal to $\exp(r)$. To relate r to the parameters of the normal distribution followed

Table 1. European call option values on dividend-paying stock; 5000 trials per estimate

S/E	Number of periods to maturity	Option values by crude Monte Carlo	Standard deviation of crude Monte Carlo estimates	Accurate option values obtained by numerical integration
0.50	2	0.004	0.002	0.003
	4	0.087	0.014	0.075
	6	0.273	0.031	0.266
	8	0.539	0.047	0.541
	10	0.879	0.066	0.867
	12	1.202	0.081	1.220
	14	1.547	0.094	1.587
	16	1.897	0.108	1.957
	18	2.281	0.130	2.327
	20	2.783	0.157	2.691
1.00	2	5.121	0.114	5.028
	4	7.427	0.170	7.251
	6	9.247	0.215	9.001
	8	10.793	0.258	10.493
	10	11.923	0.297	11.811
	12	13.135	0.330	13.000
	14	14.281	0.361	14.087
	16	15.079	0.389	15.090
	18	15.945	0.428	16.022
	20	17.190	0.479	16.891
1.50	2	26.445	0.240	26.369
	4	28.250	0.333	27.819
	6	29.832	0.405	29.300
	8	31.257	0.470	30.713
	10	32.319	0.528	32.042
	12	33.491	0.578	33.289
	14	34.835	0.624	34.459
	16	35.503	0.666	35.558
	18	36.382	0.722	36.594
	20	37.989	0.798	37.571

S = 25, 50, 75 – current stock prices.
E = 50.
r = 0.015 per period.
σ^2 = 0.025 per period.
D = 0.25 per period.

by $\log[S_{t+1}/S_t]$ it is convenient to write

$$E\left(\frac{S_{t+1}}{S_t}\right) = \exp(r) = \exp\left[(r - \tfrac{1}{2}\sigma^2) + \tfrac{1}{2}\sigma^2\right] \quad (10)$$

Hence (see Aitchison and Brown, 1963) the expected value of $\log[S_{t+1}/S_t]$ is $(r - \sigma^2/2)$.

Using the properties of the lognormal distribution, this means that we can generate the distribution of stock prices one period hence by forming the random variables

$$S_{t+1} = S_t \exp\left[(r - \tfrac{1}{2}\sigma^2) + \sigma\tilde{x}\right]$$

where $\tilde{x}$ is a normally distributed random variable with zero mean and unit variance. It is assumed that time is measured in units of one quarter. Assume also that S_t represents the stock price just after the quarterly dividend D_t has been paid.

To set up the simulation method in this case a value of S_{t+1} is generated. If this value is greater than D_{t+1}, then $(S_{t+1} - D_{t+1})$ is used as the initial value at the start of the second period and the procedure is continued until a value of S_T is obtained. If at some stage S_{t+m} $(m = 1, 2, \ldots(T - t - 1)$ is less than or equal to the corresponding dividend payment D_{t+m}, the process stops. In this case another simulation trial is started again from time t. A series of simulation trials is carried out in this way and the expected value of $\text{Max}[S_T - E, 0]$ is found. This quantity is then discounted at the risk-free rate to yield an estimate of the option value $\hat{g}$. Equation (3) is used to obtain the variance of the estimate. Ninety-five per cent confidence limits are given by $\hat{g} \pm 2\hat{s}/\sqrt{n}$.

Table 1 gives the results of the simulation method for selected values of the underlying parameters. It is assumed that $r = 0.015$ per period, $\sigma^2 = 0.025$ per period, $D = 0.25$ per period and $E = 50$. The results for three different current stock prices are given corresponding to S = 25, 50 and 75. Crude Monte Carlo estimates of options with even maturities ranging from 2 to 20 are provided. Each estimate was obtained using 5000 trials where each trial corresponds to

Table 2. European call option values on dividend-paying stocks; 5000 trials per estimate

S/E	Number of periods to maturity	Option values control variate method	Standard deviation of estimate using control variate method	Accurate option values using numerical integration
0.50	2	0.003	0.00004	0.003
	4	0.075	0.0002	0.075
	6	0.266	0.001	0.266
	8	0.542	0.001	0.541
	10	0.869	0.002	0.867
	12	1.222	0.003	1.220
	14	1.591	0.004	1.587
	16	1.964	0.006	1.957
	18	2.330	0.007	2.327
	20	2.689	0.010	2.691
1.00	2	5.028	0.0003	5.028
	4	7.251	0.001	7.251
	6	9.000	0.002	9.001
	8	10.494	0.003	10.493
	10	11.815	0.004	11.811
	12	13.006	0.006	13.000
	14	14.094	0.007	14.087
	16	15.106	0.009	15.090
	18	16.038	0.010	16.022
	20	16.903	0.013	16.891
1.50	2	26.369	0.0004	26.369
	4	27.818	0.001	27.819
	6	29.299	0.002	29.300
	8	30.715	0.003	30.713
	10	30.048	0.005	32.042
	12	33.298	0.006	33.289
	14	34.465	0.008	34.459
	16	35.578	0.009	35.558
	18	36.616	0.011	36.594
	20	37.582	0.014	37.571

S = 25, 50, 75 – current stock prices.
E = 50.
r = 0.015 per period.
σ^2 = 0.025 per period.
D = 0.25 per period.

a possible stock path. For convenience it is assumed that level dividends are payable.

To provide a benchmark the accurate results obtained by numerical integration[3] are also given. All of the thirty crude Monte Carlo estimates lie within two standard deviations of the correct answers. However, the 95 per cent confidence limits are quite wide. For example, when the current stock price is 50 and the option has 20 periods to maturity the crude Monte Carlo estimate is 17.190. The 95 per cent confidence limits are 17.190 ± 0.958. To reduce the range of these confidence limits to ±0.05 would require increasing the number of trials from 5000 to 1,835,500.[4]

This indicates that an alternative method of improving the accuracy of the Monte Carlo method should be sought. The fact that the Black–Scholes solution in the case of a non-dividend-paying stock is exact permits the use of a suitable control variate. Assume the existence of another stock, S_t^*, which pays no dividends until the exercise date.[5] Thereafter S^* is identical with the dividend-paying stock. An investor should be indifferent as between (European call) options on these two stocks (assuming both options have the same expiration date and the same exercise price). Since both stocks earn the same rate of return and will have the same value at the maturity date of the option, S_t^* will be less than S_t. In the appendix the first two moments of the return on S_t^* are derived in terms of D and the return parameters of S_t. The control variate is a stock having the same expected return and variance rate as S_t^* over the period until maturity and which in addition has returns that are lognormally distributed.[6]

The Black–Scholes solution is used to obtain the value of an option on the control variate. Then both problems are attacked together using crude Monte Carlo methods as described in the second section. In each simulation run the same random normal deviates are used to generate both lognormal distributions. (These lognormal distributions correspond to the dividend-paying stock and the control variate.) The use of the same set of normal deviates during corresponding periods in each simulation run induces high

Table 3. Option values using crude Monte Carlo and antithetic variate method; 5000 trials per estimate

S/E	Number of periods to maturity	Crude Monte Carlo estimate	Standard deviation of crude Monte Carlo estimates	Antithetic variate method estimate	Standard deviation of antithetic variate estimate
1.00	2	5.121	0.114	5.093	0.060
	4	7.427	0.170	7.254	0.092
	6	9.247	0.215	9.101	0.126
	8	10.793	0.258	10.609	0.150
	10	11.923	0.297	11.716	0.170
	12	13.135	0.330	12.974	0.194
	14	14.281	0.361	14.050	0.215
	16	15.079	0.389	15.013	0.231
	18	15.945	0.428	16.018	0.260
	20	17.190	0.479	17.030	0.287

S = 50
E = 50
r = 0.015 per period.
σ^2 = 0.025 per period.
D = 0.250 per period.

positive correlation. In the case of the control variate, D, the dividend payment is of course zero. The results are displayed in Table 2 where, for comparison purposes, the same parameters are used as in Table 1. Observe that the introduction of the control variate greatly reduces the standard deviation of the estimates. For example, the range of the 95 per cent confidence limits in the case of the 20-period option with a current stock price of 50 has been reduced from ±0.958 to ±0.026. To achieve the same reduction by increasing the number of trials would require 6,788,000 trials instead of 5,000. In all 30 cases the 95 per cent confidence limits contain the true answer. The confidence limits in Table 2 would appear to be sufficiently accurate for most practical applications.

The introduction of an appropriate control variate provides a very efficient variance reduction technique in this problem. In some problems it may be difficult to find a suitable control variate. The alternative method discussed in the second section, the antithetic variate method, is often easier to apply since it concentrates on the procedure used for generating the random deviates. Essentially, this technique relies on the introduction of negative correlation between two estimates. It is of interest to examine one method of using antithetic variates in the present problem.

Note that if $\tilde{x}$ is a normally distributed random variable with zero mean and unit variance, then so also is $-\tilde{x}$. Hence, to apply the antithetic variate technique in this case use a set of random normal deviates to obtain the initial estimate. A second estimate is obtained by using the same set of random normal deviates with their signs reversed. The revised estimate will be the mean of these two estimates. The results of one such set of calculations are displayed in Table 3, and comparison with the appropriate figures in Table 1 shows that the standard deviations of the estimates have been reduced by approximately 50 per cent – a very modest gain in efficiency. This low gain in efficiency may be explained as follows. While $\tilde{x}$ and $-\tilde{x}$ have perfect negative correlation, this does not hold for the corresponding transformed lognormal variates. The variance of the revised estimate is given by an expression analogous to expression (9) in the second section, and, while the covariance term is negative, its magnitude is not large enough to effect a significant reduction in the variance of the revised estimate.

Discussion and additional applications

This section discusses some strengths and limitations of the Monte Carlo approach and identifies situations where its use would be advantageous. Some additional applications of the method are discussed. In particular, the Monte Carlo method should prove useful in dealing with options where the underlying stocks returns are generated by a mixture of both continuous and jump processes. (In general these mixed processes give rise to mixed partial differential – difference equations which are very difficult to solve.)

As was suggested in the last section, the Monte Carlo method can be used to obtain numerical estimates of a European call option on a stock which pays discrete dividends. When an appropriate control variate was used the method gave very accurate results together with confidence limits on the estimates. This method provides the option value in respect of one particular current stock price. To obtain option values corresponding to different current stock prices a set of simulation trials has to be carried out for each starting stock price. When either a numerical integration or a finite difference approach is employed the complete range of stock prices is obtained as a by-product. This suggests that the Monte Carlo method should be used for one-off situations where only a few option values are required. Since the Monte Carlo method is very simple to set up, it can play a useful role in checking the results obtained using one of the other approaches.

It is possible to modify the simulation technique to handle American call options on dividend-paying stocks. Merton (1973) has shown that such options will only be exercised (by a rational investor) just before each dividend date if the stock price exceeds a certain critical value. These critical stock prices, one corresponding to each dividend date, must be determined as part of the solution. At each dividend date a series of option prices must be obtained to locate the critical stock price. This suggests that either the numerical integration approach or the finite difference method would be more effective. Notice however that if there is just one dividend payable during the lifetime of the contract, the critical stock price can be found exactly by using the Black–Scholes solution and an iterative procedure. In such circumstances, the Monte Carlo method should prove useful.

One advantage of the Monte Carlo method is that it is very flexible with regard to the distribution used to generate the returns on the underlying stock. Changing the underlying distribution merely involves using a different process for generating the random variates employed in the method. Furthermore, the Monte Carlo method is perhaps unique in the sense that the distribution used to generate returns on the underlying stock need not have a closed-form analytic expression. This opens the possibility of deriving option values using empirical distributions of stock returns.[7] The Monte Carlo method has the advantage that a distribution can be used for any of the parameters of the problem rather than a point estimate. For example, it may be useful in some problems to regard the variance as a probability distribution since it is usually estimated from empirical data.

The simulation method will also have advantages in cases where the underlying stock returns are generated by a mixture of stochastic processes. As an example of this we consider the model proposed (1977) by Merton (1976). In this model the process generating stock returns is composed of a continuous part and a jump component. The continuous part is modelled by a Gauss-Weiner process and the jump component is modelled by an independent Poisson process. The jump corresponds to the arrival of important pieces of information which affect the stock price. The expected number of arrivals per unit time is λ. If the Poisson event occurs, the stock price will jump from S to SY, where Y is a random variable taking non-negative values. Under these assumptions the stock price dynamics become

$$\frac{dS}{S} = (\alpha - \lambda k)\,dt + \sigma d\tilde{z} + dq \tag{11}$$

where α is the expected instantaneous return on the stock, σ^2 is the instantaneous variance of the return conditional on the Poisson event not occurring, dz is a Gauss–Weiner process, and q(t) represents the Poisson process.

The expected value of Y is $(1 + k)$ and it is assumed that the Y are independently and identically distributed.

Although it is not possible to create a riskless hedge against both the continuous changes and the discrete changes implied by equation (11), Merton shows that an option pricing formula may be derived on the assumption that the risk associated with the jump is a diversifiable risk. The dynamics of the situation dictate that the distribution of stock prices at the end of a given time period will depend on

- ❑ the basic Gauss-Weiner process,
- ❑ the number of jumps that occur,
- ❑ the amplitude of these jumps.

Thus, three different stochastic processes are involved. By generating random deviates from these three distributions a set of terminal stock values can be obtained and the simulation method used to obtain estimates of the option prices.

Merton has provided a solution to this problem in the form

$$\sum_{n=0}^{\infty} e^{-\lambda t}\frac{(\lambda t)^n}{n!}\left\{\varepsilon_n\left[W(SX_n e^{-\lambda kt}, t, E, \sigma^2, r)\right]\right\}$$

where the random variable X_n is the product of n random variables each identically distributed to the random variable Y: $X_0 = 1$; and ε_n is the expectation operator over the distribution X_n. W represents the standard Black–Scholes solution.

To obtain an explicit solution Merton makes the additional assumption that the random variable Y also has the lognormal distribution. The Monte Carlo approach affords a method of providing numerical solutions in this case for different distributions of Y. In addition, the Monte Carlo method can be used in this case when it is also assumed that the stock pays discrete dividends. Of course, to improve the accuracy of the method without resorting to a large number of simulations it is necessary to incorporate a suitable variance reduction technique. While this point is not pursued here, it is suggested that the use of either control variates or antithetic variates might prove useful.[8]

To conclude this section it will be useful to examine the role of Monte Carlo calculations in general and in particular their role in option valuation problems. The crude Monte Carlo approach provides a fast and flexible method of obtaining approximate answers together with confidence limits on the results. A number of effective techniques for reducing the variance of the estimates have been developed. However, it is not possible to formulate general rules regarding the selection and implementation of the most effective technique. As Kahn[9] observes: "... the greatest gains in variance reduction are often made by exploiting specific details of the problem, rather than by the routine application of general principles".

In connection with option valuation problems it is suggested that the Monte Carlo method should prove most valuable in situations where it is difficult if not impossible to proceed using a more accurate approach. In particular we have in mind problems where the returns on the underlying stock are generated by a mixture of stochastic processes or else drawn from an empirical distribution. The Monte Carlo method provides a viable alternative in the case of European call options on dividend-paying stocks because of the introduction of a very efficient control variate. In other situations it may be viewed as a useful supplement to existing methods. Indeed, it should have some pedagogical value since it utilises the intuitive solution technique developed by Cox and Ross (1976) and shows how the process governing the returns on the underlying stock can be used to obtain option values. The present paper has drawn attention to some of the techniques available for reducing the variance of the crude Monte Carlo estimates. Since these techniques do not appear to be widely appreciated in the finance literature, it is hoped this is a useful service.

Appendix

The determination of the parameters of the control variate is described below.

Let the current price of the dividend-paying stock be S_t and assume that the maturity date of the option is m periods hence. Assume that a dividend has just been paid and that a dividend D will be payable at the end of the 1st, 2nd, 3rd $(m-1)$st period. It is also assumed that the stock returns in each period are independently and identically distributed. Thus, the stock price evolution is generated by the series of random variables $x_1, x_2, \ldots, x_m$, where all the x_i are drawn from the same distribution. At the end of the first period the stock price is $S_t x_1$. Assuming that $S_t x_1$ is greater than D, the stock price will be $(S_t x_1 - D)x_2$ at the end of the second period. After m periods under these assumptions the stock price will be

$$S_t \prod_{i=1}^{m} x_i - D \sum_{j=2}^{m} \prod_{i=j}^{m} x_i \qquad \text{(A1)}$$

Since the x_t are independent, the expected value of this quantity is

$$S_t \prod_{i=1}^{m} E(x_i) - D \sum_{j=2}^{m} \prod_{i=j}^{m} E(x_i) \qquad \text{(A2)}$$

The variance of (A1) can be expressed in terms of $E(x_i)$ and $E(x_i^2)$ with somewhat more labour. It is convenient to write it first in the form

$$S_t^2 \operatorname{var}\left[\prod_{i=1}^{m} x_i\right] + D^2 \operatorname{var}\left[\sum_{j=2}^{m} \prod_{i=j}^{m} x_i\right] - 2S_t D \operatorname{cov}\left[\prod_{i=1}^{m} x_i, \sum_{j=2}^{m} \prod_{i=j}^{m} x_i\right] \qquad \text{(A3)}$$

and then evaluate each term in turn. Since all the x_i have the same distribution, it can be assumed that

$$E(x_i) = \rho \quad \text{and} \quad E(x_i^2) = \theta$$

for all i and this simplifies the analysis.

With this notation it can be shown (Boyle, 1976) that

$$\operatorname{var}\left[\prod_{i=1}^{m} x_i\right] = \theta^m - \rho^{2m},$$

$$\operatorname{var}\left[\prod_{j=2}^{m}\prod_{i=j}^{m} x_i\right] = \frac{(\theta+\rho)}{(\theta-\rho)}\sum_{i=1}^{m-1}\theta^i$$

$$-\frac{2\theta}{(\theta-\rho)}\sum_{i=1}^{m-1}\rho^i - \left[\sum_{i=1}^{m-1}\rho^i\right]^2,$$

$$\operatorname{cov}\left[\prod_{i=1}^{m} x_i, \sum_{j=2}^{m}\prod_{i=j}^{m} x_i\right] =$$

$$\rho\theta\frac{(\theta^{m-1}-\rho^{m-1})}{(\theta-\rho)} - \rho^m\frac{(\rho-\rho^m)}{(1-\rho)}.$$

The control variate is derived using a stock, S_t^*, which pays no dividends until the option matures and thereafter is identical with the dividend-paying stock. Hence *at maturity* the distribution of S^* is identical with that of S given by equation (A1) because from this point on both securities represent identical claims on the assets of the firm. Thus, the first two moments of the distribution of S^* at the maturity date of the option are given by (A2) and (A3). Since S and S^* must earn the same return, the current value of S^*, S_t^*, is obtained by discounting (A2) at the risk-free rate. The control variate is a stock which follows a lognormal distribution and has the same first two moments as S^*. Notice that since S^* involves a linear combination of lognormal distributions, it cannot follow a true lognormal distribution although this is a very good approximation for the range of parameters normally involved. Since the control variate follows a true lognormal distribution and its first two moments are known, the mean and variance of the underlying normal distribution can be determined (cf. Aitchison and Brown, 1963). Suppose the value of mean of the underlying normal distribution is M and that of the variance V. The normal distribution used to generate the one-period returns on the control variate will have mean M/m and variance V/m. Because of the risk-neutrality assumption,

$$rm = M + \frac{V}{2}$$

and so M can be expressed in terms of r and V. These parameters are used in determining the crude Monte Carlo estimate of the option on the control variate. The same random numbers are used in these simulations as are used in obtaining the crude Monte Carlo estimates of the option on

This paper was first published in the Journal of Financial Economics *4, pp. 323–38 (1977). It is reprinted with the permission of JAI Press. The author acknowledges research support from the University of British Columbia, Humanities and Social Sciences Grant. This paper has benefited from the comments of the referee, Clifford Smith, and the comments of an anonymous referee on an earlier version. The author is grateful to A.L. Ananthanarayanan for indispensable computer assistance and valuable discussion. In addition he acknowledges useful discussions with M.J. Brennan, E.S. Schwartz and C. Sarndal and helpful comments from M. Jensen.*

1 *The material presented in this section is available elsewhere but it is repeated here for the sake of completeness.*

2 *See Smith (1976, p. 22) and Cox and Ross (1976) for a discussion of this solution technique.*

3 *To obtain very accurate option values it proved more convenient in this instance to use a numerical integration approach rather than a finite difference approach. The numerical integration method used here is based on a simple trapezoidal rule (cf. Chen, 1969). By reducing the size of the interval, results accurate to three decimal places were obtained. This degree of accuracy was verified by showing that more refined intervals (using double precision) did not affect the results. As an additional check the numerical integration method was also used to obtain option values in the no-dividend case. The results so obtained agreed exactly with the Black-Scholes results for the range of parameter values considered to at least three decimal places.*

4 *To see this note that* $\hat{s}/\sqrt{5000} = 0.479$. *It is required to find a number,* n, *so that* $\hat{s}/\sqrt{n} = 0.025$. *Hence* $n = (0.479/0.025)^2 \times 5000 = 1{,}835{,}528$.

5 *This technique of using an associated stock which pays no dividends has also been suggested by Rubinstein (1975).*

6 *A more detailed discussion of this point is given in the appendix.*

7 *The author is indebted to A. Ananthanarayanan for this observation.*

8 *The author hopes to carry out further research on this point.*

9 *See Meyer (1956, p. 146).*

BIBLIOGRAPHY

Aitchison, J., and J.M.C. Brown, 1963, *The Lognormal Distribution*, Cambridge University Press, Cambridge.

Black, F., and M.J. Scholes, 1973, "The Pricing of Options and Corporate Liabilities", *Journal of Political Economy* 81, pp. 637–59.

Boyle, P.P., 1976, "Rates of Return as Random Variables", *Journal of Risk and Insurance* 4, pp. 693–713.

Chen, H., 1969, "A Dynamic Programming Approach to the Valuation of Warrants", Unpublished PhD thesis (University of California, Berkeley, CA).

Cox, J.C., and S.A. Ross, 1976, "The Valuation of Options for Alternative Stochastic Processes", *Journal of Financial Economics* 3, pp. 145–66.

Fishman, G.S., 1973, *Concepts and Methods in Discrete Event Simulation*, Wiley, New York.

Hammersley, J.M., and D.C. Handscomb, 1964, *Monte Carlo Methods* (Methuen, London).

Merton, R.C., 1973, "The Theory of Rational Option Pricing", *Bell Journal of Mathematics and Management Science* 4(1), pp. 141–83.

Merton, R.C., 1976, "Option Pricing When Underlying Stock Returns are Discontinuous", *Journal of Financial Economics* 3, pp. 125–44.

Meyer, M.A., (ed), 1956, *Symposium on Monte Carlo Methods*, Wiley, New York.

Parkinson, M., 1977, "Option Pricing: The American Put", *The Journal of Business* 50, pp. 21–36.

Rubinstein, M., 1976, "The Valuation of Uncertain Income Streams and the Pricing of Options", *Bell Journal of Economics* 7, pp. 407–25.

Schwartz, E.S., 1977, "Valuation of Warrants Implementing a New Approach", *Journal of Financial Economics* 4, pp. 79–93.

Shreider, Y.A., (ed), 1966, *The Monte Carlo Method*, English translation. Pergamon Press, New York.

Smith, C.W., 1976, "Option Pricing: A Review", *Journal of Financial Economics* 3, pp. 3–51.

2

Monte Carlo Methods for Security Pricing

Phelim P. Boyle, Mark Broadie and Paul Glasserman
School of Accountancy, University of Waterloo; Graduate School of Business, Columbia University; Graduate School of Business, Columbia University

The Monte Carlo approach has proved to be a valuable and flexible computational tool in modern finance. This chapter discusses some of the recent applications of the Monte Carlo method to security pricing problems, with emphasis on improvements in efficiency. We first review some variance reduction methods that have proved useful in finance. Then we describe the use of deterministic low-discrepancy sequences, also known as quasi-Monte Carlo methods, for the valuation of complex derivative securities. We summarise some recent applications of the Monte Carlo method to the estimation of partial derivatives or risk sensitivities and to the valuation of American options. We conclude by mentioning other applications.

In recent years the complexity of numerical computation in the finance area has increased enormously, putting more demands on computational speed and efficiency. Numerical methods are used for a variety of purposes. These include the valuation of securities, the estimation of their sensitivities, risk analysis and stress testing of portfolios. The Monte Carlo method is a useful tool for many of these calculations, evidenced in part by the voluminous literature on successful applications. For a brief sampling, the reader is referred to the stochastic volatility applications in Duan (1995), Hull and White (1987), Johnson and Shanno (1987) and Scott (1987);[1] the valuation of mortgage-backed securities in Schwartz and Torous (1989); the valuation of exotic options in Kemna and Vorst (1990); and the valuation of interest-rate derivative claims in Jones and Jacobs (1986) and Carverhill and Pang (1995) – also Chapter 18 of this publication.

In modern finance, the prices of the basic securities and the underlying state variables are often modelled as continuous-time stochastic processes. A derivative security, such as a call option, is a security whose payout depends on one (or more) of the basic securities. Using the assumption of no arbitrage, financial economists have shown that the price of a generic derivative security can be expressed as the discounted expected value of its payouts. This expectation is taken with respect to a transformation of the original probability measure, known as the equivalent martingale measure or the risk-neutral measure. The book by Duffie (1996) provides an excellent account of this material.

The Monte Carlo method lends itself naturally to the evaluation of security prices represented as expectations. Generically, the approach consists of the following steps:

- ❑ simulate sample paths of the underlying state variables (eg underlying asset prices and interest rates) over the relevant time horizon. Simulate these according to the risk-neutral measure;
- ❑ evaluate the discounted cashflows of a security on each sample path, as determined by the structure of the security in question; and
- ❑ average the discounted cashflows over sample paths.

In effect, this method computes a multidimensional integral – the expected value of the discounted payouts over the space of sample paths. The increase in the complexity of derivative securities in recent years has led to a need to evaluate high-dimensional integrals.

Compared with other methods of numerical integration, Monte Carlo becomes increasingly attractive as the dimension of the problem increases. Consider the integral of the function $f(x)$ over the d-dimensional unit hypercube. The simple (or crude) Monte Carlo estimate of the integral is equal to the average value of the function f over n points selected at random[2] from the unit hypercube. From the strong law of large numbers this estimate converges to the true value of the integrand as n tends to infinity. In addition, the central limit theorem assures us that the standard error[3] of the estimate tends to zero as $1/\sqrt{n}$. Thus, the error convergence rate is independent of the dimension of the problem, and this is the dominant advantage of the method over classical numerical integration approaches. The only restriction on the function f is that it should be square integrable, which is a relatively mild restriction.

Furthermore, the Monte Carlo method is flexible and easy to implement and modify. In addition, the increased availability of powerful computers has enhanced the efficiency of the method. There are some disadvantages of the method, but in recent years progress has been made in overcoming them. One drawback is that for very complex problems a large number of replications may be required to obtain precise results. Different variance reduction techniques have been developed to speed up the convergence. Two of the classical variance reduction techniques are the control variate approach and the antithetic variate method. More recently, moment matching and importance sampling methods have been introduced in finance applications.

Another technique for speeding up the valuation of multidimensional integrals uses deterministic sequences rather than random sequences. These deterministic sequences are chosen to be more evenly dispersed through the region of integration than random sequences. If we use these sequences to estimate multidimensional integrals, we can improve the convergence. Deterministic sequences with this property are known as low-discrepancy sequences or quasi-random sequences. Using this approach one can in theory derive deterministic error bounds, though the practical use of the bounds remains problematic. In contrast, the standard Monte Carlo method yields probabilistic error bounds that are practically useful. Although low-discrepancy sequences are well known in computational physics, they have only recently been applied in finance problems. There are different procedures for generating such low-discrepancy sequences and these procedures are generally based on number-theory methods. We describe some of the recent developments in this area. We also discuss applications of this approach to problems in finance and conduct some comparisons between standard Monte Carlo methods and two different quasi-random approaches.

Until recently, the conventional wisdom was that American-style options could not be valued using the Monte Carlo approach. However Tilley (1993) – also Chapter 14 of this publication, Grant, Vora and Weeks (1993), Broadie and Glasserman (1997) – also Chapter 16 of this publication, and Barraquand and Martineau (1995) have proposed approaches to this problem. We provide a brief survey of the recent research progress in this area.

The layout of the paper is as follows. Variance reduction techniques are described in the next section. The ideas behind the use of low-discrepancy sequences and brief numerical comparisons with standard Monte Carlo methods are given in the third section. Risk measure estimation using simulation is discussed in section four. Various approaches to pricing American options using simulation are briefly described in the fifth section, and other issues are touched on briefly in the final section.

Variance reduction techniques

In this section, we discuss the role of variance reduction in meeting the broader objective of improving the computational efficiency of Monte Carlo simulations. We then discuss three specific variance reduction techniques and illustrate their application to pricing problems.

VARIANCE REDUCTION AND EFFICIENCY IMPROVEMENT

The reduction of variance seems so obviously desirable that the precise argument for its benefit is sometimes overlooked. We briefly review the underlying justification for variance reduction and examine it from the perspective of improving computational efficiency.

Suppose we want to compute a parameter θ – for example, the price of a derivative security. Suppose we can generate by Monte Carlo an i.i.d. sequence $\{\hat{\theta}_i, i = 1,2,\ldots\}$, where each $\hat{\theta}_i$ has expectation θ and variance σ^2. A natural estimator of θ based on n replications is then the sample mean:

$$\frac{1}{n}\sum_{i=1}^{n}\hat{\theta}_i$$

By the central limit theorem, for large n this sample mean is approximately normally distributed with mean θ and variance σ^2/n. Probabilistic error bounds in the form of confidence intervals follow readily from the normal approximation and indicate that the error in the estimator is proportional to $\sigma/\sqrt{n}$. Thus, reducing the variance σ^2 by a factor of 10, say, while leaving everything else unchanged does as much for error reduction as increasing the number of samples by a factor of 100.

Suppose, now, that we have a choice between two types of Monte Carlo estimates which we denote by $\{\hat{\theta}_i^{(1)}, i = 1,2,\ldots\}$ and $\{\hat{\theta}_i^{(2)}, i = 1,2,\ldots\}$. Suppose that both are unbiased, so that $E[\hat{\theta}_i^{(1)}] = E[\hat{\theta}_i^{(2)}] = \theta$, but $\sigma_1 < \sigma_2$, where $\sigma_j^2 = Var[\hat{\theta}^{(j)}]$, $j = 1,2$. From our previous observations it follows that a sample mean of n replications of $\hat{\theta}^{(1)}$ gives a more precise estimate of θ than does a sample mean of n replications of $\hat{\theta}^{(2)}$. But this analysis oversimplifies the comparison because it fails to capture possible differences in the computational effort required by the two estimators. Generating n replications of $\hat{\theta}^{(1)}$ may be more time-consuming than generating n replications of $\hat{\theta}^{(2)}$; smaller variance is not sufficient grounds for preferring one estimator over another.

To compare estimators with different computational requirements as well as different variances, we argue as follows. Suppose the work required to generate one replication of $\hat{\theta}^{(j)}$ is a constant b_j, $j = 1,2$. (In some problems, the work per replication is stochastic; assuming it is constant simplifies the discussion.) With computing time t, the number of replications of $\hat{\theta}^{(j)}$ that can be generated is $\lfloor t/b_j \rfloor$; for simplicity, we drop the $\lfloor \cdot \rfloor$ and treat the ratios t/b_j as though they were integers. The two estimators available with computing time t are therefore

$$\frac{b_1}{t}\sum_{i=1}^{t/b_1}\hat{\theta}_i^{(1)} \quad \text{and} \quad \frac{b_2}{t}\sum_{i=1}^{t/b_2}\hat{\theta}_i^{(2)}$$

For large t, these are approximately normally distributed with mean θ and with standard deviations

$$\sigma_1\sqrt{\frac{b_t}{t}} \quad \text{and} \quad \sigma_2\sqrt{\frac{b_2}{t}}$$

Thus, for large t, the first estimator should be preferred over the second if

$$\sigma_1^2 b_1 < \sigma_2^2 b_2 \tag{1}$$

Equation (1) provides a sound basis for trading off estimator variance and computational requirements. In the light of the discussion leading to (1), it is reasonable to take the product of variance and work per run as a measure of *efficiency*. Using efficiency as a basis for comparison, the lower-variance estimator should be preferred only if the variance ratio σ_1^2/σ_2^2 is smaller than the work ratio b_2/b_1. By the same argument, a higher-variance estimator may actually be preferable if it takes much less time to generate.

In its simplest form, the principle expressed in (1) dates at least to Hammersley and Handscomb (1964, p. 22). More recently, the idea has been substantially extended by Glynn and Whitt (1992). They allow the work per run to be random (in which case each b_j is the *expected* work per run) and also consider efficiency in the presence of bias.

ANTITHETIC VARIATES

Equipped with a basis for evaluating potential efficiency improvements, we can now consider specific variance reduction techniques. One of the simplest and most widely used techniques in financial pricing problems is the method of antithetic variates. We introduce it with a simple example, then generalise.

Consider the problem of computing the Black-Scholes price of a European call option on a no-dividend stock. Of course, there is no need to evaluate this price by simulation, but the example serves as a useful introduction. In the Black-Scholes model, the stock price follows a lognormal diffusion. Independent replications of the terminal stock price under the risk-neutral measure can be generated from the formula

$$S_T^{(i)} = S_0 e^{(r-(1/2)\sigma^2)T+\sigma\sqrt{T}Z_i}, \quad i = 1,\ldots,n \tag{2}$$

where S_0 is the current stock price, r is the riskless interest rate, σ is the stock's volatility, T is the option's maturity and the $\{Z_i\}$ are independent samples from the standard normal distribution. See, eg Hull (1997) for more details on this model. Based on n replications, an unbiased estimator of the price of an option with strike K is given by

$$\hat{C} = \frac{1}{n}\sum_{i=1}^{n} C_i \equiv \frac{1}{n}\sum_{i=1}^{n} e^{-rT}\max\{0, S_T^{(i)} - K\} \tag{3}$$

In this context, the method of antithetic variates[4] is based on the observation that if Z_i has a standard normal distribution, then so does $-Z_i$. The price $\tilde{S}_T^{(i)}$ obtained from (2) with Z_i replaced by $-Z_i$ is thus a valid sample from the terminal stock price distribution. Similarly, each

$$\tilde{C}_i = e^{-rT}\max\{0, \tilde{S}_T^{(i)} - K\}$$

is an unbiased estimator of the option price, as is,

therefore,

$$\hat{C}_{AV} = \frac{1}{n}\sum_{i=1}^{n}\frac{C_i + \tilde{C}_i}{2}$$

A heuristic argument for preferring $\hat{C}_{AV}$ notes that the random inputs obtained from the collection of antithetic pairs $\{(Z_i,-Z_i)\}$ are more regularly distributed than a collection of $2n$ independent samples. In particular, the sample mean over the antithetic pairs always equals the population mean of 0, whereas the mean over finitely many independent samples is almost surely different from 0. If the inputs are made more regular, it may be hoped that the outputs are more regular as well. Indeed, a large value of $S_T^{(i)}$ resulting from a large Z_i will be paired with a small value of $\tilde{S}_T^{(i)}$ obtained from $-Z_i$.

A more precise argument compares efficiencies. Because C_i and $\tilde{C}_i$ have the same variance,

$$\text{Var}\left[\frac{C_i + \tilde{C}_i}{2}\right] = \frac{1}{2}(\text{Var}[C_i] + \text{Cov}[C_i,\tilde{C}_i]) \qquad (4)$$

Thus, we have $\text{Var}[\hat{C}_{AV}] \leq \text{Var}[\hat{C}]$ if $\text{Cov}[C_i,\tilde{C}_i] \leq \text{Var}[C_i]$. However, $\hat{C}_{AV}$ uses twice as many replications as $\hat{C}$, so we must account for differences in computational requirements. If generating the Z_i takes a negligible fraction of the work per replication (which would typically be the case in the pricing of a more elaborate option), then the work to generate $\hat{C}_{AV}$ is roughly double the work to generate $\hat{C}$. Thus, for antithetics to increase efficiency, we require

$$2\text{Var}[\hat{C}_{AV}] \leq \text{Var}[\hat{C}]$$

which, in the light of (4), simplifies to the requirement that $\text{Cov}[C_i,\tilde{C}_i] \leq 0$.

That this condition is met is easily demonstrated. Define ϕ so that $C_i = \phi(Z_i)$; ϕ is the composition of the mappings from Z_i to the stock price and from the stock price to the discounted option payout. As the composition of two increasing functions, ϕ is monotone, so by a standard inequality (eg Section 2.2 of Barlow and Proschan, 1975)

$$E[\phi(Z_i)\phi(-Z_i)] \leq E[\phi(Z_i)]E[\phi(-Z_i)] \qquad (5)$$

ie $\text{Cov}[C_i,\tilde{C}_i] \equiv E[\phi(Z_i)\phi(-Z_i)] - E[\phi(Z_i)]E[\phi(-Z_i)] \leq 0$, and we may conclude that antithetics help.

This argument can be adapted to show that the method of antithetic variates increases efficiency in pricing a European put and other options that depend monotonically on inputs (eg Asian options). The notable departure from monotonicity in some barrier options (eg a down-and-in call) suggests that the use of antithetics in pricing these options may sometimes be less effective.

In computing confidence levels with antithetic variates, it is essential that the standard error be estimated using the sample standard deviation of the n averaged pairs $(C_i + \tilde{C}_i)/2$ and not the $2n$ individual observations $C_1,\tilde{C}_1,\ldots,C_n,\tilde{C}_n$. The averaged pairs are independent but the individual observations are not. This is a case (we will see others shortly) in which the use of a variance reduction technique affects the estimation of the standard error and, in particular, requires some "batching" of observations to deal with dependence.

It is worth noting that the method of antithetic variates is by no means restricted to simulations whose only stochastic inputs are standard normal variates. The most primitive stochastic input in most simulations is a sequence, $\{U_n\}$, of independent variates uniformly distributed on the unit interval. In this case, $1 - U_n$ has the same distribution as U_n, and the pair $\{U_n, 1 - U_n\}$ are called antithetic because they exhibit negative dependence. If the simulation output depends monotonically on the input random numbers, then the output obtained from $\{1 - U_1, 1 - U_2,\ldots\}$ will be negatively correlated with that obtained from $\{U_1,U_2,\ldots\}$, resulting in increased efficiency compared with independent replications.

For further general background on antithetic variates and other methods based on correlation induction, see Bratley, *et al.* (1987), Hammersley and Handscomb (1964), Glynn and Iglehart (1988) and references therein. For some examples of application in finance, see Boyle (1977) – also Chapter 1 of this publication, Clewlow and Carverhill (1994) and Hull and White (1987).

CONTROL VARIATES

The method of control variates is among the most widely applicable, easiest to use and effective of the variance reduction techniques.[5] Simply put, the principle underlying this technique is "use what you know".

The most straightforward implementation of control variates replaces the evaluation of an unknown expectation with the evaluation of the *difference* between the unknown quantity and another expectation whose value is known. A specific illustration can be found in the analysis of Boyle and Emanuel (1985) and Kemna and Vorst (1990) of Asian options. Let P_A be the price of an option whose payout depends on the arithmetic average of the underlying asset. Let P_G be the price of an option equivalent in every respect except that a geometric average replaces the arithmetic average. Most options based on averages use arithmetic averaging, so P_A is of much greater practical value; but whereas P_A is analytically intractable, P_G can often be evaluated in closed form. Can knowledge of P_G be leveraged to compute P_A?

It can, through the control variate method. Write $P_A = E[\hat{P}_A]$ and $P_G = E[\hat{P}_G]$, where $\hat{P}_A$ and $\hat{P}_G$ are the discounted option payouts for a single simulated path of the underlying asset. Then

$$P_A = P_G + E[\hat{P}_A - \hat{P}_G]$$

In other words, P_A can be expressed as the known price P_G plus the expected difference between $\hat{P}_A$ and $\hat{P}_G$. An unbiased estimator of P_A is thus provided by

$$\hat{P}_A^{cv} = \hat{P}_A + (P_G - \hat{P}_G) \quad (6)$$

This representation[6] suggests a slightly different interpretation: $\hat{P}_A^{cv}$ adjusts the straightforward estimator $\hat{P}_A$ according to the difference between the known value, P_G, and the observed value, $\hat{P}_G$. The known error, $(P_G - \hat{P}_G)$, is used as a *control* in the estimation of P_A.

If most of the computational effort goes to generating paths of the underlying asset, then the additional work required to evaluate $\hat{P}_G$ along with $\hat{P}_A$ is minor. It therefore seems reasonable to compare variances alone. Since

$$Var[\hat{P}_A^{cv}] = Var[\hat{P}_A] + Var[\hat{P}_G] - 2Cov[\hat{P}_A, \hat{P}_G]$$

this method if effective if the covariance between $\hat{P}_A$ and $\hat{P}_G$ is large. The numerical results of Kemna and Vorst indicate that this is indeed the case. Fu, Madan and Wang (1995) have investigated the use of other control variates for Asian options based on Laplace transform values. These appear to be less strongly correlated with the option price.

A closer examination of (6) reveals that this estimator does not make optimal use of the relation between the two option prices. Consider the family of unbiased estimators

$$\hat{P}_A^{\beta} = \hat{P}_A + \beta(P_G - \hat{P}_G) \quad (7)$$

parameterised by the scalar β. We have

$$Var[\hat{P}_A^{\beta}] = Var[\hat{P}_A] + \beta^2 Var[\hat{P}_G] - 2\beta Cov[\hat{P}_A, \hat{P}_G]$$

The variance-minimising β is therefore

$$\beta^* = \frac{Cov[\hat{P}_A, \hat{P}_G]}{Var[\hat{P}_G]}$$

Depending on the application, β^* may or may not be close to 1, the implicit value in (6). In using an estimator of the form (6), we forgo an opportunity for greater variance reduction. Indeed, whereas (6) may increase or reduce variance, an estimator based on β^* is guaranteed not to increase variance, and will result in a strict decrease in variance so long as $\hat{P}_A$ and $\hat{P}_G$ are not uncorrelated.

In practice, of course, we rarely know β^* because we rarely know $Cov[\hat{P}_A, \hat{P}_G]$. However, given n independent replications $\{(P_{Ai}, P_{Gi}), i = 1,\ldots,n\}$ of the pairs $(\hat{P}_A, \hat{P}_G)$, we can estimate β^* via regression. At this point we face a choice. Using all n replications to compute an estimate, $\hat{\beta}$, of β^* introduces a (typically small) bias in the estimator

$$\frac{1}{n}\sum_{i=1}^{n} P_{Ai} + \hat{\beta}\left(P_G - \frac{1}{n}\sum_{i=1}^{n} P_{Gi}\right)$$

and its estimated standard error because of the dependence between $\hat{\beta}$ and the P_{Gi}. Reserving n_1 replications for the estimation of β^* and the remaining $n - n_1$ replications for the sample mean of the P_{Gi} (typically with $n_1 \ll n$) eliminates the bias but may cause deterioration in the estimate of β^*. Neither issue significantly limits the applicability of the method because the possible bias vanishes as n increases and because the estimator of β^* need not be very precise to achieve a reduction in variance.

The advantage of working with (7) over (6) becomes even more pronounced when further controls are introduced. For example, when the asset price is simulated under risk-neutral probabilities, the present value, $e^{-rT}E[S_T]$, of the terminal price must equal the current price, S_0. We can therefore form the estimator

$$\hat{P}_A + \beta_1(P_G - \hat{P}_G) + \beta_2(S_0 - e^{-rT}S_T)$$

The variance-minimising coefficients (β_1^*, β_2^*) are easily found by multiple regression. This optimisation step seems particularly crucial in this case, for whereas one might guess that β_1^* is close to 1, it seems unlikely that β_2^* would be. Optimising over the βs also allows us to exploit controls that are negatively correlated with the option payoff.

For further general background on control variates see Bratley *et al.* (1987), Glynn and Iglehart (1988) and Lavenberg and Welch (1981). For examples of control variate applications in finance, see Boyle (1977), Broadie and Glasserman (1996), Carverhill and Pang (1995), Clewlow and Carverhill (1994), Duan (1995) and Kemna and Vorst (1990).

MOMENT-MATCHING METHODS

Next we describe a variance reduction technique that was proposed by Barraquand (1995), who termed it *quadratic resampling*. His technique is based on *moment matching*. As before, we introduce it with the simple example of estimating the European call option price on a single asset and then generalise.

Let Z_i, $i = 1,\dots,n$, denote independent standard normals to drive a simulation. The sample moments of the n Zs will not exactly match those of the standard normal. The idea of moment matching is to transform the Zs to match a finite number of the moments of the underlying population. For example, the first moment of the standard normal can be matched by defining

$$\tilde{Z}_i = Z_i - \tilde{Z}, \quad i = 1,\dots,n \tag{8}$$

where $\tilde{Z} = \sum_{i=1}^{n} Z_i/n$ is the sample mean of the Zs. Note that the $\tilde{Z}_i$s are normally distributed if the Z_is are normal. However, the $\tilde{Z}_i$s are not independent. As before, terminal stock prices are generated from the formula

$$\tilde{S}_T(i) = S_0 e^{(r-(1/2)\sigma^2)T + \sigma\sqrt{T}\tilde{Z}_i}, \quad i = 1,\dots,n$$

A moment-matched estimator of the call option price is the average of the n values $\tilde{C}_i = e^{-rT} \max(\tilde{S}_T(i) - K, 0)$.

In the standard Monte Carlo method, confidence intervals for the true value, C, could be estimated from the sample mean and variance of the estimator. This cannot be done here as the n values of $\tilde{Z}$ are no longer independent, and hence the values $\tilde{C}_i$ are not independent. This points out one drawback of the moment-matching method: confidence intervals are not as easy to obtain.[7] Indeed, for confidence intervals it appears to be necessary to apply moment matching to independent batches of runs and estimate the standard error from the batch means. This reduces the efficacy of the method compared with matching moments across all runs.

Equation (8) showed one way to match the first moment of a distribution with mean zero. If the underlying population does not have a zero mean, transformed Zs could be generated using $\tilde{Z}_i = Z_i - \tilde{Z} + \mu_Z$, where μ_Z is the population mean. The idea can easily be extended to match two moments of a distribution. In this case, an appropriate transformation is

$$\tilde{Z}_i = (Z_i - \tilde{Z})\frac{\sigma_Z}{s_Z} + \mu_Z, \quad i = 1,\dots,n \tag{9}$$

where s_Z is the sample standard deviation of the Z_is and σ_Z is the population standard deviation. Of course, for a standard normal, $\mu_Z = 0$ and $\sigma_Z = 1$. An estimator of the call option price is the average of the n values $\tilde{C}_i$.

Using the transformation (9), the $\tilde{Z}_i$s are not normally distributed even if the Z_is are normal. Hence, the corresponding $\tilde{C}_i$ are biased estimators of the true option value. For most financial problems of practical interest this bias is likely to be small. However, the bias can be arbitrarily large in extreme circumstances (even when only the first moment of the distribution is matched).[8] The dependence and bias in the moment-matching method make it difficult to quantify the improvement in general analytical terms.

The moment-matching method is another example of the idea to "use what you know". In this simple European option example, the mean and variance of the terminal stock price S_T are also known. So the moment-matching idea could be applied to the simulated terminal stock values $S_T(i)$. In this case, to match the first moment, define

$$\tilde{S}_T(i) = S_T(i) - \tilde{S}_T + \mu_{S_T} \tag{10}$$

where $\mu_{S_T} = S_0 e^{rT}$ and $\tilde{S}_T$ is the sample mean of the $S_T(i)$. To match the first two moments, define

$$\tilde{S}_T(i) = (S_T(i) - \tilde{S}_T)\frac{\sigma_{S_T}}{s_{S_T}} + \mu_{S_T} \tag{11}$$

where $\sigma_{S_T} = S_0\sqrt{e^{2rT}(e^{\sigma^{2T}-1} - 1)}$ and s_{S_T} is the sample standard deviation of the $S_T(i)$. Duan and Simonato (1995) use a related method. They apply a multiplicative transformation to asset prices to enforce the martingale property over a finite set of paths.[9] They apply their method to Garch option pricing.

Comparisons of various moment-matching strategies are given in Table 1. For this comparison, n = 100 simulation trials were used to

Table 1. Standard errors for European call options

σ	S_0/K	No variance reduction	MM1 Equation (8)	MM2 Equation (9)	MM1 Equation (10)	MM2 Equation (11)
0.2	0.9	0.24	0.19	0.11	0.19	0.09
	1.0	0.62	0.29	0.09	0.26	0.10
	1.1	0.93	0.19	0.09	0.15	0.11
0.4	0.9	0.80	0.55	0.24	0.51	0.17
	1.0	1.22	0.66	0.19	0.56	0.23
	1.1	1.61	0.63	0.17	0.48	0.28
0.6	0.9	1.40	0.95	0.38	0.84	0.28
	1.0	1.93	1.10	0.31	0.91	0.39
	1.1	2.38	1.13	0.25	0.85	0.49

All results are based on $n = 100$ simulation trials. The option parameters are: $K = 100$, $r = 0.10$, $T = 0.2$, with S_0 and σ varying as indicated. Standard error estimates are based on $m = 10,000$ simulations. MM = moment matching.

estimate the European call option price. Standard errors were estimated by re-simulation. That is, $m = 10,000$ simulation trials were conducted, each one based on n replications of the estimator. The sample standard deviation of the m simulation estimates gives an estimate of the standard error of a single simulation estimate. Root-mean-squared errors are not reported because they are identical to the standard errors for the number of digits reported.

The results in Table 1 show that matching two moments can reduce the simulation error by a factor ranging from two to 10. Matching two moments dominates matching one moment, but there is no clear choice between transforming the original standard normals using (9) or the terminal stock prices using (11). Further computational results, not shown in Table 1, indicate that the improvement factor with moment matching is essentially constant as n increases. This may seem counter-intuitive, as the moment matching adjustments converge to zero as n increases. But the progressively smaller adjustments are equally important in reducing the estimation error as the number of simulation trials increases. For example, the standard error for $n = 10,000$ simulation trials is one-tenth of the corresponding number for $n = 100$ reported in Table 1.

The moment-matching method can be extended to match covariances. For options that depend on multiple assets, the entire covariance structure is typically a simulation input. Barraquand (1995) suggests a method to match the entire covariance structure and reports error reduction factors ranging from two to several hundred for this method applied to pricing options on the maximum of k assets.

The moment-matching procedure could be applied to the matching of higher-order moments as well. In addition to different methods for transforming random outcomes to match specified moments, additional points could be added as another way to match moments.

Whenever a moment is known, it can be used as a control rather than for moment matching. In an appendix, we give a theoretical argument favouring the use of moments as controls rather than for matching.

STRATIFIED AND LATIN HYPERCUBE SAMPLING

Like many variance reduction techniques, stratified sampling seeks to make the inputs to simulation more regular than random inputs. In particular, it forces certain empirical probabilities to match theoretical probabilities, just as moment matching forces empirical moments to match theoretical moments.

Consider, for example, the generation of 100 normal random variates as inputs to a simulation. The empirical distribution of an independent sample $Z_1,\ldots,Z_{100}$ will look only roughly like the normal density; the tails of the distribution – often the most important part – will inevitably be underrepresented. Stratified sampling can be used to force exactly one observation to lie between the $(i - 1)$th and ith percentile, $i = 1,\ldots,100$, and thus produce a better match to the normal distribution. One way to implement this generates 100 independent random variates $U_1,\ldots,U_{100}$, uniform on [0,1], and sets $\tilde{Z}_i = N^{-1}((i + U_i - 1)/100)$, $i = 1,\ldots,100$, where N^{-1} is the inverse of the cumulative normal distribution. This works because $(i + U_i - 1)/100$ falls between the $(i - 1)$th and the ith percentiles of the uniform distribution, and percentiles are preserved by the universe transform.

Of course, $\tilde{Z}_1,\ldots,\tilde{Z}_{100}$ are highly dependent, complicating the estimation of standard errors. Computing confidence intervals with stratified sampling typically requires batching the runs. For example, with a budget of 100,000 replications, we might run 100 independent stratified samples, each of size 10,000. To estimate standard errors we must therefore sacrifice some variance reduction, just as with moment matching.

In principle, this approach applies in arbitrary dimensions. To generate a stratified sample from the d-dimensional unit hypercube, with n strata in each coordinate, we could generate a sequence of vectors $U_j = (U_j^{(1)},\ldots,U_j^{(d)})$, $j = 1,2,\ldots$, and then set

$$V_j = \frac{U_j + (i_1,\ldots,i_d)}{n}, \quad i_k = 0,\ldots,n-1, \ k = 1,\ldots,d$$

Exactly on V_j will lie in each of the n^d cubes defined by the product of the n strata in each coordinate.

The difficulty in high dimensions is that generating even a singly stratified sample of size n^d may be prohibitive unless n is very small. Latin hypercube sampling can be viewed as a way of randomly sampling n points of a stratified sample while preserving some of the regularity from stratification. The method was introduced by McKay, Conover and Beckman (1979) and further analysed in Stein (1987). It works as follows. Let $\pi_1,\ldots,\pi_d$ be independent random permutations of $\{1,\ldots,n\}$, each uniformly distributed over all n! possible permutations. Set

$$V_j^{(k)} = \frac{U_j^{(k)} + \pi_k(j) - 1}{n}, \quad k = 1,\ldots,d, \ j = 1,\ldots,n$$

The randomisation ensures that each vector V_j is uniformly distributed over the d-dimensional hypercube. At the same time, the coordinates are perfectly stratified in the sense that exactly one of $V_1^{(k)},\ldots,V_n^{(k)}$ falls between $(j-1)/n$ and j/n, $j = 1,\ldots,n$, for each dimension $k = 1,\ldots,d$. As before, the dependence introduced by this method implies that standard errors can be estimated only through batching.

These methods can be viewed as part of a hierarchy of methods introducing additional levels of regularity in inputs at the expense of complicating the estimation of errors. Some, like stratified sampling, fix the size of the sample, while others leave flexibility. The extremes of this hierarchy are straightforward Monte Carlo (completely random) and the low-discrepancy methods (completely deterministic) discussed later in this chapter. Owen (1994, 1995) discusses these and other methods and introduces a hybrid that combines the regularity of low-discrepancy methods with the simple error estimation of standard Monte Carlo. Shaw (1995) uses an extension proposed by Stein (1987) to handle dependent inputs in a novel approach to estimating value-at-risk.

SOME NUMERICAL COMPARISONS

The variance reduction methods discussed thus far are fairly generic in the sense that they do not rely on the detailed structure of the security to be priced. This contrasts with the remaining two methods that we discuss - importance sampling and conditional Monte Carlo. These methods must be carefully tailored to each application. It therefore seems appropriate to digress briefly into a numerical comparison of the generic methods on some option pricing problems.

We first examine the performance of these methods in pricing Asian options. The payoff of a discretely sampled arithmetic average Asian option is $\max(\tilde{S} - K, 0)$, where $\tilde{S} = \sum_{i=1}^{k} S_i/k$, S_i is the asset price at time $t_i = iT/k$, and T is the option maturity. The value of the option is $E[e^{-rT} \max(\tilde{S} - K, 0)]$. There is no easily evaluated closed-form expression for this option value. Various formulas to approximate the Asian option price have been developed, but simulation is usually used to test the accuracy of the approximations.

For this Asian option, k random numbers are needed to simulate one option payoff, and nk random numbers are needed in total. Moment matching (MM2, for two moments) was applied k times to the n numbers used to generate each S_i at time t_i. Latin hypercube sampling (LHS) was applied to sample n points from the k-dimensional unit cube. The discretely sampled geometric average Asian price was used as a control variate (see Turnbull and Wakeman, 1991, for a closed-form solution for this price). The results appear in Table 2.

Table 2. Standard errors for arithmetic average Asian options

σ	K/S_0	No variance reduction	Antithetic method	Control variate	MM2	LHS
0.2	0.9	0.053	0.052	0.003	0.048	0.049
	1.0	0.344	0.231	0.004	0.162	0.161
	1.1	0.566	0.068	0.006	0.052	0.058
0.4	0.9	0.308	0.297	0.014	0.240	0.248
	1.0	0.694	0.506	0.017	0.352	0.354
	1.1	1.017	0.388	0.021	0.281	0.289
0.6	0.9	0.632	0.583	0.032	0.451	0.455
	1.0	1.052	0.817	0.038	0.566	0.578
	1.1	1.443	0.759	0.047	0.539	0.560

All results are based on n = 100 simulation trials with k = 50 prices in the average. The option parameters are: K = 100, r = 0.10, T = 0.2, with S_0 and σ varying as indicated. Standard error estimates based on m = 10,000 simulations. The geometric average Asian option is used as the control variate. Moment matching (MM2) was applied to the ith price in the average, i = 1,..., across replications.

The results in Table 2 show that matching two moments can reduce the simulation error by a factor ranging from one to 10. Using the geometric average Asian option price as a control variate reduces error by a factor ranging from 20 to 100, and is consistently the most effective method. LHS and MM2 perform similarly. Antithetics are consistently dominated by the other methods.

Next, we compare these variance reduction techniques in pricing down-and-out call options with discrete barriers. The payoff of this option at expiration is the standard call option payoff if the asset price S_i exceeds the barrier H at all times $t_i = iT/k$, $i = 1,\ldots,k$, otherwise the payoff is zero. The option is knocked out if $S_i \leq H$ at any time t_i. As a control we use the Black-Scholes price of a standard call. Moment matching and LHS are implemented as with the Asian option. Results are given in Table 3. These are consistent with the pattern in Table 2, except that the superiority of the control variate method is less pronounced.

Although it is always risky to draw conclusions from limited numerical evidence, we suggest the following broad conclusions. The antithetic method is easy to implement but often leads to only modest error reductions. Moment matching is similarly easy to implement and often leads to significant error reductions, but the error estimation is more difficult and bias is a potential problem. LHS suffers from the same error estimation difficulty but does not introduce bias. The control variate technique can lead to very substantial error reductions, but its effectiveness hinges on finding a good control for each problem.

IMPORTANCE SAMPLING

This technique builds on the observation that an expectation under one probability measure can be expressed as an expectation under another through the use of a likelihood ratio or Radon-Nikodym derivative. This idea is familiar in finance because it underlies the representation of prices as expectations under a martingale measure. In Monte Carlo, the change of measure is used to try to obtain a more efficient estimator. We present some examples using this technique; for general background see Bratley *et al.* (1987) or Hammersley and Handscomb (1964).

As a simple example, consider the evaluation of the Black-Scholes price of a call option, ie the computation of $e^{-rT}E(\max\{S_T - K,0\}]$ with S_T as in (2). A straightforward approach generates samples of the terminal value, S_T, consistent with a geometric Brownian motion having drift r and volatility σ, just as in (2). But we are in fact free to generate S_T consistent with any other drift, μ, provided we weight the result with a likelihood ratio. For emphasis, we subscript the expectation operator with the drift parameter. Then

$$E_r\{\max[S_T - K,0\}] = E_\mu\{\max[S_T - K,0\}L]$$

where the likelihood ratio, L, is the ratio of the lognormal densities with parameters r and μ evaluated at S_T, given by

$$L = \left(\frac{S_T}{S_0}\right)^{(r-\mu)/\sigma^2} \exp\left(\frac{(\mu^2 - r^2)T}{2\sigma^2}\right)$$

Indeed, S_T need not even be sampled from a lognormal distribution. The only requirement is that the support of the importance sampling measure contain the support of the original measure so that the likelihood ratio is well defined; this is an *absolute continuity* requirement. In the example above, this means that any distribution for S_T whose support includes $(0,\infty)$ is admissible.

Ideally, one would like to choose the importance sampling distribution to reduce variance. In the example above, one obtains a *zero-variance* estimator by sampling S_T from the density

$$f(x) = c^{-1}\max\{x - K,0\}e^{-rT}g(x)$$

where g is the (lognormal) density of S_T and c is a normalising constant that makes f integrate to 1. The difficulty is that c is the Black-Scholes price itself, so this method requires knowledge of the solution for its implementation. Nevertheless, it gives some indication of the potential gain from importance sampling.

Table 3. Standard errors for down-and-out call options with discrete barriers

σ	K/S_0	No variance reduction	Antithetic method	Control variate	MM2	LHS
0.2	0.9	0.96	0.44	0.37	0.43	0.39
	1.0	0.62	0.44	0.13	0.31	0.30
	1.1	0.30	0.28	0.03	0.22	0.22
0.4	0.9	1.59	1.15	0.73	0.95	0.88
	1.0	1.22	1.00	0.45	0.76	0.74
	1.1	0.88	0.82	0.26	0.61	0.61
0.6	0.9	2.19	1.83	1.07	1.44	1.36
	1.0	1.86	1.62	0.80	1.25	1.23
	1.1	1.54	1.40	0.58	1.09	1.09

All results are based on $n = 100$ simulation trials. There are $k = 5$ points in the discrete barrier at 95. The other option parameters are: $S_0 = 100$, $r = 0.10$, $T = 0.2$, with K and σ varying as indicated. Standard error estimates are based on $m = 10{,}000$ simulations. The standard European call option (Black–Scholes formula) is used as the control variate. Moment matching (MM2) was applied to the ith return, $i = 1,\ldots,5$, across replications

Reider (1993) has investigated the effect of importance sampling based on a change of drift and volatility. (Changing the volatility is consistent with absolute continuity in a discrete-time approximation of a diffusion though not in the continuous-time limit.) He finds that choosing the importance sampling distribution to have higher drift and volatility provides substantial variance reduction in pricing deep out-of-the-money options. He also investigates the combination of importance sampling with antithetic variates and control variates, and the use of put–call parity for indirect estimation. Neilsen (1994) has explored some related importance sampling ideas in sampling from a binomial tree.

Andersen (1995) has developed a powerful application of importance sampling for simulating interest rates and has applied it to non-linear stochastic differential equation models. We briefly describe his approach. Let r_t be the instantaneous short rate described, for example, by a diffusion model. Then

$$B(T) = E\left[\exp\left(-\int_0^T r_t dt\right)\right]$$

is the price today of a zero-coupon bond with face value $1 maturing at time T. In, for example, the Cox–Ingersoll–Ross and Vasicek models.[10] B(T) is available in closed form. We may therefore define a new probability measure, $\tilde{P}$, by setting

$$\tilde{P}(A) = E\left[\exp\left(-\int_0^T r_t dt - \log B(T)\right)\mathbf{1}_A\right]$$

for any event A, where $\mathbf{1}_A$ denotes the indicator of the event A. Let $\tilde{E}$ denote expectation with respect to $\tilde{P}$. Then, for any random variable, X, $E[X] = \tilde{E}[XL_T]$, where the likelihood ratio, L_T, is given by

$$L_T = \exp\left(\int_0^T r_t dt + \log B(T)\right)$$

In particular, if we take $X = \exp(-\int_0^T r_t dt)$, we know that E[X] = B(T) and, therefore, B(T) is the expectation under $\tilde{E}$ of XL_T - ie of

$$\exp\left(-\int_0^T r_t dt\right) \times \exp\left(\int_0^T r_t dt + \log B(T)\right)$$

But this simplifies to B(T) itself, meaning that we obtain a zero-variance estimator of the bond price by switching to the new probability measure. Moreover, Anderson shows that sample paths of r_t can be generated under $\tilde{P}$ simply by applying a change of drift to the original process.

As described above, the method would appear to require knowledge of the solution for its implementation. Nevertheless, the method has two important applications. The first is in the pricing of contingent claims. Because $\tilde{P}$ eliminates the variance of bond prices, it should be effective in reducing variance for pricing, for example, European bond options expiring at time T. Andersen's numerical results bear this out. A second application is in the pricing of bond models with no closed-form solutions: Andersen's results show that the change of drift derived from a tractable model (like CIR or Vasicek) remains effective when applied to an intractable model, and this significantly expands the scope of the method.

Importance sampling is frequently used to make rare events less rare; this is already suggested in Reider's (1993) application to out-of-the-money options. Our next example further highlights this aspect through a new application to barrier options. We consider a knock-in option far from the barrier and use importance sampling to increase the probability of a payout.

Suppose the barrier is monitored at discrete times $n\Delta t$, $n = 0,1,\ldots,m$, with $\Delta T = T/m$. Set the barrier at $H = S_0e^{-b}$ and the strike at $K = S_0e^{c}$, with $b,c > 0$. A down-and-in call pays $S_T - K$ at time T if $S_T > K$ and $S_{n\Delta t} < H$ for some $n = 1,\ldots,m$. We can write the price of the underlying at monitoring instants as

$$S_{n\Delta t} = S_0e^{U_n}, \quad U_n = \sum_{i=1}^{n} X_i$$

with the X_i i.i.d. normal having mean $(r - \frac{1}{2}\sigma^2)\Delta t$ and variance $\sigma^2\Delta t$. Let τ be the first time U_n drops below $-b$; then the probability of a payout is $P(\tau < m, U_m > c)$. If b and c are large, this probability is small and most simulation runs return zero. Through importance sampling, we can increase this probability and thus get more information out of each run.

Consider alternative probability measures P_{μ_1,μ_2} that give U_n a drift of $\mu_1\Delta t$ until τ and then switch the drift to $\mu_2\Delta t$. Intuitively, we would like to make $\mu_1 < 0$ to drive the asset price to the barrier and then make $\mu_2 > 0$ to drive it above the strike. For any μ_1, μ_2, we have

$$P(\tau < m, U_m > c) = E_{\mu_1,\mu_2}\left[L_{\mu_1,\mu_2}\mathbf{1}_{\{\tau<m,U_m>c\}}\right]$$

The likelihood ratio is given by

$$L_{\mu_1,\mu_2} = \exp(-\theta_1 U_\tau + \psi(\theta_1)\tau - \theta_2(U_m - U_\tau) + \psi(\theta_2)(m - \tau))$$

where $\theta_i = (\mu_i - r + \frac{1}{2}\sigma^2)/\sigma^2$, $i = 1,2$, and $\psi(\theta) = r - \frac{1}{2}\sigma^2)\Delta t\theta + \frac{1}{2}\sigma^2\Delta t\theta^2$. This follows from algebraic simplification of the product of the ratios of the densities of the X_i under the original and new means.

It remains to choose μ_1, μ_2. Intuitively, most of the variability in L_{μ_1,μ_2} comes from τ (the time of the barrier crossing): for large b, c, in the event of a payout we expect to have $U_\tau \approx -b$ and $U_m \approx c$ so these terms should contribute less variability. If we choose μ_1, μ_2 so that $\psi(\theta_1) = \psi(\theta_2)$, the likelihood ratio simplifies to

$$L_{\mu_1,\mu_2} = \exp(-(\theta_1 - \theta_2)U_\tau - \theta_2 U_m + m\psi(\theta_2))$$

which depends on τ only through $U_\tau \approx -b$. The condition $\psi(\theta_1) = \psi(\theta_2)$ translates to $\mu_1 = -\mu_2 \equiv -\mu$, so it only remains to choose this drift parameter. We choose it so that the time to traverse the straight-line path from 0 to $-b$ and then to c at rate μ equals the number of steps, m:

$$\frac{b}{\mu\,\Delta t} + \frac{(b + c)}{\mu\,\Delta t} = m$$

ie $\mu = (2b + c)/T$. Interestingly, this change of drift does not depend on the original mean increment $(r - \frac{1}{2}\sigma^2)\Delta t$.

Table 4 illustrates the performance of this method. The computational effort with and without importance sampling is essentially the same, so the efficiency improvement is just the ratio of the variances. The improvement varies widely but shows the potential for dramatic gains from importance sampling, particularly when the barrier is far from the current price of the underlying.[11]

In recent work, Andersen and Brotherton-Ratcliffe (1996) and Beaglehole, Dybvig, and Zhou (1997) show how to eliminate the bias caused by using a simulation at a discrete set of times to price continuous options on extrema – for example, barrier or lookback options.

Table 4. Standard errors for down-and-in calls: importance sampling

H	K	No variance reduction	Importance sampling	Efficiency ratio
92	100	0.00309	0.00069	20
92	105	0.00129	0.00014	85
88	96	0.00110	0.00011	96
85	90	0.00084	0.00008	116
92	105	0.01418	0.00541	7
85	105	0.00328	0.00038	75
75	96	0.00030	0.00001	1124
75	85	0.00148	0.00010	222

All results are based on $n = 100{,}000$ simulation trials. The parameters are: $S_0 = 95$, $\sigma = 0.15$, and $r = 0.05$, with the barrier H and strike K varying as indicated. The first four cases have $T = 0.25$ and $m = 50$; the last four have $T = 1$ and $m = 250$.

CONDITIONAL MONTE CARLO

This approach to efficiency improvement exploits the variance reducing property of conditional expectation: for any random variables X and Y, $\mathrm{Var}[E[X \mid Y]] \le \mathrm{Var}[X]$, with strict inequality except in trivial cases.[12] In replacing an estimator by its conditional expectation, we reduce variance essentially because we are doing part of the integration analytically and leaving less to be done by Monte Carlo.

Hull and White (1987) use this idea to price options with stochastic volatilities. Consider a model in which an asset price and its variability evolve as follows:

$$dS = rSdt + vSdW_1, \quad dv^2 = \alpha v^2 dt + \xi v^2 dW_2$$

with W_1, W_2 independent. Suppose we want to price a standard European call on S. A straightforward approach simulates sample paths of v and S up to time T and averages $\max\{S_T - K, 0\}$ over all paths. An alternative notes that, conditional on the path of v_t in $[0,T]$, the asset price S_t may be treated as having a time-varying but deterministic volatility. Thus, conditional on the volatility path, the option can be priced by the Black–Scholes formula:

$$e^{-rT}E[\max\{S_T - K, 0\} \mid v_t,\ 0 \le t \le T] = BS(S_0, K, r, T, \sqrt{V_T})$$

where

$$V_T = \frac{1}{T}\int_0^T v_t^2 dt$$

is the average squared volatility over the path, and $BS(S,K,T,r,\sigma)$ is the Black–Scholes price of a call with constant volatility, σ, and the other parameters as indicated. Using this conditional expectation as the estimator is sure to reduce variance and may even reduce computational effort since it obviates simulation of S. It is worth emphasising that both straightforward Monte Carlo and conditional Monte Carlo would have to be applied to discrete-time approximations of the continuous processes above. Also, the applicability of conditional Monte Carlo in this setting relies critically on the fact that the evolution of the asset price does not influence the volatility path. See Willard (1996) for an extension to the case of correlated W_1 and W_2.

As a further illustration of the use of conditional Monte Carlo, we give a new illustration in the pricing of a down-and-in call with a discretely monitored barrier. Let $0 = t_0 < t_1 < \cdots < t_m = T$ be the monitoring instants and S_{t_i} the price of the underlying at the ith such instant. The option price is $E[e^{-rT}\max\{S_T - K,0\}\mathbf{1}_{\{\tau_H \le T\}}]$, where H is the barrier and τ_H is the first monitoring time at which the barrier is breached.

Straightforward simulation generates paths of the underlying and evaluates the estimator

$$e^{-rT}\max\{S_T - K,0\}\mathbf{1}_{\{\tau_H \le T\}}$$

Our first alternative conditions on $\{S_0,\ldots,S_{\tau_H}\}$, the path of the underlying until the barrier crossing ie

$$\begin{aligned}&E[e^{-rT}\max\{S_T - K,0\}\mathbf{1}_{\{\tau_H \le T\}}]\\ &= e^{-rT}E\big[E[\max\{S_T - K,0\}\mathbf{1}_{\{\tau_H \le T\}}] \mid S_0,\ldots,S_n]\big]\\ &= e^{-rT}E\big[BS(S_{\tau_H},K,r,T - \tau_H,\sigma)\mathbf{1}_{\{\tau_H \le T\}}\big]\end{aligned}$$

This yields the estimator

$$CMC_1 = e^{-rT}BS(S_{\tau_H},K,r,T - \tau_H,\sigma)\mathbf{1}_{\{\tau_H \le T\}}$$

This says: simulate until the barrier is crossed or the option expires; if the barrier was crossed, return the Black-Scholes price starting from price S_{τ_H} with maturity $T - \tau_H$.

Our second alternative conditions one step earlier, at each monitoring instant evaluating the probability that the barrier will be breached for the first time at the next monitoring instant:

$$\begin{aligned}&E\big[e^{-rT}\max\{S_T - K,0\}\mathbf{1}_{\{\tau_H \le T\}}\big]\\ &= e^{-rT}E\left[\max\{S_T - K,0\}\sum_{n=1}^{m}\mathbf{1}_{\{\tau_H = t_n\}}\right]\\ &= e^{-rT}E\left[\sum_{n=1}^{m}E\big[\max\{S_T - K,0\}\mathbf{1}_{\{\tau_H = t_n\}} \mid S_{t_0},\ldots,S_{t_{n-1}}\big]\right]\\ &= e^{-rT}E\left[\sum_{n=0}^{\tau_H - 1}BS2(S_{t_n},K,H,r,t_{n+1} - t_n,T - t_n,\sigma)\right]\end{aligned}$$

where $BS2(S,K,H,r,t,T,\sigma)$ is the price of a down-and-in call that knocks in only if the underlying is below H at time t. We thus arrive at the estimator

$$CMC_2 = e^{-rT}\sum_{n=0}^{\tau_H - 1}BS2(S_{t_n},K,H,r,t_{n+1} - t_n,\ T - t_n,\sigma)$$

with

$$\begin{aligned}&BS2(S,K,H,r,t,T,\sigma)\\ &= SN_2(a_1,b_1,\rho) - e^{-rT}KN_2(a_2,b_2,\rho)\end{aligned}$$

where $\rho = -\sqrt{t/T}$, N_2 is the bivariate cumulative normal distribution with correlation ρ, and

$$a_1 = \frac{\log(S/K) + (r + \frac{1}{2}\sigma^2)T}{\sigma\sqrt{T}}, \quad a_2 = a_1 - \sigma\sqrt{T}$$

$$b_1 = \frac{\log(H/S) - (r + \frac{1}{2}\sigma^2)t}{\sigma\sqrt{t}}, \quad b_2 = b_1 + \sigma\sqrt{t}$$

(The derivation of this formula is fairly standard and therefore omitted.) The CMC_2 estimator can be expected to have lower variance than the CMC_1 estimator because it conditions on *less* information and thus does *more* integration analytically. In fact, CMC_2 is not a conditional Monte Carlo estimator in the strict sense because it conditions on different information at different times, making it more precisely a *filtered* Monte Carlo estimator in the sense of Glasserman (1991b).

Because the two estimators above have the same expectation, their difference has mean 0 and can be used as a control variate to form a further estimator

$$CMC' = CMC_1 + \beta(CMC_2 - CMC_1)$$

With β optimised, this has lower variance than either individual estimator.

Numerical results appear in Table 5. As expected, each level of conditioning further reduces variance, and the combined estimator achieves the lowest standard error of all. However, repeated evaluation of the function BS2 turns out to be time-consuming, making CMC_1 overall the most efficient estimator.

Table 5. Comparison of CMC estimators for down-and-in call

Method	Standard error (s)	Computation time (t)	$s\sqrt{t}$
Base	0.108	0.133	0.039
CMC_1	0.034	0.117	0.012
CMC_2	0.021	3.233	0.038
CMC′	0.014	3.367	0.026

Results based on n = 10,000 replications with $\sigma = 0.4$, $r = 0.10$, $S_0 = K = 100$, $H = 95$, $T = 0.5$, and 10 equally spaced monitoring times.

Low-discrepancy sequences

For complex problems the performance of the basic Monte Carlo approach may be rather unsatisfactory because the error is $O(1/\sqrt{n})$. We can sometimes improve convergence by using preselected deterministic points to evaluate the integral. The accuracy of this approach depends on the extent to which these deterministic points are evenly dispersed throughout the domain of integration. *Discrepancy* measures the extent to which the points are evenly dispersed throughout a region: the more evenly the points are dispersed, the lower the discrepancy. Low-discrepancy sequences are often called "quasi-random" sequences even though they are not at all random.[13] We shall use both terms here.

Low-discrepancy methods have recently been used to tackle a number of problems in the finance area. These applications are more fully described in papers by Birge (1994); Joy, Boyle and Tan (1996) - also Chapter 24 of this publication; and Paskov and Traub (1995); the use of quasi-Monte Carlo is also proposed in Cheyette (1992) - also Chapter 19 of this publication. In this section we describe how the approach works and review some of the recent applications. The book by Press *et al.* (1992) provides an intuitive introduction to low-discrepancy sequences and quasi-Monte Carlo methods. Spanier and Maize (1994) provide a recent overview of quasi-random methods and how they can be used to evaluate integrals with medium-sized samples. Niederreiter (1992) and Tezuka (1995) provide in-depth analyses of low-discrepancy sequences. Moskowitz and Caflisch (1996) - also Chapter 27 of this publication discuss recent developments in improving the convergence of quasi-random Monte Carlo methods. In earlier work, Haselgrove (1961) describes a method for multivariate integration that can be applied to security pricing. Haselgrove's method is developed for problems of eight dimensions or less, and our numerical experiments suggest that it is competitive with the low-discrepancy sequences investigated in this section for problems of this size.

The basic idea behind the approach is quite intuitive and is readily explained in the one-dimensional case. Suppose we wish to integrate a function, $f(x)$, over the interval $[0,1]$ using a sequence of n points. Rather than pick a random sequence, suppose we pick a deterministic sequence of points that are, in some sense, evenly distributed. With this choice, the accuracy of the estimate will be higher than that obtained using the crude Monte Carlo approach. If we use an equally spaced grid we obtain the trapezoidal method of numerical integration, which has an error of $O(n^{-2})$. However, the more challenging task is to evaluate multidimensional integrals. Without loss of generality, we can assume that the domain of integration is contained in the d-dimensional unit hypercube. The advantages of the uniformly spaced grid in the one-dimensional case do not carry over to higher dimensions. The principal reason is that the error bound for the d-dimensional trapezoidal rule is $O(n^{-2/d})$. In addition, if we use an evenly spaced Cartesian grid, we would have to decide the number of points in advance to achieve uniformity. This is restrictive because, in numerical applications, we would like to be able to add points sequentially until some termination criterion is met.

Low-discrepancy sequences have the property that as successive points are added the entire sequence of points still remains more or less evenly dispersed throughout the region. Niederreiter (1992) gives a detailed analysis of the discrepancy of a sequence. Here, we just briefly recall the definition. Suppose we have a sequence of n points, $\{x_1,x_2,\ldots,x_n\}$, in the d-dimensional half-open unit cube, $I^d = [0,1)^d$, and a subset, J, of I^d. We define

$$D(J;n) = \frac{A(J;n)}{n} - V(J)$$

where $A(J;n)$ is the number of k, $1 \le k \le n$, with $x_k \in J$ and $V(J)$ is the volume of J. The discrepancy, D_n, of the sequence is defined to be the supremum of $|D(J);n)|$ over all J. The star discrepancy, D_n^*, is obtained by taking the supremum over sets J of the form

$$\prod_{i=1}^{d} [0,u_i)$$

In the one-dimensional case there is a simple explicit form for the (star)[14] discrepancy of a sequence of n points. If we label the points so that $0 \le x_1 \le \cdots \le x_n \le 1$, then the discrepancy of this sequence is

$$D_n^* = \frac{1}{2n} + \max_{k=1,\ldots,m} \left| x_k - \frac{2k-1}{2n} \right|$$

We can see that the star discrepancy is at least $1/(2n)$ and that the lowest value is attained when

$$x_k = \frac{2k-1}{2n}, \quad 1 \le k \le n$$

In higher dimensions there is no simple form for the discrepancy of a sequence.

There are several examples of low-discrepancy sequences, including the sequences proposed by Halton (1960), Sobol' (1967), Faure (1982) and Niederreiter (1988).[15] For these sequences the asymptotic form of the star discrepancy has been shown to be

$$D_n^* = O\left(\frac{(\log n)^d}{n}\right)$$

This bound for the discrepancy involves a constant which, in general, depends on the dimension, d, of the sequence. These constants are very difficult to estimate accurately in high dimensions. For large values of d the constants "are often ridiculously large for reasonable values of n" according to Spanier and Maize (1994, p. 23). Furthermore, for high dimensions it may take a long time before the discrepancy reaches its asymptotic level. Morokoff and Caflisch (1995) note that for intermediate values of n the discrepancy may be $O(\sqrt{n})$. They suggest that the transition to $O(n^{-1}(\log n)^d)$ occurs at around values of $n = e^d$. For large d this will be an enormous number.

The error in numerical integration using a low-discrepancy sequence admits a deterministic bound. The bound reflects both the discrepancy of the sequence of points used to evaluate the integral as well as the regularity of the function. The result is contained in the following theorem.

Theorem (Koksma–Hlawka). Let $I^d = [0,1]^d$ *and let* f *have bounded variation,* $V(f)$, *on* $[0,1]^d$ *in the Hardy–Krause*[16] *sense. Then, for any* $x_1, x_2, \ldots, x_n \in I^d$, *we have*

$$\left|\frac{1}{n}\sum_{k=1}^{n} f(x_k) - \int_{I^d} f(u)\,du\right| \leq V(f)D_n^*$$

The error bound provided by this theorem, while it is of theoretical interest, is of little help in most practical situations. The theoretical bound normally overestimates the actual error by a wide margin and $V(f)$ may be difficult to evaluate or even approximate. We have noted that the constants buried in the bounds for the discrepancy are large. Another reason for the coarseness of the bound is that the Koksma–Hlawka theorem does not reflect additional smoothness in f. Intuitively, we would expect the approximation to be better as f becomes smoother. In finance applications the payoffs are normally continuous functions of the variables (with some important exceptions – payoffs on digital and barrier options are discontinuous), but may not be sufficiently smooth to have finite variation because of functions like "max" embedded in the payoffs. Hlawka (1971) provides an alternative bound under weaker smoothness requirements.

To date, studies using low-discrepancy sequences in finance applications have found that the errors produced are substantially lower than the corresponding errors generated by crude Monte Carlo. Joy, Boyle and Tan (1996) used Faure sequences to price several complex derivative securities. They found that the quasi-Monte Carlo approach resulted in significantly smaller errors than the standard Monte Carlo approach. They confirmed that the actual error bound (for cases in which it could be computed precisely) was dramatically less than the bound computed from the Koksma–Hlawka inequality. Paskov and Traub (1995) used both Sobol' sequences and Halton sequences to evaluate mortgage-backed security prices. Their work involves the evaluation of integrals with dimensions up to 360; they find that Sobol' sequences are more efficient than Halton sequences and that the quasi-random approach outperforms the standard Monte Carlo approach for these types of problems.[17] Paskov and Traub's results stand in contrast to the claim that is sometimes found in the literature[18] that the superiority of low-discrepancy algorithms vanishes for intermediate values of d around 30. Bratley, Fox and Niederreiter (1992) conducted practical numerical experiments using low-discrepancy sequences and concluded that standard Monte Carlo is superior to quasi-Monte Carlo for high dimensions, say, greater than 12. They used Sobol' and Niederreiter sequences in their tests. They conclude that in high dimensions "quasi-Monte Carlo seems to offer no practical advantage over pseudo-Monte Carlo because the discrepancy bound for the former is far larger than $\sqrt{n}$ for $n = 2^{30}$, say". (In a personal communication, Fox adds that the crossover probably depends a lot on the sequence.) The reason for the difference between this verdict and the results of the finance applications may be that the integrands typically found in finance applications behave better than those used by numerical analysts[19] to compare different algorithms. Another important consideration is that financial applications typically involve discounting, and this may effectively reduce dimensionality; for example, some of the 360 months in the life of a mortgage may have little influence on the value of a mortgage-backed security. Nevertheless, the experience of Bratley, Fox and Niederreiter (1992) serves as a useful caution against assuming that quasi-Monte Carlo will outperform standard Monte Carlo in all situation.

Some theoretical differences among low-discrepancy sequences can be understood through the concepts of (t,m,s)-nets and (t,s)-sequences, which are discussed in detail in Niederreiter (1992). Briefly, an elementary interval in base b in dimension s is a set of the form

$$\prod_{j=1}^{s}\left[\frac{a_j}{b^{k_j}}, \frac{a_j+1}{b^{k_j}}\right)$$

with k_j, a_j non-negative integers and $a_j < b^{k_j}$. A (t,m,s)-net (with $0 \le t \le m$) is a set of b^m points in the s-dimensional hypercube such that every elementary interval of volume b^{t-m} contains b^t points. Speaking loosely, this means that the proportion of points in each sufficiently large box equals the volume of the box. Smaller t implies greater uniformity. An infinite sequence forms a (t,s)-sequence if for all $m \ge t$ certain finite subsequences of length b^m form (t,m,s)-nets in base b. Sobol' points are (t,s)-sequences in base 2 and Faure points are (0,s) sequences in prime bases not less than s. Thus, Faure points achieve the smallest value of t but at the expense of a large base. A smaller base implies that uniformity holds over shorter subsequences.

An important issue in the use of quasi-Monte Carlo concerns the termination criterion, since the Koksma–Hlawka bound is often of little practical value. Various heuristics are available. Birge (1994) suggests that a rough bound may be obtained by tracking the maximum and minimum values over a period that shows equal numbers of increases and decreases. For instance, the criterion could be to stop at the first set of 2,000 observations in which the numbers of increases and decreases are within 10% of each other. He suggests that the maximum and minimum realised values could be used as bounds on the true value. Fox (1986) suggests that we compare the estimate of the integral based on a sample of $2n$ points with the estimate based on n points and stop if the answer lies within some tolerance level. Paskov and Traub (1995) use a similar termination criterion based on successive errors: stop when the difference between two consecutive approximations using $10{,}000i$, $i = 1,2,\ldots,1000$, sample points falls below some threshold. Owen (1994, 1995) proposes a hybrid of Monte Carlo and low-discrepancy methods that provides error estimates and has good convergence properties. In addition to these approaches, one can also run standard Monte Carlo at the outset and use the probabilistic error term to assess when enough low-discrepancy points have been used in the quasi-random calculation. This benchmarking with standard Monte Carlo would be useful if the same set of calculations were being carried out frequently with only slightly different input values. This situation is common in finance applications. There is often a need to perform the same set of calculations frequently – for example, the risk analysis of a book of business at the end of each day. In these cases one can conduct experiments to see which sets of low-discrepancy sequences provide the best results. The *right* number of low-discrepancy points could be determined just once at the outset.

Before leaving this section, we should mention some recent advances and new techniques for improving the performance of quasi-random Monte Carlo. Niederreiter and Xing (1995) and Ninomiya and Tezuka (1996) have proposed new low-discrepancy sequences that appear to have the potential to perform substantially better than previous methods. We have noted that the efficiency of quasi-random Monte Carlo improves as the integrand becomes smoother. Moskowitz and Caflisch (1996) illustrate procedures that can be accomplished in the discretisation of a Wiener process and in the solution of the Feynman–Kac equation. This is relevant for finance applications since the prices of derivatives securities have a Feynman–Kac representation. See Acworth, Broadie and Glasserman (1997), Berman (1996), and Caflisch, Morokoff and Owen (1997) for recent work applying low-discrepancy sequences with alternative constructions of Wiener processes. Spanier and Maize (1994) discuss a battery of techniques that can be used to improve the performance of quasi-Monte Carlo methods for relatively small sample sizes.

Next we compare the Monte Carlo method using pseudo-random numbers with the Faure, Halton and Sobol' low-discrepancy methods.

NUMERICAL RESULTS

For an initial comparison, we test the methods on the problem of pricing a European option on a single underlying asset with the usual Black–Scholes assumptions. In this framework, the Black–Scholes formula can be evaluated to give the true option values in order to compare alternative methods. Rather than using a single

1. RMS relative error versus number of points

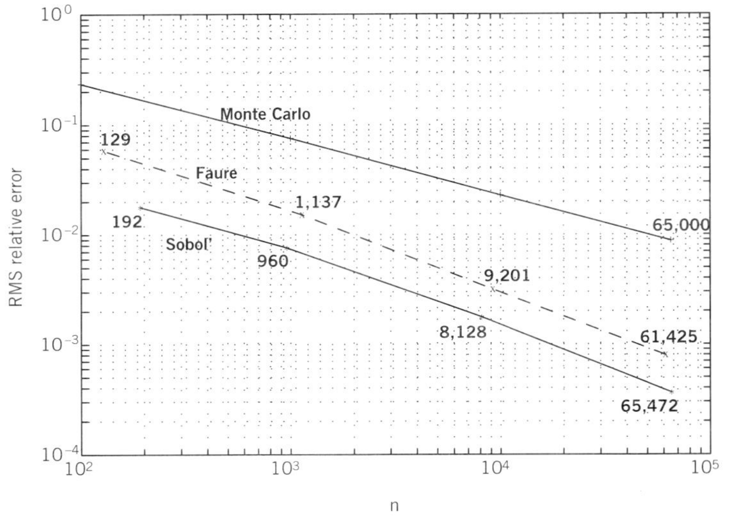

2. 1,000 two-dimensional Faure points

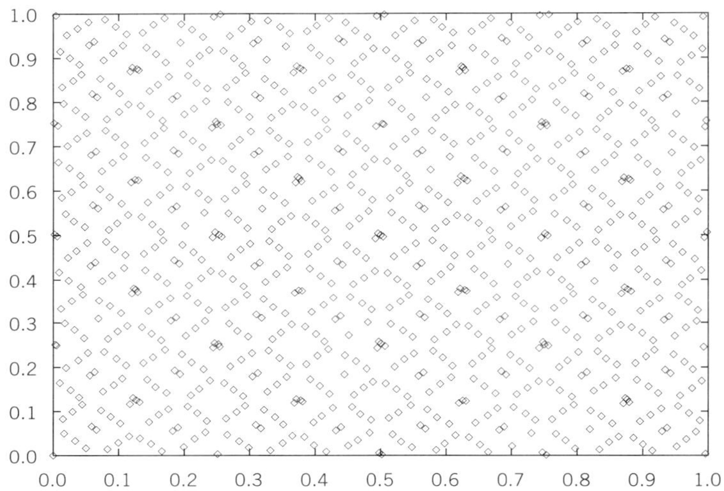

3. 1,000 two-dimensional Sobol' points

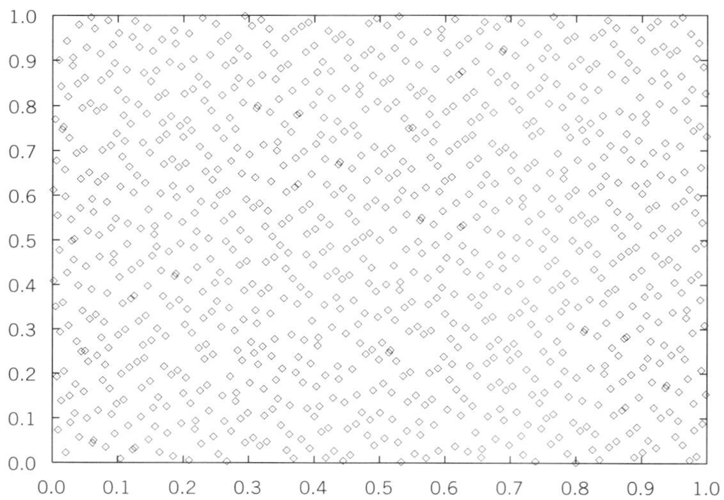

option, we evaluate the methods on a random sample of 500 options. The probability distribution of the parameters is chosen to represent a reasonable range of values in practical applications.[20] The error measure that we use is root-mean-square (RMS) relative error defined by

$$\text{RMS} = \sqrt{\frac{1}{m}\sum_{i=1}^{m}\left(\frac{\hat{C}_i - C_i}{C_i}\right)^2} \tag{12}$$

where i is the index of the $m = 500$ options in the test set, C_i is the true option value, and $\hat{C}_i$ is the estimated option value. The results are given in Figure 1.

Figure 1 plots RMS relative error against the number of points, n. The Monte Carlo method (ie using pseudo-random numbers) displays the expected $O(1/\sqrt{n})$ convergence: eg increasing n by a factor of 100 reduces the RMS error by a factor of 10. The low-discrepancy method using Faure sequences dominates the Monte Carlo method. Indeed, 129 Faure points gives an error lower than 1,000 Monte Carlo points. The Sobol' method is the best of the three methods tested. Using 192 Sobol' points gives an error lower than 10,000 Monte Carlo points.

A major consideration in the comparison of methods is the overall computation time, not just the number of points. The Sobol' sequence numbers can be generated significantly faster than Faure numbers (see, eg Bratley and Fox 1988) and faster than with most pseudo-random number methods. Hence, in the important RMS error versus computation time comparison, the relative advantage of the Sobol' method increases.

A low-discrepancy sequence will often have additional uniformity properties at certain points in the sequence (see, eg Fox, 1986, and Bratley and Fox, 1988). For example, in the Sobol' sequence the running average returns to 0.5 at the points $n = 2^k - 1$ for $k = 1,2,\ldots$. One might expect that choosing n to be one of these "favourable" points would lead to better option price estimates. For large values of n, the advantage of using favourable points becomes negligible, but for small n the effect can be quite significant. Indeed, in the experiment above, using the Sobol' points 1–254 gives an RMS error of 10%, while using the points 1–255 gives an RMS error of 4%.[21] Better results are often obtained by ignoring an initial portion of a low-discrepancy sequence. For example, using the Sobol' points 1–63 gives an RMS error of 13%, while using the Sobol' points 64–127 gives an RMS error of 2%. In the results shown in Figure 1 the Sobol' sequence was always started at point 64, so the label "192" in the figure corresponds to the 192 Sobol' points from 64 to 255. Similarly, the Faure sequence was always started at point 16, so the label "129" in the figure corresponds to the 129 Faure points from 16 to 144.

ONE-DIMENSIONAL VERSUS HIGHER-DIMENSIONAL SEQUENCES

It is sometimes asserted that low-discrepancy methods can be implemented in existing simulation programs by simply replacing the pseudo-random number generator with a low-discrepancy sequence generator. This naïve approach can lead to disastrous results, as the following example shows.

Consider pricing a European option on the maximum of two non-dividend-paying assets with the parameters: $S_1 = S_2 = K = 100$, $\sigma_1 = \sigma_2 = 0.2$, $\rho = 0.3$, $r = 0.05$, and $T = 1$. Under the usual Black–Scholes assumptions, a formula for the price of the option can be derived (see, eg Johnson, 1987, or Stulz, 1982) and gives a price of 16.442. Running one Monte Carlo simulation with 1,000 points hence 2,000 random numbers) gave an estimated price of 16.279 with a standard error of 0.533. Using 2,000 one-dimensional low-discrepancy values gave a price estimate of 4.320 using the Sobol' sequence and an estimate of 1.909 using the Faure sequence (starting at point 16). The cause of the problem can be seen by examining Figures 2–5.

Figures 2 and 3 show 1,000 two-dimensional Faure and Sobol' points, respectively. The figures illustrate how the sequences fill the two-dimensional space in regular but different ways. By contrast, Figures 4 and 5 show 2,000 one-dimensional Faure and Sobol' points, respectively, plotted in the one-dimensional sequence to be the (x,y) coordinates in two-dimensional space. In neither figure are the points filling the two-dimensional space (note that the axes do not extend from 0 to 1), and this explains why the price estimates do not converge to the correct values. Even in the quarter of the unit square where the points fall, the points do not uniformly fill the space. This problem is reminiscent of the well-known "collinearity" or "hyperplane" problem of some pseudo-random number generators, but it is even more serious with these low-discrepancy sequences.

A similar problem can occur if a high-dimensional low-discrepancy sequence is used for a problem of low dimension. Figure 6 shows the 49th and 50th dimensions of 1,000 50-dimensional Faure points. Using the last two dimensions of the

4. 2,000 one-dimensional Faure points

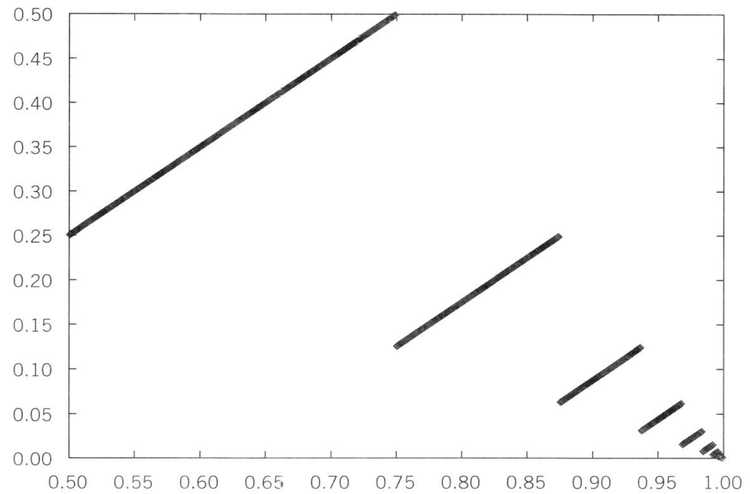

5. 2,000 one-dimensional Sobol' points

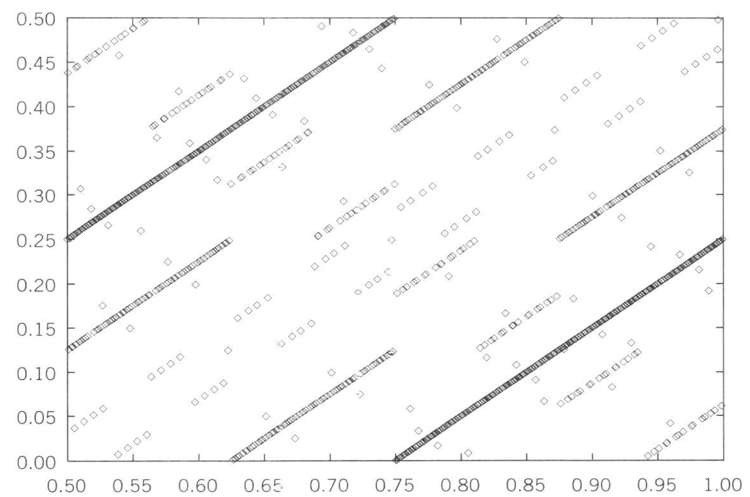

6. Coordinates 49 and 50 of 1,000 50-dimensional Faure points

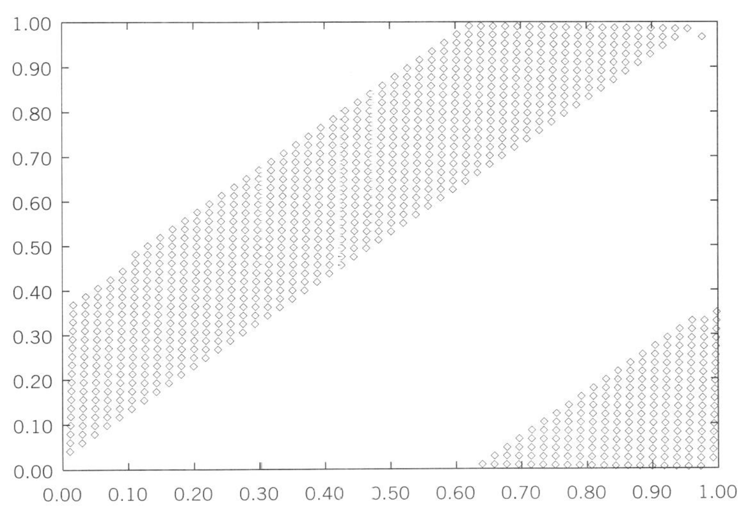

50-dimensional sequence to price a two-dimensional option will give very poor results.

HIGHER-DIMENSIONAL TEST

To test the effect of problem dimension, we price options in dimensions $d = 10, 50$, and 100. We price discretely sampled geometric average Asian options because the problem dimension is easily varied and a closed-form solution for the price is available (see Turnbull and Wakeman, 1991). The price of a geometric average Asian option is given by

$$C = E[e^{-rT}(\tilde{S} - K)^+]$$

where $\tilde{S} = (\Pi_{i=1}^{d} S_i)^{1/d}$ and S_i is the asset price at time iT/d.

7. Results with 50,000 points

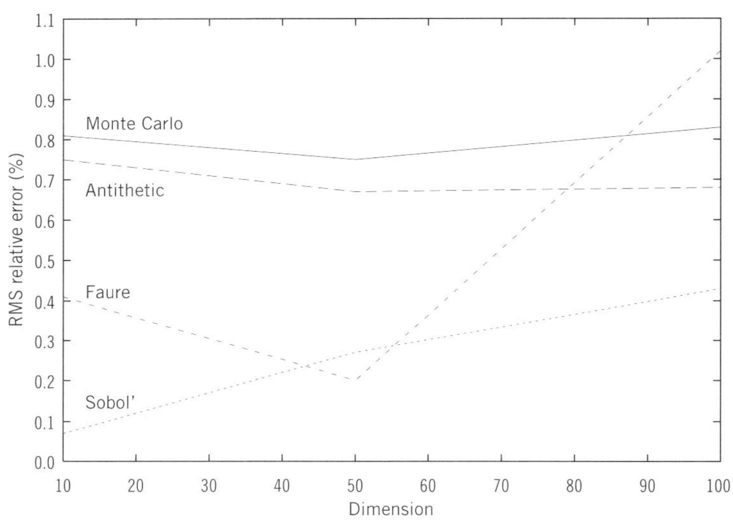

8. Results with 200,000 points

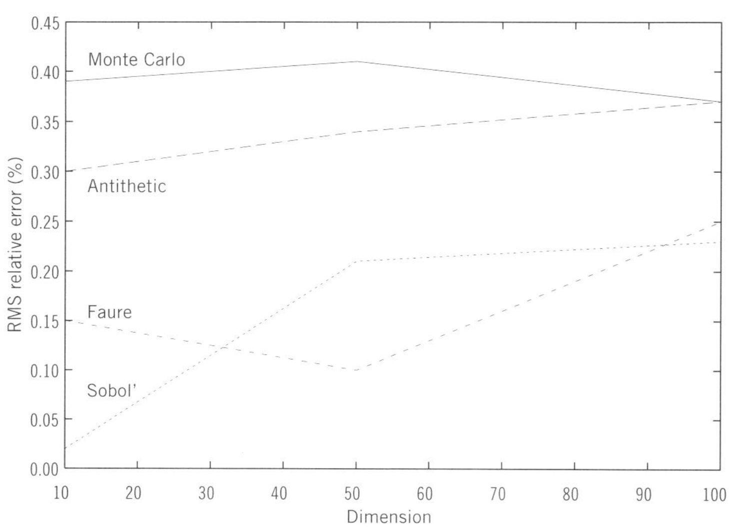

We test standard Monte Carlo, Monte Carlo with antithetic variates, and the low-discrepancy sequences of Faure, Sobol' and Halton.[22] For each dimension, we select 500 option parameters at random and compute the RMS relative error (see equation (12)) for each method.[23] Results for 50,000 and 200,000 sample points are given in Figures 7 and 8, respectively. (The antithetic method uses 25,000 and 100,000 independent pairs of points, respectively.)

Results for the Halton sequence were not competitive and are not presented. RMS error for standard Monte Carlo is nearly independent of the problem dimension. The antithetic method gives minimal variance reduction. The relative advantage, in terms of RMS error, of the low-discrepancy sequences decreases with the problem dimension. For this test problem, the crossover point is beyond dimension 100.

Estimating price sensitivities

Most of the discussion in this chapter centres on the use of Monte Carlo for pricing securities. In practice, the evaluation of price *sensitivities* is often as important as the evaluation of the prices themselves. Indeed, whereas prices for some securities can be observed in the market, their sensitivities to parameter changes typically cannot and must therefore be computed. Since price sensitivities are important measures of risk, the growing emphasis on risk management systems suggests a greater need for their efficient computation.

The derivatives of a derivative security's price with respect to various model parameters are collectively referred to as "Greeks" because several of these are commonly referred to by the names of Greek letters.[24] Perhaps the most important of these – and the one to which we give primary attention – is *delta*: the derivative of the price of a contingent claim with respect to the current price of an underlying asset. The delta of a stock option, for example, is the derivative of the option price with respect to the current stock price. An option involving multiple underlying assets has multiple deltas, one for each underlying asset.

In the rest of this section, we discuss various approaches to estimating price sensitivities, especially delta. We begin by examining finite-difference approximations and show that these can be improved through the use of common random numbers. We then discuss direct methods that estimate derivatives without requiring resimulation at perturbed parameter values.

FINITE-DIFFERENCE APPROXIMATIONS

Consider the problem of computing the delta of the Black-Scholes price of a European call – ie computing

$$\Delta = \frac{dC}{dS_0}$$

where C is the option price and S_0 is the current stock price. There is, of course, an explicit expression for delta, so simulation is not required, but the example is useful for purposes of illustration. A crude estimate of delta is obtained by generating a terminal stock price

$$S_T = S_0 e^{(r-(1/2)\sigma^2)T+\sigma\sqrt{T}Z} \qquad (13)$$

(see (2) for notation) from the current stock price, S_0, and a second, independent terminal stock price

$$S_T(\varepsilon) = (S_0 + \varepsilon)e^{(r-(1/2)\sigma^2)T+\sigma\sqrt{T}Z'} \qquad (14)$$

from the perturbed initial price $S_0 + \varepsilon$, with Z and Z′ independent. For each terminal price, a discounted payoff can be computed like this:

$$\hat{C}(S_0) = e^{-rT}\max\{0, S_T - K\},$$
$$\hat{C}(S_0 + \varepsilon) = e^{-rT}\max\{0, S_T(\varepsilon) - K\}$$

(see (3) for notation). A crude estimation of delta is then provided by the finite-difference approximation

$$\tilde{\Delta} = \varepsilon^{-1}[\hat{C}(S_0 + \varepsilon) - \hat{C}(S_0)] \qquad (15)$$

By generating n independent replications of S_T and $S_T(\varepsilon)$ we can calculate the sample mean of n independent copies of $\tilde{\Delta}$. As $n \to \infty$, this sample mean converges to the true finite-difference ratio

$$\varepsilon^{-1}[C(S_0 + \varepsilon) - C(S_0)] \qquad (16)$$

where $C(\cdot)$ is the option price as a function of the current stock price.

This discussion suggests that to get an accurate estimate of Δ we should make ε small. However, because we generated S_T and $S_T(\varepsilon)$ independently of each other, we have

$$\text{Var}[\tilde{\Delta}] = \varepsilon^{-2}(\text{Var}[\hat{C}(S_0 + \varepsilon)] + \text{Var}[C(S_0)]) = O(\varepsilon^{-2})$$

so the variance of $\tilde{\Delta}$ becomes very large if we make ε small. To get an estimator that converges to Δ we must let ε decrease slowly as n increases, resulting in slow overall convergence. A general result of Glynn (1989) shows that the best possible convergence rate using this approach is typically $n^{-1/4}$. Replacing the forward difference estimator in (15) with the central difference $(2\varepsilon)^{-1}[\hat{C}(S_0 + \varepsilon) - \hat{C}(S_0 - \varepsilon)]$ typically improves the optimal convergence rate to $n^{-1/3}$. These rates should be compared with $n^{-1/2}$, the rate ordinarily expected from Monte Carlo.

Better estimators can generally be improved using the method of *common random numbers*, which, in this context, simply uses the same Z in (13) and (14). Denote by $\hat{\Delta}$ the finite-difference approximation thus obtained. For fixed ε, the sample mean of independent replications of $\hat{\Delta}$ also converges to (16). The variance parameter is given by

$$\text{Var}[\hat{\Delta}] = \varepsilon^{-2}(\text{Var}[\hat{C}(S_0)] + \text{Var}[\hat{C}(S_0 + \varepsilon)]$$
$$- 2\text{Cov}[\hat{C}(S_0), \hat{C}(S_0 + \varepsilon)])$$

because $\hat{C}(S_0)$ and $\hat{C}(S_0 + \varepsilon)$ are no longer independent. Indeed, if they are positively correlated, then $\hat{\Delta}$ has smaller variance than $\tilde{\Delta}$. That they are in fact positively correlated follows from the monotonicity of the function mapping Z to $\hat{C}$ by the argument used in our discussion of antithetics in the third section. Thus, the use of common random numbers reduces the variance of the estimate of delta.

The impact of this variance reduction is most dramatic when ε is small. A simple calculation shows that, using common random numbers,

$$|\hat{C}(S_0 + \varepsilon) - \hat{C}(S_0)| \le |S_T(\varepsilon) - S_T|$$
$$\le \varepsilon e^{(r-(1/2)\sigma^2)T+\sigma\sqrt{T}Z}$$

Because this upper bound has finite second moment, we may conclude that

$$E||\hat{C}(S_0 + \varepsilon) - \hat{C}(S_0)|^2| = O(\varepsilon^2) \qquad (17)$$

and therefore that

$$\text{Var}[\varepsilon^{-1}\{\hat{C}(S_0 + \varepsilon) - \hat{C}(S_0)\}] = O(1)$$

ie the variance of $\hat{\Delta}$ remains *bounded* as $\varepsilon \to 0$, whereas we saw previously that the variance of $\tilde{\Delta}$ increases at rate ε^{-2}. Thus, the more precisely we try to estimate Δ (by making ε small), the greater the benefit of common random numbers. Moreover, this indicates that to get an estimator that converges to Δ we may let ε decrease faster as n increases than was possible with $\tilde{\Delta}$, resulting in faster overall convergence. An application of

Proposition 2 of L'Ecuyer and Perron (1994) shows that a convergence rate of $n^{-1/2}$ can be achieved in this case, and that is the best than can ordinarily be expected from Monte Carlo. For more on convergence rates using common random numbers, see Glasserman and Yao (1992), Glynn (1989), and L'Ecuyer and Perron (1994).

The dramatic success of common random numbers in this example relies on the fast rate of mean-square convergence of $\hat{C}(S_0 + \varepsilon)$ to $\hat{C}(S_0)$ evidenced by (17). This rate does not apply in all cases. It fails to hold, for example, in the case of a *digital option*[25] paying a fixed amount B if $S_T > K$ and 0 otherwise. The price of this option is $C = e^{-rT}BP(S_T > K)$; the obvious simulation estimator is

$$\hat{C}(S_0) = \mathbf{1}_{\{S_t > K\}} e^{-rT} B$$

Because $\hat{C}(S_0)$ and $\hat{C}(S_0 + \varepsilon)$ differ only when $S_T \le K < S_T(\varepsilon)$, we have

$$\begin{aligned} &E\left[|\hat{C}(S_0 + \varepsilon) - \hat{C}(S_0)|^2\right] \\ &= B^2 e^{-2rT} P(S_T \le K < S_T(\varepsilon)) \\ &= B^2 e^{-2rT} P(S_T \le K < (1 + \varepsilon/S_0)S_T) = O(\varepsilon) \end{aligned}$$

compared with $O(\varepsilon^2)$ for a standard call. As a result, delta estimation is more difficult for the digital option, and a similar argument applies to barrier options generally. Even in these cases, the use of common random numbers can result in substantial improvement compared with differences based on independent runs.

Table 6 compares the performance of four types of delta estimates: forward and central finite differences with and without common random numbers. The methods are compared at four values of the perturbation parameters ε, and applied to the two options discussed above. The values in the table are estimated root-mean-square errors. The numerical results substantiate the analysis above. Much lower errors are obtained for the standard call than for the digital option, allowing for smaller ε; central differences beat forward differences; common random numbers help, but they help the standard call more than the digital option. In several cases, the minimum error is obtained using a fairly large ε. This reflects the fact that the bias resulting from a large ε is sometimes overwhelmed by the large variance resulting from a small ε.

Although we have discussed common random numbers in only a limited context, the method can easily be applied to a wide range of problems. If all stochastic inputs to a simulation are samples from the normal distribution, common random numbers can be implemented by using the same samples at two different parameter settings. More generally, if the stochastic inputs are all drawn from a sequence of uniform random variates, then common random numbers can be implemented by using these variates at two different parameter settings.

Table 6. RMS errors for various delta estimation methods

		Independent		Common	
	ε	Forward	Central	Forward	Central
Standard call option	10	0.10	0.01	0.100	0.009
	1	0.18	0.09	0.012	0.006
	0.1	1.78	0.87	0.006	0.006
	0.01	7.47	8.98	0.006	0.006
Digital option	20	0.51	0.37	0.51	0.37
	10	0.22	0.11	0.21	0.10
	5	0.16	0.07	0.11	0.05
	1	0.67	0.34	0.14	0.10

Root-mean-square error of delta estimates for two options using four methods with various values of ε. Both options have $S_0 = 100$, $K = 100$, $\sigma = 0.40$, $r = 0.10$, and $T = 0.2$. The digital option has $B = 100$. Each entry is computed from 1,000 delta estimators, each estimate based on 10,000 replications. The value of delta is 0.580 for the first option and 2.185 for the second.

DIRECT ESTIMATES

Even with the improvements in performance obtained from common random numbers, derivative estimates based on finite difference still suffer from two shortcomings. They are biased (since they compute difference ratios rather than derivatives) and they require multiple resimulations: estimating sensitivities to d parameter changes requires repeatedly running one simulation with all parameters at their base values and d additional simulations with each of the parameters perturbed. The computation of 10–50 Greeks[26] for a single security is not unheard of, and this represents a significant computational burden when multiple resimulations are required.

Over the last decade, a variety of *direct* methods have been developed for estimating derivatives by simulation. Direct methods compute a derivative estimate from a single simulation, and thus do not require resimulation at a perturbed parameter value. Under appropriate conditions, they result in unbiased estimates of the derivatives themselves, rather than of a finite-difference ratio. Our discussion focuses on the use of *pathwise* derivatives as direct estimates, based on a technique generally called *infinitesimal perturbation analysis* (see, eg, Glasserman, 1991).

The pathwise estimate of the true delta, dC/dS_0, is the derivative of the sample price $\hat{C}$ with respect to S_0. More precisely, it is

$$\frac{d\hat{C}}{dS_0} = \lim_{\varepsilon \to 0} \varepsilon^{-1}\left[\hat{C}(S_0 + \varepsilon) - \hat{C}(S_0)\right]$$

provided that the limit exists with probability 1. If $\hat{C}(S_0)$ and $\hat{C}(S_0 + \varepsilon)$ are computed from the same Z, then, provided $S_T \neq K$, we have

$$\frac{d\hat{C}}{dS_0} = \frac{d\hat{C}}{dS_T}\frac{dS_T}{dS_0} = e^{-rT}\mathbf{1}_{\{S_T > K\}}\frac{S_T}{S_0} \tag{18}$$

We have used (13) to get

$$\frac{dS_T}{dS_0} = e^{(r-(1/2)\sigma^2)T+\sigma\sqrt{T}Z} = \frac{S_T}{S_0}$$

and

$$\frac{d\hat{C}}{dS_T} = e^{-rT}\frac{d}{dS_T}\max\{0, S_T - K\} = \begin{cases} e^{-rT}, & S_T > K \\ 0, & S_T < K \end{cases}$$

At $S_T = K$, C fails to be differentiable; however, since this occurs with probability zero, the random variable $d\hat{C}/dS_0$ is almost surely well defined.

The pathwise derivative $d\hat{C}/dS_0$ can be thought of as a limiting case of the common random numbers finite-difference estimator in which we evaluate the limit analytically rather than numerically. It is a direct estimator of the option delta because it can be computed directly from a simulation starting with S_0 without the need for a separate simulation at a perturbed value S_0. This is evident from the expression in (18). The question remains whether this estimator is unbiased; that is, whether

$$E\left[\frac{d\hat{C}}{dS_0}\right] = \frac{dC}{dS_0} \equiv \frac{d}{dS_0}E[\hat{C}]$$

The unbiasedness of the pathwise estimate thus reduces to the interchangeability of derivative and expectation. The interchange is easily justified in this case; see Broadie and Glasserman (1996) for this example and conditions for more general cases. Applying the same reasoning used above, we obtain the following pathwise estimators of three other Greeks for the Black–Scholes price:

Rho (dC/dr): $KTe^{-rT}\,\mathbf{1}_{\{S_T \geq K\}}$

Vega (dC/dσ):

$$e^{-rT}\mathbf{1}_{\{S_T \geq K\}}\frac{S_T}{\sigma}\left(\ln(S_T/S_0) - (r - \tfrac{1}{2}\sigma^2)T\right)$$

Theta (– dC/dT): $re^{-rT}\max\{S_T - K, 0\}$

$$-\,\mathbf{1}_{\{S_T \geq K\}}e^{-rT}\frac{S_T}{2T}\left(\ln(S_T/S_0) + (r - \tfrac{1}{2}\sigma^2)T\right)$$

Each of these estimators is unbiased.

Of course, Monte Carlo estimators are not required for these derivatives because closed-form expressions are available for each. The Black-Scholes setting is useful for illustration, but the utility of the technique lies in its applicability to more general models. In Broadie and Glasserman (1996), pathwise estimates are derived and studied (both theoretically and numerically) for Asian options and a model with stochastic volatility. For example, the Asian-option delta estimate is simply

$$e^{-rT}\frac{\tilde{S}}{S_0}\mathbf{1}_{\{\tilde{S} > K\}}$$

where $\tilde{S}$ is the average asset price used to determine the option payoff. Evaluating this expression takes negligible time compared with resimulating to estimate the option price from a perturbed initial stock price. The pathwise estimate is thus both more accurate and faster to compute than the finite-different approximation. These advantages extend to a wide class of problems.

As already noted, the unbiasedness of pathwise derivative estimates depends on an interchange of derivative and expectation. In practice, this generally means that the security payoff should be a pathwise-continuous function of the parameter in question. The standard call option payoff $e^{-rT}\max\{0, S_T - K\}$ is continuous in each of its parameters. An example where continuity fails is a digital option with payoff $e^{-rT}\mathbf{1}_{\{S_T > K\}}B$, with B the amount received if the stock finishes in the money.[27] Because of the discontinuity of $S_T = K$, the pathwise method (in its simplest form) cannot be applied to this type of option.

The problem of discontinuities often arises in the estimation of *gamma*, the second derivative of an option price with respect to the current price of an underlying asset. Consider, again, the standard European call option. We have an expression for $d\hat{C}/dS_0$ in (18) involving the indicator $\mathbf{1}_{\{S_T > K\}}$. This shows that $d\hat{C}/dS_0$ is discontinuous in S_T, preventing us from differentiating pathwise a second time to get a direct estimator of gamma.

To address the problem of discontinuities, Broadie and Glasserman (1996) constructed smoothed estimators. These estimators are unbiased but are not as simple to derive and implement as ordinary pathwise estimators. Broadie and Glasserman also investigated another technique for direct derivative estimation called the *likelihood ratio method*. This method differentiates the probability density of an asset price, rather than the outcome of the asset price itself.[28] The domains of this method and of the pathwise method overlap, but neither contains the other. When both apply, the pathwise method generally has lower variance.

Overviews of these methods can be found in Glasserman (1991a), Glynn (1987), and Rubinstein and Shapiro (1993). For discussions specific to financial applications see Broadie and Glasserman (1996) and Fu and Hu (1995).

Pricing American options by simulation

European contingent claims have cashflows that cannot be influenced by decisions of the owner. Examples include European options, barrier options, and many types of swaps. By contrast, the cashflows of American contingent claims depend both on the price path of the underlying asset or assets and the decisions of the owner. Many types of American contingent claims trade on exchanges and in the over-the-counter market. Examples include American options, American swaptions, shout options and American Asian options. They also arise in other contexts, for example as "real options" in the theory of economic investment described in Dixit and Pindyck (1994).

To be concrete, suppose that we wish to estimate the quantity $\max_T E[e^{-r\tau}h(S_\tau)]$, where r is the constant riskless interest rate, $h(S_\tau)$ is the payoff at time τ in state S_τ, and the max is taken over all stopping times $\tau \leq T$. This formulation of the American pricing problem will suffice to illustrate the major points. First, note that the state can be vector-valued and hence applies to pricing American options on multiple assets. Second, since simulation algorithms are discrete in nature, the continuous-time exercise decision must be approximated by restricting the exercise opportunities to lie in a finite set of times $0 = t_0 < t_1 < \cdots < t_d = T$. This is not always a serious restriction. For example, for a call option on a stock which pays dividends at discrete points in time, it can be shown that early exercise is only optimal just prior to the ex-dividend dates. In other cases, Richardson or other extrapolation techniques can be used to better approximate the price with exercise in continuous time from a finite set of exercise opportunities.[29] However, we now restrict attention to estimating the quantity

$$P \equiv \max_\tau E\left[e^{-r\tau}h(S_\tau)\right] \qquad (19)$$

where the max is taken over all stopping times, τ, in the set t_i, for $i = 0,\ldots,d$. The need to estimate an optimal stopping time is the crucial distinction between American and European pricing problems.

If the state space is of low dimension, say three or less, a discretisation scheme together with a dynamic programming algorithm can often be used to approximate the value in (19) numerically. Even in these cases, simulation can be used to estimate the expectation in the recursive step. Simulation-based methods become essential when the dimension of the state space is large.

An obvious simulation-based algorithm for estimating the quantity in (19) is to generate a random path of states S_{t_i}, for $i = 1,\ldots,d$, and form the path estimate

$$\hat{P} = \max_{i=0,\ldots,d} e^{-rt_i}h(S_{t_i})$$

However, this estimator corresponds to using perfect foresight, and so it is biased high. That is, $E[\hat{P}] \geq P$, which follows immediately from the inequality $\max_{i=0,\ldots,d} e^{-rt_i}h(S_{t_i}) \geq e^{-r\tau}h(S_\tau)$. A natural goal would be to develop an alternative unbiased estimator. A negative result in this regard is provided in Broadie and Glasserman (1997); among a large class of estimators, there is no unbiased estimator of P. In particular, the estimators proposed in Tilley (1993), Grant, Vora and Weeks (1994), and Barraquand and Martineau (1995) – also Chapter 15 of this publication, are all biased. Unfortunately, they provide no way to estimate the extent of the bias or to correct for the bias in a general setting. Broadie and Glasserman (1997) circumvent this problem by developing two estimators, one biased high and one biased low (but both asymptotically unbiased), which can be used together to form a valid confidence interval for the quantity P. In the remainder of this section, we give brief descriptions of the four methods mentioned and describe some strengths and weaknesses of each.

TILLEY'S BUNDLING ALGORITHM

Tilley (1993) sparked considerable interest by demonstrating the potential practicality of applying simulation to pricing American contingent claims. Tilley describes a "bundling procedure" for pricing an American option on a single underlying asset. To estimate P he suggests simulating n paths of asset prices denoted $S_{t_i}(j)$ for $i = 1,\ldots,d$ and $j = 1,\ldots,n$ in the usual way. Next, partition the asset price space and call the paths which fall into a given partition at a fixed time a "bundle". A dynamic programming algorithm is applied to bundles to estimate C. In particular, the estimated option price $P_{t_i}(j)$ at time t_i for path j is the maximum of the immediate exercise value, $h(S_{t_i}(j))$, and the present value of continuing. The latter value is defined to be the average of $e^{-r(t_{i+1}-t_i)}P_{t_{i+1}}(k)$ over all paths k which fall in the bundle containing path j at time t_i. Details of the partitioning are given in Tilley (1993).

In order to implement the algorithm, all paths must be stored so they can be sorted into bundles at each time step. Since simulation typically requires a large number of paths for good estimates, the storage and sorting requirements can be significant. More importantly, the algorithm does not easily generalise to multiple state variables. In higher dimensions, it is not clear how to define the bundles. Even then it is likely that most partitions will contain very few paths and lead to a large bias, or the partitions will be so large that the continuation values are poorly estimated.

Because Tilley's algorithm uses the same paths to estimate the optimal decisions and the value, the estimator tends to be biased high (although the bundling induces an approximation which is difficult to analyse). Tilley introduces a "sharp boundary" variant which reduces the bias, but this variant does not easily generalise to higher dimensions. Carriere (1996) contains further analysis of Tilley's algorithm and suggests a procedure based on spline functions to reduce the bias. It remains to be seen whether the spline procedure is practical for higher-dimensional problems. Nevertheless, for single state variable problems, Tilley demonstrated the potential practicality of applying simulation to American-type pricing problems.

BARRAQUAND AND MARTINEAU'S STRATIFIED STATE AGGREGATION (SSA) ALGORITHM

Barraquand and Martineau (1995) propose a partitioning algorithm but, unlike Tilley's bundling algorithm, they partition the payoff space instead of the state space. Hence, only a one-dimensional space is partitioned at each time step, independent of the number of state variables.[30] Their algorithm works as follows.

First, partition the payoff space into K disjoint cells. Then simulate n paths of asset prices denoted $S_{t_i}(j)$ for $i = 1,\ldots,d$ and $j = 1,\ldots,n$ in the usual way. For each payoff cell k at time t_i, record the number of paths, $a_{t_i}(k)$, which fall into the cell. For each pair of cells k and l at consecutive times t_i and t_{i+1}, record the number of paths, $b_{t_i}(k,l)$, which fall into both cells. Also, for each cell k at time t_i, record the sum of the payoff values, $c_{t_i}(k) = \sum h(S_{t_i}(j))$, where the sum is over all paths j which fall into cell k at time t_i. The transition probability from (t_i,k) to (t_{i+1},l) is approximated by $p_{t_i}(k,l) = b_{t_i}(k,l)/a_{t_i}(k)$. The estimated option price $P_{t_i}(k)$ at time t_i in cell k is the maximum of the immediate exercise value and the present value of continuing. The immediate exercise value is approximated by $c_{t_i}(k)/a_{t_i}(k)$. The present value of continuing is approximated by $e^{-r(t_{i+1}-t_i)}\sum_{l=1}^{K} p_{t_i}(k,l)P_{t_{i+1}}(l)$. This procedure can be applied backwards in time to determine the simulation estimate of the price P.

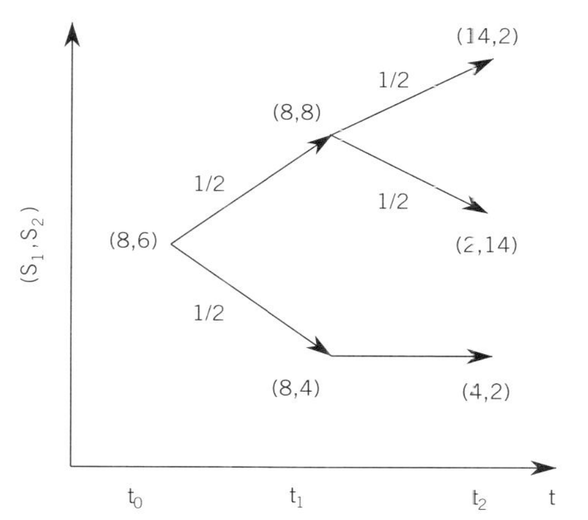

9. State evolution

Details of a payoff space partitioning scheme are given in Barraquand and Martineau (1995). Once a single path is generated and the summary information a, b and c is recorded, the path can be discarded. Hence the storage requirements with this method are modest: on the order of K^2d. One drawback of this method is a possible lack of convergence, as the following example illustrates.

Figure 9 shows the evolution of two asset prices (S_1,S_2). The option payoff is $h(S_1,S_2) = \max(S_1,S_2)$ and for convenience the riskless rate is taken to be zero. Using the risk-neutral probabilities in Figure 9, the true value of the option at time t_0 is 11, which at time t_1 involves exercise in state (8,4) but continuing in state (8,8). (Note that asset 1 pays a dividend of 4 and asset 2 a dividend of 2 after time t_1). When the states are partitioned by their payoffs, these two states are indistinguishable. As seen in the payoff evolution in Figure 10, the best strategy at time t_1 in payoff state 8 is to continue. The apparent value of the option in Figure 10 is 9 (=(1/2)14 + (1/2)4). In this example, partitioning the payoff space leads to a significant underestimate of the option value. Hence, a simulation algorithm based on partitioning the payoff

10. Payoff evolution

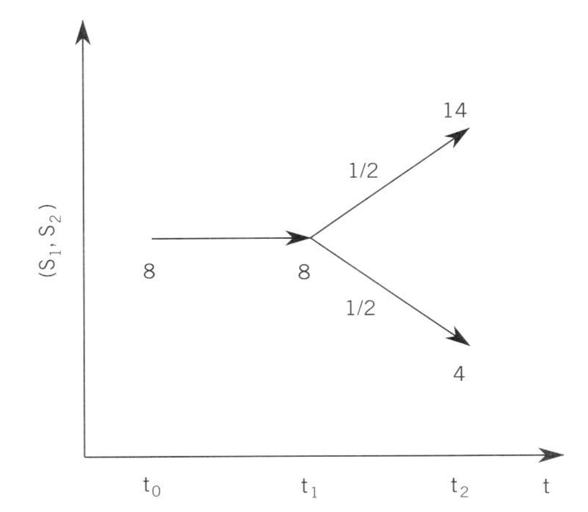

space cannot converge to the correct value. Although this example may seem contrived, Broadie and Detemple (1997) show that the payoff value is not a sufficient statistic for determining the optimal exercise decision for options on the maximum of several assets. Indeed, the payoff process $h(S_t)$ is hardly ever Markovian.

There is currently no way to bound the error in the Barraquand and Martineau method. Without an error estimate, it is difficult to determine the appropriate number of paths to simulate or the appropriate number of partitions to use. Their method can be slightly modified to generate an option price estimate which is biased low as follows. Their procedure gives an exercise strategy based on the immediate exercise payoff. Using this strategy, a new (independent) set of paths can be simulated, and an option value can be estimated under the exercise strategy previously estimated. The resulting option price estimate will be biased low because the exercise policy is not, in general, the optimal policy. With this modification, the average direction of the error is known. Raymar and Zwecher (1996) extend the Barraquand and Martineau approach by basing the exercise decision on a partition of two state variables, rather than one.

BROADIE AND GLASSERMAN'S ALGORITHM

Broadie and Glasserman (1997) propose an algorithm based on simulated trees. In order to deal with the bias problem, they develop two estimators, one biased high and one biased low, but both convergent and asymptotically unbiased as the computational effort increases. A valid confidence interval for the true value, P, is obtained by taking the upper confidence limit from the "high" estimator and the lower confidence limit from the "low" estimator. Briefly, their algorithm works as follows.

First, simulate a tree of asset prices (or, more generally, state variables) using b branches at each node. Two paths emanating from a node evolve as independent copies of the state process. The high estimator, Θ, is defined to be the value obtained by the usual dynamic programming algorithm applied to the simulated tree. Then repeat the process for n trees and compute a point estimate and confidence interval for $E[\Theta]$. A low estimator is obtained by modifying the dynamic programming algorithm at each node. Instead of using all b branches to determine the decision and value, b_1 branches are used to determine the exercise decision, and the remaining $b_2 = b - b_1$ branches are used to determine the continuation value. Their actual low estimator, θ, includes another modification of this procedure which reduces the variance of the estimate. As before, estimates from n trees are combined to give a point estimate and confidence interval for $E[\theta]$. Details of the procedure can be found in Broadie and Glasserman (1997).

For the Θ estimator, all of the branches at a given node are used to determine the optimal decision and the corresponding node value, and this leads to an upward bias, ie $E[\Theta] \geq P$. For the θ estimator, the decision and the continuation value are determined from independent information sets. This eliminates the upward bias, but a downward bias occurs, ie $E[\theta] \leq P$. The intuition for this result follows. If the correct decision is inferred at a node, the node value estimate would be unbiased. If the incorrect decision is inferred at a node, the node value estimate would be biased low because of the sub-optimality of the decision. The expected node value is a weighted average of an unbiased estimate (based on the correct decision) and an estimate which is biased low (based on the incorrect decision). The net effect is an estimate which is biased low. Both estimators are consistent and asymptotically unbiased as b increases.

The computational effort with this algorithm is order nb^d and its main drawback is that d cannot be too large for practical computations. Broadie and Glasserman (1997) give numerical results for options with $d = 4$. As mentioned earlier, to approximate option values with continuous-exercise opportunities, some type of extrapolation procedure is required. Special care is necessary in implementing extrapolation procedures within a simulation context because of the randomness in the estimates.

OTHER DEVELOPMENTS

Grant, Vora and Weeks (1994) describe a method specially designed to price American arithmetic Asian options on a single underlying asset. In this application the optimal exercise decision depends on the current asset price and the current value of the average. Using repeated simulation runs, they attempt to identify the form of an optimal exercise policy based on these two pieces of information. Once an exercise policy is specified, simulation is used to estimate the option value under this fixed policy. Since the fixed policy is a sub-optimal approximation to the optimal stopping rule, their procedure leads to a simulation estimator which is biased low.

The authors perform extensive sensitivity analysis which indicates that their option value estimate is relatively insensitive to deviations in the chosen exercise policy. So, it may be that their method gives good option price estimates relative to some accuracy level, but it is not clear how to quantify their error. It is not clear how to improve their estimates to an arbitrary accuracy level as the simulation effort increases. Their procedure is specific to the case of American Asian options and does not at this point constitute a general approach to pricing American contingent claims.

Bossaerts (1989) proposes two estimators of optimal early exercise, a moment estimator and a smooth optimisation estimator, and studies their convergence properties. His method appears to require a parametric representation of the exercise boundary and may therefore face difficulties in higher dimension. The optimisation approach described in Fu and Hu (1995) also requires a parametric representation.

Rust (1995)[31] studies the general problem of solving discrete decision problems, which include optimal stopping problems as a special case. He develops a Monte Carlo method and shows that it succeeds in breaking the "curse of dimensionality" in these problems. Rust's focus in on computational complexity, but his approach appears to provide a promising direction for finance applications.

SUMMARY

The valuation of securities with American-type features requires the determination of optimal decisions. High-dimension versions of these problems arise from multiple state variables and/or path dependencies. Although simulation is a powerful tool for solving some higher-dimensional problems, conventional wisdom was that simulation could not be applied to American-style pricing problems. The algorithms described here represent the first attempts to solve these problems, which were long thought to be computationally intractable.

Further topics

We conclude this chapter with a brief mention of two important areas of current work in the application of Monte Carlo methods to finance that have not been discussed.

A central numerical issue in simulating interest rates, asset prices with stochastic volatilities and other complex diffusions is the accurate approximation of stochastic differential equations by discrete-time processes. Kloeden and Platen (1992) discuss a variety of methods for constructing discrete-time approximations with different orders of convergence. Andersen (1995) applies some of these to interest-rate models. In general, reducing the time increment in a discrete approximation can be expected to give more accurate results but at the expense of greater computational effort. Duffie and Glynn (1995) analyse this trade-off and characterise asymptotically optimal time steps as the overall computational effort grows.

In this article we have focused almost exclusively on the use of Monte Carlo for pricing. A related, growing area of application is risk management – in particular, the use of Monte Carlo to assess value-at-risk, credit risk and related measures. For some examples of recent applications in these areas, see Iben and Brotherton-Ratcliffe (1994), Lawrence (1994) and Beckström and Campbell (1995).

Appendix: Moment controls beat moment matching asymptotically

As mentioned in the section on moment-matching methods, any time a moment is available for use with moment matching it can alternatively be used as a control variate. In this appendix, we argue that moment matching is asymptotically equivalent to a control variate technique with sub-optimal coefficients, and that it is therefore dominated by the optimal use of moments as controls. This asymptotic link applies in large samples. A related link between linear and non-linear control variates is made in Glynn and Whitt (1989), but the current setting does not fit their framework.

Let $Z_1, Z_2, \ldots$ be i.i.d. (not necessarily normal) with mean μ and variance σ^2. Let s denote the sample standard deviation of $Z_1, \ldots, Z_n$ and $\bar{Z}$ their sample mean. Suppose we want to estimate $E[f(Z)]$ for some function, f. The standard estimator is $n^{-1}\sum_{i=1}^{k} f(Z_i)$ and the moment-matching estimator is $n^{-1}\sum_{i=1}^{k} f(\tilde{Z}_i)$, with $\tilde{Z}_i$ defined in (9). For each i, the scaled difference

$$\sqrt{n}(\tilde{Z}_i - Z_i) = \sqrt{n}\left(\frac{\sigma - s}{s}\right)Z_i - \sqrt{n}\left[(\sigma\bar{Z}/s) - \mu\right]$$

converges in distribution, by the central limit theorem, for $\bar{Z}$ and s. Thus, $(\tilde{Z}_i - Z_i) = O_p(n^{-1/2})$ (see, eg Appendix A of Pollard, 1984, for O_p, o_p notation).

Suppose now that, with probability one, f is differentiable at Z_i. Then

$$f(\tilde{Z}_i) = f(Z_i) + f'(Z_i)\left[\tilde{Z}_i - Z_i\right] + o_p(n^{-1/2})$$

suggesting that up to terms $o_p(n^{-1/2})$ the moment-matching estimator and standard estimator are related via

$$\begin{aligned}
&\frac{1}{n}\sum_{i=1}^{n} f(\tilde{Z}_i) \\
&\approx \frac{1}{n}\sum_{i=1}^{n} f(Z_i) + \frac{1}{n}\sum_{i=1}^{n} f'(Z_i)\left[\tilde{Z}_i - Z_i\right] \\
&= \frac{1}{n}\sum_{i=1}^{n} f(Z_i) + \frac{1}{n}\sum_{i=1}^{n} f'(Z_i)\left[\left(\frac{\sigma}{s} - 1\right)Z_i - \frac{\sigma}{s}\bar{Z} + \mu\right] \\
&= \frac{1}{n}\sum_{i=1}^{n} f(Z_i) + \left(\frac{1}{n}\sum_{i=1}^{n} f'(Z_i)Z_i\right)\left(\frac{\sigma}{s} - 1\right) \\
&\quad + \left(\frac{1}{n}\sum_{i=1}^{n} f'(Z_i)\right)\left(\mu - \frac{\sigma}{s}\bar{Z}\right) \\
&\equiv \frac{1}{n}\sum_{i=1}^{n} f(Z_i) + \hat{\beta}_1\left(\frac{\sigma}{s} - 1\right) + \hat{\beta}_2\left(\mu - \frac{\sigma}{s}\bar{Z}\right)
\end{aligned}$$

where $\hat{\beta}_i \to \beta_i$, $i = 1,2$, as $n \to \infty$, with

$$\beta_1 = E[f'(Z)Z] \quad \text{and} \quad \beta_2 = E[f'(Z)]$$

Thus, moment matching is asymptotically equivalent to using

$$\left(\frac{\sigma}{\sigma} - 1\right) \quad \text{and} \quad \left(\mu - \frac{\sigma}{s}\bar{Z}\right) \tag{20}$$

as controls (both quantities converge to zero almost surely) with estimates of coefficients β_1, β_2. In general, these do not coincide with the optimal coefficients β_1^*, β_2^*, so moment matching is asymptotically dominated by the control variate method. In addition, the controls in (20) introduce some bias (as does moment matching itself) because, although they converge to zero, they do not have a mean of zero for finite n. In contrast, the more natural moment-control variates $(s^2 - \sigma^2)$ and $(\bar{Z} - \mu)$ have zero mean for all n and thus introduce no bias.

Reprinted from *Journal of Economic Dynamics and Control* 21(8/9), pp. 1,267–321, 1997, with permission from Elsevier Science.

The authors are grateful to Ben Fox and Leif Andersen for helpful comments, and to Ken Seng Tan and Xiaolin Gu for assistance. This research is supported in part by NSF grant DMI-94-57189.

1 *Wiggins (1987) also studies pricing under stochastic volatility but does not use Monte Carlo simulation.*

2 *In standard Monte Carlo application the n points are usually not truly random but are generated by a deterministic algorithm and are described as "pseudo-random" numbers.*

3 *We can readily estimate the variance of the Monte Carlo estimate by using the same set of n random numbers to estimate the expected value of f^2.*

4 *This method was introduced to option pricing in Boyle (1977), where its use was illustrated in the pricing of a European call on a dividend-paying stock.*

5 *The earliest application of this technique to option pricing is Boyle (1977).*

6 *To go from (6) to Boyle's (1977) example, let P_G be the price of a European call option on a no-dividend stock and let P_A be the corresponding option price in the presence of dividends.*

7 *This point is note merely a minor technical issue. The sample variance of the $\tilde{C}_i$s is usually a poor estimate of $\mathrm{Var}[\tilde{C}_i]$.*

8 *For example, let Z take the values +1 or −1 with probability one-half. Consider a security which pays +\$1 if $Z = 1$ and −\$x if $Z \neq 1$. The expected payoff of the security is $(1 - x)/2$. To estimate this expected payoff by Monte Carlo simulation, draw n samples Z_i according to the prescribed distribution. Then use equation (8) to define $\tilde{Z}_i$s which match the first moment. For almost all samples for any large n, the estimated expected payoff is $-x$ and the bias is $(1 + x)/2$. This bias does not decrease as n increases. Care must be taken when using equations (8) or (9) when the support of the random variable is not the entire real line. For example, applying (8) or (9) to uniform or exponential random variables could cause the transformed values to fall outside the relevant domain.*

9 *This is equivalent to enforcing put-call parity.*

10 *See, eg, Hull (1997, Chapter 17) for background on these models.*

11 *The standard errors in the table are all quite small, but so are the associated option values. Hence, the "relative" error without importance sampling is quite significant.*

12 *This is a direct consequence of Jensen's inequality for conditional expectations.*

13 *Thus, the name quasi-random is very misleading since these sequences are deterministic. However, it seems to be sanctioned by usage.*

14 *For the rest of the chapter we simply use the term discrepancy rather than star discrepancy to refer to* D^*_n

15 *Interestingly, linear congruential generators – frequently used to generate the pseudo-random numbers that drive ordinary Monte Carlo – produce sets of points with low discrepancy over the entire period of the generator (see Niederreiter, 1976). This suggests the possibility of choosing such a generator with period roughly equal to the total number of points required as a type of quasi-Monte Carlo method. In ordinary Monte Carlo, one prefers instead that the period be many orders of magnitude larger than the number of points required. We thank Peter Hellekalek of the University of Salzburg for this observation.*

16 *For a more complete discussion of the Hardy-Krause definition of variation and details on this theorem see Niederreiter (1992).*

17 *Bratley, Fox and Niederreiter (1992) note that the Niederreiter sequence they tested theoretically beats Sobol' sequences in dimensions higher than seven.*

18 *See, for example, Rensburg and Torrie (1993) or Morokoff and Caflisch (1995).*

19 *For example, one of the integrals used by Bratley, Fox and Niederreiter (1992) was*

$$\int_0^1 \cdots \int_0^1 \sum_{k=1}^{d} k \cos(kx_k)\, dx_1 \cdots dx_d$$

This integrand is highly periodic for large values of d.

20 *The details of the distribution are given in Broadie and Detemple (1996).*

21 *We take the first point of the Sobol' sequence to be 0.5, not 0.0.*

22 *We thank Spassimir Paskov and Joseph Traub for providing their code for the Sobol' sequences.*

23 *The details of the distribution are given in Broadie and Detemple (1996).*

24 *See, eg Chapter 14 of Hull (1997) for background.*

25 *Also called a "binary" or "cash-or-nothing" option; see Hull (1997, p. 463).*

26 *Sensitivities to various changes in the yield curve often account for several of these.*

27 *We used this example at the end of the section on low-discrepancy sequences. The settings are related: problems for which the common random numbers method is particularly effective are generally problems to which the pathwise method can be applied even more effectively.*

28 *Though not presented in a Monte Carlo context, the expressions in Carr (1993) are potentially relevant to this approach.*

29 *Geske and Johnson (1984) gave the first financial application of Richardson extrapolation. An extensive treatment of extrapolation techniques is given in Marchuk and Shaidurov (1983).*

30 *In fact, they distinguish between partitioning the state space, which they term "stratified state aggregation", and partitioning the payoff space, which they term "stratified state aggregation along the payoff". The latter method is the only one that they test or specify in detail. Hence we focus our discussion on this variant of their method.*

31 *We thank A. Dixit for pointing us to this reference.*

BIBLIOGRAPHY

Acworth, P., M. Broadie and P. Glasserman, 1997, "A Comparison of Some Monte Carlo and Quasi Monte Carlo Methods for Option Pricing", in *Monte Carlo and Quasi-Monte Carlo Methods for Scientific Computing*, H. Niederreiter, P. Hellekalek, G. Larcher, and P. Zinterhof, (eds), New York, Springer-Verlag.

Andersen, L., 1995, "Efficient techniques for simulation of interest rate models involving non-linear stochastic differential equations", Working paper, General Re Financial Products, New York, NY.

Andersen, L., and R. Brotherton-Ratcliffe, 1996, "Exact Exotics", *Risk* 9(10), pp. 85–9 (Chapter 11 of this publication).

Barlow, R.E., and F. Proschan, 1975, *Statistical Theory of Reliability and Life Testing*, Holt, Reinhart and Winston, New York.

Barraquand, J., 1995, "Numerical Valuation of High Dimensional Multivariate European Securities", *Management Science* 41, pp. 1,882–91.

Barraquand, J., and D. Martineau, 1995, "Numerical Valuation of High Dimensional Multivariate American Securities", *Journal of Financial and Quantitative Analysis* 30, pp. 383–405 (Chapter 15 of this publication).

Beaglehole, D., P. Dybvig and G. Zhou, 1997, "Going to Extremes: Correcting Simulation Bias in Exotic Option Valuation", *Financial Analysts Journal* (Jan/Feb), pp. 62–8.

Beckström, R., and A. Campbell, 1995, *An Introduction to VAR*, CATS Software, Palo Alto, CA.

Berman, L., 1996, "Comparison of Path Generation Methods for Monte Carlo Valuation of Single Underlying Derivative Securities", Research report RC-20570, IBM Research, Yorktown Heights, New York.

Birge, J.R., 1994, "Quasi-Monte Carlo Approaches to Option Pricing", Technical report 94-19, Department of Industrial and Operations Engineering, University of Michigan, Ann Arbor, MI.

Bossaerts, P., 1989, "Simulation Estimators of Optimal Early Exercise", Working paper, Carnegie-Mellon University, Pittsburgh, PA.

Boyle, P., 1977, "Options: a Monte Carlo Approach", *Journal of Financial Economics* 4, pp. 323–38 (Chapter 1 of this publication).

Boyle, P., and D. Emanuel, 1985, "The Pricing of Options on the Generalized Mean", Working paper, University of Waterloo.

Bratley, P., B.L. Fox and L. Schrage, 1987, *A Guide to Simulation*, 2nd edition, Springer-Verlag, New York.

Bratley, P., and B. Fox, 1988, "ALGORITHM 659: Implementing Sobol's Quasirandom Sequence Generator". *ACM Transactions on Mathematical Software* 14, pp. 88–100.

Bratley, P., B.L. Fox and H. Niederreiter, 1992, "Implementation and Tests of Low-Discrepancy Sequences". *ACM Transactions on Modelling and Computer Simulation* 2, pp. 195–213.

Broadie, M., and J. Detemple, 1994, "The Valuation of American Options on Multiple Assets", *Mathematical Finance* 7, pp. 241–86.

Broadie, M., and J. Detemple, 1996, "American Option Valuation: New Bounds, Approximations, and a Comparison of Existing Methods", *Review of Financial Studies* 9, pp. 1211–50.

Broadie, M., and P. Glasserman, 1996, "Estimating Security Price Derivatives by Simulation", *Management Science* 42, pp. 269–85.

Broadie, M., and P. Glasserman, 1997, "Pricing American-style Securities Using Simulation", *Journal of Economic Dynamics and Control* 21, pp. 1,323–52.

Caflisch, R.E., W. Morokoff and A. Owen, 1997, "Valuation of Mortgage-Backed Securities using Brownian Bridges to Reduce Effective Dimension", Working paper, Mathematics Department, UCLA (Chapter 26 of this publication).

Carr, P., 1993, "Deriving Derivatives of Derivative Securities", Working paper, Johnson Graduate School of Business, Cornell University.

Carriere, J.F., 1996, "Valuation of the Early-Exercise Price for Derivative Securities using Simulations and Splines", *Insurance Mathematics and Economics* 19, pp. 19–30.

Carverhill, A., and K. Pang, 1995, "Efficient and Flexible Bond Option Valuation in the Heath, Jarrow and Morton Framework", *Journal of Fixed Income* 5, pp. 70–7 (Chapter 18 of this publication).

Cheyette, O., 1992, "Term Structure Dynamics and Mortgage Valuation", *Journal of Fixed Income* 2 (March), pp. 28–41 (Chapter 19 of this publication).

Clewlow, L., and A. Carverhill, 1994, "On the Simulation of Contingent Claims", *Journal of Derivatives* 2, pp. 66–74.

Devroye, L., 1986, *Non-uniform Random Variate Generation*, New York, Springer.

Dixit, A., and R. Pindyck, 1994, *Investment Under Uncertainty*, Princeton University Press, Princeton.

Duan, J.-C., 1995, "The GARCH Option Pricing Model", *Mathematical Finance* 5, pp. 13–32.

Duan, J.-C., and J.-G. Simonato, 1995, "Empirical Martingale Simulation for Asset Prices", Working paper, Faculty of Management, McGill University, Montreal, Canada.

Duffie, D., 1996, *Dynamic Asset Pricing Theory*, 2nd edn Princeton University Press, Princeton, NJ.

Duffie, D., and P. Glynn, 1995, "Efficient Monte Carlo Simulation of Security Prices", *Annals of Applied Probability* 5, pp. 897–905.

Faure, H., 1982, "Discrépance de Suites Associées à un Système de Numeration (en Dimension s)", *Acta Arithmetica* 41, pp. 337–51.

Fox, B.L., 1986, "ALGORITHM 647: Implementation and Relative Efficiency of Quasi-random Sequence Generators", *ACM Transactions on Mathematical Software* 12, pp. 362–76.

Fu, M., and J.Q. Hu, 1995, "Sensitivity Analysis for Monte Carlo Simulation of Option Pricing", *Probability in the Engineering and Information Sciences* 9, pp. 417–46.

Fu, M., D. Madan and T. Wang, 1995, "Pricing Continuous Time Asian Options: a Comparison of Analytical and Monte Carlo Methods", Working paper, College of Business and Management, University of Maryland, College Park, MD.

Geske, R., and H.E. Johnson, 1984, "The American Put Options Valued Analytically", *Journal of Finance* 39, pp. 1511–24.

Glasserman, P., 1991a, *Gradient Estimation via Perturbation Analysis*, Kluwer Academic Publishers, Norwell, MA.

Glasserman, P., 1991b, "Filtered Monte Carlo", *Mathematics of Operations Research* 18, pp. 610–34.

Glasserman, P., and D.D. Yao, 1992, "Some Guidelines and Guarantees for Common Random Numbers", *Management Science* 38, pp. 884–908.

Glynn, P.W., 1987, "Likelihood Ratio Gradient Estimation: an Overview", in *Proceedings of the Winter Simulation Conference*, The Society for Computer Simulation, San Diego, CA, pp. 366–74.

Glynn, P.W., 1989, "Optimisation of Stochastic Systems via Simulation", in *Proceedings of the Winter Simulation Conference*, The Society for Computer Simulation, San Diego, CA, pp. 90–105.

Glynn, P.W., and D.L. Iglehart, 1988, "Simulation Methods for Queues: an Overview", *Queuing Systems* 3, pp. 221–55.

Glynn, P.W., and W. Whitt, 1989, "Indirect Estimation via $L = \lambda W$", *Operations Research* 37, pp. 82–103.

Glynn, P.W., and W. Whitt, 1992, "The Asymptotic Efficiency of Simulation Estimators", *Operations Research* 40, pp. 505–20.

Grant, D., G. Vora and D. Weeks, 1994, "Path-dependent Options: Extending the Monte Carlo Simulation Approach", Working paper, Anderson School of Management, University of New Mexico, Albuquerque, New Mexico.

Halton, J.H., 1960, "On the Efficiency of Certain Quasi-random Sequences of Points in Evaluating Multi-dimensional Integrals", *Numerische Mathematik* 2, pp. 84–90.

Hammersley, J.M., and D.C. Handscomb, 1964, *Monte Carlo Methods*, Chapman & Hall, London.

Haselgrove, C.B., 1961, "A Method for Numerical Integration", *Mathematics of Computation* 15, pp. 323–37.

Hlawka, E., 1971, "Discrepancy and Riemann Integration", in L. Mirsky (ed.), *Studies in Pure Mathematics*, Academic Press, New York, pp. 121–9.

Hull, J., 1997, *Options, Futures, and Other Derivative Securities*, 3rd edn, Prentice-Hall, Englewood Cliffs, NJ.

Hull, J., and A. White, 1987, "The Pricing of Options on Assets with Stochastic Volatilities", *Journal of Finance* 42, pp. 281–300.

Iben, B., and R. Brotherton-Ratcliffe, 1994, "Credit Loss Distributions and Required Capital for Derivatives Portfolios", *Journal of Fixed Income* 4, pp. 6–14.

Johnson, H., 1987, "Options on the Maximum or the Minimum of Several Assets", *Journal of Financial and Quantitative Analysis* 22, pp. 227–83.

Johnson, H., and D. Shanno, 1987, "Option pricing when the variance is changing", *Journal of Financial and Quantitative Analysis* 22, pp. 143–51.

Jones and Jacobs, 1986, "History Dependent Financial Claims: Monte Carlo Valuation", Working paper, Simon Fraser University.

Joy, C., P.P. Boyle and K.S. Tan, 1996, "Quasi-Monte Carlo Methods in Numerical Finance", *Management Science* 42, pp. 926–38 (Chapter 24 of this publication).

Kemna, A.G.Z., and A.C.F. Vorst, 1990, "A Pricing Method for Options Based on Average Asset Values", *Journal of Banking and Finance* 14, pp. 113–29.

Kloeden, P., and E. Platen, 1992. *Numerical Solution of Stochastic Differential Equations*, Springer, New York.

Lavenberg, S.S., and P.D. Welch, 1981, "A Perspective on the Use of Control Variables to Increase the Efficiency of Monte Carlo Simulations", *Management Science* 27, pp. 322–56.

Lawrence, D., 1994, "Aggregating Credit Exposures: The Simulation Approach", in *Derivative Credit Risk*, Risk Publications, London.

L'Ecuyer, P., and G. Perron, 1994, "On the Convergence rates of IPA and FDC Derivative Estimators", *Operations Research* 42, pp. 643–56.

Marchuk, G., and V. Shaidurov, 1983, *Difference Methods and Their Extrapolations*, Springer, New York.

McKay, M.D., W.J. Conover and R.J. Beckman, 1979, "A Comparison of Three Methods for Selecting Input Variables in the Analysis of Output from a Computer Code", *Technometrics* 21, pp. 239–45.

Morokoff, W.J., and R.E. Caflisch, 1995, "Quasi-Monte Carlo Integration", *Journal of Computational Physics* 122, pp. 218–30.

Moskowitz, B., and R.E. Caflisch, 1996, "Smoothness and Dimension Reduction in Quasi-Monte Carlo Methods", *Journal of Mathematical and Computer Modelling* 23, pp. 37–54 (Chapter 27 of this publication).

Niederreiter, H., 1976, "On the Distribution of Pseudo-random Numbers Generated by the Linear Congruential Method. III", *Mathematics of Computation* 30, pp. 571–97.

Niederreiter, H., 1988, "Low Discrepancy and Low Dispersion Sequences", *Journal of Number Theory* 30, pp. 51–70.

Niederreiter, H., 1992, *Random Number Generation and Quasi-Monte Carlo Methods*, CBMS-NSF 63, SIAM, Philadelphia, Pa.

Niederreiter, H., and C. Xing, 1995, "Low-Discrepancy Sequences and Global Function Fields with many Rational Places", Working paper, Austrian Academy of Sciences, Vienna.

Nielsen, S., 1994, "Importance Sampling in Lattice Pricing Models", Working paper, Management Science and Information Systems, University of Texas at Austin.

Ninomiya, S., and S. Tezuka, 1996, "Toward Real-time Pricing of Complex Financial Derivatives", *Applied Mathematical Finance* 3, pp. 1–20.

Owen, A., 1994, "Randomly Permuted (t, m, s)-Nets and (t, s)-Sequences", Working paper, Department of Statistics, Stanford University.

Owen, A., 1995, "Monte Carlo Variance of Scrambled Equidistribution Quadrature", in Niederreiter and P.J.S. Shiue (eds), *Monte Carlo and Quasi-Monte Carlo Methods in Scientific Computing*, Springer, Berlin.

Paskov, S., and J. Traub, 1995, "Faster Valuation of Financial Derivatives", *Journal of Portfolio Management* 22(Fall), pp. 113–20.

Pollard, D., 1984, *Convergence of Stochastic Processes*, Springer, New York.

Press, W.H., S.A. Teukolsky, W.T. Vetterling and B.P. Flannery, 1992, *Numerical Recipes in C: The Art of Scientific Computing*, 2nd edn, Cambridge University Press, Cambridge.

Raymar, S., and M. Zwecher, 1996, "A Monte Carlo Valuation of American Call Options on the Maximum of Several Stocks", Working paper, Fordham University, Graduate School of Business, New York.

Reider, R., 1993, "An Efficient Monte Carlo Technique for Pricing Options", Working paper, Wharton School, University of Pennsylvania.

Rubinstein, R., and A. Shapiro, 1993, *Discrete Event Systems*, Wiley, New York.

Rust, J., 1995, "Using Randomization to Break the Curse of Dimensionality", Working paper, University of Wisconsin, Madison, WI.

Schwartz, E.S., and W.N. Torous, 1989, "Prepayment and the Valuation of Mortgage-backed Securities", *Journal of Finance* 44, pp. 375–92.

Scott, L.O., 1987, "Option Pricing when the Variance Changes Randomly: Theory, Estimation, and an Application", *Journal of Financial and Quantitative Analysis* 22, pp. 419–38.

Shaw, J., 1995, "Beyond VAR and Stress Testing", Working paper, Global Market Risk Management Division, BZW (Chapter 21 of this publication).

Sobol', I.M., 1967, "On the Distribution of Points in a Cube and the Approximate Evaluation of Integrals", *USSR Computational Mathematics and Mathematical Physics* 7, pp. 86–112.

Spanier, J., and E.H. Maize, 1994, "Quasi-random Methods for Estimating Integrals using Relatively Small Samples", *SIAM Review* 36, pp. 18–44.

Stein, M., 1987, "Large Sample Properties of Simulations using Latin Hypercube Sampling", *Technometrics* 29, pp. 143–51.

Stulz, R.M., 1982, "Options on the Minimum or the Maximum of Two Risky Assets", *Journal of Financial Economics* 10, pp. 161–85.

Tezuka, S., 1994, "A Generalization of Faure Sequences and its Efficient Implementation", Research report RTO105, IBM Research, Tokyo Research Laboratory, Kanagawa, Japan.

Tezuka, S., 1995, *Uniform Random Numbers: Theory and Practice*, Kluwer Academic Publishers, Boston.

Tilley, J.A., 1993, "Valuing American Options in a Path Simulation Model", *Transactions of the Society of Actuaries* 45, pp. 83–104 (Chapter 14 of this publication).

Turnbull, S.M., and L.M. Wakeman, 1991, "A Quick Algorithm for Pricing European Average Options", *Journal of Financial and Quantitative Analysis* 26, pp. 377–89.

Van Rensberg, J., and G.M. Torrie, 1993, "Estimation of Multidimensional Integrals: is Monte Carlo the Best Method?", *Journal of Physics A: Mathematical and General* 26, pp. 943–53.

Wiggins, J.B., 1987, "Option values under stochastic volatility: theory and empirical evidence", *Journal of Financial Economics* 19, pp. 351–72.

Willard, G.A., 1996, "Calculating prices and sensitivities for path-dependent derivative securities in multifactor models", Working paper, Olin School of Business, Washington University, St Louis, Missouri.

Zaremba, S.K., 1968, "The Mathematical Basis of Monte Carlo and Quasi-Monte Carlo Methods", *SIAM Review* 10, pp. 310–14.

3

Monte Carlo Toolkit

Bruno Dupire
Nikko Financial Products

The aim of this short memento of algorithms is to provide the practitioner with the very basic means to start implementing Monte Carlo simulations virtually from scratch. We merely assume that he has at his disposal a good random generator to draw real samples between 0 and 1 according to a uniform law.

Gaussian samples

To obtain Gaussian samples of mean 0 and standard deviation 1 from uniform samples, two popular methods are:

BOX–MULLER TRANSFORM

Draw u and v, two uniform independent samples between 0 and 1. Compute:

$$x = \sqrt{-2\ln u}\,\cos 2\pi v$$

$$y = \sqrt{-2\ln u}\,\sin 2\pi v$$

x and y are independent Gaussian samples.

There is a way to avoid evaluating cosine and sine. Compute:

$$r = (2u - 1)^2 + (2v - 1)^2$$

If $r \leq 1$, define $\alpha = \sqrt{-2(\ln r/r)}$

$$x = u.\alpha$$
$$y = v.\alpha$$

x and y are independent Gaussian samples.

INVERSE CUMULATIVE NORMAL

Draw u, a uniform sample between 0 and 1. Compute:

$$x = N^{-1}(u) \text{ where } N(z) = \frac{1}{\sqrt{2\pi}}\int_{-\infty}^{z} e^{-v^2/2}dv$$

and x is a Gaussian sample.

To compute N^{-1}, (Beasly and Springer, 1997):

u uniform sample

$x = u - 0.5$
$y = x*x$

$$r = \frac{x*(((a_3*y + a_2)*y + a_1)*y + a_0)}{((((b_3*y + b_2)*y + b_1)*y + b_0)*y + 1)}$$

with

$$a_0 = 2.50662823884$$
$$a_1 = -18.61500062529$$
$$a_2 = 41.39119773534$$
$$a_3 = -25.44106049637$$
$$b_0 = -8.47351093090$$
$$b_1 = 23.08336743743$$
$$b_2 = -21.06224101826$$
$$b_3 = 3.13082909833$$

To improve the approximation for the tails, (Moro, 1995):

$$r = u$$
$$x = u - 0.5$$

if $|x| < 0.42$, use Beasly and Springer; else if $x > 0$, $r = 1 - u$

$$r = \log(-\log(u))$$
$$r = c_0 + r*(c_1 + r*(c_2 + r*(c_3 + r*(c_4 + r*(c_5 + r*(c_6 + r*(c_7 + r*c_8)))))))$$

if $x < 0$, $r = -r$

with

$$c_0 = 0.3374754822726147$$
$$c_1 = 0.9761690190917186$$
$$c_2 = 0.1607979714918209$$
$$c_3 = 0.0276438810333863$$
$$c_4 = 0.0038405729373609$$
$$c_5 = 0.0003951896511919$$
$$c_6 = 0.0000321767881768$$
$$c_7 = 0.0000002888167364$$
$$c_8 = 0.0000003960315187$$

To obtain a Gaussian sample y of mean m and standard deviation s from x, do $y = m + s.x$.

n-dimensional Gaussian samples

To draw a vector of Gaussian variables of covariance matrix V, a Choleski decomposition is performed.

❑ Perform a Choleski decomposition: $V = C'C$. Denote the ith row and jth column element of V as v_{ij} and that of C as c_{ij}.

For $i = 1$ to n:

For $j = i$ to n

$$c_{ij} = \frac{1}{c_{ii}}\left(v_{ij} - \sum_{k=1}^{i-1} c_{ik}c_{jk}\right)$$

$$c_{ii} = \sqrt{v_{ii} - \sum_{k=1}^{i-1} c_{ik}^2}$$

❑ Draw n independent samples.

❑ Multiply this vector of independent samples by C.

Stochastic differential equation

To simulate the stochastic differential equation

$$dx = a(x,t).dt + b(x,t).dW$$

denote $a_i = a(x_i, i\delta t)$ and $b_i = b(x_i, i\delta t)$.

Euler scheme - simplest, first-order scheme:

$$x_{i+1} = x_i + a_i\delta t + b_i\sqrt{\delta t}g_i$$

Milstein scheme - second-order scheme to account for x dependency:

$$x_{i+1} = x_i + \left(a_i + \frac{1}{2}b_i\frac{\partial b_i}{\partial x}(g_i^2 - 1)\right)\delta t + b_i\sqrt{\delta t}g_i$$

Variance reduction methods

ANTITHETIC

If a scenario is x, draw $-x$ as well. If a scenario is $(x_1,\ldots,x_n)$, draw $(-x_1,\ldots,-x_n)$ as well.

BALANCE SAMPLING

To obtain n representative samples from a Gaussian distribution, for $i = 1$ to n:

$$x_i = N^{-1}\left(\frac{(i - 0.5)}{n}\right)$$

or take x_i in each quantile.

LATIN HYPERCUBE

For each dimension, generate a random permutation of the balance sampling points. To generate a random permutation of $(x_1,\ldots,x_n)$:

For $p = n$ to 1:

Draw uniformly an integer i between 1 and p (i is the integer part of $p.u + 1$, where u is uniform).

Swap x_i and x_p

$a := x_p$

$x_p := x_i$

$x_i := a$.

Other dynamics

Allow for more realistic dynamics:

$$dx = adt + bdw + \gamma dq^{\lambda}$$

where γdq^{λ} is one if there is a jump (with probability λdt over dt) and γ is the possibly random amplitude of the jump.

For $i = 1$ to n:

Draw u uniformly.

If $u \geq \lambda dt$, $x_{i+1} = x_i + a_i\delta t + b_i\sqrt{\delta t}g_i$.

Else, draw z according to the law of γ and $x_{i+1} = x_i + a_i\delta t + b_i\sqrt{\delta t}g_i + z$.

BIBLIOGRAPHY

Beasly, J.D., and S.G. Springer, 1997, *Applied Statistics* (26), pp. 118–21.

Moro, B., 1995, "The Full Monte", *Risk* 8(2), pp. 53–7.

Press, W.H., S.A. Teukolsky, W.T. Vetterling and B.P. Flannery, 1992, *Numerical Recipes in C: The Art of Scientific Computing*, 2nd edn, Cambridge University Press, Cambridge.

PRICING

Introduction

Bruno Dupire

This section covers a wide range of contributions, including speed-up techniques, applications to specific products and calibration issues. Many of these papers have appeared previously in *Risk* magazine (Chapters 5 to 11). Others (Chapters 4, 12 and 13) are new papers written for this book.

The first chapter, by Dupire and Savine, begins by questioning the soundness of using random methods, as opposed to structured, deterministic ones. It presents a simple algorithm, "uniform with antithetic noise" (UWAN), which combines the advantages of both methods. As a variation of a regular grid, its range of applicability is confined to low dimension, which necessitates the reduction of the dimension of high-dimensional problems. This is achieved by performing a principal component analysis on Brownian paths (one asset at different dates), in contrast to classical applications of principal component analysis to various assets at one date.

Chapter 5 – "Average Intelligence", by Levy and Turnbull – reviews Asian option pricing methods. It is echoed in Chapter 6, "Beyond Average Intelligence" – a very concise paper by Michael Curran, which presents a conditioning method that computes the expected Asian payout conditional on the geometric mean.

Three other short chapters by Curran follow. "Strata Gems" proposes, in order to generate n Gaussian samples, a slicing of the real line into n intervals corresponding to the quantiles of the normal distribution and drawing the n conditional means without replacement. In dimension one, it amounts to a deterministic integration numerical scheme, but not any more than in the case of path-dependent options, where the same procedure is applied at each time step. Chapter 8, "Recovering Identity", elaborates on the previous chapter, and rescales the samples by means of a Choleski decomposition to ensure that the sample covariance structure is correct. A word of caution should apply: although the empirical covariance is zero, the samples are not independent any more! Chapter 9 explains how to compute the delta and the gamma of a derivative without having to shift the initial spot. This is done in a simple lognormal dynamics framework. It is actually possible to extend this idea to more general dynamics thanks to Malliavin calculus.

In Chapter 10, Clewlow and Carverhill present a technique they name martingale variance reduction (MVR), which uses the option hedge as a control variate. The core idea is that the hedged option exhibits much less variance than the unhedged one and the hedge, after filtering, becomes a martingale, which does not affect the expectation. They apply it to pricing lookback options under stochastic volatility assumptions, where delta, gamma and volatility hedges are computed in a Black–Scholes framework.

The last of the *Risk* magazine papers of this section is Chapter 11, on "Exact Exotics", by Andersen and Brotherton-Ratcliffe. When applying Monte Carlo simulations to pricing barrier options, time is discretised into time steps of finite size and barrier-hitting within steps has to be taken into account. They do so by computing the hitting probability over each time step conditional on the start and end points. Along each Monte Carlo path they draw, at each time step, a uniform sample and compare it with the hitting probability to decide whether or not intra-step hitting has occurred.

In other words, in the case of a knock-out option (for instance), each path sees its associated payout randomly multiplied by 1 or 0 according to its probability of survival, which is the product of the individual step probabilities of survival. Indeed, it would have been better to multiply the payout by the probability of survival, which can be interpreted as having drawn for each discrete path not one but zillions of continuous paths conditioned by the value of the discrete path. Another popular method is to shift the barrier by the expected difference between the continuous and the discrete paths (see

Broadie, M., P. Glasserman and S. Kou, 1997, "A Continuity Correction for Discrete Barrier Options", *Mathematical Finance* 7(4), pp. 325–49).

Chapter 12, a new paper by Leif Andersen, extends the same approach to double-barrier options, computing probabilities of surviving two barriers as infinite series and some truncation errors. In practice, if the time step and the volatility are not too large, multiplying the two single-barrier probabilities is more than sufficient.

The last chapter of this section, "Model Calibration in the Monte Carlo Framework", by Raphaël Douady, is also published here for the first time. It addresses the very important issue of model calibration, ie determining the parameters of a model by optimal fitting to a set of market prices, as opposed to historical estimation. It compares bootstrap (or forward, step-by-step) calibration with global calibration, and proposes remedies to the instabilities that are frequently encountered with the bootstrap method. It also establishes how to optimally distribute the computational effort (number of Monte Carlo paths) at each iteration of the optimisation algorithm. Finally, it proposes techniques to correct drift and variance to ensure that the empirical distribution has the proper moments.

4

Dimension Reduction and Other Ways of Speeding Monte Carlo Simulation

Bruno Dupire and Antoine Savine
Nikko Financial Products; General Re Financial Products

This chapter summarises research aimed at speeding up Monte Carlo simulation through improved sampling algorithms and alternative constructions of Brownian motion paths. We start with a generic approach to Monte Carlo simulation used for option pricing. Both practical and formal, it is aimed at providing an understanding of the issues involved, which in turn is the basis for algorithms and methods that ensure rapid convergence. We present a new methodology for generating and calibrating random samples and then focus on the construction of Brownian paths from such samples. We show how to optimally reduce the effective dimension of the simulation to maximise the efficiency of the sampling algorithms that we introduce.

Of all the numerical procedures available for option pricing, Monte Carlo is slow and rather inaccurate. Nonetheless, not only is Monte Carlo simulation widely used by the financial industry but it has also stimulated more papers, conferences and debates during the last few years than any other numerical topic.

This paradoxical success is probably due to the simplicity and generality of the procedure.

Simplicity because, unlike for other methods, one is not required to be an expert in stochastics and numerics to understand and run Monte Carlo simulations. It does not deal with complex analytics. Rather, it is based on the intuitive idea that the fair price of a product corresponds to its average payoff among a large number of scenarios. This intuition matches the main result of the financial theory which states that (under the standard hypotheses of non-arbitrage and completeness) derivative products can be hedged by trading the underlying asset, and that the cost of the hedge (ie the fair price of the product) can be seen as the expected payoff in a world where the average return of any asset is the risk-free interest rate. Therefore, despite its apparent simplicity, Monte Carlo is consistent with option theory.

Generality because Monte Carlo methods work however complex the product (with specific methods for American options) and the underlying assumptions of the model. Analytical approximations, trees and finite-difference methods only work in the simplest frameworks and for a limited class of products. Extension to more complex models and products is often possible but at the cost of additional difficulties – such as adding state variables, which make the numerical scheme much more complex. In contrast, Monte Carlo in its simpler form works in almost all cases.

Monte Carlo simulation has also benefited from the recent increased interest in value-at-risk analysis. Although in this chapter the concepts are presented in a pricing context (the computation of expectation), they are easily extended for risk management purposes (computations of histograms or various statistics).

Despite its apparent simplicity, Monte Carlo offers financial research one of its most challenging areas. If in its simplest form Monte Carlo works in a large number of cases, it remains slow and inaccurate. Fortunately, progress has been made recently in accelerating and optimising the method due to an improved understanding of the issues involved. Here we analyse these issues and examine state-of-the-art techniques for speeding up simulation.

The chapter is divided into four sections. The first presents a formal approach to Monte Carlo.

The second deals with sampling algorithms. In the third we demonstrate how to generate and calibrate Gaussian samples, and the fourth displays our main results with respect to spectral truncation.

All the pricing examples we give are based on the following benchmark: the initial asset price is 100, the strike is 100 and the interest rate is 0%; the model is lognormal with a volatility of 10%; and in the multi-asset case the correlation is zero.

Monte Carlo: a practical and formal approach

Monte Carlo pricing consists of simulating a large number of scenarios consistently with a given set of assumptions, and estimating an option price as the discounted average payoff among them. In a single underlying framework with a path-independent product and simple dynamics, each scenario consists of one final underlying value. Since one random number per scenario is required, the problem is set in dimension 1. If the product is path-dependent or with complex dynamics, we require the value of the underlying at a certain number, p, of points in time for each scenario. The problem is then in dimension p. If the framework consists of n random processes for which values are required at p points in time, the dimension is np.

We now pay attention to the case of stochastic differential equations, where Gaussian increments are generated at each time step. Each simulation requires the generation of a uniform random vector in the appropriate dimension, d, using a sampling algorithm. Then the uniform vector has to be transformed into a d-variate Gaussian vector. This, in turn, is used to generate n Brownian paths discretised at p points of time. Finally, the evolution of the world defined by the scenario is produced by running the stochastic differential equations defined by the model with the Brownian paths generated previously. Each scenario leads to an option payoff, and the average of these converges to the expected payoff by virtue of the strong law of large numbers. Monte Carlo simulations provide a sample of payoffs that is available for any statistical analysis: expectation, standard deviation, confidence intervals or the estimation of fractiles for value-at-risk purposes.

Let F be a technique that produces Gaussian vectors from uniform vectors.

$F: [0,1]^d$	$\rightarrow$	R^d
Uniform vector		Gaussian vector

Also, let us denote by G the chosen method that generates Brownian paths from a Gaussian vector:

$G: R^d$	$\rightarrow$	$C(R^+)^n$
Gaussian vector		n Brownian paths

Finally, let H be the payoff associated with a given scenario (H defines the product):

$H: C(R^+)^n$	$\rightarrow$	R
n Brownian paths		Option pay-off

Define $f(x) = H[G[F(x)]]$ for $x \in [0,1]^d$. Then the theoretical price of the product is $\int_{[0,1]^d} f(x)dx$ and its Monte Carlo estimate is $\frac{1}{N}\sum_{i=1}^{N} f(x_i)$, where $\{x_i, 1 \le i \le N\}$ is the sample and N is its size.

Sampling schemes

According to the above exposition, for a given set of transformations, F and G, and a given product defined by its payoff, H, the problem of Monte Carlo simulation reduces to one of approximating $\int_{[0,1]^d} f(x)dx$ by $\frac{1}{N}\sum_{i=1}^{N} f(x_i)$, where f is a real-valued d-dimensional function. A physical analogy compares this problem with that of evaluating the average heat inside a metal cube using N sensors. How should we place them to get a satisfactory approximation?

REGULAR GRID

An engineer would probably want to place the sensors uniformly throughout the cube[1] to make sure that no significant region is left out. It is a well-known result that if $\{x_i, 1 \le i \le N\}$ is

1. Regular grid in dimension 2 (25 points)

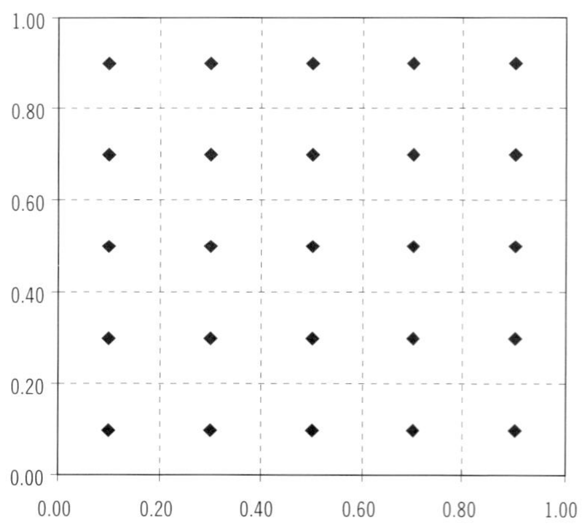

sampled through a regular scheme

$$\left|\int_{[0,1]^d} f(x)dx - \frac{1}{N}\sum_{i=1}^{N} f(x_i)\right| = O\left(\frac{1}{N^{2/d}}\right)$$

We define the convergence rate as minus the exponent of N in the expression of the error bound. The regular grid provides a convergence speed of $\frac{2}{d}$. The constant is proportional to the maximum convexity of f. The error bound corresponds to a worst case.

RANDOM SAMPLING

In contrast to this ordered approach, probability theory advises pure random sampling. If $\{x_i, 1 \le i \le N\}$ is a set of independent and identically distributed random vectors,[2] the central limit theorem states that $\frac{1}{N}\sum_{i=1}^{N} f(x_i)$ converges towards $\int_{[0,1]^d} f(x)dx$ with a speed of $\frac{1}{2}$. The constant is proportional to the L_2 norm of f. The error bound corresponds to an average case. Therefore, pure random sampling performs better than uniform sampling as soon as the dimension is greater than four.

This result is paradoxical for two reasons. First, it is disturbing that it is better to do things at random than in an ordered fashion. Second, random sampling is likely to create gaps and clusters of points within the cube, which the regular scheme avoids. And, yet, random sampling performs better.

The paradox arises from the fact that the regular grid ignores the convexity of f on the subdivisions. The error is small on each subdivision because the linear portion of f is accurately integrated. But the errors add up throughout the space. The integral of a globally convex function is undervalued, while the integral of a concave function is overvalued. This proves to be worse than the gaps and clusters created by the random scheme, in which (big) errors cancel on average.

In other words, the regular grid generates small errors that add up, whereas random sampling generates big errors that cancel on average.

UNIFORM SAMPLING WITH ANTITHETIC NOISE (UWAN)

The question naturally arises from this disappointing analysis whether we can combine the advantages of the regular grid and random sampling – that is, define a sampling scheme that generates small errors which cancel.

The answer is yes. Suppose that we cut the cube into regular sub-cubes, just as the regular grid does, to avoid gaps and clusters. Now, we want to benefit from an unbiased random sampling on each subdivision. Thus, instead of taking a sample at the centre of each subdivision, we take a random point inside it and add its symmetrical counterpart with respect to the centre as well. If we had taken only one point, big errors might arise in the case of a high gradient. The *local* antithetic, ie the symmetric point with respect to the

2. Regular grid misses convexity, generating small but additive errors

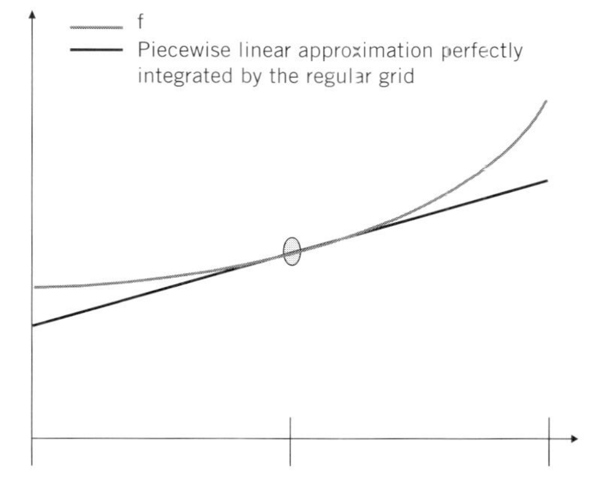

3. UWAN sampling in dimension 2 (50 points)[3]

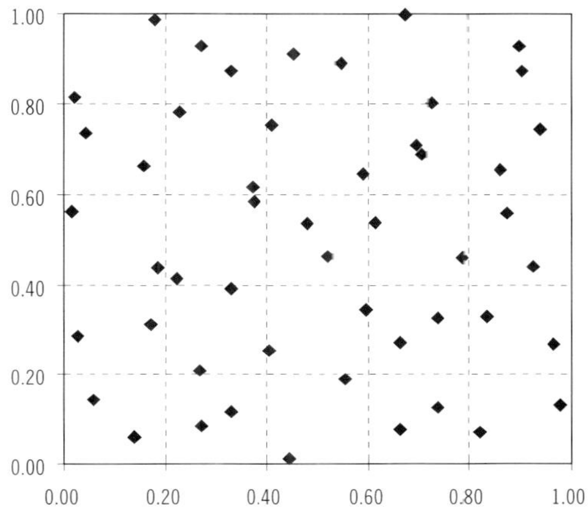

4. UWAN generates small errors that cancel

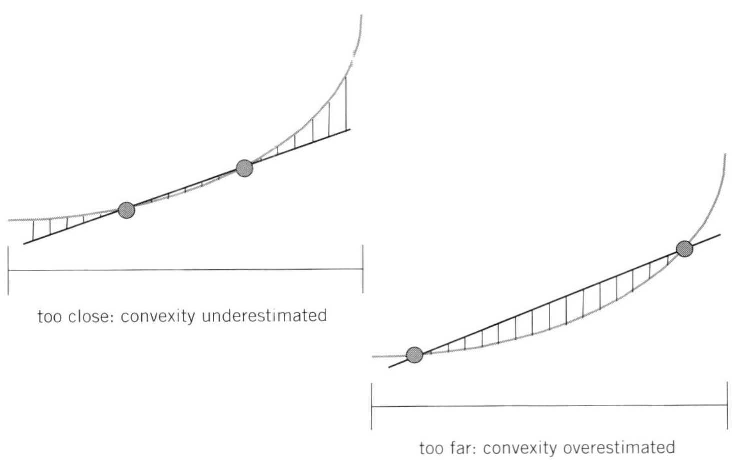

centre of the cube, removes these errors. The symmetry of these two points makes them price exactly linear functions whichever the dimension.

This method combines the benefits of the regular grid and random sampling. The regular division of the space prevents gaps and clusters and ensures a small error in each subdivision. Random sampling inside each subdivision, in turn, ensures that on average the errors cancel.

UWAN does not overestimate convexity. Depending on the random sampling on each subdivision, convexity is overvalued or undervalued. Globally, it is matched on average, as the estimate is unbiased on each interval.

In terms of convergence rate, subdivision of the space ensures a speed of $\frac{2}{d}$, while random sampling in the sub-squares independently adds $\frac{1}{2}$. As these two principles are independent, the total convergence speed is $\frac{2}{d}+\frac{1}{2}$, with an error bound corresponding to the average case. This assertion is proved in Appendix A.

As an example, Figure 5 shows the convergence of the price of a best-of-three assets option (call on the best performance as between three different financial assets). The Monte Carlo estimate is displayed as a function of the number of simulations for random sampling, the regular grid and UWAN. The random seed was reset at each simulation.

5. Monte Carlo pricing of a best-of option in dimension 3

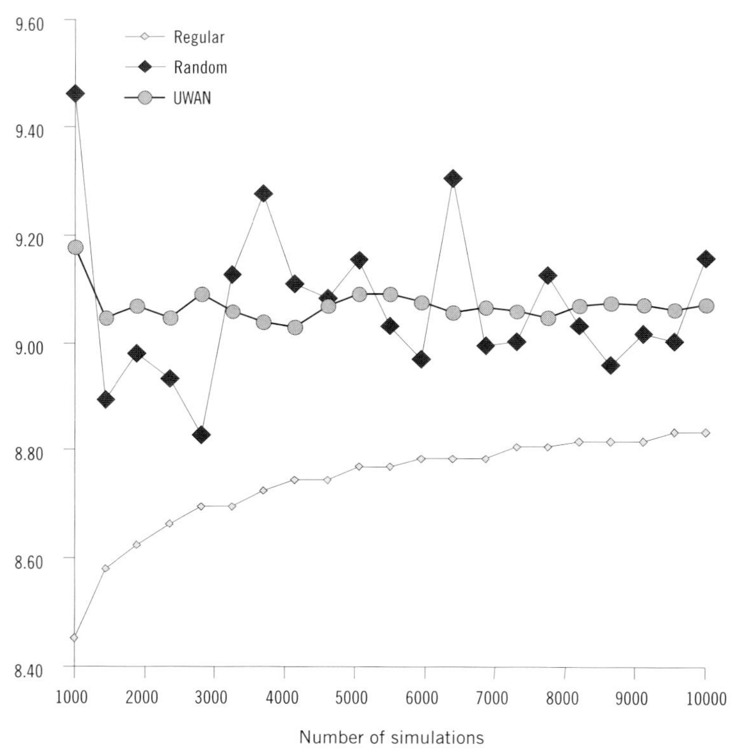

LOW-DISCREPANCY SEQUENCES

We complete this section with a discussion of low-discrepancy sequences. These are sequences of points that are chosen in a deterministic, rather than a random, fashion so as to ensure an optimal filling of the space. There are many different low-discrepancy sequences, those of Halton, Faure and Sobol' being the best known. The most widely used by the financial industry is the Sobol' sequence (described by Press *et al.*, 1992, p. 309, where an algorithm is also provided). Since much has been published on the subject (see, eg Niederreiter, 1992), here we will just summarise the principal features of such sequences.

- Low-discrepancy sequences fill the space in an additive way, which means that the $(N + 1)$th point does not modify the position of the first N points (sequences tend to place it in the centre of the biggest gap left by the previous points). This property is not satisfied by other structured schemes such as the regular grid or UWAN and presents obvious computational advantages.
- They are constructed to ensure the lowest possible discrepancy, defined as the largest error on the volume of parallelepipeds in the hypercube:

$$D(N) = \sup_{y \in [0,1]^d} \left| \prod_{j=1}^{d} y_j - \frac{1}{N} \sum_{i=1}^{N} 1_{\{\forall j \in [1,d], x_j^i \le y_j\}} \right|$$

- The Koksma–Hlawka inequality ensures that

$$\left| \int_{[0,1]^d} f(x)dx - \frac{1}{N} \sum_{i=1}^{N} f(x_i) \right| \le V(f) \cdot D(N)$$

where $V(f)$ is the variation of f. This result is more or less intuitive since the discrepancy measures the ability to integrate the volume of parallelepideds whereas the variation measures the ability to approximate f by parallelepideds.

- The ability of low-discrepancy sequences to fill the space is impressive in low dimension, but it quickly deteriorates as the dimension grows – as illustrated in Figure 6, where the number of points is 500. Note that the Sobol' sequence is the most robust in high dimension, which justifies its frequent preference by the financial community.

However, despite their ability to fill the space, low-discrepancy sequences are not able to integrate convexity correctly. Therefore, UWAN is likely to perform better on a function with high curvature, even in low dimension. This is shown in Figure 7, where $f(x) = \sqrt{x_1^2 + x_2^2 + x_3^2}$, $d = 3$.

6. Ability of low-discrepancy sequences to fill the grid space

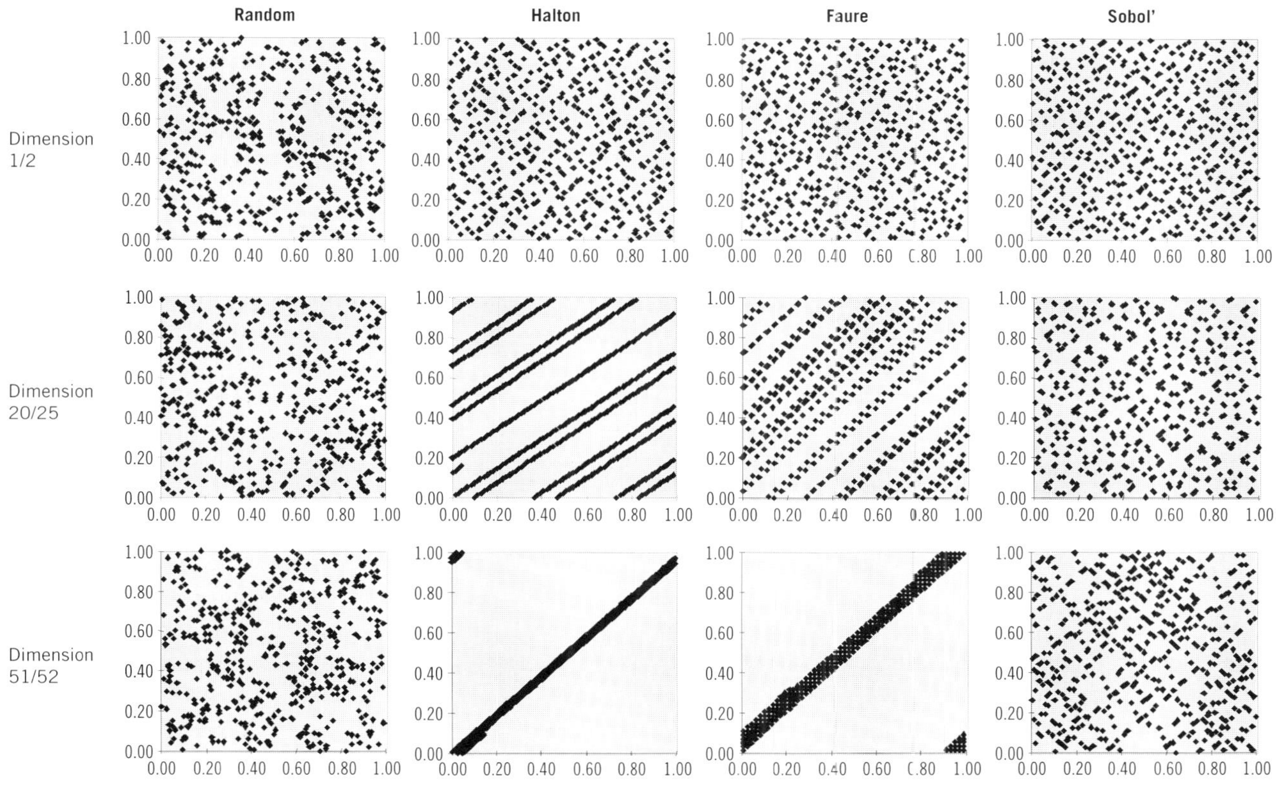

If the power of low-discrepancy sequences deteriorates in high dimension, so does the power of UWAN. Its convergence speed of $\frac{2}{d} + \frac{1}{2}$ quickly reduces to $\frac{1}{2}$ as d grows, reducing UWAN to a standard antithetic random sampling.

We therefore need to reduce the dimension of the problem (see, for instance, Caflisch, Morokov and Owen, 1997 and Chapter 26 of this publication) to provide sampling algorithms with their full strength. We raise the question in the last section after analysing Gaussian sampling and the effect of sample calibration in the next section.

7. UWAN compared with low-discrepancy sequences: norm function in dimension 3

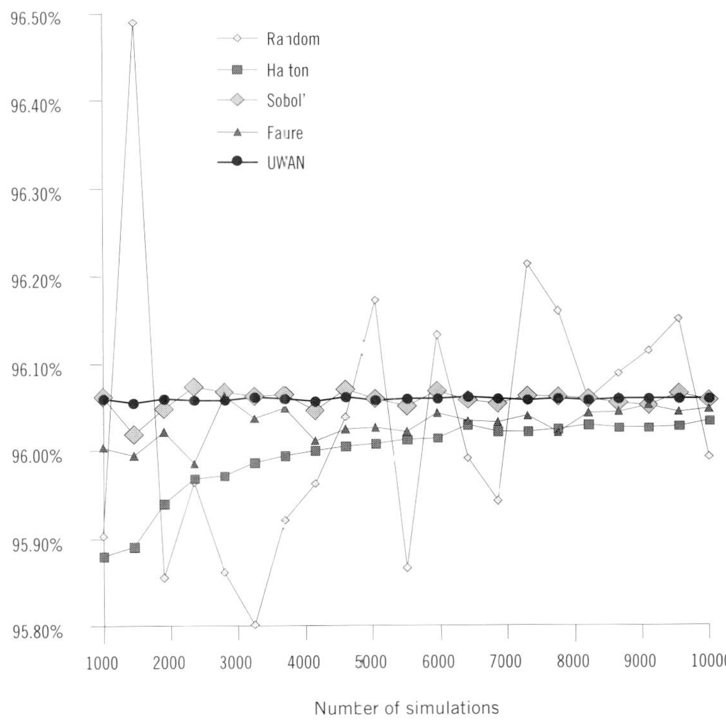

Calibrating the Gaussian sample

Once a sample, $X = \{x_i^j, 1 \leq i \leq N,\ 1 \leq j \leq d\}$, has been generated, it can be turned into a Gaussian sample, $G = \{g_i^j, 1 \leq i \leq N,\ 1 \leq j \leq d\}$, using the Box-Muller technique or a rational approximation of the inverse cumulative normal distribution (see Chapter 3).

In theory, the sample G is supposed to be generated by a d-dimensional Gaussian distribution with mean 0_d and covariance matrix I_d. In practice, it has an empirical mean $M \in R^d$, $M_j = \frac{1}{N}\sum_{i=1}^{N} g_i^j$, slightly different from 0_d and a covariance matrix, $C = G'G$, different from I_d.

This section explains why we should shift the sample slightly so as to force an empirical mean to 0_d and covariance of I_d, and how this is done.

CALIBRATING THE MEAN

Forcing the mean of the sample to 0_d ensures that all assets whose payoff is linear in terms of the state variables will be immediately matched. In addition, the linear part of non-linear assets is perfectly priced however large the number of simulations.

One way to force the mean is antithetic sampling. For each Gaussian vector, V, that is generated, add V and –V to the sample. This is all the more worth doing as the technique saves computation time, giving two samples at the cost of one.

An interesting observation on antithetic sampling as a special case of a more general framework can be found in Caflisch, Morokov and Owen (1997) and Chapter 26 of this publication.

CALIBRATING THE COVARIANCE

Calibrating the covariance matrix is also worth doing – especially if the sample was generated by some low-discrepancy sequence, which produces a correlation that deteriorates considerably in high dimension. It ensures that parabolas, and, more generally, the quadratic part of any asset, are immediately well priced. With low-discrepancy sequences the effect is remarkable – as displayed in Figure 8, which compares the convergence of Sobol' samples, with and without covariance calibration, on a one-year Asian option with weekly fixings (dimension 52).

8. Effect of covariance calibration with Sobol' samples in dimension 52 for an Asian option

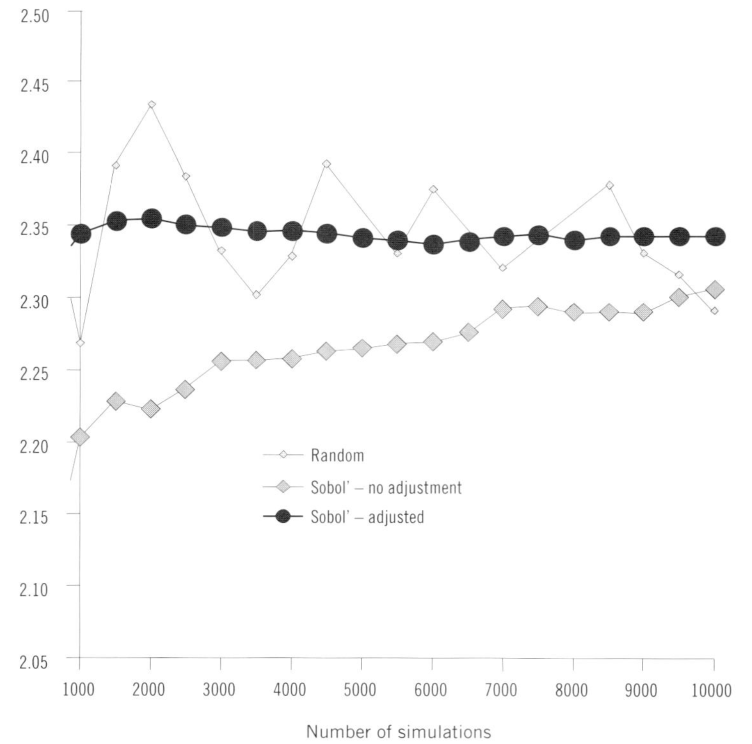

This result arises from the fact that low-discrepancy sequences generate serial correlation in high dimension. This is corrected through calibration, but the adjustment is minor enough to preserve the low-discrepancy property of the sample. The smoothness of the convergence is therefore preserved, while the bias is removed. However, after this renormalisation the samples are not exactly Gaussian any more.

The technique is quite simple and is based on the Cholesky decomposition of C: $C = cc'$. (An algorithm for the decomposition is provided in Press *et al.* (1992, p. 96). Shift the sample G by multiplying it on the right by $(c^{-1})'$. Since C is close to I_d, the adjustment is minor. Then

$$\bar{G} = G(c^{-1})' \Rightarrow C = G'G = I_d$$

Generating Brownian paths

The most efficient sampling algorithms perform better in low dimension. More precisely, they integrate better the first few dimensions of f. This section is therefore devoted to the attempt to reduce the effective dimension of f. We survey Brownian path construction techniques that allow one to concentrate the "importance" (see definition below) of the simulation on the first few dimensions of the random vectors. We also analyse the adaptability of the sampling schemes to the geometry of the problem.

We consider as given a set of N Gaussian vectors $G_i = (g^i_1, \dots, g^i_d)$, $i \leq 1 \leq N$.

FORWARD GENERATION

The standard way of generating a Brownian path is forward induction (Euler method):

$$\begin{cases} W_0 = 0 \\ W_{T_i} = W_{T_{i-1}} + \sqrt{T_i - T_{i-1}}\, g_i \end{cases}$$

This algorithm is based on

$$\text{Law}(W_{T_i} | W_{T_{i-1}}) = N(W_{T_{i-1}}, \sqrt{T_i - T_{i-1}})$$

The algorithm is attractive for its simplicity, but it does not fit our needs. Each dimension is associated with a Brownian increment. The importance of the random draws is, therefore, uniform across dimensions and we cannot concentrate our attention on the first few increments at the expense of subsequent increments.

BACKWARD GENERATION

Classical path-generation by backward induction does, however, meet our objectives. It is based on the Brownian bridge representation of the Brownian motion, which states that

$$\mathrm{Law}(W_{T_i}|W_{T_i+\alpha}, W_{T_i-\beta})$$
$$= N\left(\frac{\beta}{\alpha+\beta}W_{T_i+\alpha} + \frac{\alpha}{\alpha+\beta}W_{T_i-\beta}, \sqrt{\frac{\alpha\beta}{\alpha+\beta}}\right)$$

allowing one to proceed through backward induction.

The proof of convergence for this algorithm is also a corollary of Theorem 1. Besides, it is clear from above that the variance of the successive draws decreases exponentially. The first draws capture the essence of the variance and give the main outline of the path, on top of which some noise is superimposed. It is clear from this that the importance of the dimensions is not uniform, the first co-ordinates being used to generate an outline with high variance, the remainder being used to add some noise with little variance on top of that.

A full survey of the ability of the Brownian bridge construction to reduce the effective dimension in a particular case, as well as its application to low-discrepancy sequences, can be found in Chapter 26. A related approach to the one presented here can be found in Acworth, Broadie and Glasserman (1996).

We define the "importance" of a given dimension as the part of the variance that is explained by the direction in which the corresponding random co-ordinate is applied.

A GENERAL RESULT

Theorem 1 If (φ_k) *is a Hilbert basis of* $L^2([0,T_d])$ *and* $\phi_k(t) = \int_0^t \varphi_k(s)ds$ *are the primitives of* (φ_k) *vanishing at* 0, *then if* (g_k) *is a set of independent standard Gaussian variables,* $\sum g_k\phi_k$ *converges to a Brownian motion on* $[0,T_d]$.

Theorem 1 provides a general framework of which all path-generation techniques are special cases. It is proven in Appendix B.

Applications

Forward and backward induction fall in this framework, which mathematically justifies their use – that is, ensures that these approximations converge to a Brownian motion. In forward induction

$$\varphi_k = \frac{1_{[T_{k-1},T_k]}}{\sqrt{T_k - T_{k-1}}}$$

(see Figure 11 on next page) and in backward induction (φ_k) is the "Haar basis" (see Figure 12, also on next page). These are not Hilbert basis but can be completed to form one

Theorem 1 states that there is an equivalence between the Brownian path construction techniques and the basis of $L^2([0,T_d])$. Each new technique defines a basis, while each new basis defines a construction algorithm. In the next two subsections we define the optimal basis.

SPECTRAL GENERATION: DISCRETE CASE

From our definition of the importance of the dimensions, we call "optimal" a path construction algorithm that maximises that part of the variance explained by the first few components

9. Forward induction

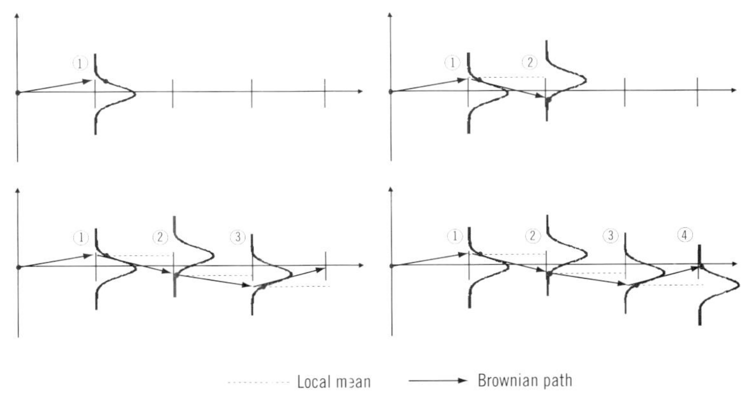

10. Backward induction

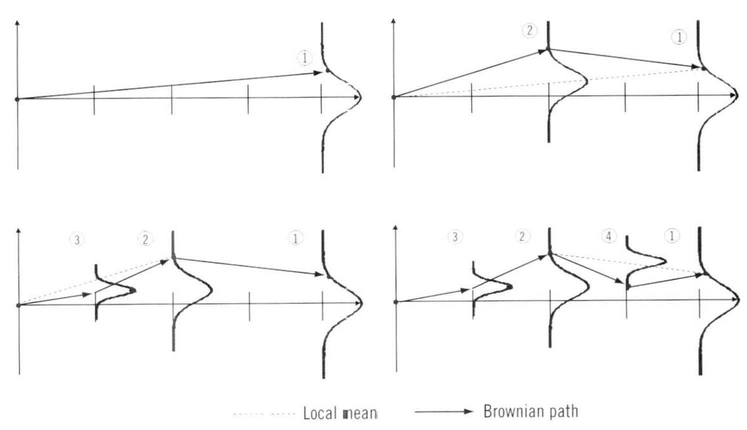

11. Forward induction as a particular case of Theorem 1

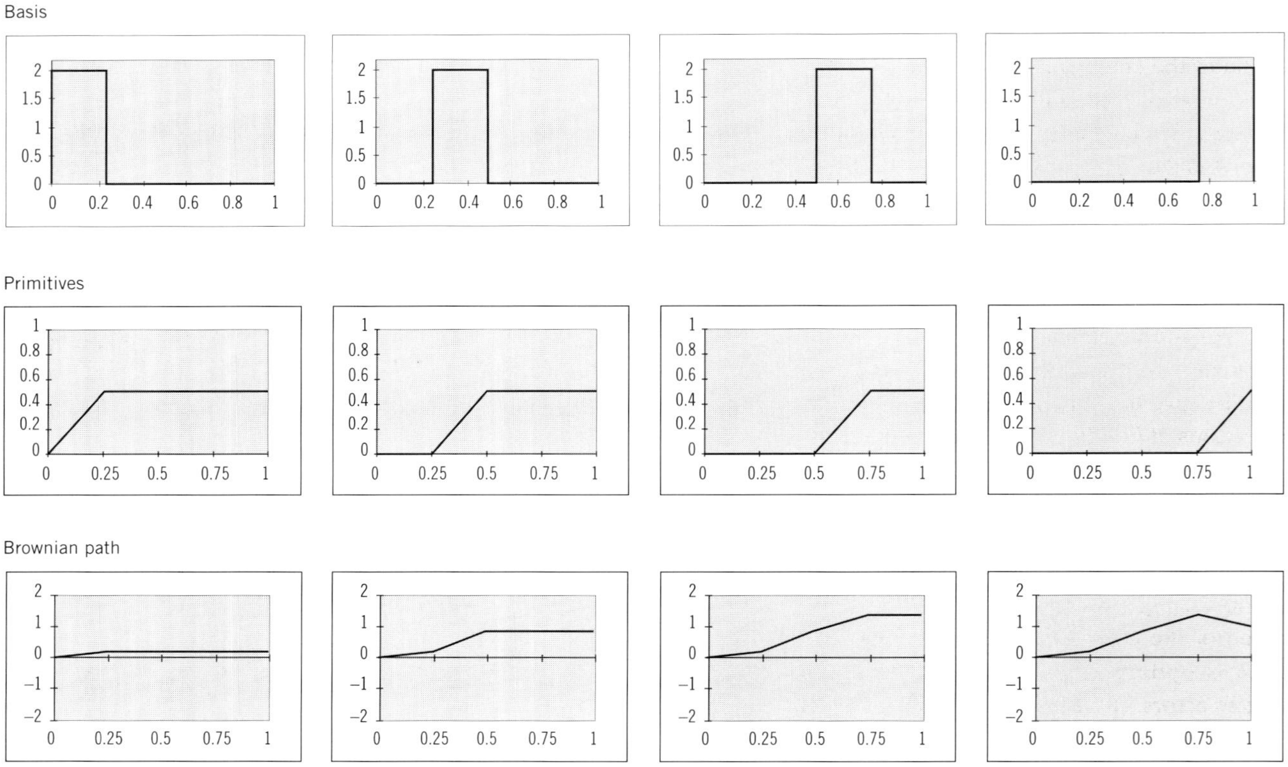

12. Backward induction as a particular case of Theorem 1

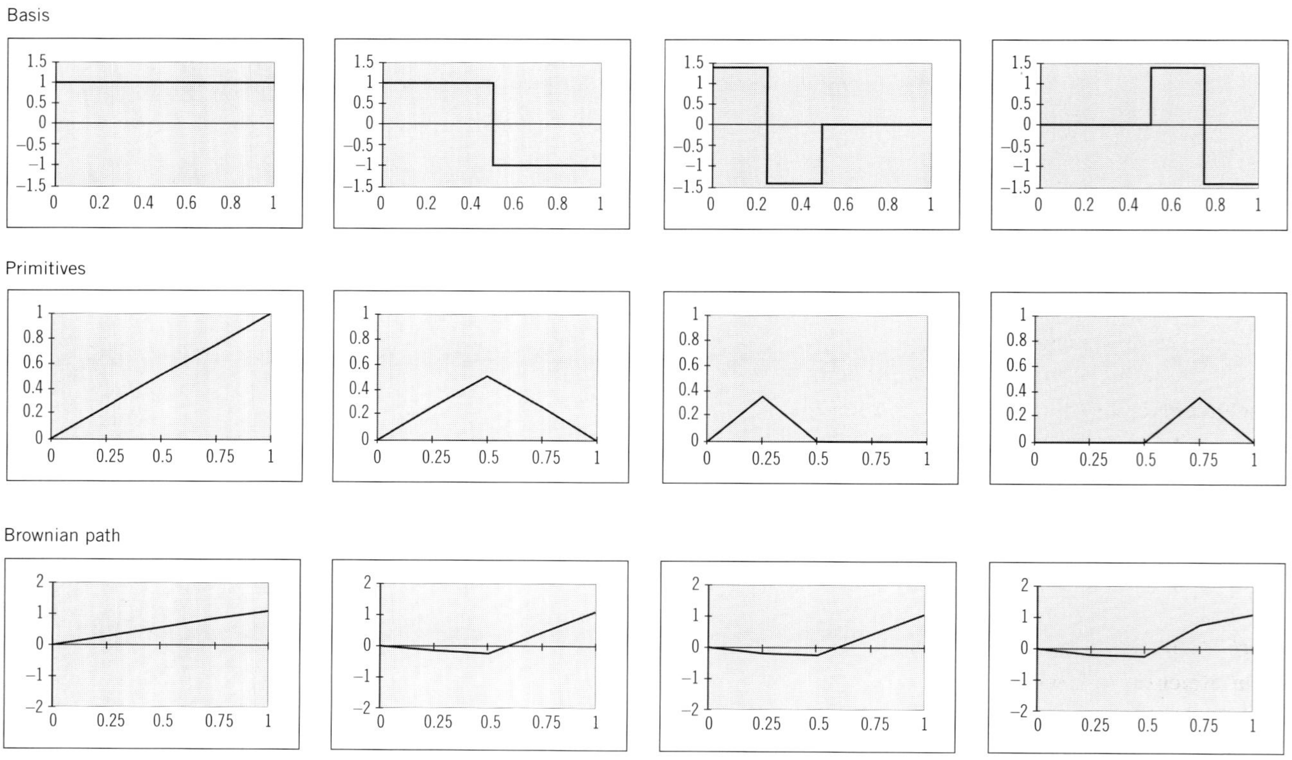

of the basis. A standard method of decomposing the variance throughout orthogonal directions of high and low variance is principal component analysis, a technique that is based on the diagonalisation of the covariance matrix. For example, consider a set of d-variate Gaussian variables, X, with covariance matrix C. Since C is a symmetric positive definite matrix, there exists a unique diagonal matrix, D, and an orthogonal matrix, P, such that $C = PDP'$. Denote by $\sigma_i^2 = D_{ii}$ the eigenvalues and by $(P_1,\ldots,P_d)$ the columns of P or eigenvectors[4] of C. The eigenvectors provide an orthogonal decomposition of the multivariate Gaussian distribution, while the corresponding eigenvalues give the variances in the corresponding directions. If $G = (g_1,\ldots,g_d)$ is an independent standard Gaussian vector, then X and the random vector Y, defined by $Y_k = \Sigma_{i=1}^{d}\sigma_i g_i P_{ik}$, have the same probability distribution. In addition, this decomposition is optimal in the sense that (provided that the eigenvectors are sorted by decreasing value of the corresponding eigenvalues) the variance is optimally concentrated on the first few eigenvectors. If we are to simulate X we can use this decomposition, which allows us to concentrate attention on the first co-ordinates of G. Typically, we would generate the first components of G through UWAN and fill in the rest at random.

A Brownian motion seen at discrete times $(T_1,\ldots,T_d)$ is a set of Gaussian variables with covariance matrix $C = [\min(T_i,T_j)]$. The above analysis applies and

$$W_{T_k} = \sum_{i=1}^{d} \sigma_i g_i P_{ik}$$

The first five components of a Brownian path are shown in Figure 13. The first eigenvectors represent the main outline of the path, capturing a large part of the total variance, while the remaining eigenvectors represent a little noise to be added to the outline. It is therefore worth sampling the first eigenvectors, m,[5] with a structured algorithm such as UWAN, while sampling the remaining $d - m$ eigenvectors at random:

$$W_{T_k} = \sum_{i=1}^{m} \sigma_i g_i^* P_{ik} + \sum_{i=1}^{d-m} \sigma_{m+i} g_i P_{m+i,k}$$

where g^* is a Gaussian sample generated by UWAN in dimension m, while g is a sample generated at random in dimension $d - m$.

The results are remarkable, as shown in Figure 14 for the pricing of an Asian call option.

13. Orthogonal decomposition of a Brownian path: the first five components
Total length, 1 year, with weekly discretisation (dimension 52)

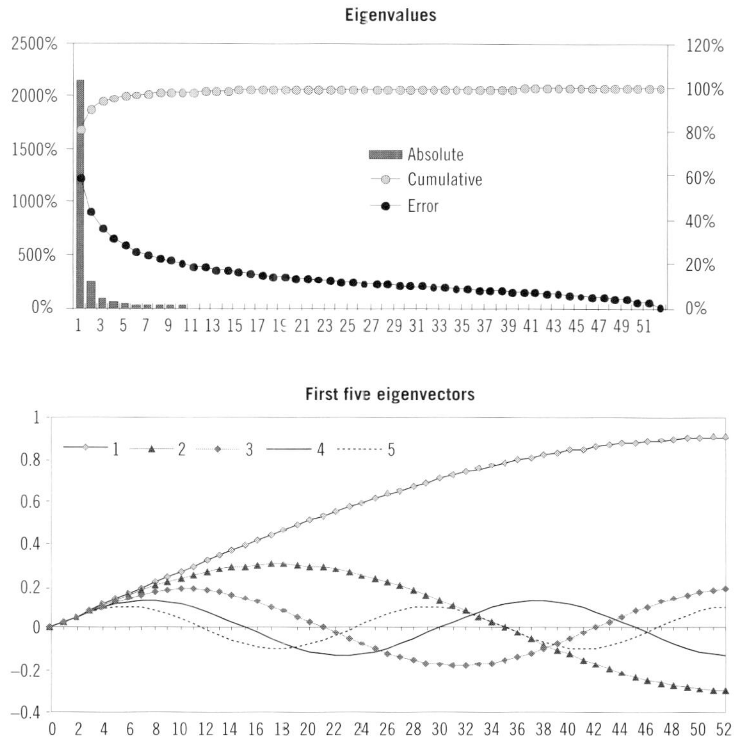

14. Effect of dimension reduction in the pricing of an Asian call
Asian call option, one-year maturity with weekly fixing (dimension 52)

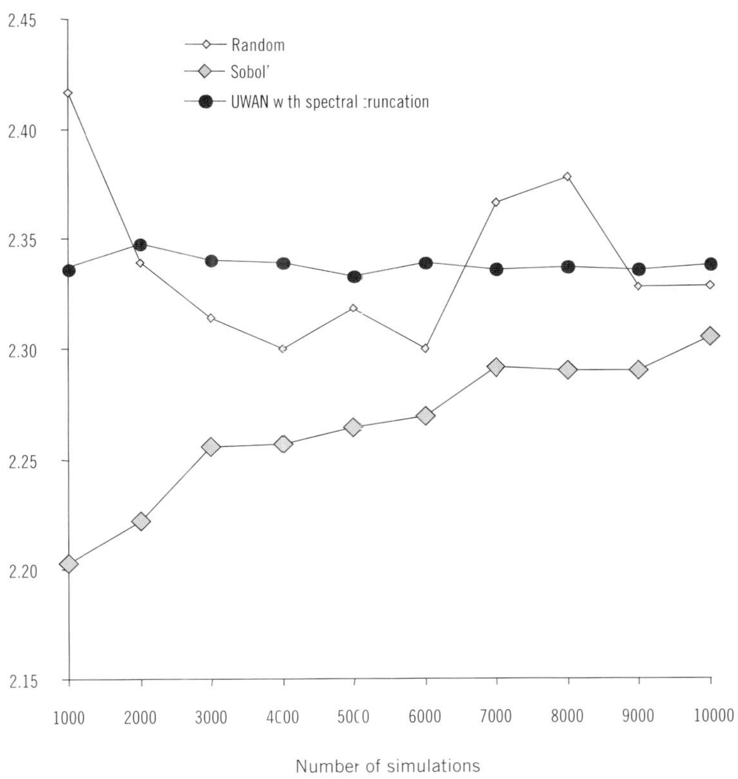

Remark This analysis is directly extendible to problems requiring more than one Brownian motion. For example, two Brownian motions with correlation ρ seen at times $(T_1,\ldots,T_p)$ are a set of $2p$ Gaussian variables with covariance matrix

$$C = \begin{bmatrix} C & \rho C \\ \rho C & C \end{bmatrix}$$

to which the same analysis can be applied.

SPECTRAL GENERATION: CONTINUOUS CASE

Here we justify the use of the technique presented above by replacing it in the general framework given in Theorem 1 and reveal the analytical form of the eigenvectors of Brownian paths so as to avoid calling diagonalisation routines.

We shall describe as the "Brownian covariance operator" the operator A on $L^2([0,T])$, defined by

$$Af(t) = \frac{1}{T}\int_0^T \min(s,t)f(s)ds$$

A is the continuous equivalent of the discrete Brownian covariance matrix C. We shall denote by (h_k) the eigenfunctions of A and by (λ_k) the corresponding eigenvalues.

Theorem 2.1

1. *Eigenvalues and eigenfunctions of* A *are the solutions of*

$$\begin{cases} \lambda T h'' + h = 0 \\ h(0) = 0 \\ h'(T) = 0 \end{cases}$$

2. $\sqrt{\lambda_k}h_k = \sqrt{\frac{2}{T}}\frac{\sin(\omega_k t)}{\omega_k}$ *where* $\omega_k = \frac{2k+1}{2T}$.
3. $(\sqrt{\lambda_k}h_k')$ *is a Hilbert basis of* $L^2([0,T])$.
4. *If* (g_k) *is a series of orthonormal Gaussian variables,* $\sum g_k\sqrt{\lambda_k}h_k$ *is a Brownian motion.*

The proof of the theorem is given in Appendix C.

Recall the spectral generation algorithm:

$$W_{T_k} = \sum_{i=1}^{m} \sigma_i g_i^* P_{ik} + \sum_{i=1}^{d-m} \sigma_{m+i} g_i P_{m+i,k}$$

In substance, Theorem 2.1 says that

$$\sigma_i P_{ik} = \sqrt{\frac{2}{T}}\frac{\sin(\omega_k t)}{\omega_k}$$

which provides an analytical form of the eigensystem of the Brownian paths and removes the need to call time-consuming diagonalisation algorithms in the pricing routines. The eigenvectors are shown in Figure 15. So far, all dates have been given the same importance. However, an option on the maximum, for instance, will require time to be more finely sampled around the most likely dates the maximum should reach, as given by the arcsine law.

Suppose that the "importance" of time is represented by a density φ such that $\phi(T)=\int_0^T\varphi(s)ds = 1$. In Theorem 2.1, time is uniformly important and $\varphi \equiv \frac{1}{T}$. If we observe the Brownian motion at a discrete set of times, $\varphi(t) = \frac{1}{n}\sum_{i=1}^{n}\delta_{T_i}(t)$. We consider the operator B on $L^2([0,T])$ defined by $Bf(t) = \int_0^T \min(\phi^{-1}(s),\phi^{-1}(t))f(s)ds$.

15. Spectral generation as a special case of Theorem 2.1

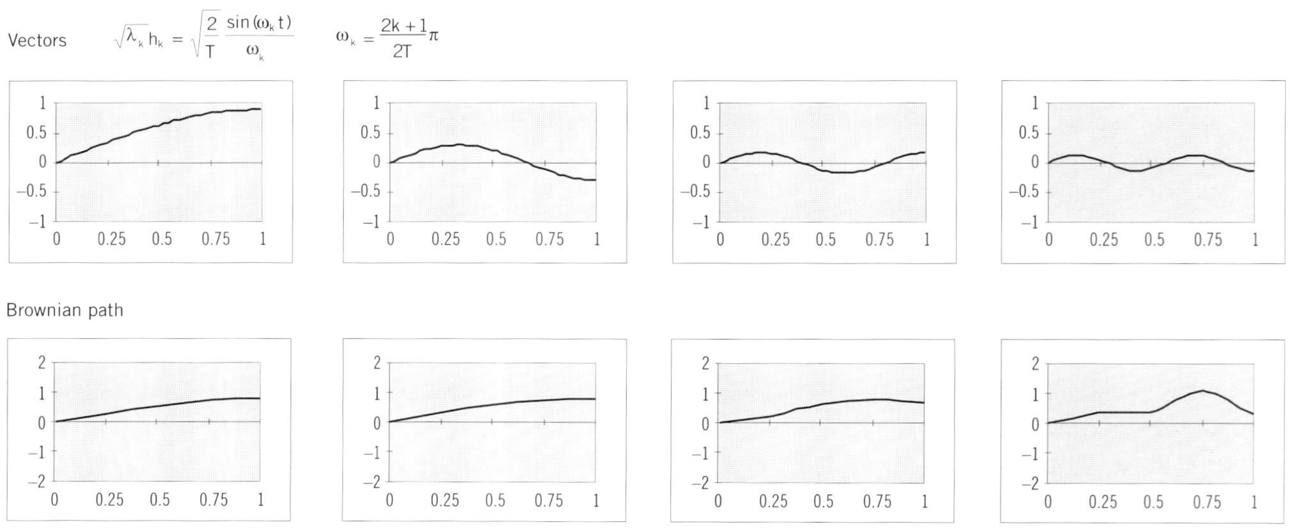

Theorem 2.2

1. *Eigenvalues and eigenfunctions of* B *are the solutions of*

$$\begin{cases} \lambda T h'' + \varphi h = 0 \\ h(0) = 0 \\ h'(T) = 0 \end{cases}$$

2. $(\sqrt{\lambda_k} h_k')$ *is a Hilbert basis of* $L^2([0,T])$.
3. *If* (g_k) *is a series of orthonormal Gaussian variables,* $\sum g_k \sqrt{\lambda_k} h_k$ *is a Brownian motion.*

The proof of this theorem is available on request.

SPECTRAL TRUNCATION

The spectral decomposition of Brownian paths has two major implications for simulation. First, it allows the use of UWAN to sample the main harmonics. This ensures that the directions explaining the largest part of the variance - the main outlines of the paths - are generated with the appropriate technique, all the more so as UWAN is highly efficient in low dimension. Second, it enables a finer discretisation of the most significant directions, as shown in Figure 16.

16. Spectral truncation

PCA
- ❑ Rotate co-ordinates (sample eigenvectors)
- ❑ Rescale axis (sample proportionally to the square root of eigenvalues)

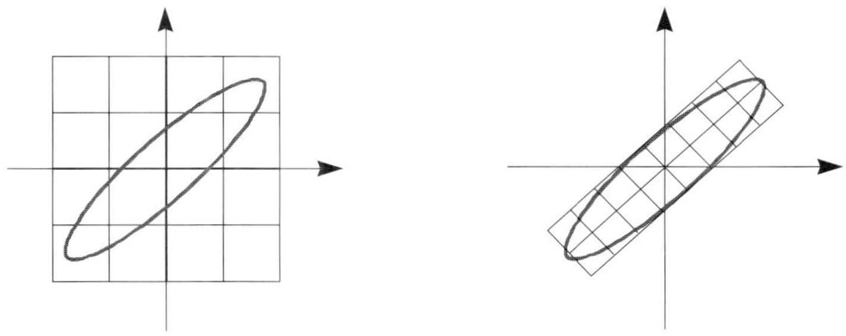

Classical generation
Ignores the geometry of the problem

Spectral generation
Adapts to the geometry of the problem
- ❑ Sample eigenvectors
- ❑ Scale the grid with reference to eigenvalues

Let us take an example in dimension 2. We generate two Gaussian variables that sample, respectively, the first and second eigenvectors. Suppose for the sake of simplicity that we use a regular grid, that the first eigenvalue is 16 times higher than the second one and that the total number of scenarios is 100. We could use a 10×10 grid, but it would be sub-optimal, as the dimension should be sampled according to the square root of the eigenvalues. That is, we need to discretise the first axis more finely, using a 20×5 grid.

In the Brownian case, it is clear from Figure 13 that among the five main harmonics "importance" is clearly not uniform. The drop between the first and second eigenvalues (from 81% to 9%) is dramatic. Thus, it is a natural requirement that the structured algorithm, even when truncated to the first two-to-five dimensions, should respect this disproportion and sample the most important dimensions more finely, in proportion to the square root of the eigenvalues.

ROBUSTNESS OF THE ALGORITHM

In contrast to other variance reduction techniques, particularly those based on low-discrepancy sequences, the spectral-truncated UWAN algorithm is robust in the sense that it is superior to random sampling whatever the scenario.

With our algorithm, the most significant harmonics are sampled through UWAN, whereas the less significant ones are sampled with pure noise. The worst case is a product whose payoff depends only on the less significant harmonics. This corresponds to a volatility swap with payoff

$$\frac{1}{d}\sqrt{\sum_{i=0}^{d-1}\left(\frac{S_{i+1}-S_i}{S_i}\right)^2}$$

and daily fixing. This payoff is linked to the quadratic variation of the stock, which does not depend on the main outline of the path but only on the daily movements of the stock price. In this case our algorithm performs with exactly the same efficiency as random sampling because the only relevant harmonic is sampled at random in both methods. At best, the algorithm performs dramatically better than standard Monte Carlo, and is, at worst, equivalent.

A worst-case analysis with low-discrepancy sequences can give alarming results. Since these sequences are deterministic and generate gaps in high dimension, it is possible to define a product with a payoff that occurs only in scenarios that are mapped to areas of the space which are left blank by such sequences for a reasonable value of N. Such a product would be priced at zero by low-discrepancy Monte Carlo.

Appendix A – UWAN: convergence order

UWAN divides the hypercube into $N/2$ cubes with side length $\ell = (N/2)^{-(1/d)}$. Let us look at the integral over one of the sub-cubes. We call $x_0 \in [0,1]^d$ the centre of the sub-cube. As linear functions are properly priced by UWAN, the error on f is the same as the error on

$$g(x) = f(x) - f(x_0) - x - (x_0).f'(x_0)$$

For all x in the sub-cubes,

$$|g(x)| \le \frac{M}{2}|x - x_0|^2 \le \frac{M}{2}\left(\frac{\ell}{2}\right)^2 d$$

where M is the maximum of the second derivative of f.

The variance of the sum of two samples in a sub-cube is then lesser than

$$\frac{M^2}{4}.\ell^4.d^2$$

As the errors are independent, their variances add-up over the $N/2$ sub-cubes and the total variance of the average of the N samples is bounded from above by

$$\frac{1}{8.N}M^2.\ell^4.d^2$$

and the standard deviation of the total error is of order $\ell^2/\sqrt{N}$, which is order $N^{-(1/2+2/d)}$.

This proves that the convergence order of UWAN is $\frac{1}{2} + \frac{2}{d}$.

Appendix B – Proof of Theorem 1

Consider the transformation, I, from $L^2([0,T])$ into the Gaussian space on a given probability space (Ω,A,P) (ie the set of all Gaussian variables, such as any finite-dimensional sample, from this set is a Gaussian vector):

$$\begin{array}{ccc} L^2([0,T]) & \rightarrow & G \\ I:f & \alpha & N\left(0,\int_0^T f^2(s)ds\right) \end{array}$$

I is obviously linear.

I is an isomorphism at the condition to set G as the set of equivalence classes for the almost sure equality of centred variables.

Clearly

$$\|I(f)\|_{L^2(\Omega)} = \|f\|_{L^2([0,T])}$$

Therefore I is an isometry.

For $t \in [0,T]$, denote $\chi_t(s) = 1_{[0,T]}(s)$; for $\chi_t \in L^2([0,T])$, denote $w(t) = I(\chi_t)$. w is a Gaussian process by property of the Gaussian space. Since χ is continuous in $L^2([0,T])$, w is continuous by isometry.

$$E[w(t)w(s)] = \langle w(t),w(s)\rangle_{L^2(\Omega)} = \langle \chi_t,\chi_s\rangle_{L^2([0,T])} = \inf(s,t)$$

Therefore w is a Brownian motion on (Ω,A,P).

Denote by (φ_k) a Hilbert basis of $L^2([0,T])$; then, for any function, f, in $L^2([0,T])$, $f = \Sigma a_k(f)\varphi_k$, where $a_k(f) = \int_0^T f(s)\varphi_k(s)ds$.

In particular, $a_k(\chi_t) = \int_0^t \varphi_k(s)ds = \varphi_k(t)$.

$w(t) = I(\chi_t) = \Sigma a_k(\chi_t)I(\varphi_k) = \Sigma\phi_k(t)\beta_k$ and, finally, $w = I(\chi) = \Sigma a_k(\chi)I(\varphi_k) = \Sigma\phi_k\beta_k$, where β_k is a series of orthonormal Gaussian variables.

Appendix C – Proof of Theorem 2.1

Part 1 comes from the double differentiation of $Ah = \lambda h$ (ie $\frac{1}{T}\int_0^T \inf(s,t)h(s)ds = \lambda h(t)$) with respect to t.

Part 2 solves the differential equation in part 1.

Part 3 is immediate.

Part 4 comes from Theorem 1.

1 *To produce a regular grid, denote* $m = \lceil \sqrt[d]{N} \rceil$, $N^* = m^d$, *and set the* j*th co-ordinate of the* i*th point to*

$$x_i^j = \frac{\frac{1}{2} + [\frac{i-1}{m^{j-1}}] \bmod m}{m}$$

for $1 \le i \le N^*$ *and* $1 \le j \le d$.

2 *Uniform random vectors have independent uniform random co-ordinates. Efficient independent uniform random number generators are implemented in Press et al. (1992), chapter 7. We recommend the ran3 algorithm (rather than the usual reference ran2) as its slight bias is more than compensated by its low complexity.*

3 *In practice, generate a regular grid of points* $\frac{N}{2}\{x_i, 1 \le i \le \frac{N}{2}\}$ *and select the points*

$$\left\{x_i + \frac{1}{2}\left(\frac{N}{2}\right)^{-1/d}U_i,\ x_i - \frac{1}{2}\left(\frac{N}{2}\right)^{-1/d}U_i\right\}$$

where $\{U_i, 1 \le i \le \frac{N}{2}\}$ *is a set of* d*-dimensional vectors following a uniform distribution in* $[-1,1]^d$.

4 *Press* et al. *(1992) provide two ways of computing eigenvectors and eigenvalues: the Jacobi algorithm (p. 463) and the Householder method (p. 470) which is more robust in high dimension.*

5 *Typically* m *is small, ranging from 2 to 5 depending on the problem.*

BIBLIOGRAPHY

Acworth, P., M. Broadie and P. Glasserman, 1997, "A Comparison of Some Monte Carlo and Quasi Monte Carlo Methods for Option Pricing", in *Monte Carlo and Quasi-Monte Carlo Methods for Scientific Computing*, H. Niederreiter, P. Hellekalek, G. Larcher, and P. Zinterhof, (eds), New York, Springer-Verlag.

Caflisch, R.E., W. Morokov and A.B. Owen, 1997, "Valuation of Mortgaged-backed Securities Using Brownian Bridges to Reduce Effective Dimension", *Journal of Computational Finance* 1, 27–46 (Chapter 26 of this volume).

Niederreiter, H., 1992, *Random Number Generation and Quasi-Monte Carlo Methods*, Philadelphia, SIAM.

Press, W.H., S.A. Teukolsky, W.T. Vetterling and B.P. Flannery, 1992, *Numerical Recipes in C*, 2nd edition, Cambridge, Cambridge University Press.

5

Average Intelligence

Edmond Levy and Stuart Turnbull
HSBC MIDLAND and Queens University, Canada

In a relatively short time, the average price or "Asian" option has become an established hedging instrument, regularly offered by financial institutions to their clients. As with most financial innovations, the market for these instruments is predominantly over-the-counter, specified to suit client needs.

Average price options were developed more or less in parallel in the foreign exchange, interest rate and commodity markets. In the foreign exchange market, the option enables a corporate treasurer to cover a series of foreign currency flows with one hedging instrument whose exercise is cash-settled. For commodities, such as metals or fuels, averaging is also useful as a means to remove extreme sensitivity of the option's expiration value to the underlying cash price on a particular day.

Emerging interest in these options has provoked curiosity as to just how they are being valued. Although most specifications of average price options are European - that is, exercise takes place only on the expiration date - it is known that no closed-form solution for their value exists. The terminal value of these contracts is determined by the arithmetic average of past prices; consequently, valuation is made difficult by the choice of geometric Brownian motion for the underlying asset price.

Under this description, the underlying asset price at any future time is described by the lognormal density function. The valuation of a conventional European option is feasible and given by the celebrated Black-Scholes formula. Similarly, because the product of lognormal prices is itself lognormal, the valuation of options determined by a geometric average of prices is relatively simple. However, the distribution of the *sum* of lognormal components has no explicit representation and thus it is not possible to derive a closed-form solution for average price options.[1]

In this chapter we will review various methods that have emerged for valuing average price options. Many institutions have proprietary models (either developed in-house or provided by "exotic option" software), which are thus unavailable for discussion. A number of papers on this subject have surfaced recently (and others are no doubt in progress), but very few have been published. Our purpose will be to outline some important developments in pricing methodologies that yield insights into the problem at hand. The emphasis is on methods that can be easily implemented and are computationally efficient for the purposes of portfolio risk evaluation. For these reasons we cannot promise to be exhaustive in our coverage.

The average price option

Let us introduce some notation so that we can formally describe the average price option contract. Let $S(t)$ denote the current spot (or cash) price at time t. We suppose that the average is calculated over the time interval $[t_0,t_N]$ and at points on this interval $t_i = t_0 + ih$ for $i = 0,1,\ldots,N$ where $h = (t_N - t_0)/N$. We will denote as $A(t)$ the "running average" to date, which is defined for the current time point, t, $t_m \leq t < t_{m+1}$, by

$$A(t) = \frac{1}{m+1}\sum_{i=0}^{m} S(t_i)$$

for a corresponding integer $0 \leq m \leq N$, and $A(t) = 0$ for $t < t_0$. Thus $A(t_N)$ represents the simple arithmetic average of $N + 1$ prices taken at equal intervals of time, h, between t_0 and t_N. Typically, this time interval is specified to be a day, a week, or a month.[2] Many contracts have fixing dates which are not equally spaced and some allow for a weighted average specification rather than the simple average. These developments, though important to the armoury of financial instruments offered by banks, can be easily represented as a modification to the solution to the basic problem.

The option we focus on here is characterised by the payoff function at time t_N given by $\max[A(t_N) - K,0]$ for a call option, or $\max[K - A(t_N),0]$ for a put option. K denotes the specified strike price of the option. The market has also introduced an average strike or "floating average" option having payoff $\max[A(t_N) - S(t_N),0]$. We merely make the observation that once a successful methodology to the initial problem is found, the average strike option becomes an application of the exchange of asset option problem examined in Margrabe (1978). Finally, define r as the domestic (continuously compounding) interest rate and y as the continuous yield on the asset. It is assumed that r and y are constant over the life of the option. (A deterministic term structure for interest rates, yields and volatilities can be incorporated into the analysis without too much trouble but at the cost of increasing notation.)

Some observations

Whilst there is no simple analytical solution to the problem of valuing average price options, there are some basic observations to be made concerning the behaviour of a desirable answer. Firstly, for an option structure in which the averaging period is a relatively small proportion of the option maturity horizon (eg a one-year option in which payoff is determined by the average of daily fixings over the final month), we should expect an answer close to the price of a conventional European option for the corresponding period. Indeed, this European option can be viewed as the limit of an average price option in which the averaging period is an infinitesimal time period prior to expiry.

Secondly, at the other extreme, with one fixing remaining, the average price option has its terminal value determined by a single spot price. Thus we can immediately value it as $1/(N + 1)$ times a European option on $S(t_N)$ with strike $K^{**} = [(N + 1)K - A(t)N]$. With this in mind, it will always pay to redefine the option payoff as $\max[M(t) - K^*,0]$, where $M(t) = [A(t_N) - A(t) \times (m + 1)/(N + 1)]$ denotes the undetermined component of the average and $K^* = [K - A(t) \times (m + 1)/(N + 1)]$ denotes the effective strike price of the option. Because prices are assumed always positive, whenever K^* is negative exercise on a call option at t_N is guaranteed.

Thirdly, as with conventional European options, a put-call parity condition exists for average price options. To avoid arbitrage, once a call option valuation method is adopted, put options should be valued through this condition unless it is implicit in the pricing methodology. For $t_m \leq t < t_{m+1}$, given the average call option price $C[A(t),K,t]$, the value for the corresponding average put option, $P[A(t),K,t]$, can subsequently be found using the put-call parity relation:

$$P[A(t),K,t] = C[A(t),K,t] - e^{-r\tau} \times \left\{ \frac{A(t)(m + 1)}{(N + 1)} + \frac{S(t)e^{(gh-\xi)}}{N + 1} \frac{[1 - e^{g(N-m)h}]}{1 - e^{gh}} - K \right\}$$

where $g = (r - y)$, $\xi = (t - t_m)$ and $\tau = (t_N - t)$. For the instance of $t < t_0$ use

$$P[A(t),K,t] = C[A(t),K,t] - e^{-r\tau}\left\{ \frac{S(t)}{N + 1} \frac{[e^{g(t_0-t)} - e^{g[t_0-t+h(N+1)]}]}{1 - e^{gh}} - K \right\}$$

These conditions can be derived by evaluating all the unknown future spot prices at their respective arbitrage-free forward prices and discounting back to time t.

Pricing: initial ideas

In option-pricing theory, the spot price process assumed is usually the familiar geometric diffusion

$$dS(t) = \mu S(t)dt + \sigma S(t)dz \tag{1}$$

where dz is a Wiener process (that is, it is distributed as normal with mean zero and unit standard deviation, or $N(0,1)$ for short), and μ and σ are, respectively, the constant drift and volatility parameters. As the payoff on average price options is based on the future path of spot prices, we should keep to the same assumption. Under (1) we can express $S(t_j)$ in terms of $S(t_{j-1})$ as

$$S(t_j) = S(t_{j-1})e^{(\mu-0.5\sigma^2)h+\sigma\sqrt{h}Y_j} \tag{2}$$

where Y_j is $N(0,1)$. For $t_j > t$, $\log_e S(t_j)$ is normally distributed with mean $\log_e S(t_j) + (m - \frac{1}{2}\sigma^2) \times (t_j - t)$ and standard deviation $\sigma\sqrt{t_j - t}$.

A frequently used approach to valuing options is to adopt the risk-neutrality transformation of Cox and Ross.[3] This enables us to characterise the answer to the average price option as

$$C[S(t),K,t] = e^{-r\tau}E^*\max[M(t_N) - K^*,0] \tag{3}$$

where E^* is the expectation operator conditioned on $[A(t),S(t)]$ at time t under the risk-adjusted density function. The process for $S(t)$ is now described by the diffusion (1) with μ

replaced by g:

$$dS(t) = gS(t)dt + \sigma S(t)dz \qquad (4)$$

We can see clearly now why knowledge of the distribution for M(t) (under (4)) is critical to the determination of the average option price. Suppose we denote its density function by $f^*(.)$, then the expectation term in (3) can be written as

$$E^*\max|M(t_N) - K^*,0| = \int_{K^*}^{\infty} |M(t) - K^*|\, f^*(\omega)d\omega \qquad (5)$$

Because the function $f^*(.)$ is non-standard, to evaluate this integral it is necessary to adopt numerical procedures. Two approaches have been proposed. Firstly, Monte Carlo techniques can be employed (as demonstrated by Kemna and Vorst, 1990). Returning to expression (2), if we draw outcomes of Y_j, then a sequence of $S(t_j)$ $(j = i,...,N)$ can be constructed and the payoff $\max[M(t_N) - K^*,0]$ evaluated (i is suitably defined, depending on whether we are before or into the averaging period). Repeating this several times and averaging the results will give a good estimate for (5).

A second numerical approach is to concentrate on M(t) directly and evaluate the convolution of density functions. This was proposed by Carverhill and Clewlow (1990).

Although these ideas provide a solution to valuing average price options, both approaches are time-consuming.[4] This is especially true for options with a large number of fixings remaining and for the risk-sensitivity analysis of large portfolios.

Pricing: analytical approximations

An alternative approach to pricing average price options is to approximate the density function $f^*(.)$ and solve (5). There are several studies supporting the choice of a lognormal function as a good approximation. This would seem to make life very simple, for if $\log_e M(t)$ is $N(\alpha,\nu)$, then

$$E^*\max|M(t_N) - K^*,0| = e^{\alpha+\frac{1}{2}\nu^2}\,\Phi(d_1) - K^*\Phi(d_2) \qquad (6)$$

where $d_1 = (\alpha - \log_e K^* + \nu^2)/\nu$, $d_2 = d_1 - \nu$ and $\Phi(.)$ is the standard normal distribution function. However, calculating α and ν is not a straightforward matter.

Suppose we treat M(t) as distributed according to its corresponding geometric average, $G(t) = [S(t_i) \times S(t_{i+1})....S(t_N)]^{1/(N+1)}$. Thus the parameters α and ν become, respectively, the mean and standard deviation of $\log_e G(t)$. This essentially gives the formula presented in Ruttiens (1990). For a large number of fixings (large N), ν is $\sigma\sqrt{\frac{1}{3}(t_n - t)}$ for $t > t_0$ and $\delta\sqrt{\frac{1}{3}(t_n - t_0) + (t_0 - t)}$ for $t \le t_0$, which explains the use of the "$1/\sqrt{3}$ rule" that some have mentioned.[5]

Tables 1a, 1b and 1c present results for the geometric solution in the columns headed GS.[6] The benchmark values are given in the columns headed MC, which contain Monte Carlo estimates (columns Std Err report the standard errors of these estimates). The options valued have $(t_N - t_0)$ = one year and N = 52. We have computed prices for this structure for various strikes and volatilities at three points in time: (a) 20 weeks prior to t_0, (b) at t_0, and (c) 20 weeks into the averaging period.

It is clear from the tables that use of the geometric average distribution tends to undervalue

Table 1a. Call option values 20 weeks prior to averaging period

$N = 52$, $S(t) = 100$, $r = 0.09$, $y = 0$, $t_n - t_0 = 1$, $A(t) = 0$

σ	K	Std Err	MC	GS	VC	MG	WD	EM	TW	LT
	95	0.00	11.76	11.71	11.76	11 76	11.76	11.76	11.76	11.76
0.0	100	0.00	7.39	7.34	7.39	7 39	7.39	7.40	7.39	7.39
	105	0.00	3.48	3.43	3.47	3 47	3.48	3.53	3.47	3.47
	95	0.00	11.96	11.85	11.96	11 96	11.97	11.97	11.96	11.96
0.1	100	0.00	8.07	7.97	8.06	8 07	8.08	8.09	8.07	8.07
	105	0.00	4.87	4.78	4.85	4 87	4.87	4.90	4.87	4.87
	90	0.01	19.17	18.55	19.10	19 16	19.21	19.20	19.10	19.16
0.3	100	0.01	13.42	12.90	13.34	13 41	13.47	13.46	13.41	13.43
	110	0.01	9.06	8.61	8.93	9 00	9.07	9.06	9.06	9.06
	90	0.03	24.17	22.61	23.79	24 10	24.24	24.17	23.58	24.02
0.5	100	0.03	19.38	18.02	19.00	19 34	19.50	19.42	19.26	19.35
	110	0.03	15.44	14.26	15.06	15 40	15.57	15.49	15.63	15.49

MC = Monte Carlo; GS = geometric solution VC = Vorst correction;
MG = modified geometric; WD = Wilkinson discrete; EM = exact moments;
TW = Turnbull/Wakeman; LT = Levy/Turnbull.

Table 1b. Call option values at the averaging period

$N = 52$, $S(t) = 100$, $r = 0.09$, $y = 0$, $t_n - t_0 = 1$, $m = 0$, $A(t) = 0$

σ	K	Std Err	MC	GS	VC	MG	WD	EM	TW	LT
	95	0.00	8.81	8.76	8.81	8.81	8.81	8.80	8.81	8.81
0.0	100	0.00	4.31	4.26	4.31	4.31	4.31	4.34	4.31	4.31
	105	0.00	0.95	0.92	0.94	0.94	0.95	1.06	0.95	0.95
	95	0.00	8.91	8.80	8.91	8.91	8.92	8.92	8.91	8.91
0.1	100	0.00	4.91	4.81	4.90	4.90	4.92	4.94	4.91	4.91
	105	0.00	2.06	1.98	2.03	2.04	2.06	2.10	2.06	2.06
	90	0.01	14.96	14.38	15.04	14.98	15.04	15.03	14.91	15.00
0.3	100	0.01	8.81	8.31	8.72	8.77	8.86	8.84	8.78	8.84
	110	0.01	4.68	4.27	4.53	4.58	4.67	4.66	4.69	4.69
	90	0.03	18.14	16.75	18.00	18.20	18.38	18.29	17.66	18.13
0.5	100	0.03	12.98	11.76	12.71	12.94	13.16	13.06	12.86	13.00
	110	0.03	9.10	8.00	8.71	8.93	9.17	9.07	9.22	9.12

MC = Monte Carlo; GS = geometric solution; VC = Vorst correction;
MG = modified geometric; WD = Wilkinson discrete; EM = exact moments;
TW = Turnbull/Wakeman; LT = Levy/Turnbull.

call options (and consequently, using put–call parity, put options will also be undervalued). This is largely explained by the fact that $G(t) \le M(t)$, and so the central tendency of the distribution is biased downwards. For example, at $t = t_0$ and 30% volatility, $E^*[G(t)] = 103.81$ and $E^*[A(t)] = 104.64$. In noting this bias, Vorst (1990) suggests correcting the strike price of the option by this difference so that K is replaced by $K^v = K + E^*[G(t)] - E^*[M(t)]$. This procedure is reported in the columns headed VC.[7]

An alternative and more accurate approach is to correct for the mean directly by observing that if $\log_e M(t)$ is $N(\alpha, v)$, then $E^*[M(t)] = \exp(\alpha + \frac{1}{2} v^2)$ - hence we can choose $\alpha = \log_e E^*[M(t)] - \frac{1}{2} v^2$. A convenient feature of choosing the parameters in this manner, implicit in the pricing methodology, is the fact that put–call parity is ensured. That is, put options could be valued by directly solving $E^*\max[K^* - M(t),0]$ with this choice for $f^*(.)$. Call option values under this "modified geometric" approach are denoted in the tables as MG. The tables reveal the significance of the adjustment, leading to much-improved option values. Unfortunately, a tendency to undervalue call options persists, especially for out-of-the-money options and in the mid-to-higher volatility regions.

Recall that α and v denote the mean and standard deviation for $\log_e M(t)$. Although determining the moments for $\log_e M(t)$ is not a trivial exercise, we may estimate them by "discovering" these parameters through equating the true moments for M(t) with those implied by a lognormal fit. The relationship between these moments is given by the equation

$$E^*\left[M(t)^n\right] = e^{n\alpha + \frac{1}{2}n^2\nu^2}$$

Table 1c. Call option values at the averaging period

N = 52, S(t) = 100, r = 0.09, y = 0, $t_n - t_0 = 1$, m = 0, A(t) = 0

σ	K	Std Err	MC	GS	VC	MG	WD	EM	TW	LT
	95	0.00	6.39	6.38	6.39	6.39	6.39	6.38	6.39	6.39
0.0	100	0.00	1.73	1.72	1.73	1.73	1.73	1.80	1.73	1.73
	105	0.00	0.01	0.01	0.01	0.01	0.01	0.03	0.01	0.01
	95	0.00	6.40	6.36	6.40	6.40	6.40	6.40	6.40	6.40
0.1	100	0.00	2.11	2.07	2.10	2.10	2.11	2.15	2.10	2.10
	105	0.00	0.20	0.19	0.19	0.19	0.20	0.23	0.20	0.20
	90	0.01	11.32	11.07	11.33	11.33	11.34	11.34	11.32	11.32
0.3	100	0.01	4.12	3.94	4.10	4.11	4.14	4.14	4.12	4.12
	110	0.01	0.91	0.82	0.87	0.88	0.90	0.90	0.93	0.93
	90	0.02	12.30	11.70	12.29	12.33	12.38	12.36	12.17	12.29
0.5	100	0.02	6.23	5.78	6.15	6.22	6.29	6.26	6.21	6.24
	110	0.02	2.73	2.45	2.63	2.69	2.76	2.74	2.83	2.79

MC = Monte Carlo; GS = geometric solution; VC = Vorst correction; MG = modified geometric; WD = Wilkinson discrete; EM = exact moments; TW = Turnbull/Wakeman; LT = Levy/Turnbull.

for any integer n. Given $E^*[M(t)]$ and $E^*[M(t)^2]$, we can determine α and v to fit the mean *and* variance of M(t). That is, solve the above equations using n = 1 and n = 2 to give

$$\alpha = 2\log_e E^*\left[M(t)\right] - \tfrac{1}{2}\log_e E^*\left[M(t)^2\right] \qquad (7a)$$

$$\nu = \sqrt{\log_e E^*\left[M(t)^2\right] - 2\log_e E^*\left[M(t)\right]} \qquad (7b)$$

This idea underlies the methods in a number of papers and is referred to here as the "Wilkinson estimates".[8] To calculate the moments $E^*[M(t)^n]$ use the algorithm in Turnbull and Wakeman (1991) or, alternatively, for n = 1 and n = 2, the closed-form expressions in Levy (1991). The results with α and v chosen in this way are shown in columns headed WD. These results taken together with further simulations (not reported here) demonstrate little or no evidence of bias up to 20% volatility. This makes the procedure acceptable for most currency and interest rate markets. However, for, say, commodity contracts, where volatility is largely in the 30% and upward range, use of the procedure may result in a tendency to overvalue average rate options.

To measure the extent to which this bias is due solely to inaccuracies in determining α and v, we can adopt an alternative method to refine these estimates. By extending the approach in Schwartz and Yeh (1982) to allow for correlated random variables, we can determine reasonably accurate estimates for α and v over a wider range of volatilities. These results are shown in the column headed EM. Up to 30% volatility there is little improvement over WD, but above this level the methodology yields significant benefits. There remains a tendency to overvalue options and a slight downward bias for deep-out-of-the-money options, but it is not as dramatic.

We should remind ourselves that the lognormal distribution is only a first approximation for the true distribution of M(t). Although we may be able to capture the mean and variance for M(t) perfectly, we will inevitably encounter differences between the skewness and kurtosis implied by the lognormal fit and those for M(t). These differences will become increasingly apparent as we move to higher levels of volatility and/or longer option structures. A method, implemented by Turnbull and Wakeman (1991), to counter these effects is to expand the distribution for M(t) around the lognormal. Let a(.) be the lognormal distribution. Then it can be shown

PANEL 1

ASIAN ARITHMETIC

Valuation formulae for average rate (or Asian) currency options have been presented using adaptations of the formula for valuing an option on the geometric average. The market, however, is concerned with an instrument whose payoff is determined by the arithmetic average of past exchange rates.

The difference between the two forms of averaging is important in at least two respects. Firstly, the put–call parity relation for arithmetic average rate options is not satisfied by the geometric-based formula, and as a consequence it will tend to overprice put options and underprice call options. Secondly, the time-decay profile, as valued by this formula, will not correctly reflect the necessary dynamic hedging requirements. Undoubtedly these inconsistencies diminish as the option reaches maturity, but we can do better than rely on a modified geometric answer.

The formula below overcomes these anomalies by taking into account known features of the distribution of the evolving average when the spot exchange rate process is assumed to follow the conventional geometric diffusion. This formula assumes continuous daily fixings, but the approach can be extended to allow for (say) monthly or quarterly fixings (Levy, 1990). The resulting generalised formula, though not exact, represents a closed-form solution with an accuracy which is acceptable relative to those provided by numerical methods and considerably easier to implement.

Let T be the original maturity of the option with remaining maturity $\tau = (T - t)$. Define the following: S_{AV} = the arithmetic average of the known spot rate fixings; S = the spot rate at time t; r_d = the domestic interest rate; r_t = the foreign interest rate; $g = (r_d - r_t)$; σ = the assumed volatility; and N(.) is the cumulative standard normal distribution function. Then the value of a call option with strike K is calculated as

$$C_A = S_A N(d_1) - e^{-r_d\tau} K^* N(d_2)$$

where

$$S_A = \frac{S}{Tg}\left(e^{-r_t\tau} - e^{-r_d\tau}\right)$$

$$d_1 = \frac{1}{\sqrt{V}}\left(\frac{1}{2}\log_e D - \log_e K^*\right)$$

$$d_2 = d_1 = \sqrt{V}$$

$$V = \log_e D - 2(r_d\tau + \log S_A)$$

$$D = \frac{M}{T^2} \qquad K^* = K - \frac{t}{T} S_{AV}$$

$$M = \frac{2S^2}{(g + \sigma^2)}\left\{\frac{\left[e^{(2g+\sigma^2)\tau} - 1\right]}{(2g + \sigma^2)} - \frac{\left[e^{g\tau} - 1\right]}{g}\right\}$$

Put options (P_A) can be valued through the put–call parity relation

$$(C_A - P_A) = S_A - e^{-r_d\tau} K^*$$

The table provides examples of option prices obtained using the above formula. The options are on the daily average of rates over one year with $t = 0$, $\sigma = 12\%$, $r_d = 5.5\%$, $r_t = 9.5\%$ and $S = 1.78$. C_M and P_M are the call and put prices from a Monte Carlo approach with 10,000 replications representing the "true" prices. For comparison, prices from the geometric-based formula (C_G and P_G) are also included. All prices are expressed as a percentage of the base (or foreign) currency amount.

K	C_M	C_A	C_G	P_M	P_A	P_G
1.80	1.389	1.386	1.357	4.319	4.318	4.406
1.74	2.676	2.681	2.630	2.416	2.422	2.489
1.68	4.585	4.598	4.523	1.136	1.149	1.192

that we can expand $f^*(.)$ as

$$f^*(\omega) =$$

$$a(\omega) + E_1\frac{da(\omega)}{d\omega} + E_2\frac{d^2a(\omega)}{d\omega^2} + E_3\frac{d^3a(\omega)}{d\omega^3} + \ldots \quad (8)$$

where E_i are terms involving the difference between cumulants (or semi-invariants) implied by the lognormal fit and the true cumulants for M(t). For example, E_1 is the difference between the mean of M(t) implied by $a(\omega)$ and the true mean of M(t), and E_2 is the difference between the variances. If we substitute (8) into (5) and evaluate the integral, (6) becomes augmented by further terms having E_i $(i = 1,2,\ldots)$ as coefficients. When α and ν are chosen as (7), the first two terms are zero (by design), but E_i $(i = 3,4,\ldots)$ are non-zero. The Turnbull and Wakeman procedure is to expand (8) to E_4 using estimates from (7), thereby correcting for differences in skewness and kurtosis.

The benefit from including these "correction" terms can be seen by comparing columns headed WD and TW with the Monte Carlo estimates. Though contributing very little in the low-volatility region (between 20% and 30%) the correction terms are significant, so that quite accurate values are achieved. Above this level, the numbers reveal a tendency to undervalue in-the-money call options and overvalue out-of-the-money calls.

Casual empiricism of simulated distributions for the arithmetic average at high volatilities, when compared with their lognormal fit, reveals the slight presence of further skewness and kurtosis not accounted for. These are partly, but clearly unsatisfactorily, catered for in the correction terms in TW. However, the parameters there were determined by the first two cumulants and hence distort the necessary weighting appropriate to terms involving higher cumulants. An obvious solution is to choose to fit the lognormal by using information on the higher-order moments of $M(t)$, when skewness and kurtosis are relatively more pronounced, thereby diminishing the resulting pricing bias at high volatilities. For instance, in Table 1, columns headed LT display call option values when α and ν are chosen along these lines (applying these values to the correction terms in the Edgeworth expansion).

The improvement in the results given by LT over those given by TW is significant. For volatilities between 30% and 50%, this approach, although still short of ideal, on the whole performs better than other methods considered in this chapter. A surprising feature in the tables is the relatively poor performance of LT at 50% volatility where the current time point is 20 weeks prior to the averaging period. We can explain this by noting that the shape of the distribution of $M(t)$ in these instances is more lognormal than not (see our earlier remarks). This would also explain why the exact moments results here can be particularly good even though it fails to incorporate any information about the true higher-order cumulants for $M(t)$.

Conclusions

This article has reviewed the successes and failures of recent attempts to value average price options. In the main, we can say that analytical approximations along the lines of fitting a lognormal distribution can yield acceptable closed-form analytical formulae for these option contracts. Our analyses lead to a number of conclusions. Firstly, users should be aware of biases that may exist within "black-box" algorithms which do not correct for the mean and variance of the arithmetic average. For asset volatilities up to 20%, the Wilkinson procedure is simple to implement and is quite adequate. If the underlying asset volatility generally trades beyond this level and up to 30%, augmenting by the more general Turnbull and Wakeman (1991) procedure is highly recommended.

Beyond this level there is no clear favourite. Generalising Turnbull and Wakeman and incorporating information on higher cumulants into the procedure does a great job when valuing average price options at or into the averaging period. On the other hand, more precise calculation of the first two moments of the average is more important when valuing prior to the averaging period. What we have is a general framework for developing good solutions to the average price option problem. Unfortunately, in applying this framework we find good answers for certain instances and other good answers for others. As with all option valuation models, the old adage applies – ***user beware!***

1 *Eric Reiner has developed a representation for the distribution under continuous fixing for the average. In practice, the average price option is dependent on fixings at discrete points in time.*

2 *Ritchken, Sankarasubramanian and Vijh (1990) and Carverhill and Clewlow (1990) define their average as determined in the interval* (t_0, t_N) *but at points on this interval* $t_i = t_0 + ih$ *for* $i = 1,\ldots,N$. *Thus the first fixing is at time* t_1 *with* $N - 1$ *remaining.*

3 *Cox and Ross (1976).*

4 *There are a number of techniques that could be adopted to limit the necessary number of replications in the Monte Carlo procedure – such as the use of control variates or antithetic variates. This greatly reduces the number of replications, but one would still require several thousand for an acceptable degree of accuracy.*

5 *See, for example, John Hull's comment quoted in Krzyzak (1990).*

6 *When pricing in the averaging period note that the distribution of* $\log_e M(t)$ *is approximated by that for* $\log_e G^*(t) + \log_e(N - m) - \log_e(N + 1)$, *where* $G^*(t) = [S(t_{m+1}),S(t_{m+2}),\ldots,S(t_n)]^{1/(N-m)}$.

7 *Our numbers differ from Vorst (1990) as we modified* K *only by the unknown component of* $A(t_N)$. *This minor change yields slightly better estimates.*

8 *Papers which have independently adopted this method include Levy (1990, 1991), Ritchken, Sankarasubramanian and Vijh (1990) and Turnbull and Wakeman (1991).*

BIBLIOGRAPHY

Carverhill, Andrew P., and Les J. Clewlow, 1990, "Flexible Convolution", *Risk* 3(4), pp. 25–9.

Cox, John C., and Stephen A. Ross, 1976, "The Valuation of Options for Alternative Stochastic Processes", *Journal of Financial Economics*, September, pp. 145–66.

Kemna, A.G.Z., and A.C.F. Vorst, 1990, "A Pricing Method for Options Based on Average Asset Values", *Journal of Banking and Finance*, March, pp. 113–29.

Krzyzak, Krystyna, 1990, "Asian Elegance", *Risk* 3(1), pp. 30–4.

Levy, Edmond, 1990, "Asian Arithmetic", *Risk* 3(5), pp. 7–8.

Levy, Edmond, "Pricing European Average Rate Currency Options", *Journal of International Money and Finance*.

Margrabe, William, 1978, "The Value of an Option to Exchange One Asset for Another", *Journal of Finance*, March, pp. 177–96.

Ritchken, Peter L., L. Sankarasubramanian and Anand M. Vijh, 1990, "The Valuation of Path Dependent Contracts on the Average", *unpublished manuscript*, School of Business Administration, University of Southern California.

Ruttiens, Alan, 1990, "Classical Replica", *Risk* 3(2), pp. 33–6.

Schwartz, Stuart C., and Y.S. Yeh, 1982, "On the Distribution Function and Moments of Power Sums with Lognormal Components", *Bell System Technical Journal*, September, pp. 1,441–62.

Turnbull, Stuart M., and Lee Macdonald Wakeman, 1991, "A Quick Algorithm for Pricing European Average Options", *Journal of Financial and Quantitative Analysis*, September, pp. 377–89.

Vorst, Ton, 1990, "Analytic Boundaries and Approximations of the Prices and Hedge Ratios of Average Exchange Rate Options", *unpublished manuscript*, Econometric Institute, Erasmus University, Rotterdam, February.

6

Beyond Average Intelligence

Michael Curran
RiskCare Ltd

Computational approaches to pricing Asian, or average price, options have been reviewed by Edmond Levy and Stuart Turnbull (see Chapter 5). The methods they covered all revolve around attempts to approximate the risk-neutral probability distribution of the arithmetic mean price, which has no closed-form representation under the standard assumptions for asset price dynamics. The price of the Asian option is then estimated by integrating the payoff function with respect to this approximating distribution. This article presents an alternative approach giving more accurate estimates, according to the numerical results given in Levy and Turnbull.

This approach uses the mathematical technique known as "conditioning" to compute the expected discounted risk-neutral payoff. The basic idea is simple. The integration is done across all possible geometric mean prices, for which the probability density is easily obtainable, and the expected Asian payoff conditional on that geometric mean price is computed.

Performing the integration results in a formula which resembles a sum of terms similar to the first term in the Black–Scholes formula, and a single term similar to the second Black–Scholes term:

$$P \approx \exp(-rT)\left\{(1/n)\sum_{i=1}^{n}\exp(\mu_i + \sigma_i^2/2)\right.$$
$$\left.\Phi\left[(\mu - \ln\hat{K})/\sigma_x + \sigma_{Xi}/\sigma_X\right] - K\Phi\left[(\mu - \ln\hat{K})/\sigma_X\right]\right\}$$

where

r = the risk-free rate,
d = the dividend payout on the asset,
T = time of payoff,
K = strike price,
S_0 = initial asset price,
n = number of averaging points,
t_1 = time of first averaging point,
Δt = time between averaging points,
σ = volatility of asset,
$\mu_i = \ln S_0 + (r - d - \sigma^2/2)t_i$,
$\sigma_i^2 = \sigma^2(t_i + (i-1)\Delta t)$,
$\Phi(\cdot)$ = the standard normal distribution function,
$\sigma_{Xi} = \sigma^2\{t_i + \Delta t[(i-1) - i(i-1)/(2n)]\}$,
$\mu = \ln S_0 + (r - d - \sigma^2/2)(t_i + (n-1)\Delta t/2)$,
$\sigma_X^2 = \sigma^2[t_i + \Delta t(n-1)(2n-1)/6n]$,

and

$$\hat{K} = 2K - \frac{1}{n}\sum_{i=1}^{n}\exp\left[\mu_i + (\sigma_{Xi}/\sigma_X^2)(\ln K - \mu)\right.$$
$$\left.+(\sigma_i^2 - \sigma_{Xi}^2/\sigma_X^2)/2\right]$$

Table 1 on the next page reproduces the Monte Carlo (MC) simulation results and the prices given by Levy and Turnbull's own technique, which appears to be the best among the distribution-approximating (DA) methods. The geometric conditioning (GC) technique gives superior estimates for almost all cases, according to the MC simulations.

The GC method requires an amount of time that is roughly proportional to n. Therefore, if n is large an improvement in efficiency may be achieved by valuing a few Asians with smaller n and using Richardson extrapolation, or some other extrapolation technique, to estimate the prices for larger n.

Table 1. Call option values

20 weeks before averaging period

σ	K	MC (std err)	DA	GC
	95	11.76 (0.00)	11.76	11.76
0.05	100	7.39 (0.00)	7.39	7.39
	105	3.48 (0.00)	3.47	3.47
	95	11.96 (0.00)	11.96	11.96
0.10	100	8.07 (0.00)	8.07	8.07
	105	4.87 (0.00)	4.87	4.87
	90	19.17 (0.01)	19.16	19.17
0.30	100	13.42 (0.01)	13.43	13.43
	110	9.06 (0.01)	9.06	9.05
	90	24.17 (0.03)	**24.02**	24.10
0.50	100	19.38 (0.03)	19.35	19.37
	110	15.44 (0.03)	15.49	15.47

$n = 53$, $S_0 = 100$, $r = 0.09$/year, Δt = one week, $d = 0$, $t_n = 72$ weeks.

At averaging period

σ	K	MC (std err)	DA	GC
	95	8.81 (0.00)	8.81	8.81
0.05	100	4.31 (0.00)	4.31	4.31
	105	0.95 (0.00)	0.95	0.95
	95	8.91 (0.00)	8.91	8.91
0.10	100	4.91 (0.00)	4.91	4.91
	105	2.06 (0.00)	2.06	2.06
	90	14.96 (0.01)	**15.00**	14.96
0.30	100	8.81 (0.01)	**8.84**	8.80
	110	4.68 (0.01)	4.69	4.67
	90	18.14 (0.03)	18.13	18.14
0.50	100	12.98 (0.03)	13.00	12.98
	110	9.10 (0.03)	9.12	9.07

$n = 53$, $S_0 = 100$, $r = 0.09$/year, Δt = one week, $d = 0$, t_n = one year.

20 weeks into averaging period

σ	K	MC (std err)	DA	GC
	95	6.39 (0.00)	6.39	6.39
0.05	100	1.73 (0.00)	1.73	1.73
	105	0.01 (0.00)	0.01	0.01
	95	6.40 (0.00)	6.40	6.40
0.10	100	2.11 (0.00)	2.10	2.11
	105	0.20 (0.00)	0.20	0.20
	90	11.32 (0.01)	11.32	11.32
0.30	100	4.12 (0.01)	4.12	4.12
	110	0.91 (0.01)	0.93	0.92
	90	12.30 (0.02)	12.29	12.29
0.50	100	6.23 (0.02)	6.24	6.23
	110	2.73 (0.02)	**2.79**	2.77

$n = 53$, $S_{20} = 100$, $A_{20} = 100$, $r = 0.09$/year, Δt = one week, $d = 0$, $t_n = 32$ weeks.

Estimates three standard errors or more from Monte Carlo estimate in bold.

7

Strata Gems

Michael Curran
RiskCare Ltd

The use of Monte Carlo simulation to price path-dependent derivative instruments has increased as products have become more complex. This is because, for many path-dependent and other complex products, closed-form solutions for prices under standard assumptions do not exist, and it is not feasible to apply other numerical methods given the multi-dimensional nature of the problems.

The relatively slow convergence of simulation puts a premium on so-called variance reduction techniques (methods of adjusting the random sampling procedure so that convergence is accelerated). Two widely known and used techniques of this kind are the antithetic variate and control variate methods.[1] This article discusses a variance reduction method known as stratified sampling and explains how it can be adapted to cases when sequences of normally distributed variates are to be generated.

The benefit of stratified sampling can be appreciated by first considering the antithetic variate method as applied to the normal distribution. Normal variates are required to simulate the most common models used in financial modelling: Brownian motion and geometric Brownian motion. In this technique, if a vector of normal variates, $\underline{z}$, is used to generate a price or interest rate path, then $-\underline{z}$ is used to generate another path. This ensures symmetry of the sampling distribution, so that the odd sample moments are equal to their population values of zero. However, the even sample moments, which really give the normal distribution its shape, can still stray substantially from their population values. Stratified sampling addresses the issue of generating representative even sample moments.

The usual approach to stratified sampling is to divide the probability distribution into ranges and to sample from each range according to its probability. For example, if 10 equal probability ranges were chosen, and if 10n normal variates were desired, then n variates would be generated from each decile of the distribution. If, instead, we use a sufficiently large number of ranges, we can justify specifying a single value that is used repeatedly, which represents each interval. This would obviate the time-consuming generation of new normal variates.

To see how this might be accomplished, suppose that we divide the normal distribution into 1,000 intervals and use the mean conditional on being in each interval as our representative variate for that interval. These values are easy to compute given the appropriate quantiles of the normal distribution, which can be obtained from a table or via a numerical approximation of the inverse of the normal distribution function. The first few are ±3.367090, ±2.973103, ±2.808998, etc.

A uniform distribution on these values constitutes a discrete distribution with a variance of 0.9998 and kurtosis of 2.989, very close to the normal population parameters of 1 and 3. We will refer to this vector of variates as $\underline{z}$. (It may be desired to divide $\underline{z}$ by its sample standard deviation, $\sqrt{0.9998}$, to make the variance $\underline{z}$ equal to unity, but this minuscule adjustment has little practical effect.)

To illustrate the use of these variates, consider how they can be employed to price a one-year European call option with 20% volatility and an underlying price of 100, where the risk-free interest rate is 5%, under the standard assumption of geometric Brownian motion. We price this option as if we were using Monte Carlo simulation and have generated one of each element of $\underline{z}$:

$$\hat{C} = \frac{\exp(-0.05)}{1000} \sum_{i=1}^{1000} \max(0, f(z_i) - K)$$

where

$$f(z) = \exp(\ln(100) + 0.05 - (0.2)(0.2)/2 + 0.2z)$$

There is nothing probabilistic about this

Table 1. Pricing European calls

K	Black–Scholes	Numerical integration
70	33.54	33.54
80	24.59	24.62
90	16.70	16.72
100	10.45	10.47
110	6.04	6.05
120	3.25	3.25
130	1.64	1.64

Table 2. Pricing Asian call and put options

	Asian call		Asian put	
K	Stratified	Simple	Stratified	Simple
80	21.71	21.91	0.070	0.069
	(0.0003)	(0.0145)	(0.00002)	(0.00003)
90	12.96	12.85	0.779	0.817
	(0.0005)	(0.0119)	(0.00070)	(0.00040)
100	6.072	6.160	3.54	3.48
	(0.0016)	(0.0074)	(0.00088)	(0.00260)
110	2.286	2.311	9.12	9.27
	(0.0014)	(0.0030)	(0.00055)	(0.00720)
120	0.671	0.673	17.06	17.14
	(0.0002)	(0.0009)	(0.00035)	(0.01220)

method. It is essentially numerical integration where intervals of equal area under the density curve, instead of equal widths, are used. The results of pricing with this formula are given in Table 1, with the Black–Scholes values given for comparison. Observe that we incur very little bias from using the discrete distribution we have created rather than sampling from a continuous normal distribution.

To price path-dependent options we simply sample without replacement by randomly shuffling the vector z for each time period. The increment of the underlying price or rate is then determined by the z value at the corresponding index, where the first index will correspond to the first path, the second to the second path, etc. Summing z values from period to period, as is done in generating normal or lognormal variates, will help to mitigate the (slight) bias incurred from the deviation of the discrete distribution from the normal. The number of possible outcomes clearly increases very quickly, making the resulting distribution less granular. Also, the discrete distributions resulting from summing elements of the original vector converge to the normal, by the central limit theorem. For example, the kurtosis of the population of sums of two shuffled zs is 2.994, significantly closer to the normal population value of 3 than that of the distribution of z. In practice, it may be desirable to generate the first increment in the path by summing a few shuffled z vectors to accelerate the convergence to normality.

To demonstrate the efficiency of this method, one-year Asian call and put options were priced using both simple sampling and the stratified sampling method described above. (Accurate closed-form approximations exist for these options. We use them simply as a convenient test case.) Averaging is done on a monthly basis. Other parameters are as above. For the stratified results 10 sets of 1,000 trials were run, and for comparison 10,000 trials were simulated in the standard way to get the simple sampling estimates (see Table 2, where the sample variances are in parentheses).

The stratification method gives a substantial increase in efficiency. The ratios of sample variances give the acceleration factor. Also, the task of generating new normal variates and the associated transcendental function evaluations have been eliminated. Interestingly, stratification appears to increase efficiency most for in-the-money options. The intuition here is that, since the overall probability distribution for the future value of the underlying is very well approximated, the options for which most of the distribution is relevant benefit most.

It is well known that for problems in one dimension, standard Monte Carlo simulation is inefficient relative to numerical integration, while the reverse is the case for problems in higher dimensions. The method described here reconciles these facts and is efficient for all cases. The method is easy to implement and can save considerable computer time when using Monte Carlo simulation to price path-dependent derivative instruments.

1 *Bratley, P., B. Fox and L. Schrage, 1987,* A Guide to Simulation, *Springer-Verlag.*

8

Recovering Identity

Michael Curran
RiskCare Ltd

The standard method of generating Brownian paths for use in Monte Carlo simulation results, due to sampling error, in paths that are not particularly representative. There are at least two reasons for this. First, the marginal distributions are not normally distributed, and second, the sample covariance matrix does not equal the identity. The technique in this article remedies this situation and greatly reduces the computation needed for an acceptable estimate.

Mathematical basis

Suppose that we wish to simulate the value of Brownian paths at n points in time. Since we want to generate representative paths, we would like the marginal distributions of our paths at each of the n time points to be very nearly normally distributed. We can accomplish this with a technique which selects a single value from each quantile of the normal distribution to represent that quantile (Curran, 1994) - also Chapter 7 of this publication. For example, if we choose 1,000 divisions of the normal distribution we then select a single z value from each of the 1,000 quantiles. A uniform distribution over this set of values is a good (discrete) approximation to the normal distribution. At each of the n time points we shuffle this vector in order to generate the increments of each of the m paths. These values can sensibly be chosen to be the mean of the normal distribution provided they lie in each of the quantiles.

Another desirable feature of a set of simulated Brownian paths is an $(n \times n)$ covariance matrix which equals the identity. After all, these paths should be independent. Being able to generate them with the identity as a covariance matrix is a step in that direction. Specifically, we know from multivariate statistics that any multivariate normal distribution with the identity as a covariance matrix has independent components. Since the components of our paths are nearly normal, we assume that if we can transform them so that they have the identity as a covariance matrix, they will be very close to independent.

Now, suppose that we have generated m simulated Brownian paths, each of length n, using the shuffling procedure described above, and that the covariance matrix of the n components is

$$V = \mathrm{Cov}(X)$$

where X is an n vector representing the increments of the simulated Brownian paths. To determine a linear transformation of X that has a covariance matrix equal to the identity, we seek a matrix C such that $\mathrm{Cov}(CX) = I$. So we must have

$$C\,V\,C^{T} = I, \text{ or}$$
$$V = C^{-1}(C^{T})^{-1} = C^{-1}(C^{-1})^{T} = DD^{T}$$

where $D \equiv C^{-1}$. We can find D using Cholesky decomposition and then set $C = D^{-1}$ and $Y' = CY$, where Y is the initially generated set of paths. That is, each of the m rows of Y is a sample path of length n. We then use Y' as our set of pricing paths. Since the columns of Y were structured to be close to normal, it is reasonable to invoke the following result for normal distributions. This result states that for a multivariate normal vector X and matrix C, CX is also multivariate normal. Therefore, we expect the columns of Y' also to be approximately normally distributed.

If we follow this procedure once for a large number of columns, N, we need not repeat it every time we perform a new simulation. We simply shuffle the (orthogonal) columns of Y' for each new set of m paths. There are then $N!/(N-n)!$ ways to permute the columns of Y' in generating paths at n time points, so there should be plenty of sets of potential paths to generate using a reasonable value of N.

Computational results

The method described above was applied to Asian call options with monthly averaging over

Table 1. Results of 50,000 paths

Strike	Antithetic	40-column run	12-column run	Accurate value
90	12.984	13.031	13.009	13.016
	(0.028)	(0.007)	(0.006)	
100	7.348	7.307	7.320	7.323
	(0.031)	(0.011)	(0.007)	
110	3.741	3.734	3.780	3.754
	(0.026)	(0.011)	(0.010)	

one year. The initial underlying price was set to 100, the risk-free interest rate and dividend yield were fixed at zero, and a volatility of 30% was used. The results are given in the table. The first column is the result of using the standard antithetic method. Since antithetic paths were used for each of the 50,000 runs, a total of 100,000 paths were evaluated. The standard error is in parentheses below the price estimates. The second column shuffles the full 40 columns of X' and uses the first 12 columns to price each set of $m = 1{,}000$ runs. The third column shuffles only the first 12 columns of X' and reuses these, shuffling for each new set of 1,000 paths. There seems to be some advantage in using the same 12 columns again and again. Possibly there is a further stratification effect going on. The last column gives an accurate value for each option.

The square of the ratios of the standard errors gives the acceleration factor for the new method over the standard antithetic variate method on a path basis. The new method, however, takes approximately one-third the time for these paths. The total acceleration factor varies from about 20 to 60.

Summary

The method described here generates Brownian paths that are, in important ways, more representative of the population of independent Brownian paths. In particular, the components of the paths are very close to normally distributed and the sample covariance matrices are exactly equal to the identity. The method yields dramatic improvements in pricing path-dependent securities such as Asian options.

BIBLIOGRAPHY

Curran, M., 1994, "Strata Gems", *Risk* 7(3), pp. 70–1 (Chapter 7 of this publication).

Johnson, R.A., and D.W. Wichern, 1982, *Applied Multivariate Statistical Analysis*, Prentice Hall.

9

Greeks in Monte Carlo

Michael Curran
RiskCare Ltd

Traders who want to use Monte Carlo simulation to estimate derivatives' hedge parameters (Greeks) are often advised to use finite difference approximations. With a path-dependent derivative, for example, delta is calculated by incrementing the initial value of the underlying asset and computing a new value for the option using the same paths.

There is a more efficient technique, however: it is often possible to compute analytically the sensitivities of a particular path or set of paths without re-summing the paths with the shifted initial value.

Let us suppose that an asset price follows geometric Brownian motion, and let B represent the vector-random variable generating the risk-adjusted asset price path. Let the payoff of a call option be based on a function $f(B{:}S)$ of the path when the initial asset price is S. Virtually all such functions of the path of the underlying price will be homogeneous of degree one in the initial asset price. Technically this means that $f(B;S) = Sf(B;1)$. More simply, we may speak of the option value as a proportion of the value of the underlying asset, as is standard practice.

Now, suppose that we have generated a path B and wish to estimate the delta of the option using this path. Such a path will contribute to the delta only if it finishes in the money, so we focus on this case. The payoff of the option is $f(B{:}S) - K$, where K is the strike price. From the assumption above, the payoff is equal to $Sf(B{:}1) - K$. The delta estimate from this path is clearly just $\delta f(B{:}1)$, where δ is the present value factor. We then estimate delta by averaging these path-by-path estimates, where out-of-the-money paths are assigned a delta of zero. More generally, this approach can be used for estimating other hedge ratios such as theta and vega and can also be applied to interest rate-dependent instruments.

The calculation of the derivative's gamma poses other problems, however. The gamma of an option can be expressed as $\Gamma = \delta\Delta/\delta S$. If we differentiate our expression for the delta of a particular path, we get an intuition-defying function known as the Dirac function, which is defined to be zero everywhere except at the strike price, where it takes an infinite value.

We can get a handle on the problem by noting that in the finite difference method the delta will be increased when more price paths end up in the money as a result of incrementing S. At the margin, these paths will contribute with terms of the delta of an at-the money path. Since, for an at-the-money path, $Sf(B{:}1) = K$, the delta of an at-the-money path is $\delta K/S$. The rate at which more paths finish in the money as S is increased is also determined by the delta of at-the-money paths.

We can therefore estimate gamma with $\delta(K/S)^2$ multiplied by an estimate of the density function of $f(B{:}S)$ at the value K. This is easily done by counting the number of paths such that $|f(B;S) - K| < \varepsilon$ for some ε. If there are m such paths, then the estimate of the density is $m/(2\varepsilon n)$, where n is the total number of trials.

This technique can substantially reduce the computation time associated with Monte Carlo simulation.

10

Quicker on the Curves

Les Clewlow and Andrew Carverhill
Lacima Consultants Ltd and Warwick Business School;
University of Science and Technology, Hong Kong

The Monte Carlo method of valuing options is extremely flexible and easy to implement. It can be used for virtually any European option, and handling special features such as random volatility or interest rates is not a problem. It was first introduced to the option valuation literature by Boyle (1977) - also Chapter 1 of this publication.

However, in its naive form the method has an overwhelming disadvantage: it is so inefficient in terms of computer time as to be impracticable for many purposes. In this article, we show how to make the Monte Carlo method many orders of magnitude faster by a special implementation of the martingale variance reduction (MVR) technique.[1] Kemna and Vorst (1990) have proposed a similar technique for the valuation of average-rate ("Asian") options based on the analytic formula for the geometric average option. Their control variate differs from ours since it is not a martingale variate; therefore their estimates are likely to be significantly biased.

We will illustrate the MVR technique by using it to value a European lookback call option with stochastic volatility. We will first apply the "naive" Monte Carlo method to this option, and then contrast its performance to that using the MVR technique. With the MVR technique we can get fast, accurate answers for this option. Furthermore, this technique simultaneously provides a measure of the effectiveness of the hedge portfolio.

Naive valuation

The Monte Carlo method is simply to simulate the (risk-neutrally adjusted) random path of the underlying instrument over the lifetime of the option, and thence the option payoff. The option value is just the average over these simulated payoffs, discounted back at the riskless rate, over the lifetime of the option.

For our lookback, we will assume that the behaviour of the underlying asset S_t and its volatility σ_t are given by[2]

$$dS_t = \mu S_t dt + \sigma_t S_t dz_t^S \qquad (1)$$

$$d\sigma_t = \alpha(\bar{\sigma} - \sigma_t)dt + \theta\sigma_t dz_t^{\sigma} \qquad (2)$$

that is, we take the volatility to be stochastic but mean-reverting. If the underlying is an exchange rate and the domestic and foreign interest rates are r_d and r_f, respectively, then we should take the drift to be $r_d - r_f$.

To simulate the underlying, we first discretise the time: if the option lives over the time interval $[0, T]$, then take the time grid $0 \equiv t_0 < t_1 < \ldots < t_N \equiv T$, with

$$t_i - t_{i-1} = \varepsilon \equiv \frac{T}{N}$$

for each i. This time grid should correspond to the fixing times of the lookback, that is, the times at which the price is recorded for the calculation of the option payoff. Then replace (1) and (2) by

$$S_{t_{i+1}} = S_{t_i} + \mu S_{t_i}\varepsilon + \sigma_{t_i} S_{t_i} \Delta z_{t_i}^S \qquad (3)$$

$$\sigma_{t_{i+1}} = \sigma_{t_i} + \alpha(\bar{\sigma} - \sigma_{t_i})\varepsilon + \theta\sigma_{t_i}\Delta z_{t_i}^{\sigma} \qquad (4)$$

where $\Delta z_{t_i}^S$ and $\Delta z_{t_i}^{\sigma}$ are independent random normal variables with mean zero and standard deviation $\sqrt{\varepsilon}$.

Now, to value the lookback call, develop the simulation as in (3) and (4), starting from the known initial values (S_0, σ_0), and then calculate the simulated payoff:

$$P_T = S_T - \min_{i=1\ldots,N} (S_{t_i}) \qquad (5)$$

Then we average over many simulations, and multiply by $e^{-r_d T}$ for the discounting, to arrive at the option value.

The Monte Carlo method does not guarantee an accurate answer; in fact it gives an "estimate", which is itself random - that is, if we run the simulation on a different computer, with a different random number generator, we will obtain a slightly different answer. However, it is easy to obtain the standard error of the Monte Carlo estimate, and this estimate is very unlikely to be more than, say, four standard errors away from the true value. Also, the standard error decreases in proportion to one over the square of the number of simulations taken. Therefore to halve the standard error, for example, one should take four times as many simulations.

Taking our option to be a foreign exchange lookback with a lifespan of six months and 128 fixings, and with realistic parameters,[3] we obtained a standard error of about 10% using 1,000 simulations. Such a standard error is clearly unacceptable, but the calculation took about 30 seconds on a 486 PC. A standard error of 1% would perhaps be marginally acceptable, but this would have taken $10^2 = 100$ times as long. Note that in some cases we could save time by taking fewer time-steps. However, in our case this would reduce the accuracy because we would not be correctly reflecting the stochastic volatility and the frequency of fixing is quite critical to the price of the lookback.

The MVR technique

The MVR technique is to generate, simultaneously with the simulated path, a related collection of martingales - that is, random paths with zero expectation; these are the MVR variates. Then a linear combination of these martingale variates is subtracted from the simulated option payoff. These variates will not affect the final average that gives the Monte Carlo valuation, because they will themselves average to zero. If the linear combination is chosen carefully, however, the standard error will be reduced.

For our lookback, we took variates $(x^1_{t_i}, x^2_{t_i}, x^3_{t_i})$ defined by

$$x^1_{t_{i+1}} = x^1_{t_i} + \frac{\partial}{\partial S} L(t_i, S_{t_i}, \sigma_{t_i})(\Delta S_{t_i} - E[\Delta S_{t_i}]) \qquad (6)$$

$$x^2_{t_{i+1}} = x^2_{t_i} + \frac{\partial^2}{\partial S^2} L(t_i, S_{t_i}, \sigma_{t_i})(\Delta S^2_{t_i} - E[\Delta S^2_{t_i}]) \qquad (7)$$

$$x^3_{t_{i+1}} = x^3_{t_i} + \frac{\partial}{\partial \sigma} L(t_i, S_{t_i}, \sigma_{t_i})(\Delta \sigma_{t_i} - E[\Delta \sigma_{t_i}]) \qquad (8)$$

in which, for example, $\Delta S_{t_i} = S_{t_{i+1}} - S_{t_i}$, and starting from $x^1_0 = x^2_0 = x^3_0 = 0$. Note that the expressions such as $(\Delta S_{t_i} - E[\Delta S_{t_i}])$ are easily computed from (3) and (4), and have expectation zero; and so $x^1_{t_i}, x^2_{t_i}, x^3_{t_i}$ are indeed martingales. Also in (6), (7) and (8) we take L to be the formula for valuing a lookback with constant volatility and continuous fixings.[4] That is:

$$L(t,S,\sigma) = Se^{r_f(T-t)}N(x + \sigma\sqrt{T-t}) - m_{0,t}e^{r_d(T-t)}N(x) + \frac{S}{B}\left(e^{r_d(T-t)}(S/m_{0,t})^{-B} N(y + \sigma\sqrt{T-t}) - e^{r_f(T-t)}N(y)\right) \qquad (9)$$

where

$$B = 2(r_f - r_d)/\sigma^2$$

$$x = \frac{\log(S/m_{0,t}) + (b - \frac{1}{2}\sigma^2)(T-t)}{\sigma\sqrt{T-t}}$$

$$y = \frac{-\log(S/m_{0,t}) - (b + \frac{1}{2}\sigma^2)(T-t)}{\sigma\sqrt{T-t}} \qquad (10)$$

and $m_{0,t}$ is the minimum underlying value up to time t.

We now explain our choice of the MVR variates $x^1_{t_i}$, $x^2_{t_i}$ and $x^3_{t_i}$. Variate $x^1_{t_i}$ corresponds to delta hedging the option, except that this hedging is only approximate because the formula for L fails to account for the stochastic volatility and the discrete fixings. But, in spite of this, one expects this hedge to be reasonably good, and this is reflected in the variance reduction. Variate $x^2_{t_i}$ corresponds to gamma hedging, and $x^3_{t_i}$ to hedging the volatility.

In general, one should choose MVR variates to capture as much as possible of the randomness in the system. It is a beautiful fact that the MVR variates cannot really do any harm. The worst that could happen would be to choose a variate which was completely uncorrelated to the underlying. This would fail to reduce the variance, but calculating it would take up computer time. For our lookback, we have used a convenient analytic formula for a very similar option; however, we would get considerable benefit even if we used the elementary Black-Scholes formula instead.

The variance-reduced Monte Carlo estimate (let us call this $\hat{P}$) is the discounted expectation of

$$P_t - (\beta_1 x^1_T + \beta_2 x^2_T + \beta_3 x^3_T) \qquad (11)$$

and it remains to explain how to choose the coefficients β_1, β_2, β_3 and thence $\hat{P}$. In fact we can get $\hat{P}$ directly by a least-squares estimation procedure.

Putting $\underline{\beta} = (\hat{P}, \beta_1, \beta_2, \beta_3)$, we have

$$\underline{\beta} = (X'X)^{-1}X'P$$

where X is the matrix whose rows correspond to

each of the simulations and are $(1,x_T^1,x_T^2,x_T^3)$ and "dash" denotes transpose, and P is the vector of simulated payoff values.[5] Moreover, we can calculate $X'X$ and $X'P$ efficiently as the simulation proceeds via the formulae

$$(X'_{n+1}X_{n+1})_{j,k} = (X'_nX_n)_{j,k} + x_j^{n+1}x_k^{n+1}$$
$$(X'_{n+1}P_{n+1})_j = (X'_nP_n)_j + x_j^{n+1}P^{n+1}$$

in which $x_0^n = 1$, and x_k^n and P^n are the values, returned by the nth simulation, of X_T^k and the option payoff, respectively.

For our lookback option, using the three MVRs reduced the standard error for a given number of simulations by a factor of about 14. This means that one can get a given accuracy by taking fewer simulations by a factor $14^2 \approx 200$. Also, each simulation took longer by a factor of about three, and so the MVR technique makes the option valuation about 70 times faster![6]

1 *We are grateful to Professor Stewart Hodges for originally suggesting the techniques of this article.*

2 *Note that since we are simulating we can choose virtually any stochastic processes.*

3 *Namely* $S_0 = 100$, $\sigma_0 = 6\%$, $r_d = 5\%$, $r_f = 7\%$, $\alpha = 1.35$, $\theta = 0.085$.

4 *See Goldman, Sosin and Gatto (1979).*

5 *For unbiased least-squares estimates the payoffs and control variates must be multivariate normal, but they are not. However, for typical cases with good hedging control variates and small time-steps the bias will be insignificant.*

6 *A further improvement in efficiency can be obtained by applying the antithetic variate technique as well. Details of all these techniques can be found in Clewlow and Carverhill (1992).*

BIBLIOGRAPHY

Boyle, P.P., 1977, "Options: a Monte Carlo Approach", *Journal of Financial Economics* 4, pp. 323-38 (Chapter 1 of this publication).

Clewlow, L.J., and A.P. Carverhill, 1992, "Efficient Monte Carlo Valuation and Hedging of Contingent Claims", Financial Options Research Centre Preprint 92/30, Warwick Business School, Warwick University.

Goldman, M.B., H. Sosin and M. Gatto, 1979, "Path-dependent Options: 'Buy at the Low, Sell at the High'", *Journal of Finance* 34 (5), pp. 1,111-27.

Kemna, A.G.Z., and A.C.F. Vorst, 1990, "A Pricing Method for Options on Average Asset Values", *Journal of Banking and Finance* 14, pp. 113-29.

11

Exact Exotics

Leif Andersen and Rupert Brotherton-Ratcliffe
General Re Financial Products

Pricing formulas for various barrier and lookback options have been available for more than 20 years. Starting with Merton's 1973 formula for a knock-out call and Goldman, Sosin and Gatto's (1979) formula for lookback options, researchers have steadily developed analytical formulas for a variety of related structures, eg barrier puts and calls, digital options, ladder options, partial lookback options, options with rebates or contingent premiums, and barrier options triggered by a variable other than the underlying. Many of these pricing formulas have appeared in *Risk* magazine (eg Rubinstein and Reiner, 1991a and 1991b, and Heynen & Kat, 1994a and 1994b).

Generally, these analytical formulas hold only in a Black–Scholes framework. In particular, they require both the discount rate and volatility to be constant. If we wish to relax these assumptions or to price more complex structures – eg step-up or step-down barrier options, curved barrier options or options with discretely monitored trigger dates – a numerical approach is required.

In this article, we will discuss the application of the Monte Carlo method, a numerical technique known for its high flexibility but also its slow convergence rate (generally $O(N^{-1/2})$ where N is the number of simulations). Fortunately, many variance reduction schemes are capable of significantly improving the convergence of the Monte Carlo method (eg Boyle, 1977 – also Chapter 1 of this publication, and Brotherton-Ratcliffe, 1994 – also Chapter 29 of this publication). In the case of barrier and lookback options, however, a potentially more significant problem is that Monte Carlo methods yield biased results for options which depend on the continuously monitored path of some stochastic variable (such as an option which can be knocked out at any time over its life).

In a Monte Carlo simulation, where we can only sample the value of a stochastic variable at discrete times, we lose information about the parts of the continuous-time path that lie between the sampling dates. It is easy to see and verify that, with a probability equal to one, the discretely observed minimum will be too high and the discretely observed maximum too low compared with the real extremes of the continuous-time process. For, say, a knock-out option, this will mean we would tend to underestimate the likelihood of the option being knocked out and hence overestimate the option's value (and vice versa for knock-in options).

Of course, this bias does not apply to options with discrete monitoring of the underlying random variable since the simulation can be devised to match the actual observation dates. But, as calculation time is proportional to the number of observation dates, simulation might become prohibitively lengthy if there are too many observation dates, such as for a long-term knock-out option with a trigger variable monitored daily.

So, on the one hand we must eliminate the bias that discrete sampling yields in continuously monitored barrier and lookback options and, on the other, find an efficient way of pricing discretely monitored options when there are many observation dates. We will see that the solutions to these two problems are closely related.

First, consider the value of an up-and-out call option on a stock that pays no dividends. We will make the usual Black–Scholes assumptions, so we can compare Monte Carlo prices with the exact price of the continuously monitored knock-out option. Table 1 shows the Monte Carlo prices of one-year options as a function of the number of discrete knock-out dates, M, when the initial stock price $S = \$100$, strike price $K = \$100$, volatility $\sigma = 25\%$, the discount rate $r = 5\%$ and the knock-out strike $H = \$150$. The exact price of the continuously monitored option is \$6.442. In each case, the value is estimated by simulating 10,000 and 100,000 stock price paths

Table 1. Up-and-out call prices

Knock-out dates (M)	10,000 paths Value	(Error)	100,000 paths Value	(Error)
1	8.214	(0.122)	8.365	(0.039)
2	8.046	(0.121)	8.231	(0.039)
4	7.772	(0.118)	7.922	(0.038)
8	7.426	(0.114)	7.595	(0.037)
16	7.154	(0.112)	7.294	(0.036)
32	6.979	(0.110)	7.078	(0.035)
64	6.773	(0.108)	6.904	(0.035)
128	6.631	(0.106)	6.769	(0.034)
256	6.542	(0.105)	6.667	(0.034)

The exact value of the continuously monitored option is \$6.442.

of M sampling dates each, using

$$S_i = S_{i-1}e^{(r-\frac{1}{2}\sigma^2)\delta + W_i\sigma\sqrt{\delta}}$$
$$\text{for } i = 1,2,\ldots,M \text{ and } S_0 = S$$

Here, W_i are independent draws from N(0,1), the standard normal distribution. Also, the time step $\delta = T/M$, where T is the life of the option (here T = 1). For the option with M knock-out dates, the payout of the option at time T is calculated as

$$\max(S_M - K,0) \text{ if } S_i < H \text{ for } i = 1,2,\ldots,M$$

otherwise, the payout is zero.

Table 1 also gives the standard errors of the Monte Carlo estimates, which approximately equal \$0.036 with 100,000 paths, so the 95% confidence intervals are roughly ±\$0.07.

In compiling Table 1, we have applied a useful trick: the sample paths generated for M = 256 simulation times, corresponding to the highest sampling frequency, have been reused to price all the other discretely monitored options, ie for M = 128, say, we have used every second stock price drawn from the full path of M = 256 simulations, and so on. As well as speeding up the simulation, this ensures that the correlations between simulation errors are high and the standard deviation of the difference between any two option prices is low, around \$0.003. Had we used different random paths to price these options, the standard errors of the differences would have been much larger, roughly \$0.05, and the shape of the convergence profile less clear.

The sampling frequency of the knock-out has a significant effect on the option's price. In valuing the continuous knock-out, even daily sampling, ie M = 256, overestimates its value, by \$0.225. The difference between the daily and continuously monitored options is almost seven times the random error for 100,000 simulations and hence is highly significant.

A more accurate estimate of the value of the continuous knock-out can be found by taking into account the non-zero probability of a barrier breach between the simulation dates. Consider the interval $[T_{i-1},T_i]$ and suppose $\tau_{i-1}(H)$ is the random time-to-first-hit for the barrier H after time T_{i-1}, ie, $\tau_{i-1}(H) = \inf\{t > T_{i-1}:S_t = H\}$. We are interested in the distribution of $\tau_{i-1}(H)$ given S_{i-1} and S_i, and it can be shown (Andersen, 1996) that, for $0 < u < \delta$:

$$\text{Prob}\{\tau_{i-1}(H) - T_{i-1} < u \mid S_{i-1},S_i\} = N(\phi w_i) + N(\phi\hat{w}_i)\xi_i \tag{1}$$

where N denotes the normal distribution and

$$\phi = \text{sign}(H - S_{i-1})$$
$$w_i = \frac{u\ln(S_i/S_{i-1}) - \delta\ln(H/S_{i-1})}{\sigma\sqrt{u\delta(\delta-u)}}$$
$$\hat{w}_i = \frac{u\ln(H^2/S_{i-1}S_i) - \delta\ln(H/S_{i-1})}{\sigma\sqrt{u\delta(\delta-u)}}$$
$$\xi_i = \exp\left(\frac{2\ln(H/S_{i-1})\ln(S_i/H)}{\sigma^2\delta}\right)$$

Notice that this result is independent of the risk-neutral drift rate of the stock price. This is because the distribution is conditional upon both S_{i-1} and S_i; the distribution of S_i, however, does depend on the drift rate. The conditional distribution of the time-to-first-hit remains the same as long as the drift rate is constant over the interval $[T_{i-1},T_i]$. By taking the limit of equation (1) as $u \to \delta$ we obtain

$$\text{Prob}\{\tau_{i-1}(H) \leqslant T_i \mid S_{i-1},S_i\} = \begin{cases} \xi_i & \text{if } \phi(S_i - H) < 0 \\ 1 & \text{otherwise} \end{cases} \tag{2}$$

That is, if both S_{i-1} and S_i are on the same side of the barrier, either above or below H, then the probability that the stock price hits the barrier in the interval (T_{i-1},T_i) is ξ_i.

To use equation (2) in Monte Carlo simulation of the up-and-out call, we simply have to draw M random uniformly distributed variables X_i, $i = 1,2,\ldots,M$ from the uniform distribution U(0,1) and calculate the payout as

$$\max(S_M - K,0) \text{ if}$$
$$S_i < H \textit{ and } X_i \geq \xi_i \text{ for } 1,2,\ldots,M$$

In Table 2, we give the Monte Carlo estimates of the continuously monitored knock-out option using this method for 10,000 and 100,000 simulations and from 1 to 256 simulation dates. The

Table 2. Revised up-and-out call prices

Simulation dates (M)	10,000 paths Value	(Error)	100,000 paths Value	(Error)
1	6.605	(0.106)	6.449	(0.033)
2	6.431	(0.105)	6.501	(0.033)
4	6.508	(0.105)	6.436	(0.033)
8	6.428	(0.104)	6.477	(0.033)
16	6.267	(0.102)	6.415	(0.033)
32	6.417	(0.105)	6.473	(0.033)
64	6.255	(0.103)	6.456	(0.033)
128	6.275	(0.103)	6.421	(0.033)
256	6.511	(0.105)	6.446	(0.033)

The exact option value is \$6.442.

errors are independent here, and the bias has been removed: all these estimates are within the 95% confidence interval of the exact value.

A significant feature of these results is that the estimate from just one time jump ($M = 1$) is as good as the estimate from 256 time jumps. There is no need to simulate the stock price prior to option expiry. Crude Monte Carlo simulation with just one time step gave a value of \$8.365 ± 0.076 (at the 95% confidence level) using 100,000 simulations; here, one time step gives a value of \$6.449 ± 0.065, the exact value being \$6.442.

The application of equation (2) to eliminate bias for the simple knock-out option is quite general and can be applied to a large class of barrier option payouts. For example, we can apply the method to cases where there are separate trigger and payout variables, and we can also use it for non-constant barriers as long as the barriers are piecewise flat as a function of time. As any smooth function can be approximated to an arbitrary degree of precision by a piecewise flat "staircase" function, the method is also useful in pricing options with barriers that vary smoothly with time. A similar argument shows that the method can also be used where the volatility and risk-neutral drift of the underlying variables are not constant.

In practice, many barrier-style options also include rebates or contingent payments which occur only if the barrier is breached. As for the payment's timing, there are two types of contingent payments: those which occur on a fixed date, typically at the option expiry, and those which occur when the barrier is first breached. The first type of payment can be priced in the same way as above but, for the second type, we also need to know when the barrier is hit. Using equation (1), we can easily generate bias-free simulations of the time of hit. Specifically, remember that we simulate a hit in the interval $[T_{i-1},T_i]$ whenever we draw X_i from $U(0,1)$ with $X_i < \xi_i$. If we now write

$$F(u) = \mathrm{Prob}\{\tau_{i-1}(H) - T_{i-1} < u \mid S_{i-1},S_i\}$$

then, when $X_i < \xi_i$, the simulated time of hit is simply

$$t_H = T_{i-1} + F^{-1}(X_i)$$

where the inverse function $F^{-1}(\cdot)$ can be computed from equation (1) by, for example, a simple Newton–Raphson search routine. The present value of a cash-at-hit rebate can then be calculated as $R\exp(-rt_H)$, where R is the rebate amount. This expression can easily be generalised to a case where the discount rate is not constant.

As a second example, consider lookback options on the stock S. Although lookback options and barrier options may appear quite different at first, they are closely related as they can both be considered options on the maximum or minimum of a stochastic process. We know that, for any interval $[T,u]$, if $\tau(H) = \inf\{t > T : S_t = H\}$ is the time-of-first-hit after T, we have

$$\begin{aligned}&\mathrm{Prob}\{\tau(H) < u \mid S_T,S_u\}\\ &= \mathrm{Prob}\{\max(\phi S_t : t \in [T,u]) \geq \phi H \mid S_T,S_u\}\end{aligned}$$

where, as before, $\phi = \mathrm{sign}(H - S_T)$. Therefore, the probability distributions of the maximum and minimum of the stock price over the interval are directly related to the probability distribution of the time of first hit. For the interval $[T_{i-1},T_i]$, we have

$$\begin{aligned}&\mathrm{Prob}\{\max(\phi S_t : t \in [T_{i-1},T_i]) \geq \phi H \mid S_{i-1},S_i\}\\ &= \xi_i(H)\end{aligned}$$

when $\phi H \geq \phi S_{i-1}$ and $\phi H \geq \phi S_i$. Note that, in the case of barrier options, ξ_i is a function of the fixed value of H. But in the case of lookback options, we consider H to be a variable. We then immediately obtain the following:

$$\begin{aligned}&\mathrm{Prob}\{\max(S_t : t \in [T_{i-1},T_i]) < H \mid S_{i-1},S_i\}\\ &= 1 - \xi_i(H) \text{ for } H \geq \max(S_{i-1},S_i)\end{aligned}$$

$$\begin{aligned}&\mathrm{Prob}\{\min(S_t : t \in [T_{i-1},T_i]) < H \mid S_{i-1},S_i\}\\ &= \xi_i(H) \text{ for } H \leq \min(S_{i-1},S_i)\end{aligned}$$

Thus, to simulate the minimum of the stock price over the interval $[T_{i-1},T_i]$, given S_{i-1} and S_i, we draw X_i from $U(0,1)$ and set

$$S_i^{min} = \xi_i^{-1}(X_i)$$

Table 3. Lookback put prices

Simulation dates (M)	Discrete lookback Value	Discrete lookback (Error)	Revised lookback Value	Revised lookback (Error)
1	7.47	(0.03)	18.72	(0.04)
2	9.65	(0.04)	18.70	(0.04)
4	11.77	(0.04)	18.76	(0.04)
8	13.39	(0.04)	18.66	(0.04)
16	14.86	(0.04)	18.75	(0.04)
32	15.81	(0.04)	18.65	(0.04)
64	16.68	(0.04)	18.73	(0.04)
128	17.32	(0.04)	18.80	(0.04)
256	17.72	(0.04)	18.77	(0.04)

The exact value of the continuous lookback is $18.72.

Also, $\xi_i^{-1}(1 - X_i)$ generates the maximum S_i^{max}, which seems paradoxical at first, but this is because $\xi_i^{-1}(X)$ has two solutions: the lower-branch solution gives the minimum, the upper-branch the maximum.

For example, consider the value of a lookback put, which pays the difference between the maximum stock price over the life of the option and its final price. Using crude Monte Carlo simulation, we would calculate the payout along one price path as

$$\max\{S_i : i = 0,1,\ldots,M\} - S_M$$

while using the current approach, we calculate the payout as:

$$\max\{S_i^{max} : i = 1,\ldots,M\} - S_M$$

Table 3 presents the results for a one-year lookback put, with the same parameters used earlier. The lookback with strike set at the maximum stock price over 256 discrete points during its life has a value of about \$17.72, compared with the exact value of \$18.72 for the continuous lookback. Thus, the daily lookback underestimates the value of the continuous lookback by more than 5%. Again, by sampling the maximum stock price directly, we can obtain an unbiased estimate of the continuous lookback with just a single simulation step.

1. Comparison of up-and-out call option values
S = \$100, K = \$100, H = \$150, r = 5%, σ = 25%

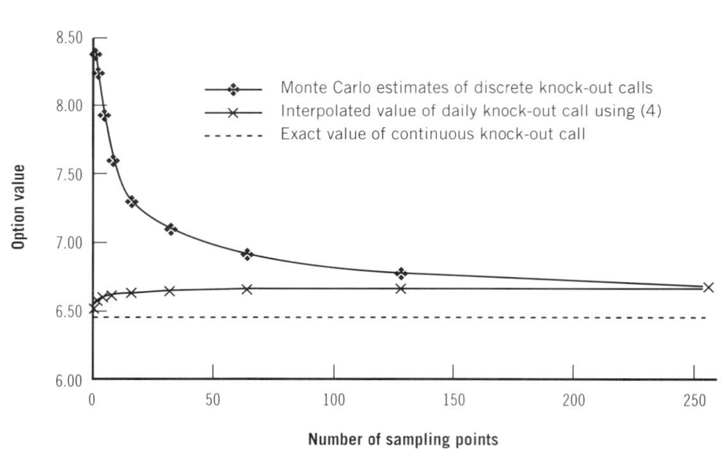

Finally, we turn our attention to efficient valuation of discretely monitored barrier and lookback options with many observation dates. In principle, it is easy to use Monte Carlo simulation to value these options by simulating the underlying stochastic variable at every future observation date. However, this method will become computationally expensive if there are too many observation dates.

We have already seen that even options with a large number of observation dates can differ significantly from those which are continuously monitored. We can, however, estimate that the difference between a discretely and continuously sampled option decreases as $\delta^{1/2}$, approximately, where δ is the time between the discrete observation dates. Consider, for example, the value of a lookback put. Here, discrete sampling will underestimate the true maximum of the stock price on every path. When the discrete sampling dates are close together ($\delta \ll 1$), the stock price can change by an amount $O(\delta^{1/2})$ between the sampling dates, and so the maximum stock price must be around $O(\delta^{1/2})$ greater than the discretely sampled maximum.

This, in turn, gives a systematic $O(\delta^{1/2})$ underestimate in the continuous lookback option payout along each path, which therefore gives a net $O(\delta^{1/2})$ error in its computed price. A similar argument can be made for the difference between continuous and discrete barrier options. (See also Broadie *et al*, 1997.)

We can use this result to obtain a relatively efficient pricing method for a discrete option with a large number of observation dates. If V_C and V_δ are the values of the continuously and discretely monitored options, respectively:

$$V_\delta = V_C + \lambda\delta^{1/2} + O(\delta^{1/2}) \text{ when } \delta \ll 1 \qquad (3)$$

where the coefficient λ does not depend on δ. Equation (3) allows us to estimate V_δ from the value V_{δ^*} of an otherwise identical option but with a lower monitoring frequency, δ^*, using interpolation:

$$V_\delta \approx V_C + (\delta/\delta^*)^{1/2}(V_{\delta^*} - V_C) \qquad (4)$$

Table 4. Interpolated up-and-out call prices

Simulation dates (M)	Time between observations (δ^*)	V_δ with $\delta = 1/256$ using equation (4)	Random error
1	1	6.520	0.033
2	1/2	6.559	0.033
4	1/4	6.588	0.033
8	1/8	6.609	0.033
16	1/16	6.622	0.033
32	1/32	6.638	0.033
64	1/64	6.650	0.033
128	1/128	6.660	0.034
256	1/256	6.667	0.033

Table 4 gives the value of the daily monitored (ie $\delta = 1/256$) up-and-out call studied earlier, using equation (4) in combination with Monte Carlo estimates for V_C and V_{δ^*}. The last row in the table gives the price computed directly, without using equation (4). Note that the errors reported in this table are only the usual non-systematic Monte Carlo errors, and these are sometimes smaller than the systematic error induced by equation (4). If necessary, the error can be reduced by including higher-order terms in the asymptotic expansion (3).

Figure 1 shows the rapid convergence of the interpolated option values V_δ with $\delta = 1/256$, as a function of the number of observation dates used in equation (4). The graph also shows the exact value of the continuous knock-out call and the values of the various discrete knockout calls V_{δ^*} with δ^* ranging from 1 to 1/256. Note that all the interpolated values of V_δ are closer to the brute-force Monte Carlo estimate of V_δ than the continuous knock-out value.

In conclusion, the current methodology extends the usefulness of the Monte Carlo method to the valuation of lookback and barrier options with both continuous and discrete monitoring of the underlying asset. Our approach provides unbiased price estimates for continuous knock-out, contingent premium and lookback options. We have also shown how these continuous prices can be used to improve the efficiency of pricing certain discretely monitored options. Needless to say, the real power of our method lies in its application to far more complex structures than those discussed here, and in situations where the dynamics of the underlying variables are also more complex.

BIBLIOGRAPHY

Andersen, L., 1996, "Monte Carlo Simulation of Lookback and Barrier Options with Continuous or High-Frequency Monitoring of the Underlying Asset", *working paper*, General Re Financial Products Corp.

Boyle, P., 1977, "Options: A Monte Carlo Approach", *Journal of Financial Economics* 4, pp. 323–38 (Chapter 1 of this publication).

Broadie, M., P. Glasserman and S. Kou, 1997, "A Continuity Correction for Discrete Barrier Options", *Mathematical Finance* 7(4), pp. 325–49.

Brotherton-Ratcliffe, R., 1994, "Monte Carlo Motoring", *Risk* 7(12), pp. 53–7 (Chapter 29 of this publication).

Goldman M., H. Sosin and M. Gatto, 1979, "Path Dependent Options: Buy at the Low, Sell at the High", *Journal of Finance* 34, pp. 1,111–27.

Heynen R., and H. Kat, 1994a, "Crossing Barriers", *Risk* 7(6), pp. 46–51.

Heynen R., and H. Kat, 1994b, "Selective Memory", *Risk* 7(11), pp. 73–6.

Merton R., 1973, "A Theory of Rational Option Pricing", *Bell Journal of Economics and Management Science* 4, Spring, pp. 141–83.

Rubinstein M., and E. Reiner, 1991a, "Breaking Down the Barriers", *Risk* 4(8), pp. 28–35.

Rubinstein M., and E. Reiner, 1991b, "Unscrambling the Binary Code, *Risk* 4(9), pp. 75–83.

12

Monte Carlo Simulation of Options on Joint Minima and Maxima

Leif Andersen
General Re Financial Products

From humble beginnings in the 1970s (Merton, 1973, Goldman, Sosin and Gatto, 1979), barrier and lookback options have become increasingly popular and today can be found embedded in many popular structured derivatives in stock, foreign exchange and interest rate markets. Working mostly under stylised Black–Scholes assumptions[1] with constant interest rates and volatilities, researchers have been able to devise analytical or semi-analytical pricing formulas for some of these new structures, including, for example, double barrier options (eg Kunitomo and Ikeda, 1992, Geman and Yor, 1996, Jamshidian, 1997), partial barrier options (Heynen and Kat, 1994a), bivariate barrier options (Andersen, 1994, Heynen and Kat, 1994b), double-lookback options (He, Keirstad and Rebholz, 1998), and partial lookback options (Heynen and Kat, 1995). In practice, however, the Black- Scholes process assumptions are rarely realistic and, in particular, real contracts always seem to have at least one feature which is beyond any analytical approach. In either case, numerical methods are needed to price the contract. As the inherent path-dependence of barrier and lookback options often makes lattice and tree approaches quite cumbersome, the method of Monte Carlo simulation[2] appears to be a good candidate for a general pricing tool. For options with continuous or high-frequency sampling of the underlying asset, however, Monte Carlo simulation of process extremes might, at first glance, appear infeasible as an impossibly large number of simulation dates would be necessary to generate a single path.

As demonstrated by Beaglehole, Dybvig and Zhou (1997) and by Andersen and Brotherton-Ratcliffe (1996) - also Chapter 11 of this publication, this problem can in many instances be overcome by relatively straightforward conditioning techniques; we will review the basic idea of the method in the fourth section of this chapter. The approach taken in Andersen and Brotherton-Ratcliffe (1996) can be applied to a large class of barrier and lookback options, but is - in theory at least - limited to payouts not depending on the joint values of asset minima and maxima over shared time intervals. This restriction excludes certain contracts that are of considerable practical importance, eg double barrier options. In this paper we will discuss how the basic idea in Andersen and Brotherton-Ratcliffe (1996) can be extended to cover options that depend on joint maxima and minima. We will proceed as follows. In the next section we provide notation, list our process assumptions and firm up the objective of the paper. In the third section we list certain useful probabilistic results that form the basis of the proposed technique. The fourth section outlines the application of the results obtained in the third section to option pricing problems, and the fifth section contains some examples. The sixth section briefly discusses some possible extensions of the results in the chapter. The final section presents our conclusions.

Setup

We consider a horizon, $[0,T]$, that is subdivided into a finite, sorted set of dates, $t_0,t_1,t_2,\dots,t_N$, with $t_0 = 0$ and $t_N = T$. An asset, S, is assumed to follow geometric Brownian motion under the risk-neutral probability measure Q (which we assume exists and is unique):

$$dS_t/S_t = \mu(t)dt + \sigma(t)dW_t,\ S_0 = \text{const.} > 0 \qquad (1)$$

The functions $\mu(.)$ and $\sigma(.)$ are assumed to be deterministic, bounded, and piecewise constant on the sub-intervals[3] $[t_i,t_{i+1}]$, $i = 0,1,2,\dots,N-1$. Under these assumptions, (1) can be solved explicitly as

$$S_t = S_0 \exp\Big(\sum_{i \geq 1: t_i < t} (\mu_{i-1} - \tfrac{1}{2}\sigma_{i-1}^2)(t_i - t_{i-1}) + (\mu_\alpha - \tfrac{1}{2}\sigma_\alpha^2)(t - t_\alpha)\Big) \times \exp\Big(\sum_{i \geq 1: t_i < t} \sigma_{i-1}(W_{t_i} - W_{t_{i-1}}) + \sigma_\alpha(W_t - W_{t_\alpha})\Big) \quad (2)$$

where μ_i denotes the constant drift over $[t_i, t_{i+1}]$, σ_i denotes the constant volatility over $[t_i, t_{i+1}]$, and $\alpha = \max\{i : t_i < t\}$. The assumption of piecewise constant volatility and drift is necessary for the results that follow; we notice, however, that (2) can approximate smooth term structures to arbitrary precision. The assumption of deterministic drift and volatility can be relaxed (see the sixth section, "Extensions").

We define the continuous-time extrema of S on an arbitrary time interval $[a, b]$ as

$$M_{[a,b]} = \max\{S_t : t \in [a, b]\}$$
$$m_{[a,b]} = \min\{S_t : t \in [a, b]\}$$

We now consider the class of contingent claims V with time payouts at time T of the form

$$V_T = f(S_{t_0}, S_{t_1}, \ldots, S_{t_N}, m_{[t_0,t_1]}, \ldots, m_{[t_{N-1},t_N]}, M_{[t_0,t_1]}, M_{[t_1,t_2]}, \ldots, M_{[t_{N-1},t_N]}) \quad (3)$$

where $f(.)$ is some well-behaved real function. Well-known examples of such options include double knock-out call options:

$$V_T = (S_T - K)^+ \text{ iff } M_{[0,T]} < H \text{ and } m_{[0,T]} > h$$

and double-lookback calls:

$$V_T = (M_{[0,T]} - m_{[0,T]} - K)^+$$

As an example of a more complicated, non-standard payout, consider, for example, a partial double lookback, partial double knock-in call option, with a forward-starting Asian strike:[4]

$$V_T = [M_{[t_1,t_2]} - m_{[t_1,t_2]} - (S_{t_3} + S_{t_4})/2]^+ \text{ iff } M_{[t_5,t_6]} > H \text{ or } m_{[t_5,t_6]} < h$$

Some conditional distributions

We now focus on a specific time bucket $[t_i, t_{i+1}]$ and consider the problem of estimating the distributions of the process extremes $M_{[t_i,t_{i+1}]}$ and $m_{[t_i,t_{i+1}]}$, *conditional* upon knowledge of S_{t_i} and $S_{t_{i+1}}$. For the marginal conditional distributions of $M_{[t_i,t_{i+1}]}$ and $m_{[t_i,t_{i+1}]}$, we have the following result.

PROPOSITION 1
For $h < \min(S_{t_i}, S_{t_{i+1}})$ *and* $H > \max(S_{t_i}, S_{t_{i+1}})$

$$P\{M_{[t_i,t_{i+1}]} > H | S_{t_i}, S_{t_{i+1}}\} = \xi_i(H) \quad (4a)$$
$$P\{m_{[t_i,t_{i+1}]} \leq h | S_{t_i}, S_{t_{i+1}}\} = \xi_i(h) \quad (4b)$$

where $\delta_i \equiv t_{i+1} - t_i$, *and*

$$\xi_i(x) = \exp\left(\frac{2\ln(x/S_{t_i})\ln(S_{t_{i+1}}/x)}{\sigma_i^2 \delta_i}\right)$$

PROOF
This is a standard result. It is also a limiting case of the main result in Andersen and Brotherton-Ratcliffe (1996) (proven in Appendix A of Andersen, 1996).

Notice that the drift, μ_i, appears nowhere in (4a–b), a result of conditioning upon the start- and end-points of S and assuming a constant drift over the time bucket $[t_i, t_{i+1}]$. The drift affects the distributions only indirectly through its effects on the law of $S_{t_{i+1}}/S_{t_i}$. Suitably normalised, the log-increments of S, conditional on S_{t_i} and $S_{t_{i+1}}$, become a so-called "Brownian bridge" (or "tied-down Brownian motion") on $[t_i, t_{i+1}]$. Not surprisingly, (4a–b) can alternatively be proven using results for extremes of Brownian bridges; see, for example, Karatzas and Shreve (1991, p. 265).

We now extend Proposition 1 to include conditioning on both start- and end-point of the stock as well as information about the maximum attained on $[t_i, t_{i+1}]$. Here and in the following we will use $N(.)$ to denote the standard Gaussian cumulative distribution function and $n(.)$ to denote the standard Gaussian density.

PROPOSITION 2
For $h < \min(S_{t_i}, S_{t_{i+1}})$, $H > \max(S_{t_i}, S_{t_{i+1}})$, *and an integer* $l \geq 0$

$$P\{m_{[t_i,t_{i+1}]} \geq h | M_{[t_i,t_{i+1}]} < H, S_{t_i}, S_{t_{i+1}}\} = \frac{1}{\lambda_i(1 - \xi_i(H))} \sum_{j=-l}^{l} \{n(A_i(j)) - n(B_i(j))\} + e_1(l) \quad (5a)$$

$$P\{m_{[t_i,t_{i+1}]} \geq h | M_{[t_i,t_{i+1}]} \geq H, S_{t_i}, S_{t_{i+1}}\} = \frac{1 - \xi_i(h)}{\xi_i(H)} - \frac{1}{\lambda_i \xi_i(H)} \times \sum_{j=-l}^{l} \{n(A_i(j)) - n(B_i(j))\} + e_2(l) \quad (5b)$$

$$P\{m_{[t_i,t_{i+1}]} \geq h | M_{[t_i,t_{i+1}]} = H, S_{t_i}, S_{t_{i+1}}\} = \frac{\sigma_i\sqrt{\delta_i}\sum_{j=-l}^{l}\{jA_i(j)n(A_i(j)) - jB_i(j)n(B_i(j))\}}{\lambda_i\xi_i(H)\ln(H^2/(S_{t_{i+1}}S_{t_i}))} + e_3(l) \tag{5c}$$

where $\xi_i(.)$ *and* δ_i *are as in Proposition 1, and*

$$A_i(j) = \frac{\ln(S_{t_{i+1}}/S_{t_i}) - 2j\ln(H/h)}{\sigma_i\sqrt{\delta_i}}$$

$$B_i(j) = \frac{\ln(S_{t_{i+1}}S_{t_i}/h^2) - 2j\ln(H/h)}{\sigma_i\sqrt{\delta_i}}$$

$$\lambda_i = n\left(\frac{\ln(S_{t_{i+1}}/S_{t_i})}{\sigma_i\sqrt{\delta_i}}\right)$$

The error terms satisfy $e_1(\infty) = e_2(\infty) = e_3(\infty) = 0$ *and are bounded as follows:*

$$|e_1(l)| \leq \frac{\sigma_i\sqrt{\delta_i}\,|N(B_i(l)) - N(A_i(l+1)) + N(-A_i(-l)) - N(-B_i(-l-1))|}{2\ln(H/h)\lambda_i(1-\xi_i(H))} \tag{6a}$$

$$|e_2(l)| \leq \frac{\sigma_i\sqrt{\delta_i}\,|N(B_i(l)) - N(A_i(l+1)) + N(-A_i(-l)) - N(-B_i(-l-1))|}{2\ln(H/h)\lambda_i\xi_i(H)} \tag{6b}$$

$$|e_3(l)| \leq \frac{\sigma_i\sqrt{\delta_i}\left\{2 + \frac{\sigma_i^2\delta_i}{2\ln(H/h)^2} - \frac{\ln(S_{t_{i+1}}/h)}{\ln(H/h)} + F_+^{A_i}(l) + F_+^{B_i}(l) - F_-^{A_i}(-l) - F_-^{B_i}(-l)\right\}}{\lambda_i\xi_i(H)\ln(H^2/(S_{t_{i+1}}S_{t_i}))} \tag{6c}$$

where the functions $F_\pm^{A_i}(.)$ *and* $F_\pm^{B_i}(.)$ *are defined in Appendix A, equations (A1) and (A2).*

PROOF

For brevity we only provide an outline of the proof. We first notice that $Y_t \equiv \ln(S_t/S_{t_i})/\sigma_i$ is a standard Brownian motion on $[t_i,t_{i+1}]$ with drift $\mu_i/\sigma_i - \frac{1}{2}\sigma_i$ and initial value $Y_{t_i} = 0$. As Y_t is strictly increasing in S_t, the extreme values of S_t on $[t_i,t_{i+1}]$ are uniquely given by the extremes of Y_t. The result in the proposition for $l = \infty$ then follows from Bayes' rule, Girsanov's theorem (to remove the drift), and the known infinite series for the joint distribution of maximum, minimum and end-point of a Brownian motion (see, eg, Borodin and Salminen, 1996, or Karatzas and Shreve, 1991). To determine the truncation errors for a finite value of $l \geq 0$, notice that

$$\left|\sum_{j=l+1}^{\infty}\{n(A_i(j)) - n(B_i(j))\}\right| = \sum_{j=l+1}^{\infty}\{n(B_i(j)) - n(A_i(j))\} \leq \int_l^{\infty}\{n(B_i(x)) - n(A_i(x+1))\}dx = \frac{\sigma_i\sqrt{\delta_i}}{2\ln(H/h)}|N(B_i(l)) - N(A_i(l+1))| \tag{7a}$$

In (7a), the first equality follows from the characteristics of the Gaussian density function and $A_i(j) < B_j(j) < 0$ for $j \geq 1$. The inequality in (7a) follows from an obvious geometric argument. We also have

$$\left|\sum_{j=-\infty}^{-l-1}\{n(A_i(j)) - n(B_i(j))\}\right| = \sum_{j=-\infty}^{-l-1}\{n(A_i(j)) - n(B_i(j))\} \leq \int_{-\infty}^{-l}\{n(A_i(x)) - n(B_i(x-1))\}dx = \frac{\sigma_i\sqrt{\delta_i}}{2\ln(H/h)}|N(-A_i(-l)) - N(-B_i(-l-1))| \tag{7b}$$

Combining (7a) and (7b) leads to (6a) and (6b). Equation (6c) is proven in Appendix A.

As we shall see shortly, under certain circumstances another representation of the distribution functions in Proposition 2 might be easier to implement numerically:

PROPOSITION 3

For $h < \min(S_{t_i},S_{t_{i+1}})$, $H > \max(S_{t_i},S_{t_{i+1}})$, *and an integer* $l \geq 1$

$$P\{m_{[t_i,t_{i+1}]} \geq h | M_{[t_i,t_{i+1}]} < H, S_{t_i}, S_{t_{i+1}}\} = \frac{\sigma_i\sqrt{8\pi\delta_i}\sum_{j=1}^{l} n(C_i(j))\sin D_i(j)\sin E_i(j)}{\lambda_i(1-\xi_i(H))\ln(H/h)} + e_4(l) \tag{8a}$$

$$P\{m_{[t_i,t_{i+1}]} \geq h | M_{[t_i,t_{i+1}]} \geq H, S_{t_i}, S_{t_{i+1}}\} = \frac{1-\xi_i(h)}{\xi_i(H)} - \frac{\sigma_i\sqrt{8\pi\delta_i}\sum_{j=1}^{l} n(C_i(j))\sin D_i(j)\sin E_i(j)}{\lambda_i\xi_i(H))\ln(H/h)} + e_5(l) \tag{8b}$$

$$P\{m_{[t_i,t_{i+1}]} \geq h | M_{[t_i,t_{i+1}]} = H, S_{t_i}, S_{t_{i+1}}\} = \frac{\sigma_i^3\delta_i^{3/2}\sqrt{2\pi}\sum_{j=1}^{l}\Psi_i(j)n(C_i(j))\sin D_i(j)\sin E_i(j)}{\lambda_i\xi_i(H)\ln(H/h)^2\ln(H^2/(S_{t_{i+1}}S_{t_i}))} + e_6(l) \tag{8c}$$

where λ_i, $\xi_i(.)$ *and* δ_i *are as in Proposition 2, and*

$$C_i(j) = \frac{j\pi\sigma_i\sqrt{\delta_i}}{\ln(H/h)},\quad D_i(j) = \frac{j\pi\ \ln(S_{t_i}/h)}{\ln(H/h)}$$

$$E_i(j) = \frac{j\pi\ \ln(S_{t_{i+1}}/h)}{\ln(H/h)}$$

$$\Psi_i(j) = C_i(j)^2 - 1 - D_i(j)\mathrm{ctg}D_i(j) - E_i(j)\mathrm{ctg}E_i(j)$$

The error terms satisfy $e_4(\infty) = e_5(\infty) = e_6(\infty) = 0$ *and are bounded as follows:*

$$|e_4(I)| \le \frac{\sqrt{8/\pi}}{\lambda_i(1-\xi_i(H))}\,N(-C_i(I)) \qquad (9a)$$

$$|e_5(I)| \le \frac{\sqrt{8/\pi}}{\lambda_i\xi_i(H)}\,N(-C_i(I)) \qquad (9b)$$

$$|e_6(I)| = \frac{\sigma_i^3\delta_i^{3/2}\sqrt{2\pi}\,\{N(-C_i(I))\,\Omega_i(h,H) + n(C_i(I))(I+2+\Omega_i^*(h,H))\}}{\lambda_i\xi_i(H)\,\ln(H/h)^2\,\ln(H^2/(S_{t_{i+1}}S_{t_i}))} \qquad (9c)$$

In (9c)

$$\Omega_i(h,H) = \frac{\frac{2}{\pi}\ln(H/h) + \ln(S_{t_{i+1}}S_{t_i}/h^2)}{\sigma_i\sqrt{\delta_i}} + \frac{\pi\sigma_i\sqrt{\delta_i}}{\ln(H/h)}$$

$$\Omega_i^*(h,H) = \frac{\ln(H/h)\,\ln(S_{t_{i+1}}S_{t_i}/h^2)}{\pi\sigma_i^2\delta_i}$$

PROOF

When $I = \infty$, (8a–c) are just the Fourier sine representations of (5a–c). As for the truncation errors, notice that

$$\left|\sum_{j=I+1}^{\infty} n(C_i(j))\sin D_i(j)\sin E_i(j)\right| \le \sum_{j=I+1}^{\infty} n(C_i(j)) \le \int_I^{\infty} n(C_i(x))dx = \frac{\ln(H/h)}{\pi\sigma_i\sqrt{\delta_i}}N(-C_i(I)) \qquad (10)$$

(9a–b) follow directly from (10) and (8a–b). Using arguments similar to those in Appendix A, (9c) follows from the triangle inequality applied to each of the four terms of $\Psi_i(j)$ in (8c).

In both propositions 2 and 3 we have provided convenient[5] bounds on the errors involved in terminating otherwise infinite series after a finite number of terms. A closer analysis of (6a–c) and (9a–c) reveals that (6a–c) tend to be large when $\ln(H/h)$ is small relative to $\sigma_i\sqrt{\delta_i}$, whereas the exact opposite is true for (9a–c). This is a convenient circumstance as we can switch back and forth between the representations (5a–c) and (8a–c) whenever errors start getting out of hand in a particular application. As an example, set $S_{t_i} = 1{,}000$, $\sigma_i = 0.3$ and $\delta_i = 0.25$, and consider the comparison of (6a) and (9a). In Table 1 we have illustrated how, for this example, the bounds on the $I = 3$ truncation errors behave as a function of H, h, and $S_{t_{i+1}}$. Notice that the largest errors tend to happen for $S_{t_{i+1}}$ very close to the upper barrier, H. For the example used, $I = 3$ is sufficient to keep the smallest of (6a) and (9a) negligible in almost all situations. In fact, for most applications it is possible to pick the series representation in such a way that very few terms of the series need to be included. So, while the formulas in propositions 2 and 3 might appear slow and difficult to evaluate, this is normally not the case in practice.

Applications to option pricing

Consider now the problem of determining the time 0 price of an option with time T payout V_T of the form (3). Standard arguments (eg Harrison and Pliska, 1981) equate the arbitrage-free option price with an expectation under Q:

$$V_0 = E^Q\{V_T e^{-\int_0^T r_u du}|\mathfrak{I}_0\} \qquad (11)$$

Table 1. Upper bounds on truncation errors $|e_1(3)|$ and $|e_4(3)|$

$S_{t_{i+1}}$	Bound (6a)	Bound (9a)
(a) H = 1150, h = 850, S_{t_i} = 1000, σ_i = 0.3, δ_i = 0.25		
850.001	0*	1.07e–5
900	0*	7.84e–6
1000	0*	7.08e–6
1100	0*	1.68e–5
1149	0*	8.33e–4
1149.9	0*	8.33e–3
1149.999	0*	0.833
(b) H = 1100, h = 900, S_{t_i} = 1000, σ_i = 0.3, δ_i = 0.25		
900.001	3.55e–13	5.85e–12
950	2.33e–12	5.57e–12
1000	3.01e–11	6.74e–12
1050	4.62e–10	1.21e–11
1099	1.54e–7	5.93e–10
1099.9	1.59e–6	5.93e–9
1149.999	1.60e–4	5.93e–7
(c) H = 1050, h = 950, S_{t_i} = 1000, σ_i = 0.3, δ_i = 0.25		
950.001	1.4e–3	0*
975	2.1e–3	0*
1000	4.8e–3	0*
1025	1.6e–2	0*
1045	0.12	0*
1049	0.65	0*
1049.9	6.62	0*

*To 15-digit machine precision.

Here, r_t is the instantaneous risk-free rate and $\mathfrak{I}_0$ the filtration representing all information available at time 0. For processes of the form (1) and general payouts of the form (3), (11) is typically not analytically tractable and we must resort to numerical methods. As mentioned in the opening section, we will focus here on applications of the Monte Carlo simulation method. Under this approach, we use a random number generator and (2) to generate[6] L simulated paths $S_{t_0}, S_{t_1}, \ldots, S_{t_N}$ of the asset S. Assuming deterministic interest rates, the Monte Carlo estimate of (11) takes the form of an average:

$$\hat{V}_0 = \frac{e^{-\int_0^T r_u du}}{L} \sum_{\ell=1}^{L} f(S^\ell_{t_0}, S^\ell_{t_1}, \ldots, S^\ell_{t_N}, m^\ell_{[t_0,t_1]}, \ldots, m^\ell_{[t_1,t_2]}, \ldots, m^\ell_{[t_{N-1},t_N]}, M^\ell_{[t_0,t_1]}, M^\ell_{[t_1,t_2]}, \ldots, M^\ell_{[t_{N-1},t_N]}) \qquad (12)$$

where $S^\ell_{t_i}$ is the ℓth draw of S_{t_i}, and similarly for m and M. In the evaluation of (12), the question remains how to use Monte Carlo techniques to generate the various continuous-time extremes $m^\ell_{[t_i,t_{i+1}]}$ and $M^\ell_{[t_i,t_{i+1}]}$. A naïve approach would be to subdivide the interval $[t_i, t_{i+1}]$ into a finer time grid and estimate $m^\ell_{[t_i,t_{i+1}]}$ and $M^\ell_{[t_i,t_{i+1}]}$ as the minimum and maximum of a discrete-time Monte Carlo path on this grid. Obviously this approach would, with probability one, understate the true continuous-time process extremes, but one could hope that the resulting option price biases tend to be low. Unfortunately, as demonstrated by many authors (eg Andersen and Brotherton-Ratcliffe, 1996, Andersen, 1996, Geman and Yor, 1996,[7] Kat and Verdonk, 1995), even with daily sampling this naïve approach can lead to significant price errors. Besides potential bias problems from the discretisation of time, evaluation of (12) also inevitably involves statistical errors stemming from the finite sample size L. In basic applications of the Monte Carlo method this error approaches 0 quite slowly (as $O(L^{-1/2})$), but, fortunately, a variety of so-called variance-reduction techniques can be applied to speed up convergence (see Boyle, Broadie and Glasserman, 1997, for an excellent survey – also Chapter 2 of this publication). Being mainly interested in methods to reduce biases – which, by definition, are unaffected by L – we shall not pursue variance reduction further but will just point out that any realistic, time-efficient application of our results should be complemented by appropriate variance-reduction techniques.

As shown by Andersen and Brotherton-Ratcliffe (1996), the conditional distributions (4a–b) can be used to remove the naïve Monte Carlo bias for certain simpler forms of (3), such as single-barrier options and regular lookback options. To summarise their approach, consider simulation path ℓ and an option dependent on, say, $M^\ell_{[t_i,t_{i+1}]}$. We assume that $S^\ell_{t_i}$ and $S^\ell_{t_{i+1}}$ have already been generated in a Monte Carlo scheme, and let $\bar{u}^\ell_i$ be a draw from a uniform distribution $U(0,1)$ on $[0,1]$; $\bar{u}^\ell_i$ must be independent of the random numbers used to generate the W's in (2). For a barrier option with an upper barrier at H, we immediately conclude that the barrier was breached if $\max(S^\ell_{t_i}, S^\ell_{t_{i+1}}) \geq H$. If $\max(S^\ell_{t_i}, S^\ell_{t_{i+1}}) < H$, the decision of whether H was touched or not can, from (4a), be based on the value of $\bar{u}^\ell_i$ as follows:

$$\begin{aligned} \bar{u}^\ell_i \leq \xi_i(H) &\Rightarrow M^\ell_{[t_i,t_{i+1}]} \geq H \\ \bar{u}^\ell_i > \xi_i(H) &\Rightarrow M^\ell_{[t_i,t_{i+1}]} < H \end{aligned} \qquad (13)$$

(13) only gives us information about whether a barrier was breached or not. For lookback options, where a specific estimate of $M^\ell_{[t_i,t_{i+1}]}$ is required, we draw directly from the conditional distribution of $M_{[t_i,t_{i+1}]}$ by inverting (4a):

$$M^\ell_{[t_i,t_{i+1}]} = \xi_i^{-1}(\bar{u}^\ell_i) \qquad (14)$$

The inversion of ξ_i^{-1} can be performed analytically.[8] Using (4b), a scheme similar to (13) and (14) can be constructed for sampling the minimum, $m_{[t_i,t_{i+1}]}$. It is important to notice, however, that (4a) and (4b) only work "marginally" – that is, we *cannot* use these equations to simultaneously generate unbiased samples of *both* the maximum and the minimum over a shared time-interval $[t_i, t_{i+1}]$.

Let us now consider extending the scheme above to options on the joint minimum and maximum of the underlying asset. As an example, consider an option that knocks in if and only if the asset breaches either an upper barrier, H, or a lower barrier, h, in the time interval $[t_i, t_{i+1}]$. We focus on a particular path ℓ and assume that $S^\ell_{t_i}$ and $S^\ell_{t_{i+1}}$ have already been generated. If either $\max(S^\ell_{t_i}, S^\ell_{t_{i+1}}) \geq H$ or $\min(S^\ell_{t_i}, S^\ell_{t_{i+1}}) \leq h$, we conclude that the knock-in option has been activated and can proceed to compute its final payout. If, instead, $\max(S^\ell_{t_i}, S^\ell_{t_{i+1}}) < H$ and $\min(S^\ell_{t_i}, S^\ell_{t_{i+1}}) > h$, we first draw a $\bar{u}^\ell_i \in U(0,1)$ and apply (13). If (13) makes us conclude that $M^\ell_{[t_i,t_{i+1}]} \geq H$, the option has been knocked in; otherwise, we still need to check whether the lower barrier was breached. For a simple approach to this problem, we could first attempt to just ignore any dependence between the minimum and the maximum of S.

This approach will obviously introduce a bias, but at least it is simple: we can use (4b) and proceed to check whether the lower barrier has been touched in much the same way as the maximum. That is, we would draw a new uniformly distributed number $\bar{v}_i^\ell \in U(0,1)$, independent of $\bar{u}_i^\ell$, and check whether this number exceeded $\xi_i(h)$ or not. In general it is, of course, inappropriate to ignore the information contained in the knowledge of $M^\ell_{[t_i,t_{i+1}]} < H$. A more correct approach to checking for breach of the lower barrier is as follows:

$$\bar{v}_i^\ell \geq P\{m_{[t_i,t_{i+1}]} \geq h | M_{[t_i,t_{i+1}]} < H, S_{t_i}, S_{t_{i+1}}\} \Rightarrow m^\ell_{[t_i,t_{i+1}]} \leq h$$
$$\bar{v}_i^\ell < P\{m_{[t_i,t_{i+1}]} \geq h | M_{[t_i,t_{i+1}]} < H, S_{t_i}, S_{t_{i+1}}\} \Rightarrow m^\ell_{[t_i,t_{i+1}]} > h \quad (15)$$

The probability in (15), $P\{m_{[t_i,t_{i+1}]} \geq h \,|\, M_{[t_i,t_{i+1}]} < H, S_{t_i}, S_{t_{i+1}}\}$, can be computed to within a given precision by (5a–6a) or (8a–9a). If (15) leads us to conclude that $m^\ell_{[t_i,t_{i+1}]} > h$, the option did not knock in along path ℓ, and we assign a value of 0 to the option. Otherwise, we can proceed with the computation of the final payout.

Using (5b) and (8b), (15) can easily be modified to cover double knock-out options and other types of double barrier options (for example, options that knock in or out if and only if *both* barriers have been breached). For the case of double-lookback options that require specific estimates of $M^\ell_{[t_i,t_{i+1}]}$ and $m^\ell_{[t_i,t_{i+1}]}$, we proceed as follows. First, we use (14) to sample $M^\ell_{[t_i,t_{i+1}]}$. The distribution of the minimum, $m^\ell_{[t_i,t_{i+1}]}$, conditional upon $M^\ell_{[t_i,t_{i+1}]}$ as well as $S^\ell_{t_i}$ and $S^\ell_{t_{i+1}}$ is given by the infinite series (5c) or (8c); for convenience we define

$$G_i(x) = P\{m_{[t_i,t_{i+1}]} \geq x | M_{[t_i,t_{i+1}]} = H, S_{t_i}, S_{t_{i+1}}\} \quad (16)$$

We then proceed to draw a uniform number $\bar{v}_i^\ell \in U(0,1)$ and sample the minimum $m^\ell_{[t_i,t_{i+1}]}$ by inverting (16):

$$m^\ell_{[t_i,t_{i+1}]} = G_i^{-1}(\bar{v}_i^\ell) \quad (17)$$

Table 2. Double knock-out call option pricing
$S_0 = 1000$, $K = 1000$, $T = 0.25$, $\sigma = 0.3$, $r = 0.05$, $N = 1$, $I = 3$, $L = 100,000$, antithetic sampling

h	H	KI price	BBMC price	SD	e_{max}	Series
400	1600	64.77	64.744	0.16	0*	(5a)
600	1400	55.72	55.601	0.14	0*	(5a)
800	1200	19.31	19.314	0.077	0*	(5a)
850	1150	8.60	8.630	0.050	0*	(5a)
900	1100	1.29	1.296	0.017	1.10E–05	(8a)
930	1070	0.05	0.0516	0.0030	0*	(8a)
950	1050	0.00	4.30E–05	4.3E–05	0*	(8a)

*To 15-digit machine precision.

The inversion of (17) can be done by standard numerical root-search methods; the function $G_i(x)$ can be computed to given precision by either (5c–6c) or (8c–9c). While (17) is a "correct", bias-free approach, it is likely to be slow due to the need to find roots in a function that is computationally expensive to evaluate. To speed things up we could – as for double barrier options – instead envision a simplified approach where we effectively assume that the minimum and the maximum are independent. That is, we sample the minimum directly from the marginal distribution (4b), ie

$$m^\ell_{[t_i,t_{i+1}]} = \xi_i^{-1}(\bar{v}_i^\ell) \quad (18)$$

(see footnote 8). As ξ_i can be inverted analytically, this approach is much faster than (17). The errors committed in this simplified approach depend on the application; for straightforward double-lookback put and call options the simplified approach can normally be set up to work well (see "Double-lookback options" in the next section for details).

Examples

In this section we will test the algorithms described in the previous section by applying them to simple options for which closed-form results are known and have been reported in the literature. Obviously, the applications chosen here are for test purposes only; in practice, simulation should only be applied to securities for which no closed-form expressions are available.

DOUBLE KNOCK-OUT CALL OPTION

Let us consider the pricing of a double knock-out call option with flat upper and lower barriers H and h, respectively (see the second section of this chapter for discussion of the exact payout). We will assume that the option has maturity $T = 0.25$ and a strike of $K = 1000$. The underlying asset, S, pays no dividends, has a spot price of $S_0 = 1000$, and a constant volatility of $\sigma = 0.3$. We also assume a constant discount rate of $r = 0.05$. Given that S pays no dividends, the drift of S under the risk-neutral measure is constant and equal to the interest rate, $\mu = r$. To use the scheme (13) and (15) discussed above, we first notice that the lack of term structure in the drifts, volatilities and barriers allows us to simulate the option in a single time step, ie, $N = 1$. To compute $P\{m_{[0,T]} \geq h \,|\, M_{[0,T]} < H, S_0, S_T\}$ we use $I = 3$ series terms and the experimental results in Table 1 as a guide to selection between representations (5a) and (8a).

Table 3. Double knock-out call option pricing
$I = 3$, $L = 100{,}000$, antithetic sampling

N	NMC Price	NMC SD	SMC Price	SMC SD	r_{comp}	BBMC Price	BBMC SD	r_{comp}
(a) $S_0 = 1000$, $K = 1000$, $h = 800$, $H = 1200$, $T = 0.25$, $\sigma = 0.3$, $r = 0.05$								
1	33.105	0.095	19.330	0.077	2.0	19.314	0.077	5.0
2	30.954	0.093	19.275	0.077	2.1	19.306	0.077	5.3
4	28.878	0.092	19.305	0.077	2.3	19.368	0.077	5.8
8	26.632	0.089	19.330	0.077	2.4	19.342	0.077	6.8
16	24.761	0.086	19.321	0.077	2.6	19.326	0.077	7.2
32	23.321	0.084	19.359	0.077	2.8	19.287	0.077	7.4
64	22.215	0.083	19.292	0.077	2.9	19.325	0.077	7.7
Analytical value: 19.31.								
(b) $S_0 = 1000$, $K = 1000$, $h = 930$, $H = 1070$, $T = 0.25$, $\sigma = 0.3$, $r = 0.05$								
1	5.828	0.031	0.349	0.0079	1.6	0.0516	0.077	2.5
2	3.657	0.026	0.132	0.0049	1.2	0.0532	0.077	1.6
4	2.079	0.020	0.0601	0.0032	1.1	0.0556	0.077	1.3
8	1.165	0.015	0.0582	0.0032	1.1	0.0542	0.077	1.2
16	0.647	0.011	0.0540	0.0031	1.0	0.0514	0.077	1.1
32	0.394	0.0089	0.0525	0.0031	1.0	0.0497	0.077	1.1
64	0.251	0.0070	0.0534	0.0031	1.0	0.0554	0.077	1.1
Analytical value: 0.052.								

In Table 2 we have used $L = 100{,}000$ Monte Carlo trials with antithetic sampling to compute the price of the knock-out call for various values of h and H. Besides the estimated option price ("BBMC price"), the table lists the choice of series representation ("Series") used ((5a) or (8a)); the largest truncation error bound ("e_{max}") encountered in the simulation (computed by (6a) or (9a)); the sample standard deviation on the price estimate ("SD"); and analytical results from Table 3.1 in Kunitomo and Ikeda, 1992 ("KI price").

As all Monte Carlo estimates are within one sample standard deviation of the analytical price, we can be confident that the Monte Carlo method works as intended. We also note that our crude way of determining series representation and truncation limits worked well as all truncation errors were negligible. (Indeed, the truncation errors were so small relative to the Monte Carlo sample error that we probably could have got away with fewer series terms). In practice, more efficient "adaptive" schemes for the truncation limit and the series representation can be implemented with little difficulty.

Having verified that the Brownian bridge approach works as intended, we now focus on two options ($H = 1200$, $h = 800$ and $H = 1070$, $h = 930$) to compare our results with: a naïve Monte Carlo scheme ("NMC"), where no adjustments for "overshoot" between sampling points are applied; and a simplified Brownian bridge scheme ("SMC"), where the test for breach of the lower barrier is done using the marginal distribution (4b) (see discussion in the preceding section, "Applications to option pricing"). Table 3 lists the NMC, SMC and BBMC prices obtained for various numbers of (equidistant) sampling points, N. We also list the ratio of computing times for the adjusted Monte Carlo methods relative to naïve Monte Carlo ("r_{comp}") (on a DEC Alpha 8400 minicomputer).

We notice that the convergence of the NMC method to the continuous-time limit is very slow. In the first part of Table 3, for instance, even with $N = 64$ (or roughly daily sampling), the NMC price estimate is around 15 per cent higher than the continuously sampled value of 19.31 reported in Kunitomo and Ikeda (1992). In both parts of the table the NMC price error for $N = 64$ is more than 25 sample standard deviations and, hence, highly significant. As expected, N has no influence on the BBMC price, which in both parts stays within two standard deviations of the theoretical option value for all N. The added complexity of the BBMC method and the need to draw three times as many random numbers per time step as in the NMC approach obviously makes the former approach more computer-intensive: in part (a) the computation times of the BBMC method vary between five and eight times those of the NMC method; in part (b), where the barriers are hit more frequently, the computation time of the BBMC method is gener-

Table 4. Double lookback call option pricing
$S_0 = 1000$, $T = 1.0$, $\sigma = 0.3$, $r = 0.05$
$N = 1$, $I = 3$, $L = 100{,}000$, antithetic sampling

K	HKR price	BBMC price	SD	e_{max}
0	470.9	471.06	0.46	4.2e–7
450	84.6	84.86	0.37	4.2e–7
500	64.2	64.44	0.33	4.2e–7
1000	6.0	6.08	0.11	4.2e–7

ally less than twice that of the NMC method. As a convergence analysis (see the next section) reveals that more than $N = 1000$ sampling points would be needed even to make the NMC price less than 20.0 (which still would constitute a more than 3 per cent bias), the increase in computation time of the Brownian bridge technique is easily justified.

As for the SMC method, the results in Table 3, part (a) ($H = 1200$, $h = 800$), are remarkably competitive: the computation times are more than twice as fast as for the BBMC approach and, for all values of N, the computed prices are within a single standard deviation of the exact price. Although we know that the SMC method will induce a bias, for the option considered in part (a) this bias is so small that it is not possible to separate it from the random noise. In part (b), however, the bias is detectable for N less than 16, but eventually vanishes for higher values of N. As the SMC method is only marginally faster than the BBMC method for the option in part (b), the need to take 16 steps or more with the SMC method makes the BBMC method the most efficient approach here.

The decrease of the induced bias of the SMC method when N is increased is a general phenomenon. In general, by splitting a time interval $[t_i, t_{i+1}]$ up into progressively smaller sub-intervals, it becomes less and less likely that $M_{[t_i, t_{i+1}]}$ and $m_{[t_i, t_{i+1}]}$ will occur within the same sub-interval. This "decoupling" of the maximum and minimum will, in the limit, justify the SMC method – that is, the simultaneous application of the marginal distributions (4a) and (4b). As even naïve Monte Carlo has the property that the price bias vanishes for large N, the question of whether the SMC technique is worthwhile is an empirical one, and will typically (as we saw above) depend heavily on the particular application. If one can accept the additional computational overhead, the BBMC method based on (15) is a "safer" approach for general applications as it by construction is bias-free and does not necessitate the insertion of sub-intervals.

DOUBLE-LOOKBACK OPTIONS

We now turn to the pricing of the lookback options mentioned in the second section. He, Keirstad and Rebholz (1998) discuss this class of options in detail, and we will use their analytical results to test our method. As a specific example, consider a one-year ($T = 1$) double-lookback call on an asset, S, that pays no dividends, has a spot price, S_0, of 1000, and has a constant volatility of $\sigma = 0.3$. We will assume that the spot rate, r, is constant and equal to 0.05.

As in the previous example, we set the number of series terms in the computation of (17) to $I = 3$. To choose between series representations (5c) and (8c), we use the somewhat crude rule of always picking the series that has the lowest error bound. Obviously, more efficient schemes based on, say, predefined tolerances are possible and, indeed, recommended for real applications.

For various values of the double-lookback call strike, K, Table 4 lists the results of applying (17) with $L = 100{,}000$ Monte Carlo paths, one time step per path ($N = 1$), and antithetic sampling. Along with the Monte Carlo price ("BBMC price"), we also list the largest truncation error bound ("e_{max}") encountered in the simulation (computed by (6c) or (9c)); the sample standard deviation on the price estimate ("SD"); and analytical prices ("HKR price") from Table 1 in He, Keirstad and Rebholz (1998).[9]

All simulated price estimates are within one standard deviation of the analytical prices, so we conclude that the method works as intended.

As noted earlier, the root-search algorithm needed to invert (16) can be slow. A faster, but biased, algorithm can be based on the "simplified" approach. To compare the simplified approach ("SMC") to the proper Brownian bridge adjustment ("BBMC") and naïve Monte Carlo ("NMC"), consider the double-lookback option above with strike $K = 500$. For various equidistant sampling points (N), Table 5 contains prices, standard deviations ("SD"), as well as computing time ratios ("r_{comp}") relative to naïve Monte Carlo.

As before, all prices generated by the BBMC approach are within two sample errors whereas the naive Monte Carlo approach leads to significant biases: for the finest discretisation in the grid ($N = 64$), the NMC price is more than 20 per cent below the exact value. The bias of the SMC approach is visible for $N < 4$ but is indistinguishable from noise for higher values of N. For the example above, the SMC method is around 20 times faster to execute than the BBMC method and is, despite its bias, the most efficient approach here.

Table 5. Double-lookback call option pricing

1000, K = 500, T = 1.0, σ = 0.3, r = 0.05, l = 3, L = 100,000, antithetic sampling

	NMC		SMC			BBMC		
N	Price	SD	Price	SD	r_{comp}	Price	SD	r_{comp}
1	21.160	0.22	72.216	0.337	3.0	64.550	0.331	52.2
2	23.218	0.22	65.165	0.333	3.0	64.324	0.333	59.3
4	27.532	0.23	64.407	0.334	3.0	64.462	0.334	64.4
8	32.951	0.25	64.029	0.337	3.0	64.049	0.337	65.9
16	38.773	0.27	64.317	0.338	2.9	64.315	0.333	64.7
32	44.523	0.29	64.086	0.338	2.9	64.081	0.333	65.0
64	49.201	0.30	64.154	0.338	2.9	64.157	0.333	64.9

Analytical value: 64.2.

Extensions

The techniques discussed so far are easily extended to more complicated payouts of the form (3) and also allow for certain classes of multi-asset barrier options (see Panigirtzoglou, 1997, for a specific application). In this section we will briefly survey some non-trivial extensions, most of which are discussed in more detail in Andersen and Brotherton-Ratcliffe (1996) or Andersen (1996).

DISCRETE OBSERVATIONS

Many lookback and barrier options are based on daily or weekly, rather than continuous, sampling of the underlying asset. While it is common practice to equate daily and continuous sampling for pricing purposes, the examples in the preceding section illustrate that this practice might lead to significant pricing errors. One way to devise more accurate methods for options with discrete observations is based on the observation in Broadie, Glasserman and Kou (1996) that discretely sampled barrier and lookback options can be approximated by continuously sampled options after appropriate adjustments of the barrier or process extremes. By incorporating these adjustments in the Brownian bridge simulations, all techniques described so far become valid for discretely monitored options as well. An alternative technique described in Andersen and Brotherton-Ratcliffe (1996) and Andersen (1996) involves writing down an expansion of the discretely sampled option price in terms of the sampling interval and using interpolation methods to compute the discretely monitored option price from the price of the continuously monitored option and a coarsely sampled option. As argued in Andersen and Brotherton-Ratcliffe (1996), for sufficiently small sampling periods the option price can often be described as a linear function of the square root of the sampling period. This, for instance, holds for both the double knock-out and double-lookback options considered in the preceding section. For example, Figure 1 graphs the data for the NMC method in Table 3, part (a), as functions of $\sqrt{\delta}$, $\delta = T/N$; for low δ the relationship is, as predicted, close to linear. Further discussion and results for first- and second-order interpolation in $\sqrt{\delta}$ can be found in Andersen (1996).

STATE-DEPENDENT DRIFT AND VOLATILITY OF THE UNDERLYING ASSET

Consider the general diffusion process

$$dS_t/S_t = \mu(S_t,t)dt + \sigma(S_t,t)dW_t \tag{18}$$

for drift and volatility functions $\mu(.)$, $\sigma(.)$, sufficiently well-behaved for (18) to have a unique, non-explosive solution. In general the SDE (18) will not allow for an explicit solution (such as (2)); instead, we need to introduce an approximate discretisation scheme such as those described in detail in the monograph by Kloeden and Platen (1992). For instance, the so-called log-Euler scheme for advancing S from time t_i to t_{i+1}

1. Naïve Monte Carlo price of double knock-out call vs $\sqrt{\delta}$, $\delta = T/N$

$S_0 = 1000$, K = 1000, h = 800, H = 1200, T = 0.25, σ = 0.3, r = 0.05
L = 100,000, antithetic sampling

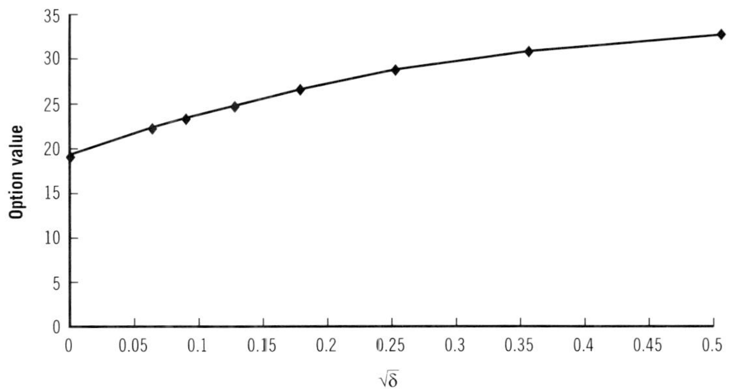

takes the form

$$S_{t_{i+1}} = S_{t_i} \exp\left(\left[\mu(S_{t_i},t_i) - \tfrac{1}{2}\sigma(S_{t_i},t_i)^2\right](t_{i+1} - t_i) + \sigma(S_{t_i},t_i)\sqrt{t_{i+1} - t_i}\, Z_i\right) \qquad (19)$$

where $Z_i \sim N(0,1)$. In effect, the log-Euler scheme freezes the arguments of $\mu(.)$ and $\sigma(.)$ to the left point of each simulation step. As such, the Brownian bridge schemes discussed earlier are still applicable as the requirement of piecewise flat drift and volatility is satisfied. While the Brownian bridge schemes remove the bias that is due to underestimation of continuous-time extremes, option prices based on the log-Euler scheme will still exhibit a first-order bias stemming from the inexact dynamics of the approximation (19). Further details on this topic can be found in, for instance, Kloeden and Platen (1992) and Andersen (1995).

OPTIONS WITH PAYMENTS AT HITTING TIMES

Many barrier options include rebates paid at the moment a barrier is hit. In simulating such options, it becomes necessary not only to determine *if* a barrier was breached but also *when* the barrier was hit. To incorporate this timing feature into Brownian bridge simulations, Andersen and Brotherton-Ratcliffe (1996) work with the distribution of the hitting time conditional upon the start and end values of the stock. Inverting this distribution leads to a bias-free scheme to sample the hitting time for single-barrier options. For double barrier options the exact conditional hitting-time distributions get very convoluted and are probably not worth the effort. In many cases it will suffice to insert additional time steps and use the hitting-time distribution in Andersen and Brotherton-Ratcliffe (1996) in the usual "simplified" approach assuming independence of minimum and maximum.

Conclusions

In this chapter we have discussed the application of Brownian bridge simulation techniques to a general class of path-dependent options dependent on local or global asset extremes. In particular, we worked out the necessary conditional distribution functions to allow for bias-free simulation of options dependent on the joint evolution of minimum and maximum. As these distributions take the form of infinite series, we also supplied analytic bounds for the error associated with truncation of the series after a finite number of terms. The techniques presented in the chapter were demonstrated on double barrier and double-lookback options. In our numerical examples we showed that the bias associated with the application of the simpler marginal distributions in Andersen and Brotherton-Ratcliffe (1996) can sometimes be quite small and manageable through the insertion of extra simulation dates. Finally, we briefly discussed possible extensions of the proposed techniques to more complicated asset dynamics and to options with discrete-time sampling and pay-at-hit rebates.

Appendix A – Proof of (6c)

Routine application of integration by substitution and by parts yields

$$\int (x \pm 1)A_i(x \pm 1)n(A_i(x))dx$$

$$= N(A_i(x))\left[1 \pm \frac{\ln(S_{t_{i+1}}/S_{t_i})}{2\ln(H/h)} + \frac{\sigma_i^2\delta_i}{4\ln(H/h)^2}\right]$$

$$+ \frac{n(A_i(x))\sigma_i\sqrt{\delta_i}(x \pm 2)}{2\ln(H/h)} \equiv F_{\pm}^{A}(x) \qquad (A1)$$

$$\int (x \pm 1)B_i(x \pm 1)n(B_i(x))dx$$

$$= N(B_i(x))\left[1 \pm \frac{\ln(S_{t_{i+1}}S_{t_i}/h^2)}{2\ln(H/h)} + \frac{\sigma_i^2\delta_i}{4\ln(H/h)^2}\right]$$

$$+ \frac{n(B_i(x))\sigma_i\sqrt{\delta_i}(x \pm 2)}{2\ln(H/h)} \equiv F_{\pm}^{B}(x) \qquad (A2)$$

A little thought shows that

$$\int_{l}^{\infty} (x + 1)A_i(x + 1)n(A_i(x))dx \leq \sum_{j=l+1}^{\infty} jA_i(j)n(A_i(j))$$

$$\leq \int_{l+1}^{\infty} (x - 1)A_i(x - 1)n(A_i(x))dx \leq 0$$

$$\int_{l}^{\infty} (x + 1)B_i(x + 1)n(B_i(x))dx \leq \sum_{j=l+1}^{\infty} jB_i(j)n(B_i(j))$$

$$\leq \int_{l+1}^{\infty} (x - 1)B_i(x - 1)n(B_i(x))dx \leq 0,$$

$$\int_{-\infty}^{-l} (x - 1)A_i(x - 1)n(A_i(x))dx \leq \sum_{j=-\infty}^{-l-1} jA_i(j)n(A_i(j))$$

$$\leq \int_{-\infty}^{-l-1} (x + 1)A_i(x + 1)n(A_i(x))dx \leq 0$$

$$\int_{-\infty}^{-l} (x - 1)B_i(x - 1)n(B_i(x))dx \leq \sum_{j=-\infty}^{-l-1} jB_i(j)n(B_i(j))$$

$$\leq \int_{-\infty}^{-l-1} (x + 1)B_i(x + 1)n(B_i(x))dx \leq 0$$

From the triangle inequality and (5c), we get

$$|e_3(I)| \le \frac{-\sigma_i\sqrt{\delta_i}}{\lambda_i\xi_i(H)\ln(H^2/(S_{t_{i+1}}S_{t_i}))}$$

$$\left\{\int_I^{\infty}(x+1)A_i(x+1)n(A_i(x))dx\right.$$

$$+\int_I^{\infty}(x+1)B_i(x+1)n(B_i(x))dx$$

$$+\int_{-\infty}^{-I}(x-1)A_i(x-1)n(A_i(x))dx$$

$$\left.+\int_{-\infty}^{-I}(x-1)B_i(x-1)n(B_i(x))dx\right\}$$

Inserting (A1) and (A2) leads to (6c).

1 *Two interesting exceptions are the papers by Roberts and Shortland (1994) and Bhagavatula and Carr (1996), both of which are set in an extended Black–Scholes economy with time-varying drift and volatility. The first paper discusses single-barrier options, the second double-barrier options.*

2 *Throughout this paper we will assume familiarity with basic Monte Carlo techniques. An excellent survey of the Monte Carlo method for securities pricing can be found in Boyle, Broadie and Glasserman (1997).*

3 *Strictly speaking, we should of course here refer to either* $(t_i, t_{i-1}]$ *or* $[t_i, t_{i-1})$. *For our purposes, whether or not the time buckets include their end-points is, however, immaterial.*

4 *To the uninitiated, this payout might perhaps appear too complicated to ever be of practical interest. In reality, however, this product is downright straightforward when compared to some of the more recent innovations in retail equity derivative markets.*

5 *While the error bounds are easier to compute than they might appear, in some circumstances it might be sufficient simply to monitor the absolute values of each term in the series and terminate when the contribution appears small enough. Care must clearly be taken when the series display oscillatory characteristics. At some lack of accuracy, the error bounds might also be simplified through order-type analysis; see, eg, Bhagavatula and Carr (1996), who also discuss problems with Fourier-series representations for very short maturities (Gibb's phenomenon).*

6 *In (2), each increment* $W_{t_i} - W_{t_{i-1}}$ *can be generated by making a pseudo-random draw from a normal distribution with mean* 0 *and standard deviation* $\sqrt{t_i - t_{i-1}}$.

7 *The authors specifically warn of "the danger inherent in Monte Carlo methods applied to barrier and double barrier options". As we shall see, the situation is not quite this dire.*

8 *Notice that* $\xi_i^{-1}(x)$ *is a two-valued function of* x. *The upper branch solution gives the maximum of* S *whereas the lower branch gives the minimum.*

9 *The table in He, Keirstad and Rebholz is based on an asset price of 100. We have scaled their results by a factor of ten.*

BIBLIOGRAPHY

Andersen, L., 1994, "Bivariate Binary Options", Working paper, General Re Financial Products Corp.

Andersen, L., 1995, "Efficient Techniques for Simulation of Interest Rate Models Involving Non-Linear Stochastic Differential Equations", Working paper, General Re Financial Products Corp.

Andersen, L., and R. Brotherton-Ratcliffe, 1996, "Exact Exotics", *Risk* 9(10), pp. 85–9 (Chapter 11 of this publication).

Andersen, L., 1996, "Monte Carlo Simulation of Barrier and Lookback Options with Continuous or High-Frequency Monitoring of the Underlying Asset", Working paper, General Re Financial Products Corp.

Beaglehole, D., P. Dybvig and G. Zhou, 1997, "Going to Extremes: Correcting Simulation Bias in Exotic Option Valuation", *Financial Analyst Journal* January/February, pp. 62–8.

Bhagavatula, R., and P. Carr, 1996, "Valuing Double Barrier Options with Time-Dependent Parameters", Working paper, Morgan Stanley.

Borodin, A., and P. Salminen, 1996, *Handbook of Brownian Motion – Facts and Formulae*. Basel, Birkhauser Verlag.

Boyle, P., M. Broadie and P. Glasserman, 1997, "Monte Carlo Methods for Security Pricing", *Journal of Economic Dynamics and Control* 21, pp. 1267–321 (Chapter 2 of this publication).

Broadie, M., P. Glasserman and S. Kou, 1996, "Connecting Discrete and Continuous Path-Dependent Options", Working paper, Columbia University.

Geman, H., and M. Yor, 1996, "Double-Barrier Options: A Probabilistic Approach", *Mathematical Finance* 6, pp. 365–78.

Goldman, M., H. Sosin and M. Gatto, 1979, "Path-Dependent Options: Buy at the Low, Sell at the High", *Journal of Finance* 34, pp. 1111–27.

Harrison, M., and S. Pliska, 1981, "Martingales and Stochastic Integrals in the Theory of Continuous Trading", *Stochastic Processes and their Applications* 11, pp. 215–60.

He, H., W. Keirstad and J. Rebholz, 1998 "Double Lookbacks", *Mathematical Finance* 8(3), pp. 201-28.

Heynen R., and H. Kat, 1994a, "Partial Barrier Options", *Journal of Financial Engineering* 3/4, pp. 253-74.

Heynen R., and H. Kat, 1994b, "Crossing Barriers", *Risk* 7(6), pp. 46-51.

Heynen R., and H. Kat, 1995, "Lookback Options with Discrete and Partial Monitoring of the Underlying Price", *Journal of Applied Mathematical Finance* 4, pp. 273-84.

Jamshidian, F., 1997, "A Note on Analytical Valuation of Double Barrier Option", Working paper, Sakura Global Capital.

Karatzas, I., and S. Shreve, 1991, *Brownian Motion and Stochastic Calculus*, Springer Verlag.

Kat, H., and L. Verdonk, 1995, "Tree Surgery", *Risk* 8(2), pp. 53-6.

Kloeden, P., and E. Platen, 1992, *Numerical Solution of Stochastic Differential Equations*, Springer Verlag.

Kunitomo, N., and M. Ikeda, 1992, "Pricing Options with Curved Barriers", *Mathematical Finance* 4, pp. 275-98.

Merton, R., 1973, "Theory of Rational Option Pricing", *Bell Journal of Economics and Management Science* 4, pp. 141-83.

Panigirtzoglou, N., 1997, "Investigation of the Pricing of Outside Barrier Options on LME Forwards using Monte Carlo Simulation, Lattice Models, and Analytical Approximations", M.Sc. thesis, Warwick Business School.

Roberts, G., and C. Shortland, 1994, "Pricing Barrier Options with Time-Dependent Coefficients", Working paper, University of Cambridge.

13

Model Calibration in the Monte Carlo Framework

Raphaël Douady
Ecole Normale Supérieure, CMLA

Financial models always depend on one or several parameters. The problem of calibration is to provide the value of these parameters with the two constraints of matching the market price of liquid instruments and of fitting certain statistics. The Monte Carlo framework is sometimes the only available numerical method when the model involves a large number of variables. In this case, calibration can become very time-consuming because it requires a lot of option calculations.

In this chapter, we study the complexity of the problem and some techniques to overcome it. We first compare the pros and the cons of the so-called "bootstrap" calibration with the global parametric approach. In particular, we assess the stability issue. Then we study what should be the optimal number of paths and present some easy to implement, quick and efficient variance reduction tricks. In Panel 1, we study some issues related to the infinite dimensional limit and their implications for finite dimensional models.

A simple stochastic volatility model is used throughout the chapter to illustrate theoretical observations.

The calibration problem

Mathematical models of financial markets, which are widely used to price and hedge portfolios of derivative assets, can mainly be calibrated in two ways. The first, which we shall call *statistical* calibration (or parameter estimation), relies on the study of statistical properties of past series and on the assumption that the future will be like the past. The second, called *implied* calibration, uses the information given by the market price of sufficiently liquid standard options to price exotic ones. The aim of these models is to help providers of derivative securities reduce the volatility of their wealth. Apart from the numerical issue, statistical calibration is not limited only by the fact that the future may differ from the past, but also because even if one's estimation is correct the market may think otherwise, leading to possible losses due to implied parameter changes. On the other hand, unless the model is simple – too simple – implied calibration does not have a unique solution. Moreover, even if one can reduce the model to achieve uniqueness, the problem is generally ill-posed, which implies that the result is highly unstable. An oversimple or unstable model can be very dangerous in practice because it may give rise to illusory hedges that will actually increase the wealth variance. Practical solutions usually involve a mixture of both procedures (statistical and implied), with some parameter estimations left to the trader's "guess". In fact, most recent models require the estimation of functions (curves or surfaces, or even functions of more variables), and that is often the reason for model complexity. A subsidiary question is now to choose between parametric and non-parametric estimation of these functions, the answer depending on the smoothness of the particular function.

In the Monte Carlo framework, which, when the model has a large number of independent random variables, is the only available numerical method, the model parameters are directly linked to the diffusion coefficients of the variables. It thus seems natural to slice the calibration problem into time steps. One begins to compute the coefficients in the first slice so as to match the market price of short maturity options and, possibly, certain statistical data. Then we continue with the diffusion and adjust parameters in the second slice to match the second set of option prices, and so on. This technique, which we call *bootstrapping*,[1] looks attractive and is, indeed, efficient in terms of computation time. However we shall see that it gives rise to high numerical instability, for which the only solution is a parameterisation of the time behaviour of coefficients and the minimisation of a global least-square criterion (LSC). Similarly, parameterisation with respect to other variables is sometimes one of the only ways to avoid the oscillations due to the fact that the problem is ill-posed (see Panel 1 on page 114).

Bootstrapping is the fastest algorithm when there are not too many time slices and when one can rely on good precision (sufficiently large number of paths, variance reduction, etc). We shall see that the required precision depends exponentially on the number of time slices, as in the Prigogine paradox of the flap of the butterfly's wing that can, eventually, induce a hurricane.[2] One needs to switch to a global calibration if oscillations appear: it would be hopeless to try to kill them by increasing the number of paths. Stabilisation techniques exist for the bootstrap method, but they are less efficient in terms of mean square error and can lead to systematic error (ie errors in the same sense).

In this chapter, we shall first describe how to implement both the bootstrap and the global calibration, then we shall discuss the criteria for the choice of one or the other. In the first case, stabilisation techniques are described. In the second case, different methods of improving the computation time are mentioned. The Monte Carlo framework provides a very simple link between precision and computation time. We shall see how to optimise the number of paths so as not to be too precise when this is not needed. It will also be shown how to organise the calculation of the LSC to avoid performing the same computation several times.

In most models the drift of a certain number of variables is implied by the absence of arbitrage opportunity. Although this drift is usually given in a closed form, it is better in a Monte Carlo model to adjust it numerically to ensure the value of risk-neutral expectations. We shall see how to perform this adjustment efficiently and how to overcome the difficulty of a drift parameter that depends on the value of the variables.

In the last part, we shall list some variance reduction techniques that are not specific to calibration but which, in this setting, become crucial. These are higher-order schemes (allowing longer time steps), variance/covariance correction, fast Gaussian simulation, control variate techniques and payoff smoothing.

The panel on pages 114–5 considers the infinite dimensional limit of complex models and the properties of operators which can be responsible for the instability of finite dimensional approximations.

Techniques for inferring diffusion parameters from historical data are not considered in this chapter, and statistical properties of historical data are used only to overcome the lack of information from the market. The choice of these statistics is left to the user, as well as how to compute them. Most of the time these are simple variance and correlation calculations, performed with very standard formulas. The only common sense remark is that statistics that one will "calibrate" must be computed in the same way from historical data and from Monte Carlo simulation.

General setting of a Monte Carlo model

Let us consider a model which describes the behaviour of a financial market where assets $S_1,\ldots,S_m$ are traded. We assume that the values of $S_1,\ldots,S_m$ are entirely determined by market variables $x_1,\ldots,x_n$:

$$S_i = \phi_i(t;x_1,\ldots,x_n) \quad i = 1,\ldots,m$$

where t is the current date. Market variables could be the asset prices themselves, interest rates, exchange rates, volatilities, etc. Assets could be underlyings as well as options. The market variables follow a joint random process, which we shall suppose is a multidimensional Itô process:

$$dx_i = \mu_i(t;x_1,\ldots,x_n)dt + \sum_{j=1}^{n} \sigma_{ij}(t;x_1,\ldots,x_n)dW_j \qquad i = 1,\ldots,n \tag{1}$$

where Brownian motions W_j are assumed independent. This is the most general shape, but, in practice, the coefficients μ_i and σ_{ij} will depend on only a few variables. The total number of Brownian motions cannot exceed the number of variables – also called the *dimension* of the model – but it could be less (as would be the case for a non-Markov Heath–Jarrow–Morton model).

Some of the functions ϕ_i may be explicitly computable (underlying assets, simple options), but others may require a numerical procedure (more complicated options): a tree or other PDE-solving algorithm, or a Monte Carlo simulation. We here assume that we are in the latter case.

Similarly, some of the drift coefficients, μ_i, are implied by the theory of arbitrage, whereas others can take any value.[3] Generally speaking, the arbitrage theory has no bearing on the volatility coefficients, σ_{ij}.

The coefficient functions μ_i and σ_{ij} depend on one or several parameters $\alpha_1,\ldots,\alpha_p$, $p \geq 1$. These are the *model parameters*. The total number of parameters, p, can be much larger than the number of variables, n, especially if one wishes to reproduce whole curves or surfaces (ie coefficients depending on one or two, or even more, variables, including time, t). We say that they are in a *parametric shape* if they are given by a formula that depends on parameters (for instance a linear combination of basic functions such as polynomials or exponentials). They are in a *non-parametric shape* if their value at grid points is given and if an interpolating procedure is defined (eg piecewise constant or linear, splines). In both cases, the entire function is determined by a finite list of numbers, which we try to make as short as possible while keeping flexibility in the shape of the graph.

Implied calibration consists of finding the values of parameters such that when evaluating the functions ϕ_i (those given in closed form and those which require a simulation), one exactly matches market prices, $\hat{S}_i$. The series of options that are used for calibration is called the *calibrating set*.

If there are more parameters than options ($p > m$), one usually finds families of solutions. Other constraints then have to be added. The most natural ones are that in simulations market variables have the same statistics as historical data. This type of requirement is called *statistical calibration* (parameter estimation from historical data; a broad literature is available on this topic). We now have too many constraints, so we must relax some of the statistics - mainly those which are redundant with implied calibration - and perform a *hybrid calibration*. This choice is an art left to the individual model user.

In our hybrid framework the statistical part can be achieved in two ways. The first way is to infer from historical data the value of some parameters and obtain others by implied calibration. The second way is to compute various statistical characteristics of simulations, such as correlations, volatilities of non-traded variables, etc, and to make them match historically observed values of the same characteristics.

There are two main issues in calibration:

- ❑ the accuracy of implied calibration - the error should never exceed the bid/offer spread; and
- ❑ the stability of parameters - if the model assumes that parameters are fixed, they should not change by a large amount every day and become real random processes (although a slight adjustment to new market conditions should regularly be performed).

The next section describes an example of a Monte Carlo model and its calibration problem.

A stochastic volatility model

Let S_t be the price of a stock which, for the sake of simplicity, delivers no dividend. We assume that for any strike K and any maturity date $T > t$, the standard European call $C(K,T)$ is traded and its value given by the market (as well as the put, the value of which is given by the put–call parity). We denote by r the interest rate, supposed constant. In the risk-neutral probability, the underlying process follows a diffusion equation:

$$\frac{dS_t}{S_t} = r\,dt + \sigma_t\,dW$$

where the volatility σ_t is itself a random process:

$$\sigma_t = \lambda \exp(x_t) \qquad dx_t = -ax_t\,dt + \nu\,dW'$$

Parameters λ, α and ν are here taken to be constant. In other words, σ_t is the exponential of an Ornstein–Uhlenbeck process. Its logarithm, x_t, has a Gaussian distribution, centred at $\bar{x}_t$ and with variance V_t defined by

$$\bar{x}_t = x_0 \exp(-at) \qquad V_t = \frac{\nu^2}{2a}(1 - \exp(-at))$$

If the two Brownian motions W and W′ are independent, one can find very good analytical approximations for the joint distribution of S_t and σ_t. If not, Monte Carlo simulation is a good technique for approximating this joint distribution. We denote by ρ the correlation between dW and dW′.

Given the values of the parameters r, λ, a, ν and ρ, one can compute the price, $C(t\ S_t,\sigma_t;K,T)$, of a call with strike K and maturity T at time t, assuming the stock and its spot volatility are at levels S_t and σ_t. The *implied volatility* is the parameter $\bar{\sigma}(t,S_t,\sigma_t;K,T)$ to input in the Black–Scholes formula in order to get the price $C(t,S_t,\sigma_t;K,T)$.

This model is arbitrage-free as soon as the drift of S_t equals r. The volatility process can be anything (provided that it stays positive and its size remains controlled so that S_t is still an integrable process). The shape of the "volatility surface" $\bar{\sigma}(K,T)$ at date t is then inferred from the process followed by the spot volatility as soon as S_t and σ_t are given. This is a two-factor Markov model (in the Dupire stochastic volatility framework – see Dupire, 1995 – the volatility surface is driven by only one factor).

Assuming r is given by the market, the "implied calibration" problem is to find λ, a, ν and ρ so as to match the market prices of European calls:

$$C(0,S_0,\sigma_0;K,T) = \hat{C}(K,T) \quad \forall K,T$$

One can see that four parameters are not enough to match an "infinite" number of prices. Therefore, one has to make parameters λ, a, ν and ρ depend on time and/or variables, namely S_t and σ_t. More precisely, one samples a series of options $(K_1,T_1),\ldots,(K_n,T_n)$, and then makes *functions* λ, a, ν and ρ entirely determined by the parameters $\alpha_1,\ldots,\alpha_p$ (in a parametric or non-parametric shape – see the explanation of these terms given earlier). These parameters are now the new parameters of the model. Then one must solve a (usually non-linear) system of equations:

$$C(0,S_0,\sigma_0;K_i,T_i;\alpha_1,\ldots,\alpha_p) = \hat{C}(K_i,T_i) \quad i = 1,\ldots,n$$

The function $\hat{C}(K,T)$ depends on two variables, so if we make coefficient functions λ, a, ν and ρ depend on all three variables, it is very likely (unless we adopt a rather rigid shape) that $p > n$ and that solutions of the system will not be unique. The model is said to be *underspecified* by the market. In our particular case, underspecification cannot be avoided by a tighter sampling of strikes and maturities (see the panel on infinite dimensions). One has to assume that coefficient functions depend on only one or two variables, but not on all three.

Implied calibration: bootstrapping

ONE-DIMENSIONAL BOOTSTRAP

The one-dimensional bootstrap applies to a situation in which the maturities of options in the calibrating set are sorted, say $T_1 < \ldots < T_n$, with no two options of the same maturity. We also assume that the parameter functions depend only on time and that all parameter functions but one are inputs, the last one being the only one to be calibrated. In this case, we shall adopt for it a non-parametric shape.

In the case of our stochastic volatility model, assume for instance that a(t), ν(t) and ρ(t) are given and that we want to calibrate λ(t). For each T_i, we assume that the strike K_i is at-the-money, or close to, and we set $\lambda_i = \lambda(T_i)$. Let λ be, for instance, linearly interpolated between T_i and T_{i+1}. The value of the option of maturity T_i depends only on $\lambda_1,\ldots,\lambda_i$ but not on the sequel. Assume that these values are already known. Then $C(K_{i+1},T_{i+1})$ depends only on λ_{i+1} and it is not difficult to solve for λ_{i+1} to match the market price $\hat{C}(K_{i+1},T_{i+1})$. This is done by any algorithm which solves a non-linear equation in one variable (eg the Newton method, secant method, dichotomy, etc, or a mixture of them). These algorithms can find a very precise solution with only a small number of evaluations.

MULTIDIMENSIONAL BOOTSTRAP

The multidimensional bootstrap is similar to the one-dimensional bootstrap, but one has a fixed number of options for each maturity, say m. In this case, we can calibrate m time-dependent parameters. In the stochastic volatility model we know that the volatility of volatility creates a "smile", whereas the correlation, ρ, affects the slope of the implied volatility with respect to the strike. Let us take m = 3 and sort the calibrating set in the following way:

$$(K_{11},T_1), (K_{12},T_1), (K_{13},T_1),\ldots,(K_{n1},T_n), (K_{n2},T_n), (K_{n3},T_n)$$

where, for every i, the strike K_{i1} is in-the-money, K_{i2} is at-the-money and K_{i3} is out-of-the-money. For each i we shall solve a system of three non-linear equations to match the market prices $\hat{C}(K_{i1},T_i)$, $\hat{C}(K_{i2},T_i)$ and $\hat{C}(K_{i3},T_i)$. The three unknowns of the system are $\lambda(T_i)$, $\nu(T_i)$ and $\rho(T_i)$. The mean reversion, a(t), still has to be an input. There are again several efficient algorithms to solve such systems, such as the Newton- Raphson or Broyden (a generalisation of the secant method – see Broyden, 1965, or Press *et al.*, p. 389), and embedded dichotomy (ie solving one equation under the constraint of others), etc.

Note that one may add some statistical constraints, such as the level of some correlation or volatility, while also adding the same number of unknowns – that is, of parameter functions to be calibrated. Usually, one matches the number of unknowns to the number of constraints (or vice versa). Nevertheless, it is possible to have more constraints than unknowns and seek at each step an approximate solution by minimising a least-square criterion. Conversely, one can have more unknowns than constraints and remove the problem degeneracy (families of solutions) by adding penalising terms to the criterion.

MONTE CARLO IMPLEMENTATION

The bootstrap calibration is particularly adapted to Monte Carlo models. One just needs to keep track of the values of the market variables for each path at date T_i. Then the market value of T_{i+1} options is obtained by adjusting the diffusion parameters on the interval $[T_i,T_{i+1}]$. We then only need to compute options of maturity T_{i+1} and, moreover, we only have to diffuse variables over a short period of time.

INSTABILITY

There is, however, a major drawback to this method. It concerns stability. Suppose that there is an error in parameter values at the first date T_1 - for instance, too high a volatility. Note that such an error may come from the Monte Carlo diffusion, but it could also be an uncertainty on a market price (bid/offer) or simply a rounding error. As variables at date T_1 have a large variance, one needs to input a lower volatility in the second period, $[T_1,T_2]$, to get correct prices at date T_2. If the sensitivity of option prices to this local volatility (the local vega) is not strong enough, the compensation will have to be bigger than the error at the previous step, typically by a multiplicative factor $\mu > 1$. After n time steps the compensation is multiplied by a factor μ^n (with alternative signs) and calibration may become impossible because it would seek negative volatilities.

At the ith step μ is the ratio between partial derivatives $\partial C_{i+1}/\partial\lambda_i$ and $\partial C_{i+1}/\partial\lambda_{i+1}$. This ratio exceeds one when time step i is shorter than time step $i-1$, or when some "elasticity" in the model lowers the effect of parameters on option prices. This is the case in the stochastic volatility model when the mean reversion, a, is low and the volatility of volatility, v, is high. Huge variations in the parameter λ are required to produce small changes in prices. The butterfly flap is the initial error, the hurricane happens on λ.

The linear interpolation of parameters with respect to time is perhaps more aesthetic, but it is exactly twice as unstable as piecewise constant parameters because the "compensation" has to be doubled for the same effect. Avoiding instability is not an easy task. We now show how to reduce the oscillation by accepting a non-exact calibration.

STABILISATION

Oscillations can be identified as an excessive variation of parameters from one period to the next. By penalising this variation, one can find a trade-off between the accuracy of calibration and the quadratic variation of parameters (in mathematical terms, the Sobolev H^1-norm of the parameters with respect to time). Returning to the section on one-dimensional bootstrapping, suppose that we do not want the series $\lambda_1,\ldots,\lambda_n$ to oscillate too much. One first calibrates the value of λ_1 in the usual way. Then, instead of solving at each step i the equation $C(K_i,T_i,\lambda_i) = \hat{C}(K_i,T_i)$, we seek that value of λ_i which minimises the criterion

$$\Phi_c(\lambda_i) = (C(K_i,T_i,\lambda_i) - \hat{C}(K_i,T_i))^2 + c(\lambda_i - \lambda_{i-1})^2$$

where c is a constant. The bigger the constant c, the more regular the series of λ_is but also the less accurate the calibration. If, in the opposite extreme case $c = 0$, we fall into the previous situation of an exact calibration but with no control of the oscillations. This methodology is easily extended to the multidimensional case, where the penalisation criterion involves the sum of the squared variations of each coefficient. Note that this regularisation procedure is achieved at almost no extra cost in terms of computation time.

The weakness of this approach, compared to the global one described below, is that if the first parameter λ_1 is misplaced (because, for instance, the first market price is erroneously quoted), or simply if the series has a strong slope at the beginning, then all subsequent parameters λ_i will be shifted in the same direction, thus creating a systematic error. Moreover, one will end with a higher mean square error.

It is possible to apply this technique to control higher-order Sobolev norms, H^s, of the parameters with respect to time. Assume for example that $s = 2$. One must calibrate the first two options in the usual way. Then, at step $i \geq 3$, one is led to reconsider the value of λ_{i-1} and minimise with respect to the pair $(\lambda_{i-1},\lambda_i)$ the criterion

$$\begin{aligned}\Psi_c(\lambda_{i-1},\lambda_i) = {} & (C(K_{i-1},T_{i-1},\lambda_{i-1}) - \hat{C}(K_{i-1},T_{i-1}))^2 \\ & + (C(K_i,T_i,\lambda_{i-1},\lambda_i) - \hat{C}(K_i,T_i))^2 \\ & + c(\lambda_i - 2\lambda_{i-1} + \lambda_{i-2})^2\end{aligned}$$

Again, this algorithm is quite fast, though not as fast as when $s = 1$ (it requires a multidimensional minimisation even in the one-dimensional case), but the mean square error is not much higher than with a global calibration procedure.

Implied calibration: global approach

GENERAL IDEA

Global calibration is rather easy to explain, although its implementation is an art and requires, first, a close geometrical analysis of the model. The idea is to build a "black box" which, for any set of parameter[4] values $(\alpha_1,\ldots,\alpha_p)$, provides the model price of any calibrating option:

$$(\alpha_1,\ldots,\alpha_p) \mapsto F_i(\alpha_1,\ldots,\alpha_p) \quad i = 1,\ldots,n$$

Let P_i be the market price of the ith option and define the least-square criterion:

$$\Psi(\alpha_1,\ldots,\alpha_p) = \sum_{i=1}^{n} w_i(F_i(\alpha_1,\ldots,\alpha_p) - P_i)^2$$

where the coefficients w_i are positive weights. Several numerical algorithms can be used to find the minimum of the function Ψ, subject to possible constraints on the parameters. The weights are very important to ensure a uniform precision of the result over the various parameters.

This technique is extremely flexible. The Monte Carlo framework allows easy computation of market variable statistics (variance, correlation of variables, etc). One may add corresponding terms to the LSC in order to impose some target value on those statistics. It is also possible to penalise the use of some parameters by adding terms of the form α_i^2 for undesired indexes i. Suppose for instance that one of the functions one wants to calibrate is given in a polynomial form. Instead of limiting the degree of the polynomial, which could make the model provide imprecise results in certain market situations, one can allow a higher degree but penalise the use of high-degree coefficients.

Generally speaking, there is no reason why Ψ should be convex. In particular, there is nothing to prevent the existence of local minima that could mislead the minimisation algorithm. A geometrical analysis of the behaviour of the criterion is also useful to overcome problems due to restricted parameter domains (constraints) or almost flat areas (which make optimisers unstable). Solutions are to change variables to avoid constraints and flatness and the incorporation of extra regularisation terms to avoid local minima, multiple solutions and also flatness.

The stability of the solution – that is, the continuity of the minimum point with respect to inputs (market prices) – is ensured by the non-degeneracy of the Hessian (ie its positive definiteness).

In this approach, the number of options, n, and the number of parameters, p, need not be equal. However, the total number of terms of the LSC, including penalisations and regularising terms, should exceed p; if not it is likely that the Hessian criterion will be degenerate, or nearly so.

OPTIMISATION ALGORITHMS

At this stage, there is nothing specific to Monte Carlo models except the computation of statistics. Let us list below the standard optimisation algorithms (see Press, *et al.*, 1996, chapters 10 and 15):

- steepest descent: $x_{n+1} = x_n - \lambda \text{ grad } \Psi(x_n)$, where λ is the gain;
- Newton method: $x_{n+1} = x_n - (\text{Hess } \Psi(x_n))^{-1} \times \text{grad } \Psi(x_n)$ seeks a zero of the gradient (not necessarily a minimum);
- Levenberg–Marquardt: a mixture of both; descent is ensured;
- Powell: a steepest descent with optimal gain;
- Davidson–Fletcher–Powell and Broyden–Fletcher–Goldfarb–Shanno: a pseudo-Newton method which is a clever adaptation of the secant method to several dimensions; and
- simulated annealing: to step out from local minima.

Conjugate gradient and the Generalised Minimum Residual method are specific to quadratic minimisation and cannot be applied here. Similarly, the simplex method is well suited to constrained linear problems.

In fact, due to the general behaviour of option prices, the plot of the LSC usually has the shape of a "child's cube", that is, wide flat areas and round angles. These kinds of shapes are a nightmare for optimisers. One of the most efficient techniques for overcoming the difficulty is a general multidimensional algorithm, such as Newton or BFGS, coupled with a line search algorithm, which minimises the criterion along a given line (or half-line). There is a whole literature on the optimal precision of the line search and on different ways of ensuring a minimum descent at each step.

The drawback of the Newton method is that it requires the computation of the Hessian. Levenberg–Marquardt propose an approximation, specific to LSC, that does not take longer to compute than the gradient and is close to the real Hessian if we approach the solution (see Press *et al.*, 1996, p. 687 or Marquardt, 1963). Moreover, it is always positive definite (unlike the real Hessian). It consists of writing

$$\frac{d^2}{dx^2}(f^2) = 2(ff'' + f'^2)$$

and of ignoring the first term, so that only the first derivative is necessary. If we are close to a solution, f is small and, if f'' is bounded, the first term is small with respect to the second.

MONTE CARLO IMPLEMENTATION

Let us first make an observation about the computation of the LSC. Options should be sorted by maturity so that each new "slice" does not require rediffusion of the variables from the beginning but only on the interval $[T_i, T_{i+1}]$.

The usual remarks on the computation of sensitivities in the Monte Carlo setting apply here. The gradient and the Hessian (if needed) are computed by adjusting the parameters and launching a new Monte Carlo with the same seed (if the seed is different, the result will not even be close to the real value). One may also use Malliavin calculus, which consists of differentiating the probability measure on the Wiener set of paths (see Mikhailov, 1995). In fact this makes no improvement for the gradient, for each partial derivative would need the computation of an expectation, which is not less than what is required by finite difference methods. Nonetheless, it appears to be an efficient method for computing Hessians.

ADAPTIVE NUMBER OF PATHS

Calibration is an iterative process which requires option computations at each iteration step. In this section, we show how to distribute the Monte Carlo efforts across the steps. Indeed, high precision is useless at the beginning.

It is well known that the accuracy of an expectation computed by means of a Monte Carlo simulation is of the order of $c_0/\sqrt{N}$, where N is the number of paths. Variance reduction techniques may improve the constant c_0 but, in general, not the order of convergence.[5] The gradient and the Hessian behave in the same way. Their errors are also of the order of $c_1/\sqrt{N}$ and $c_2/\sqrt{N}$, with constants c_1 and c_2 that might differ. Note that if p is the total number of parameters to calibrate (that is, the dimension of the optimisation problem), the number of first derivatives is p and that of second derivatives is $p(p+1)/2$.

Assume constants c_0, c_1 and c_2 are known. Then one can more or less control the size of the error as well as the computation time, which is strictly proportional to N. We here show how to optimise the number of paths at each iteration step, knowing that it is useless to have too high an accuracy when we are far from the solution. *A priori*, the rate of convergence of a pseudo-Newton method is super-exponential: the nth step error is of the order of $\exp(-\lambda\varphi^n)$, where $\varphi = 2$ for the true Newton method and φ is the golden ratio $(1+\sqrt{5})/2$ for secant-type methods, such as BFGS. However, this is only valid close to the minimum, where, in any case, one uses the maximum number of steps. When one is further away experience shows that the rate of convergence is simply exponential. Let $\tau(\varepsilon_1,\varepsilon_2)$ be the minimum number of steps from the distance ε_1 to the distance ε_2 from the solution, with $\varepsilon_1 > \varepsilon_2$. On the one hand, $\tau(\varepsilon_1,\varepsilon_2) \sim \gamma\log(\varepsilon_1/\varepsilon_2)$. On the other hand, the required number of paths is not larger than c/ε_2^2. Let us define an optimal sequence $\varepsilon_1,\ldots,\varepsilon_k,\ldots$. The computing time required by this sequence (which we suppose is finite) is of the order of

$$T = \gamma c \sum_k \frac{\log(\varepsilon_{k-1}/\varepsilon_k)}{\varepsilon_k^2}$$

One has

$$\frac{\partial T}{\partial \varepsilon_k} = \gamma c \left(\frac{1}{\varepsilon_k\,\varepsilon_{k+1}^2} - \frac{1-2\log\varepsilon_k}{\varepsilon_k^3}\right)$$

Hence, the sequence is optimal if

$$\varepsilon_{k+1} = \frac{\varepsilon_k}{\sqrt{1-2\log\varepsilon_k}}$$

which is equivalent to

$$N_{k+1} = N_k\left(1 + \log\frac{N_k}{c}\right)$$

One should take care that N_k is not the number of paths attached to the kth step. The idea is that we set the optimal sequence, (ε_k); then, as long as the error (which can be estimated from the gradient and the Hessian, or an approximation of it) is between ε_k and ε_{k+1}, one launches only N_k paths. When we reach the distance ε_{k+1}, it is time to launch N_{k+1} paths, etc.

The sequence (N_k) will be capped at a value N_{max} corresponding to the required accuracy of the calibration. Again, it is indeed useless to compute option prices to six digits accuracy when only two decimal places of market prices will be matched.

One may adapt this computation of the optimal number of paths to a heuristically observed rate of convergence.

Drift correction

As mentioned in our discussion of the general setting of a Monte Carlo model, the drift of certain market variables is implied by arbitrage theory. Although it is not difficult to generate a set of Brownian increments with zero average (for instance, with antithetic noise), when these increments are input into the diffusion formulas, often non-linear, their average no longer exactly matches its theoretical value. To show this, let us consider the stochastic volatility model. Assume that the Monte Carlo diffusion procedure applies to the logarithm of the exchange rate and of the volatility:

$$\log X(t_{i+1}) = \log X(t_i) + (r^d - r^f - \tfrac{1}{2}\sigma(t_i)^2)\delta t + \sigma(t_i)\delta W$$
$$\log\sigma(t_{i+1}) = \log \sigma(t_i) + (b - a \log \sigma(t_i))\delta t + \nu\delta W'$$

where δt, δW and $\delta W'$ denote finite increments of time and Brownian motions. Up to a first-order error, one can check that

$$E(X(t_i)) = X(t_0) \exp((r^d - r^f)(t_i - t_0)) \tag{2}$$

for any time index i. This is the equation that ensures the put/call parity. Even a small error on this expectation can lead, for instance, to prices of in-the-money options below the intrinsic value. The fact that $\log X(t_i)$ has a correct expectation, and even a correct variance, does not imply that its exponential, $X(t_i)$, has the correct expectation. It is thus necessary to adjust the drift term $r^d - r^f - \frac{1}{2}\sigma(t_i)^2$ so that equation (2) is satisfied with accuracy.

In fact not only does the expectation at the origin of time, t_0, have to match the forward value, but all intermediate conditional expectations must also match the corresponding forward exchange rate:

$$\mathbf{E}_{t_i}(X(t_j)) = X(t_i) \exp((r^d - r^f)(t_j - t_i)) \quad t_i < t_j \tag{3}$$

where we denote by $\mathbf{E}_{t_i}(X(t_j))$ the conditional expectation of $X(t_j)$, conditional on the value of $X(t_i)$. The difficulty lies in the fact that different trajectories involve different values of $X(t_i)$, and the Monte Carlo simulation is unable to provide a group of trajectories starting from $X(t_i)$ at date t_i.

This example is rather simple. The theoretical drift term does not depend on $X(t_i)$ and one can assume that nor will the correction depend on $X(t_i)$. Let us set

$$X_{i+1} = X_i \exp(\mu\delta t + \sigma_i\delta W) \quad \mu \simeq r^d - r^f - \tfrac{1}{2}\sigma_i^2$$

One must have

$$\mathbf{E}_{t_i}(X(t_{i+1})) = X(t_i) \exp((r^d - r^f)\delta t)$$

which is equivalent to

$$\mathbf{E}_{t_i}(\exp(\sigma_i\delta W)) = \exp((r^d - r^f - \mu)\delta t) \tag{4}$$

This is a conditional expectation, which normally depends on market variables at date t_i. However, one can assume that μ differs from its theoretical value $r^d - r^f - \frac{1}{2}\sigma_i^2$ by a correction term that is independent of $X(t_i)$ and σ_i. Equation (4) will then be satisfied "on average" and the put/call parity will occur for any European option priced at date t_0. The Bayesian rule shows that this correction "on average" is equivalent to satisfying equation (2). In this example, the drift correction factor is given by an explicit calculation. In general, as we already start from a good approximation of the solution, the Newton–Raphson method provides a very precise expectation in just a few steps.

In certain interest rate models, such as the Brace–Gatarek–Musiela lognormal model (see Brace, Gatarek and Musiela, 1997), the drift term is rate-dependent. This is due to changes of forward-neutral probabilities. We now face the following problem. The equivalent of equation (3) for interest rates is

$$\mathbf{E}_t\left(\exp\left(-\int_t^T r(s)\,ds\right)\right) = B(t,T) \tag{5}$$

where $r(s)$ is the actual spot rate at date s (known only at this date) and $B(t,T)$ is the discount factor from date t to date T as known at date t. In their article, Brace, Gatarek and Musiela also propose approximating the drift value by its expectation at date t_0. Performing the same approximation in the Monte Carlo setting would mean making the same drift correction "on average". However, the rate-dependency of the drift is an essential feature of the process as this is nothing other than its mean reversion (positive or negative), and cancelling out mean reversion can incur serious errors in the final variance of rates and, hence, on option prices.

In this particular case, one knows that if rates were not stochastic there would also not be any mean reversion, but in a lognormal interest rate model the volatility vanishes with the rates. Therefore, one can assume that the drift correction is proportional to the rates. Again, this assumption, which is less rough than the previous one, leaves only one parameter to be adjusted, and that can be performed by matching expectations at date t_0.

When drift terms depend on market variables, one should take care to respect (at least "on average") the mean reversion. The general idea is to make an assumption that leaves only one degree of freedom per time and per expectation to match and adjust the level of drift on the value of expectations at date t_0, thus avoiding the problem of conditional expectations.

If such a reduction of the problem is impossible, we advise the use of higher-order approximations, such as the Runge–Kutta method (for the pure drift part) or the Milstein scheme (for the whole diffusion - see the section below on higher-order schemes). This will at least cancel the second-order systematic error.

Variance reduction

Some of the techniques presented below are discussed in other chapters in this volume. Nevertheless, we prefer to repeat these studies with a particular emphasis on calibration, so that we can compare their efficiency. Some techniques, which can be applied independently of others (such as low discrepancy, etc), will not be mentioned. Other techniques, such as importance sampling, are just not suited for calibration. Indeed, calibration requires the simulataneous computation of a series of options. This is incompatible with focusing trajectories towards a given strike (except in very special examples).

HIGHER-ORDER SCHEMES

Most of the time, Monte Carlo simulations are adopted either because of the dimension or because a closed-form formula for the pdf of market variables cannot be found. The basic Euler scheme consists of simulating the diffusion process (1) by simply adding the differential terms to the market variables:

$$x_i(t_{k+1}) = x_i(t_k) + \mu_i(t_k;x_1(t_k),\ldots,x_n(t_k))\delta t + \sum_{j=1}^{n} \sigma_{ij}(t_k;x_1(t_k),\ldots,x_n(t_k))\delta W_j$$

where $\delta t = t_{k+1} - t_k$ and where δW_j are pseudo-random numbers with zero mean and variance δt (there is no lack of generality in assuming no correlations).

This scheme leaves out a second-order error. Indeed, on the interval $[t_k,t_{k+1}]$ the values of variables x_i change, and this affects the levels of drift and volatility. For the drift part, the error is of the order of δt^2, but for the stochastic part the order of magnitude of δW_j is $\sqrt{\delta t}$ and the error becomes of the order of δt (although, on average, this is still a second-order error).

The principle of higher-order schemes is well described in Kloeden and Platen (1995). It consists of writing $x_i(t_{k+1})$ as a stochastic integral:

$$x_i(t_{k+1}) = x_i(t_k) + \int_{t_k}^{t_{k+1}} \mu_i(t;x_1(t),\ldots,x_n(t))dt + \sum_{j=1}^{n} \int_{t_k}^{t_{k+1}} \sigma_{ij}(t;x_1(t),\ldots,x_n(t))dW_j$$

Then, within the integral, each $x_i(t)$ is replaced by a value obtained from a lesser-order scheme. The values of the drift and volatility are approximated by a Taylor expansion. The second-order approximation is called a Milstein scheme (see Milstein, 1995). The following schemes are due to Pardoux and Talay (1985).

We show here the example of the Milstein scheme. One first picks $x_i(t)$ from the Euler scheme:

$$x_i(t) = x_i(t_k) + \mu_i(t;x_1(t_k),\ldots,x_n(t_k))(t - t_k) + \sum_{j=1}^{n} \sigma_{ij}(t;x_1(t_k),\ldots,x_n(t_k))(W_j(t) - W_j(t_k))$$

then performs a first-order Taylor expansion of μ_i:

$$\mu_i(t;x_1(t),\ldots,x_n(t)) = \mu_i(t;x_1(t_k),\ldots,x_n(t_k)) + \sum_{j=1}^{n} \frac{\partial \mu_i}{\partial x_j}(x_j(t) - x_j(t_k))$$

and does the same for volatilities σ_{ij}. In the end, we are left with a combination of integrals of one of the following kinds:

$$\int_{t_k}^{t_{k+1}} f(t)\,dt$$

$$\int_{t_k}^{t_{k+1}} f(t)\,dW_i$$

$$\int_{t_k}^{t_{k+1}} f(t)W_i(t)\,dt$$

$$\int_{t_k}^{t_{k+1}} (W_i(t) - W_i(t_k))\,dW_i$$

$$\int_{t_k}^{t_{k+1}} (W_i(t) - W_i(t_k))\,dW_j$$

The first integral is deterministic. The second and third are Gaussian variables. The computation of their variance and covariance is left to the reader. For the fourth, we observe that

$$d(\tfrac{1}{2}W^2) = dt + WdW$$

hence

$$\int_{t_k}^{t_{k+1}} (W_i(t) - W_i(t_k))dW_i$$

$$= \tfrac{1}{2}(W_i(t_{k+1}) - W_i(t_k))^2 - t_{k+1} + t_k$$

The last integral is the most difficult to simulate. It can be seen as the product of two independent variables. For each choice of a trajectory for W_i, the stochastic integral is a Gaussian variable with variance

$$V_i = \int_{t_k}^{t_{k+1}} (W_i(t) - W_i(t_k))^2 dt$$

One now has to make this choice in accordance with the previous integrals. A detailed computation of the joint distribution of these stochastic integrals is given in Milstein's book (Milstein, 1995). Note that sampling W_i on the interval $[t_k, t_{k+1}]$ is a solution that has to be compared, in terms of numerical complexity, to a simple Euler scheme with a shorter time step.

VARIANCE AND COVARIANCE CORRECTION

According to its definition, a Brownian process is a "martingale with independent increments". When one launches a simulation, the actual expectation and variance need not be equal to their theoretical values. The standard deviation of the expectation of δW_t is $\delta t/\sqrt{N}$ whereas that of the variance is $\delta t^2\sqrt{2/N}$. The expectation can be made rigorously zero by antithetic sampling and will anyway be adjusted by the drift correction (see the earlier section on drift correction). In order to correct the variance, one could also compute the actual variance of the sample and resize it to match the theoretical value. However, this is not sufficient and one has to take care of correlations. Let

$$W(t_k) = \sum_{i=1}^{k} \delta W(t_i) \qquad \delta W(t_i) = W(t_i) - W(t_{i-1})$$

The variance of the "integrated" Brownian $W(t_k)$ is given by

$$\mathrm{Var}W(t_k) = \sum_{i=1}^{k} \mathrm{Var}\delta W(t_i) + \sum_{i,j=1}^{k} \mathrm{Cov}(\delta W(t_i), \delta W(t_j))$$

Even if the increments $\delta W(t_i)$ have a correct variance, the global variance of $W(t_k)$ will be corrupted if they are not independent.

Similarly, in multi-variable models it is often the case that an asset has a price which is given by a combination of various market variables. Moreover, the diffusion of each market variable is a combination of supposedly uncorrelated Brownian motions. If this is not the case, the global variance of market variables and asset prices will be incorrect, as will be the price of options.

To reduce this error, one can first launch all the increments that will be needed in the simulation. This is a (rather big) three-dimensional array, $\delta W_i(t_j, k)$, where k is the index of the Monte Carlo path. Then one computes its covariance matrix, the size of which is the total number of increments – that is, the product pm, where p is the number of Brownian motions and m is the number of sample dates:

$$\Gamma = (\gamma_{(i,j)(i',j')}) \text{ where } \gamma_{(i,j)(i',j')} = \mathrm{Cov}(\delta W_i(t_j), \delta W_{i'(t_{j'})})$$

The matrix Γ is (hopefully) positive definite,[6] so the Gram-Schmidt orthonormalisation algorithm[7] provides a triangular matrix S such that

$${}^tS\Gamma S = \delta t.\mathrm{Id}$$

Applying this matrix to the global set of increments provides a set of truly independent increments with exact variance δt.

The main problems are computation time and accuracy. The numerical complexity of the orthonormalisation is proportional to the square of the matrix size – in particular, to the square of the number of dates. Moreover, Gram-Schmidt orthonormalisation becomes very imprecise when the dimension exceeds 30 (as in a matrix inversion by the standard Gauss method).[8] A four-dimensional model could not afford more than six sample dates!

One way to pull complexity back to order 1 with respect to the number of dates and reduce the matrix size to twice the number of Brownian motions is to relax useless covariance checking. In practice, what we need is that independent Brownian motions be truly independent and that the variance of $W_i(t_k)$ be really equal to t_k. It then suffices to check that $\delta W_i(t_k)$ and $\delta W_j(t_k)$ be orthogonal if $i \neq j$ and that each $\delta W_i(t_k)$ be orthogonal to all $W_j(t_{k-1})$, including $W_i(t_{k-1})$. In this case, one has

$$\begin{aligned}\mathrm{Var}W_i(t_k) &= \mathrm{Var}W_i(t_{k-1}) + \mathrm{Var}\delta W_i(t_k)\\ &\quad + \mathrm{Cov}(W_i(t_{k-1}),\delta W_i(t_k))\\ &= t_{k-1} + \delta t = t_k\end{aligned}$$

This is slightly less precise than checking all correlations, but the significant time saved allows a larger number of paths. For a given computational time, the accuracy is much better.

FAST GAUSSIAN SIMULATION

The advantage of simulating a Gaussian variable rather than any distribution with the correct mean and variance is that it reproduces the exact kernel of the heat equation (with constant coefficients) – that is, the distribution of a Brownian motion. Thanks to the central limit theorem, any distribution would converge (at least weakly) to the Gaussian one. In other words, like higher-order schemes, a Gaussian distribution allows longer time steps (it would allow a single time step if the coefficients were constant). Therefore, one must control the time spent in simulating the Gaussian distribution.

The usual random generators are based on Birkhoff ergodic theorem and provide a number which is uniformly distributed on the interval [0,1]. The two most common methods for simulating a Gaussian distribution are Box–Muller and the inverse normal cumulative function. The Box–Muller technique generates two independent Gaussian variables by a change into polar co-ordinates. It requires the computation of a logarithm, a square root and two trigonometric functions. The inverse normal cumulative technique relies on the fact that if x is uniformly distributed on [0,1], then $N^{-1}(x)$ is Gaussian, where

$$N(x) = \int_{-\infty}^{x} g(s)\,ds$$

$$g(s) = \frac{1}{\sqrt{2\pi}} \exp\left(-\frac{s^2}{2}\right)$$

There are various approximations of N and N^{-1} in closed form, which require computation of exponentials (for N) and logarithms and square roots (for N^{-1}). We here show a piecewise linear approximation of N and N^{-1} based on moment matching, so that the rate of convergence of the central limit theorem is maximised. This technique resembles Gauss–Legendre integration procedure. We first approach the Gaussian density $g(s)$ by a step function h_n, which we shall suppose even. Let $0 = \alpha_0 < \alpha_1 < \ldots < \alpha_n$ and $\lambda_1,\ldots,\lambda_n > 0$. We define the function h_n by

$$\begin{aligned} h_n(s) &= \lambda_i \qquad \text{if } s \in [\alpha_{i-1},\alpha_i) \quad i = 1,\ldots,n\\ h_n(s) &= 0 \qquad \text{if } s \geq \alpha_n\end{aligned}$$

The even moments of h are

$$\begin{aligned}M_{2k}(\vec{\alpha},\vec{\lambda}) &= \int_{-\alpha_n}^{\alpha_n} s^{2k}\, h_n(s)\,ds\\ &= \frac{2}{2k+1}\sum_{1}^{n} \lambda_i\,(\alpha_i^{2k+1} - \alpha_{i-1}^{2k+1})\end{aligned}$$

Those of g are

$$\bar{M}_{2k} = 1 \times \ldots \times (2k-1)$$

As we have $2n$ degrees of freedom, we can match moments until the order $4n - 2$ after solving a non-linear system of $2n$ equations:

$$M_{2k}(\vec{\alpha},\vec{\lambda}) = \bar{M}_{2k} \qquad k = 0,\ldots,2n-1$$

The cumulative distribution, N, is approximated by the primitive H_n of h_n:

$$H_n(x) = \int_{-\infty}^{x} h_n(s)\,ds$$

This is a piecewise linear function defined by its values at points α_i and $-\alpha_i$. It satisfies the same symmetry property as N:

$$H_n(-x) = 1 - H_n(x)$$

and one has

$$H_n(\alpha_i) = \mu_i \quad H_n(-\alpha_i) = 1 - \mu_i \quad i = 0,\ldots,n$$

where

$$\mu_0 = \frac{1}{2} \quad \mu_i = \frac{1}{2} + \sum_{j=1}^{i} \lambda_j(\alpha_j - \alpha_{j-1}) \quad i = 1,\ldots,n$$

It is obvious that

$$\begin{aligned}0 &= 1 - \mu_n < \ldots < 1 - \mu_1 < \mu_0\\ &= \tfrac{1}{2} < \mu_1 < \ldots < \mu_n = 1\end{aligned}$$

The reciprocal function H_n^{-1} is also a piecewise linear function defined on the interval [0,1] by its values at points μ_i and $1 - \mu_i$, $i = 1,\ldots,n$:

$$H_n^{-1}(\mu_i) = \alpha_i \quad H_n^{-1}(1 - \mu_i) = -\alpha_i$$

Note that the linear interpolation between $1 - \mu_1$ and μ_1 matches the value $H_n^{-1}(\frac{1}{2}) = 0$.

PANEL 1

ON THE ROAD TO INFINITE DIMENSIONS

Financial models are more and more complex and, with the increasing power of computers, involve more and more dimensions, both in the number of market variables and Brownian motions and in the number of parameters. This increasing complexity is mainly due to attempts to reproduce functional spaces which are, effectively, infinite dimensional. Functional space approximation by finite subspaces (so-called *Galerkin subspaces*) is the topic of a large literature. It is not possible to summarise this topic in a few lines, and we shall only make a couple of observations.

❑ To simulate the behaviour of a fluid, one may individualise each molecule and thus obtain a "finite" dimensional model. The quotes on the word "finite" come from the fact that the dimension, several billions of billions, is completely out of reach of any computer. It is much more efficient first to model the fluid with continuous functions (temperature, pressure, currents, etc) and then to discretise the Navier–Stokes equation. The same holds in financial modelling. Although the complexity is not as high, it is often better to consider a large class of assets (for instance interest rate contracts or options on one given underlying) as a continuous, hence infinite, family, and then consider this family as a mathematical object that will be discretised for numerical purposes.

❑ Smoothness is an essential issue in approximation theory. Indeed, one knows that the speed of convergence of Fourier series is directly linked to the function smoothness. Financial data are often provided with a given number of decimal places ("tick data"), and this creates an artificial noise which in fact has no financial meaning. It is always useful to smooth data (yield curves, implied volatility surfaces) to recover that "portion of a tick" which has been destroyed by rounding. There are two main ways of smoothing data: convolution with a "bump" function and projection on to a Galerkin subspace spanned by smooth functions, or a mixture of both.

When performing any kind of calibration, one should always check its infinite dimensional limit. If the problem is unsolvable at the limit, it is likely that numerical results will be very unstable and, in practice, unusable. Let us give an example in which infinity depends on the number of different traded assets: the famous Black–Karasinski lognormal model of interest rates (Black *et al.* 1991). As observed by Musiela and Rutkowski (1998) and Goldys, Musiela and Sondermann (1994), if the spot rate is strictly lognormal then, because of the Itô term, it will take an infinite value with a positive probability. The practical implementation does not face this problem because of its time step. Instead of the pure spot rate, the rate whose tenor is equal to the time step is the lognormal variable, making the model rather of the BGM type. In fact, only bonds with a maturity that is a multiple of the time step can be priced by the model. Their number tends to infinity as the time step tends to zero. As the diffusion depends highly on the time step, so will the price of options. By a feedback effect, the parameters provided by the calibration will also depend strongly on the time step. The relation between standard and exotic options and the corresponding hedge ratios will not be reliable.

In a general infinite dimensional functional setting, the calibrating set consists of one or several functions $F_1 \in \mathcal{F}_1,\ldots,F_m \in \mathcal{F}_m$, depending on variables $(\xi_1,\ldots,\xi_q) \in \mathbb{R}^q$ such as strike and maturity. In our stochastic volatility model, these are the function $C(K,T)$ and the term structure of forward prices, $F(T)$, of the underlying at date T. Parameters to calibrate are, again, one or several functions $f_1 \in \mathcal{E}_1,\ldots,f_n \in \mathcal{E}_n$ depending on variables $(x_1,\ldots,x_p) \in \mathbb{R}^p$. Consider the operator

$$\Phi\colon \mathcal{E} = \mathcal{E}_1 \times \ldots \times \mathcal{E}_n \to \mathcal{F} = \mathcal{F}_1 \times \ldots \times \mathcal{F}_m$$
$$(f_1,\ldots,f_n) \mapsto (F_1,\ldots,F_m)$$

Two problems may occur. First, if Φ is not injective, then there will be several solutions (if any). We call such a model *truly underspecified*. Even if the numerical procedure provides only one solution, this choice can change drastically after a small change in the parameters (as in catastrophe theory). Stability at least requires uniqueness of the solution. A typical case where injectivity cannot occur is when $p > q$. For instance, any function of two variables can be displayed as an infinite sum of products of functions of one variable:

$$f(x,y) = \sum_{i=1}^{\infty} \phi_i(x)\psi_i(y)$$

but finite sums do not span the whole space of functions of two variables. Even in finite dimensional approximations, m has to be substantially larger than n and, usually, the market is not that rich in liquid assets. In other words, arbitrage theory will never avoid bets to tarders, or the "absence of arbitrage opportunity" hypothesis would be violated.

The second problem is more vicious. The operator Φ could be injective, but, as we deal with infinite dimensions, its reciprocal Φ^{-1} may not be continuous. This is the definition of an *ill-posed* problem. Let us give as an example Rubinstein's "implied binomial tree" (see Rubinstein, 1995). It consists of finding the value of the local volatility, $\sigma(S,t)$, from that of options C(K,T) assuming that the drift is known by arbitrage.[1] Dupire (1995) observed that local volatilities are given by the forward Kolmogorov equation and gave a formula that depends on $\frac{\partial C}{\partial T}$ and $\frac{\partial^2 C}{\partial K^2}$.

If the function C is considered in the C^0 topology, the inverse operator is not continuous. For instance, if one slightly modifies the value of one option, while leaving the others unchanged, the effect of the local volatility can potentially be infinite.

This example is, however, "weakly" ill-posed: it becomes well posed in some stronger topologies. Some problems involving "deconvolution" are strongly ill-posed, such as finding the profile of a payoff from the value of the option with respect to the spot today.

The first problem requires more information, such as statistical data. The second can be solved by first smoothing the data and then adding a regularisation term to the LSC. Indeed in this situation the criterion is only weakly convex. By adding small but strongly convex terms, the minimum depends continuously on inputs. This technique is also valid for the first problem

Calibration will always be a trade-off between the adaptability of the model and the lack of information from the market.

[1] This is only valid for futures. For stocks and exchange rates, one cannot predict the behaviour of interest rates with respect to the underlying. In this case the model is truly underspecified.

Here are numerical values of α_i, λ_i and μ_i for n = 1, 2 and 3:

❑ n = 1:

$$\alpha_1 = \sqrt{3} \qquad \lambda_1 = \frac{1}{2\sqrt{3}} \qquad \mu_1 = 1$$

❑ n = 2:

$$\alpha_1 = 1.355626 \quad \lambda_1 = 0.333214 \quad \mu_1 = 0.951714$$
$$\alpha_2 = 2.856970 \quad \lambda_2 = 0.032162 \quad \mu_2 = 1$$

❑ n = 3:

$$\alpha_1 = 1.154406 \quad \lambda_1 = 0.354107 \quad \mu_1 = 0.906409$$
$$\alpha_2 = 2.366759 \quad \lambda_2 = 0.074851 \quad \mu_2 = 0.997155$$
$$\alpha_3 = 3.750439 \quad \lambda_3 = 0.002056 \quad \mu_3 = 1$$

Figure 1 shows the approximate inverse Gaussian cumulative normal, H_n^{-1}, for n = 1, 2 and 3.

CONTROL VARIATE APPROACH

Usually, control variate techniques are used to improve the accuracy of an option price computed through a Monte Carlo simulation by comparing it with that of another option for which a closed-form formula exists. One removes from the Monte Carlo price of the exotic the error that appears with respect to the closed form in the price of the standard option.

Calibration mostly deals with standard options. However, instead of comparing options, we shall compare models. The need for a Monte

1. Piecewise linear approximations of N^{-1} matching two, six and 10 moments

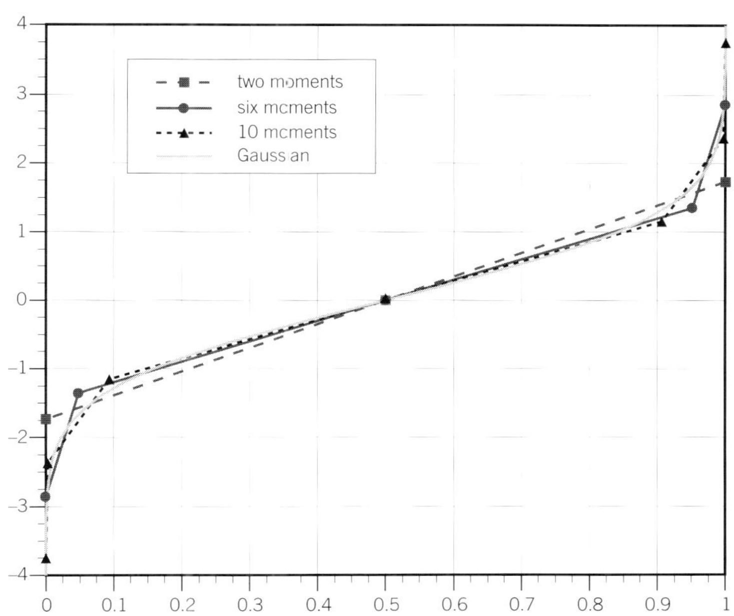

Carlo simulation arises when it is impossible to compute the exact kernel of the diffusion equation (ie the Green function of the process). When the model is just a perturbation of another simpler model, the kernel of which is computable, one can compute the Monte Carlo price of calibrating options in both models, then check the error on the simple model and remove it from the complex one.

2. Comparison of H¹-smoothing, H²-smoothing and global calibration: flat volatility term structure = 20%

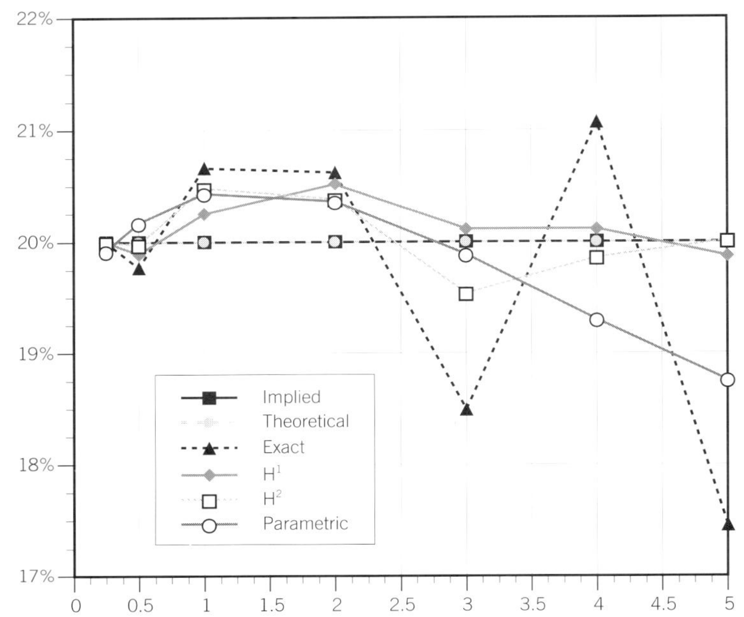

3. Comparison of H¹-smoothing, H²-smoothing and global calibration: increasing volatility term structure: 20% → 26%

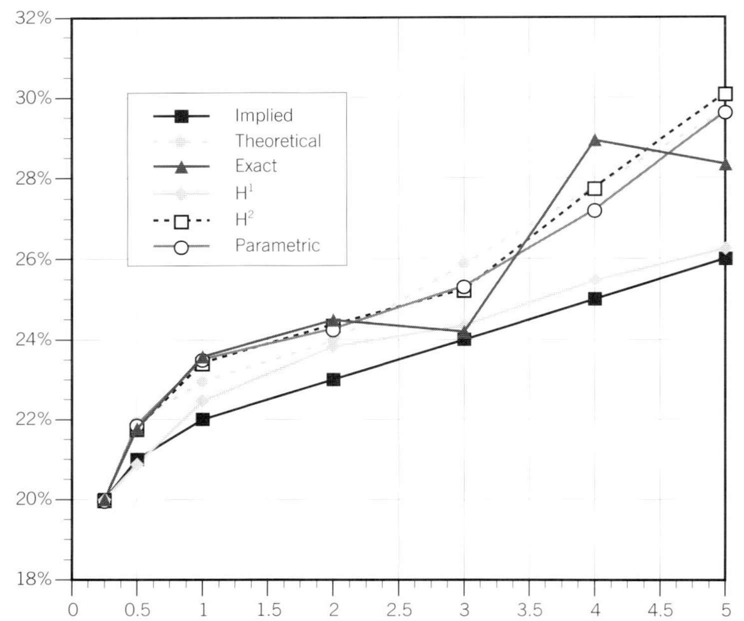

In the case of the stochastic volatility model, one can take as a simple base a standard Black–Scholes model in which the stochastic volatility has been replaced by its expectation.

PAYOFF SMOOTHING

Singularities of boundary conditions – especially of option payoffs – are responsible for a large fraction of the pricing error, both with Monte Carlo and with other techniques (trees, finite differences, etc). Their effect on sensitivities is even stronger. For European options, these singularities appear at the strike, which shows an angle. Other options, such as American, digitals or barriers can even have discontinuities at dividend or coupon dates, or at maturity (digitals and irregular barriers). These options are usually not used for calibration, so we shall focus on the angle at the strike.

The principle of payoff smoothing is to replace the very last step by a closed-form formula. Most of the time the payoff depends on only one of the variables. During the last step on a Euler scheme, diffusion parameters (drift and volatility) are known and do not change further. Hence the underlying follows a pure Brownian motion and it is possible to explicitly compute the value of the option at the one but final date as soon as all market variables are known at this date. This value depends smoothly on all variables, including the underlying.

In the stochastic volatility model, the last step – once the volatility is known – is nothing but a Black–Scholes model. Thus, one replaces the payoff of the option at date t_n by the Black–Scholes formula for an option of maturity $\delta t = t_n - t_{n-1}$.

The longer the last step, the better the smoothing but, also, the bigger the error. If the size of this step can be modified, its optimal value is a trade-off between these two issues. It is possible to stop several steps before the end when a higher-order approximation of the Green function can be found (see the earlier analysis of higher-order schemes).

Graphs

Figures 2–5 compare the bootstrap method and the parametric approach. The model is a standard Garman–Kohlagen foreign exchange diffusion:

$$\frac{dX}{X} = (r^d - r^f)dt + \sigma(t)dW$$

where we suppose that the domestic and foreign rates r^d and r^f are constant and that the volatility

$\sigma(t)$ is a deterministic function of time. We fix a series of maturities, $T_1,\ldots,T_n$, and of corresponding implied volatilities, $\bar\sigma_1,\ldots,\bar\sigma_n$. We know in this case that, for the model to match market prices, one must have, for every i:

$$\int_0^{T_i} \sigma(t)^2\,dt = T_i\,\bar\sigma_i^2$$

If we assume that $\sigma(t)$ is constant on each interval $[T_i,T_{i+1}]$, there is a unique solution to the calibration problem, given by

$$\sigma(t) = \sqrt{\frac{T_{i+1}\,\bar\sigma_{i+1}^2 - T_i\bar\sigma_i^2}{T_{i+1} - T_i}} \qquad t \in [T_i,T_{i+1}] \qquad (6)$$

We mean to choose an explicitly computable setting so that it is possible to compare Monte Carlo results with theoretical values.

In this experiment, we set

$$X_{i+1} = X_i \exp((r^d - r^f - \tfrac{1}{2}\sigma_i^2)(T_{i+1} - T_i) + \sigma_i\sqrt{T_{i+1} - T_i}z_i)$$

where the z_is are independent normalised Gaussian variables. In our example the calibrating set is composed of seven options with maturities: three months, six months, one year, two years, three years, four years and five years. The number of paths is 100 (deliberately low to enlarge instabilities), Brownian innovations have been orthonormalised and call values are computed by diffusing the exchange rate to the last but one date; then the last period is replaced by a closed-form formula (see the section on payoff smoothing). Gaussian variables are simulated by a piecewise linear approximation of N^{-1} matching 10 moments (see the section on fast Gaussian simulation). The parametric shape of σ_i is

$$\sigma_i = a\exp(-\alpha T_i) + b\exp(-\beta T_i) + c + dT_i$$

Optimisation is performed on the six variables a, b, c, d, α and β. The H^1-smoothing penalises $(\sigma_i - \sigma_{i-1})^2$ and the H^2-smoothing penalises $(\sigma_{i+1} - 2\sigma_i + \sigma_{i-1})^2$. The weights are chosen to provide reasonable results. The "theoretical" value is that given by equation (6), and its error is the pure Monte Carlo error. The "exact" value is given by the bootstrap calibration with no penalisation at all. By definition, it creates no calibration error (but some error of the order of magnitude of the theoretical error will appear if one changes the seed of the Monte Carlo sample).

We studied two examples, one where all implied volatilities, $\bar\sigma_i$, have the same value, 20%. In this case, the H^1-smoothing provides good results. In the second example, $\bar\sigma_i$ increases. As expected, the H^1-smoothing induces a systematic error because it has difficulty following the variation of the volatility. In both cases, we see that the H^2-smoothing does not give a higher relative error than the parametric approach although there are almost as many parameters as calibrating options.

4. Comparison of H^1-smoothing, H^2-smoothing and global calibration: relative error (flat volatility term structure)

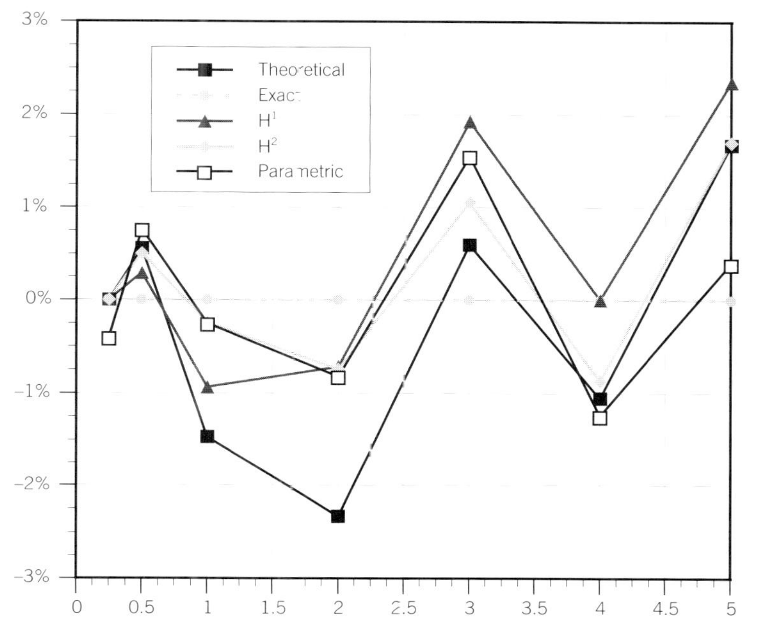

5. Comparison of H^1-smoothing, H^2-smoothing and global calibration: relative error (increasing volatility term structure)

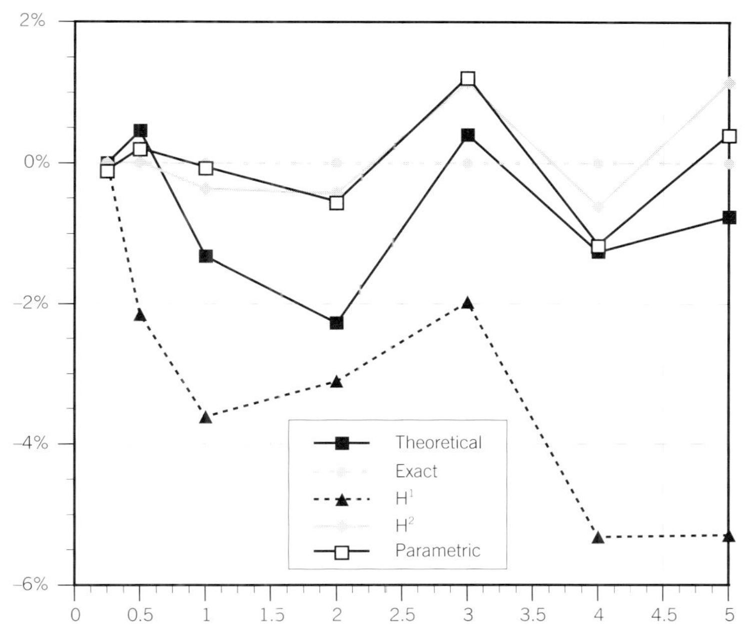

The author wishes to thank the research team of CIBC New York for implementing most of the techniques described in this chapter, and thus providing him with precious information on their practical use.

1 *The term "bootstrapping" is meant here in the sense that is used by market practitioners. This meaning has very little in common with the usual bootstrap sampling technique known in statistics.*

2 *This is a famous consequence of chaos theory. The imprecision in weather prediction grows exponentially with the length of the forecast. If the forecast horizon is too far away, an uncertainty in the data as small as that caused by a butterfly flicking its wings can lead, after running the model, to two different situations, one with a storm and the other without it.*

3 *This depends on whether the variable represents an asset price (stock, exchange rate, discount factor), in which case* μ_i *is given by arbitrage, or an observed, but not traded, quantity (spot interest rate, spot volatility), in which case* μ_i *can be customised.*

4 *We mean here model parameters, ie those which describe the behaviour of coefficient functions (drifts and volatilities) in a parametric or non-parametric shape.*

5 *Low-discrepancy series and other techniques, like UWAN, improve the order of convergence to better than* $\frac{1}{2}$.

6 *Definiteness requires that the sample size* N *is larger than the total number of increments.*

7 *A dual version of Cholesky decomposition, which directly creates the inverse of the Cholesky matrix.*

8 *If one really wishes to orthonormalise a huge matrix, we recommend a "singular value decomposition" (SVD), see Press* et al., *(1996, p. 66).*

BIBLIOGRAPHY

Press, W., S. Teukolsky, W. Vetterling and B. Flannery, 1996, *Numerical Recipes in C: The Art of Scientific Computing*, second edition, Cambridge, Cambridge University Press.

Brace, A., D. Gatarek and M. Musiela, 1997, "The Market Model of Interest Rate Dynamics", *Mathematical Finance* 7, pp. 127–54.

Black, F., E. Derman and W. Toy, 1990, "A One-Factor Model of Interest Rates and its Application to Treasury Bond Options", *Financial Analysts Journal* January–February, pp. 33–9.

Black, F., and P. Karasinski, 1991, "Bond and Option Pricing when Short Rates are Lognormal", *Financial Analysts Journal* July/August, pp. 52–9.

Broyden, C.G., 1965, *Mathematics of Computation* vol. 19, pp. 577–93.

Dupire, B., 1995, "A Unified Theory of Volatility", preprint, Paribas Capital Markets.

Goldys, B., M. Musiela and D. Sondermann, 1994, "Log-Normality of Rates and the Term Structure of Interest Rates", preprint, School of Mathematics, University of New South Wales, Sydney.

Kloeden, P., and E. Platen, 1995, *Numerical Solutions of Stochastic Differential Equations*, Berlin, Springer Verlag.

Marquardt, D.W., 1963, *SIAM Journal* 11, pp. 431–41.

Mikhailov, G.A., 1995, "New Monte Carlo Methods with Estimating Derivatives", Utrecht, VSP

Milstein, G.N., 1995, *The Numerical Integration of Stochastic Differential Equations*, Dordrecht, Kluwer Academic Press.

Musiela, M., and M. Rutkowski, 1998, *Arbitrage Pricing*, Berlin, Springer Verlag.

Pardoux, E., and D. Talay, 1985, "Discretization and Simulation of Stochastic Differential Equations", *Acta Appl. Math.* 3, pp. 23–47.

Rubinstein, M., 1995, "As Simple as One, Two, Three", *Risk* 8(1), pp. 44–7.

AMERICAN-STYLE

Introduction

Bruno Dupire

For a long time Monte Carlo simulations have been considered as a powerful tool for pricing options with complex payouts (path-dependent) or complex dynamics (stochastic volatility, multi-asset, etc). However, the nature of the path-by-path generation of Monte Carlo simulations seemed to make them unable to deal with American-style options, whose early exercise features requires knowledge of the option price at any intermediate date for any value of the spot. In fact, some structuring of the path generation makes it possible.

The interest in American Monte Carlo methods has been spurred by the Tilley paper on "Valuing American Options in a Path Simulation Model" (Chapter 14), which, appropriately, starts this section. His main idea is to try to mimic, in a Monte Carlo simulation, the dynamic programming procedure classically implemented in trees.

Dynamic programming proceeds backwards from the option maturity and compares, at each time step and for several values of the stock, the "live" value of the option with its "dead" value. The live value is the holding value, computed as the discounted expectation one step ahead conditioned by the current stock price. The dead value is the intrinsic value, corresponding to immediate exercise. The dynamic programming algorithm defines the option value as the greater of those two quantities, and carries on recursively backwards until the initial date when the desired option premium is collected.

The non-trivial step in a Monte Carlo approach is the computation of the conditional expectation, which requires knowledge of the option price for several paths emanating from the same conditioning value. As the paths are generated independently, at a given time step no two paths will generally lead to the same stock price. Tilley proposes a remedy through an approximation of the conditional expectation, by making use of neighbouring points. More precisely, he generates classical Monte Carlo paths and, at each date, reorders the stock prices and partitions them into bundles. Within a bundle, the holding value is constant, computed as the discounted average of the option value at the successors of this bundle one step ahead. The approximation of the option value by a piecewise constant function can easily be improved.

Indeed, the real interest of American Monte Carlo methods resides in American path-dependent options, such as American Asian or American look-backs. In these cases, the optimal exercise policy does not depend only on the current stock price but also on a second state variable, such as the current average or the current minimum. This requires a two-dimensional bucketing, which makes matters considerably more complex.

Chapter 15, by Barraquand and Martineau, is also oft-quoted. It deals with American options on several underlying assets, such as options on the maximum of several assets. They introduce a state variable which dictates the exercise policy. Unfortunately, it is only a sub-optimal policy, which induces a bias that does not disappear when the number of simulations increases to infinity. Moreover, this technique is not generic and the case considered is of limited practical application.

The last chapter of this section, by Broadie and Glasserman, is an excellent overview of the subject. They mainly review the first two chapters of this section and summarise two approaches of their own. The first one is the "simulated tree method", which is a random non-recombining multinomial tree scheme. The natural way to account for early exercise in such a tree is to proceed backward, and to take at each node the highest between the discounted expectation and the exercise value. However, the discrete expectation is random because the nodes are, while taking the maximum with the intrinsic value creates a positive convexity bias. They build a lower estimate which makes use only of a subset of daughter nodes to decide on the exercise. The sub-optimality of this exercise policy biases the estimate low. The problem of the whole approach, however, is that its complexity is exponential in the number of exercise dates.

The second approach, coined "stochastic mesh", contains this exponential explosion, by generating Monte Carlo paths and connecting each point of date n to each point of date $n + 1$. The connections are computed to reflect the conditional probabilities, and the pricing is performed in a backward sweep as usual. This method is promising, but its computational complexity is of the order of the square of the number of paths, which quickly becomes prohibitive.

Any approach to pricing American options requires the ability to value the option at any intermediate time. This means that the scheme must be able to estimate at each point the expected value of the option one step ahead. It is a difficult task in high dimension if one wishes to prevent the exponential proliferation of points (simulated trees) or full connectivity (stochastic mesh). The way for the future is probably in recombining the scheme, together with a good pruning methodology.

14

Valuing American Options in a Path-Simulation Model

James A. Tilley
Morgan Stanley

The goal of this paper is to dispel the prevailing belief that American-style options cannot be valued efficiently in a simulation model, and thus remove what has been considered a major impediment to the use of simulation models for valuing financial instruments. I present a general algorithm for estimating the value of American options on an underlying instrument or index for which the arbitrage-free probability distribution of paths through time can be simulated. The general algorithm is tested by an example for which the exact option premium can be determined.

Mathematicians seem to resort to simulation models to analyse a problem only when all other models fail to yield a solution. In financial economics, evidence of this tendency to avoid simulation models is found in the proliferation of published binomial and multinomial lattice solutions (or their equivalent) to the problem of valuing instruments with cashflows or payoffs contingent on interest rates or stock prices. The standard approach to valuing an American option is to utilise a one-factor (continuous) model of the stochastic price behaviour of the option's underlying asset, then create a binomial or multinomial (discrete) connected lattice representation of that stochastic process, and finally solve the valuation problem by backward induction on the lattice. Market-makers who deal in today's complicated financial instruments and investors who buy and sell them are beginning to sense a need for more realistic multifactor models of the stochastic dynamics of interest rates, foreign exchange rates, stock prices and commodity prices. These more complex models demand analysis by simulation because constructing approximate solutions (whether by means of lattices or otherwise) to the non-linear differential and integral equations to which they give rise is extremely difficult.

In general, the use of simulation models for valuing financial instruments has been restricted to assets that have path-dependent cashflows or payoffs, for example, mortgage-backed securities, including collateralised mortgage obligations (CMOs), and esoteric derivative instruments, such as "look-back" options (see Dyer and Jacob, 1991; Hull, 1989. (An exception, at least in the academic literature if not in practice, is the paper by Boyle (1977) – also Chapter 1 of this publication, which examines how Monte Carlo simulation can be used to value European-style options). Indeed, it has been thought that simulation models could not be used to value American-style options efficiently, if at all (Dyer and Jacob, 1991; Geske and Shastri, 1985; and Pedersen's discussion of Tilley, 1992). Ideally, a broker–dealer would like to be able to use a *single* method to value its entire book, and a financial intermediary would like to be able to use a *single* method to analyse its entire asset-liability condition. I believe that simulation models offer this possibility.

Simulation models consume large amounts of computer processing time, and some problems have heretofore required too much execution time to be handled practically by simulation. However, the arrival of powerful workstations, servers and parallel processors has rendered simulation feasible in many situations where it previously was not, a condition that can only improve with time as the pace of major technological advances accelerates. In many situations, a *single sample* of paths can be generated and then *used repeatedly* to value many different instruments, for example, a dealer's entire book of interest rate swaps, caps, floors and swaptions; a dealer's entire book of stock index derivatives or of currency swaps and options; or a financial intermediary's entire portfolio of fixed-income securities. Simulation may not be the best method when each financial instrument must be valued on the basis of its own random sample of paths, but this situation can often be avoided by designing the simulation properly.

The algorithm for valuing American options is described in the second section and tested by an example in the section after that. The issue of bias in the estimator of the option premium is then examined, after which the example is revisited in the next section. A summary and some concluding remarks are presented in the final section.

The valuation algorithm

A textbook by Cox and Rubinstein (1985) provides a comprehensive treatment of the subject of options. I assume that the reader is familiar with the general subject area, including various models for pricing options. For convenience, the option's underlying asset is referred to as a "stock," but the entire development in this section applies to any type of asset or index for which the arbitrage-free probability distribution of paths through time can be simulated. In an earlier paper (Tilley, 1992) I discussed what is meant by "arbitrage-free" and showed how arbitrage-free paths of interest rates can be sampled stochastically. The example in the third section of this chapter utilises paths of stock prices that are sampled from a probability distribution that is arbitrage-free because its mean has been adjusted appropriately.

Let us consider how to evaluate put and call options on a stock. The options are exercisable only at specified epochs $t_1, t_2, \ldots, t_N$, which are indexed $1,2,\ldots,N$ for convenience. The origin of time is $t = 0$, which is indexed as epoch 0. The options can be considered to be first exercisable at epoch 0 or at epoch 1, as appropriate. A path of stock prices is a sequence $S(0), S(1), S(2)\ldots, S(N)$, in which the arguments of S refer to the epoch indexes at which the stock prices occur. *All* paths of stock prices emanate from the initial stock price, $S(0)$. The simulation procedure involves the random generation of a finite sample of R such paths and the estimation of option prices from that sample. The kth path in the sample is represented by the sequence $S(0), S(k,1), S(k,2), \ldots, S(k,N)$, in which the first index refers to the path and the second index refers to the epoch. Two paths of stock prices are represented in Figure 1. Let $d(k,t)$ be the present value at epoch t on path k of a \$1 payment occurring at epoch $t+1$ on path k. Let $D(k,t)$ be the present value at epoch 0 of a \$1 payment

1. Two illustrative paths of stock prices sampled from a discrete-time continuous-state model of stock price movements

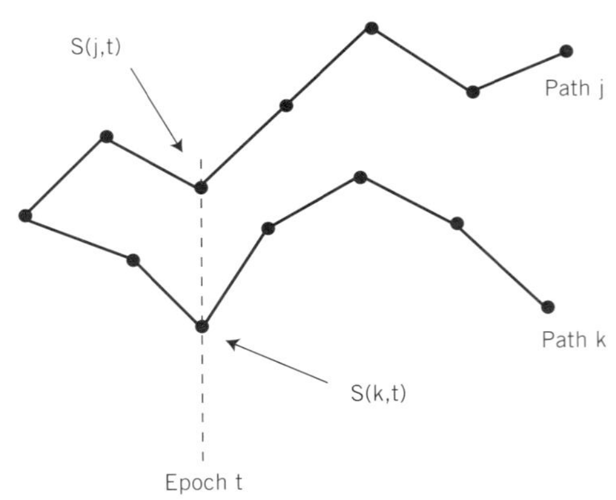

occurring at epoch t on path k, computed as the product of the discount factors $d(k,s)$ from $s = 0$ to $s = t-1$.

Assume that the option has strike prices that can depend on the date of exercise but not on the stock price at the time of exercise. Let $X(1), X(2),\ldots,X(N)$ denote the sequence of exercise prices at epochs $1,2,\ldots,N$, respectively. Typical stock options have a constant strike price X independent of date of exercise, but typical call options in private placement bonds do not. The intrinsic value, $I(k,t)$, of the option on path k at epoch t can now be found as

$$I(k,t) = \begin{cases} \max\,|0, S(k,t) - X(t)| & \text{for a call option} \\ \max\,|0, X(t) - S(k,t)| & \text{for a put option} \end{cases}$$

Finally, let $z(k,t)$ be the "exercise-or-hold" indicator variable, which takes the value 0 if the option is not exercised at epoch t on path k and which takes the value 1 if the option is exercised at epoch t on path k. Clearly, either $z(k,t) = 0$ at all epochs t along path k or $z(k,t_*) = 1$ at one and only one epoch t_* along path k. If such a t_* exists, it is the date at which the option is exercised on path k.

The price of any asset is known at epoch 0 if its cashflows are known at all epochs along all possible paths. That price is calculated in two steps: first, compute for each path k the present value at epoch 0 of the asset's cashflows along that path using the path-specific discount factors $D(k,t)$, and second, average across all paths the present values computed in the first step. The paths must be drawn from the appropriate arbitrage-free distribution. More details on this general valuation procedure can be found in Tilley

(1992). On a given stock price path, the "cashflow" for an option is 0 at every epoch other than the one at which the option is exercised. At exercise the option's "cashflow" is equal to its payoff, which is its intrinsic value. Assuming the usual situation that all randomly sampled stock price paths are equally likely with probability weight R^{-1}, we can express the option premium estimator by the following equation:

$$\text{Premium estimator} = R^{-1} \sum_{\substack{\text{all}\\ \text{paths}\\ k}} \sum_{\substack{\text{all}\\ \text{epochs}\\ t}} z(k,t)D(k,t)I(k,t)$$

Thus, to estimate the price of the option, we need to estimate the exercise-or-hold indicator function $z(k,t)$, given a finite sample of R paths drawn from an arbitrage-free distribution of paths. The algorithm presented in this section for estimating $z(k,t)$ mimics the standard backward induction algorithm implemented on a connected lattice for estimating the value of an American option. A discussion of this standard technique can be found in the textbook by Cox and Rubinstein (1985).

The backward induction is begun at the latest epoch at which the option can be exercised, that is, at its expiration date. On that date, represented by epoch N, the option, if it is still "alive" on path k (that is, not previously exercised), will be exercised if and only if $I(k,N) > 0$. The general step of the backward induction performed at an arbitrary epoch t involves determining whether it is optimal to hold the option for possible exercise beyond epoch t or to exercise the option immediately at epoch t. This decision is made by comparing the option's "holding value" to its "exercise value." The option's exercise value is equal to its intrinsic value and can be directly calculated for each path because the price of the underlying stock is known at each epoch on each path. The option's holding value on any path is calculated as the present value of the expected one-period-ahead option value.

Many believe that utilising the path structure illustrated in Figure 1 precludes estimation of an option's holding values because the only point from which *many* paths emerge is epoch 0. On any particular path, at any epoch $t > 0$, only a *single* path is simulated. One might think that many paths would need to be simulated from each such point to estimate closely the mathematical expectation of the one-period-ahead option value. Unfortunately, such an approach would lead to a multinomial "tree" in which the number of paths grows exponentially with the number of epochs – a computational infeasibility. Instead, computational feasibility can be achieved by utilising the path structure illustrated in Figure 1 and then estimating the option's holding value by means of a distinct partitioning at each epoch of the R paths into Q bundles of P paths each. The hope is that the paths within a given bundle are sufficiently alike that they can be considered to have the same expected one-period-ahead option value; in other words, Q must not be too small. The mathematical expectation of the one-period-ahead option value is estimated as an average over all the paths in the bundle. Thus, the estimate of the option holding value will be good only if there are sufficiently many paths in the bundle; in other words, P must not be too small.

In general, there is at least one bundle in which the decision for some paths is to hold the option, while the decision for the rest of the paths in the bundle is to exercise the option. Such a bundle generally has a "transition zone" in stock price from a decision to hold the option to a decision to exercise the option. Specifically, for a call option, there exist stock prices $S_L(t)$ and $S_U(t)$ at epoch t, with $S_L(t) < S_U(t)$, such that the decision is to hold for $S \leq S_L(t)$ and to exercise for $S \geq S_L(t) < S_U(t)$. However, for $S_L(t) < S < S_U(t)$, the decision is "inconsistent"; that is, there exist stock prices $S_t(t)$ and $S_u(t)$ such that $S_L(t) < S_l(t) < S_u(t) < S_U(t)$, yet the decision is to exercise at $S = S_l(t)$ and to hold at $S = S_u(t)$! The transition zone from an unambiguous hold decision to an unambiguous exercise decision often extends across several consecutive bundles. The algorithm can be refined to eliminate the transition zone by creating a "sharp boundary" at $S = S_*(t)$ such that the decision is to hold for $S < S_*(t)$ and to exercise for $S \geq S_*(t)$.

The general step that is performed at epoch t in the backward induction algorithm includes eight substeps, as follows:

1. Reorder the stock price paths by stock price, from lowest price to highest price for a call option or from highest price to lowest price for a put option. Re-index the paths from 1 to R according to the reordering.
2. For each path k, compute the intrinsic value $I(k,t)$ of the option.
3. Partition the set of R ordered paths into Q distinct bundles of P paths each. Assign the first P paths to the first bundle, the second P paths to the second bundle, and so on, and finally the last

P paths to the Qth bundle. We assume that P and Q are integer factors of R.

4. For each path k, the option's holding value, H(k,t), is computed as the following mathematical expectation taken over all paths in the bundle containing the path k:

$$H(k,t) = d(k,t)P^{-1} \sum_{\substack{\text{all } j \\ \text{in bundle} \\ \text{containing } k}} V(j,t+1)$$

The variable V(k,t) is fully defined in substep 8 below. At epoch N, V(k,N) = I(k,N) for all k.

5. For each path k, compare the holding value H(k,t) to the intrinsic value I(k,t), and decide "tentatively" whether to exercise the option or to hold it. Define an indicator variable, x(k,t) as follows:

$$x(k,t) = \begin{cases} 1 & \text{if } I(k,t) > H(k,t) \quad \text{Exercise} \\ 0 & \text{if } H(k,t) \geq I(k,t) \quad \text{Hold} \end{cases}$$

6. Examine the sequence of 0s and 1s {x(k,t); k = 1,2,...,R}. Determine a sharp boundary between the hold decision and the exercise decision as the start of the first string of 1s, the length of which exceeds the length of every subsequent string of 0s. Let $k_*(t)$ denote the path index (in the sample as *ordered* in substep 1 above) of the leading 1 in such a string. The transition zone between *hold* and *exercise* is defined as the sequence of 0s and 1s that begins with the first 1 and ends with the last 0. An example is given below:

Boundary
↓
0 0 ... 0 1 1 0 0 0 1 1 1 1 1 0 0 1 ... 1 1

7. Define a new exercise-or-hold indicator variable, y(k,t), that incorporates the sharp boundary as follows:

$$y(k,t) = \begin{cases} 1 & \text{for } k \geq k_*(t) \\ 0 & \text{for } k < k_*(t) \end{cases}$$

8. For each path k, define the current value, V(k,t), of the option as follows:

$$V(k,t) = \begin{cases} I(k,t) & \text{if } y(k,t) = 1 \\ H(k,t) & \text{if } y(k,t) = 0 \end{cases}$$

After the algorithm has been processed backwards from epoch N to epoch 1 (or epoch 0 if immediate exercise is permitted), the indicator variable z(k,t) for t ≤ N is estimated as follows:

$$z(k,t) = \begin{cases} 1 & \text{if } y(k,t) = 1 \text{ and } y(k,s) = 0 \text{ for all } s < t \\ 0 & \text{otherwise} \end{cases}$$

This completes the description of the algorithm for valuing an American option.

The partition of R paths into Q bundles or P paths each can be characterised by defining a "bundling parameter", α, by means of the equation $Q = R^{\alpha}$, and, therefore, $P = R^{1-\alpha}$. It is clear that $0 \leq \alpha \leq 1$. The value α = 0 corresponds to the partition into a single bundle of R paths, and the value α = 1 corresponds to the partition into R bundles of one path each. A particular American option valuation algorithm can now be fully described by the sample size, R, the technique used to sample paths, and the bundling parameter, α. If α is restricted to rational numbers, we can fix α and take sensible limits as $R \to \infty$ to investigate the convergence properties of the option premium estimator. For example, with α = 2/5, we can examine sample sizes equal to $2^5, 3^5, 4^5, \ldots$ paths for which we can study the estimators associated with the partitions $Q = 2^2, 3^2, 4^2, \ldots$ bundles and $P = 2^3, 3^3, 4^3, \ldots$ paths per bundle, respectively.

For any exercise–hold decision algorithm with α fixed and $0 < \alpha < 1$, it can be proved that the option premium estimate must converge to the proper result as $R \to \infty$. This follows from the observation that the algorithm for determining the exercise–hold decision variable is based on the standard backward induction algorithm for valuing American options and that all sources of error arise from P, Q and R being finite. For finite R, imprecision in the premium estimates arises because: the continuous distribution of stock prices at each epoch is not sampled finely enough; and the mathematical expectation in substep 4 above is approximated by an average over a finite number of paths. Imprecision of the first type can be reduced by increasing Q, the number of bundles. Imprecision of the second type can be reduced by increasing P, the number of paths per bundle. For fixed R, increasing Q means reducing P, and vice versa, implying a tradeoff between the first and second types of imprecision. However, if α is held constant at some value in the interval (0,1), then both types of imprecision are eliminated simultaneously as $R \to \infty$ because then *both* $Q \to \infty$ *and* $P \to \infty$.

The distinction between the variables y(k,t) and x(k,t) disappears as $R \to \infty$ and α is held constant at a value other than 0 or 1. As $R \to \infty$, the boundary between a decision to exercise the option and a decision to hold the option becomes sharper and sharper; that is, at each epoch the transition zone with alternating strings of 1s and 0s occurs over a smaller and smaller

Table 1. Bundling parameter, α, for various partitions of 5,040 paths

Partition			Partition		
Number of bundles	Paths per bundle	α	Number of bundles	Paths per bundle	α
1	5040	0.00000	72	70	0.50165
2	2520	0.08131	80	63	0.51401
3	1680	0.12887	84	60	0.51973
4	1260	0.16261	90	56	0.52783
5	1008	0.18879	105	48	0.54591
6	840	0.21017	112	45	0.55348
7	720	0.22825	120	42	0.56157
8	630	0.24392	126	40	0.56730
9	560	0.25773	140	36	0.57965
10	504	0.27009	144	35	0.58296
12	420	0.29148	168	30	0.60104
14	360	0.30956	180	28	0.60913
15	336	0.31765	210	24	0.62721
16	315	0.32522	240	21	0.64288
18	280	0.33904	252	20	0.64860
20	252	0.35140	280	18	0.66096
21	240	0.35712	315	16	0.67478
24	210	0.37279	336	15	0.68235
28	180	0.39087	360	14	0.69044
30	168	0.39896	420	12	0.70052
35	144	0.41704	504	10	0.72991
36	140	0.42035	560	9	0.74227
40	126	0.43270	630	8	0.75608
42	120	0.43843	720	7	0.77175
45	112	0.44652	840	6	0.78983
48	105	0.45409	1008	5	0.81121
56	90	0.47217	1260	4	0.83739
60	84	0.48027	1680	3	0.87113
63	80	0.48599	2520	2	0.91869
70	72	0.49835	5040	1	1.00000

interval of stock prices. Defining a sharp boundary by means of substep 6 above generally improves the convergence of the algorithm considerably for any α in the interval $(0,1)$ and also generally broadens considerably the interval of α over which the option premium estimates are good. In general, the option premium estimate based on a given sample size, sampling technique and bundling parameter is more accurate when a sharp exercise-hold boundary is determined than when it is not. However, the *ultimate* convergence of the exercise-hold decision algorithm to the exact option premium does *not* depend at all on whether substeps 6 and 7 above are implemented. If substeps 6 and 7 were omitted from the algorithm, $x(k,t)$ would be used in lieu of $y(k,t)$, both in substep 8 and in the calculation of $z(k,t)$.

An example

To test the algorithm presented in the preceding section, we consider the situation of a non-dividend-paying stock. Let $S(t)$ denote the price of the stock at time t. We assume that the random variable $\ln[S(t)/S(0)]$ is normally distributed with mean μt and variance $\sigma^2 t$. We further assume that the yield curve is flat and that interest rates are constant over time at an annual effective rate r. For the distribution of stock price movements to be arbitrage-free over time, it must be true that $\mu = \ln[1 + r] - \sigma^2/2$. Refer to the textbook by Cox and Rubinstein (1985) for a proof of this statement.

When a non-dividend-paying stock is the underlying asset, the price of an American call option must be exactly the same as the price of an otherwise identical European call option. The price of an American put option must be no less than the price of an otherwise identical European put option, but the former will in general exceed the latter (Cox and Rubinstein, 1985). Therefore, we test the valuation algorithm on a *put option* that is first exercisable in one quarter of a year and is exercisable every quarter of a year thereafter until its expiration in three years. The stock price has logarithmic volatility, σ, equal to 30%. The initial price of the stock, $S(0)$, is 40; the strike price, X, of the option is 45 at all epochs; and the annual effective interest rate, r, is 7%. Paths of stock price movements are generated randomly by stratified sampling of the standard normal density, as described in Tilley (1992). Random samples of size $7! = 5,040$ are used so that many different partitions can be examined. Table 1 lists the values of the bundling parameter α that correspond to each of the 60 different partitions of 5,040 paths into equal bundles.

2. Premium estimates for three-year American put option (5,040 paths partitioned 40 ways into exercise-decision bundles)

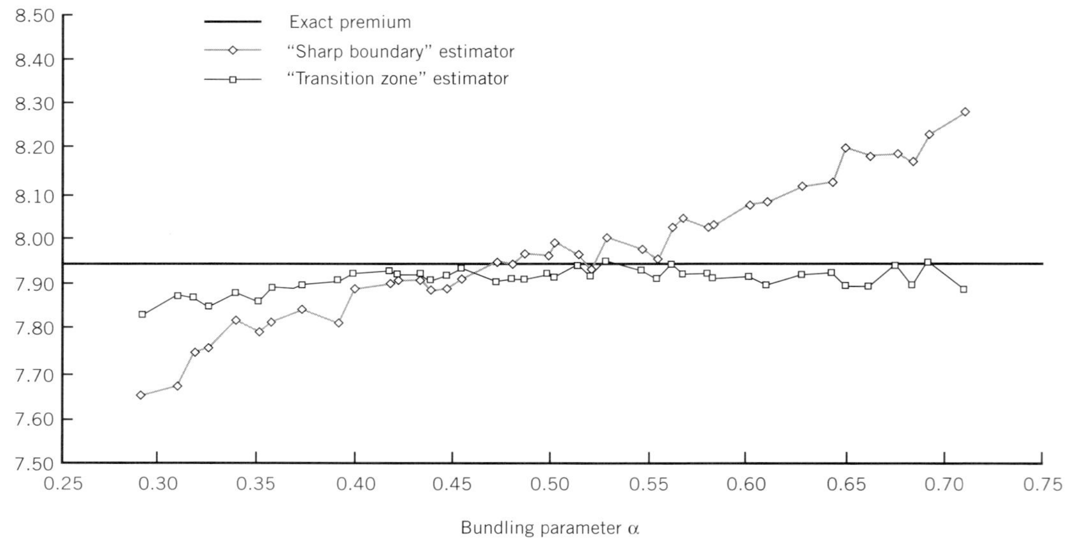

The exact price of the three-year American put option with quarterly exercise intervals was determined to be 7.941 by using a binomial lattice with 1,200 time periods constructed according to the procedure described in Cox and Rubinstein (1985). This is approximately 1.61 higher than the price of the corresponding three-year European put option. Using a *single* sample of 5,040 paths, the exercise-decision algorithm described in the preceding section was tested for all partitions having at least 12 bundles but no more than 420 bundles. The results are displayed in Figure 2.

In Figure 2, the solid line connecting the diamonds corresponds to application of the algorithm without substeps 6 and 7 – that is, with a *transition zone* from hold to exercise, not a sharp boundary between hold and exercise. The broken line connecting the squares corresponds to application of the algorithm with substeps 6 and 7 included – that is, with a *sharp boundary* between hold and exercise. The horizontal line across the graph at a vertical axis value of 7.941 marks the exact option premium. Figure 2 clearly demonstrates the importance of including substeps 6 and 7 in the algorithm. When a sharp boundary is determined the option premium estimates are essentially flat across an interval from $\alpha = 0.29$ to $\alpha = 0.71$ and cover a range of only 12 cents. However, when only a transition zone is utilised the option premium estimates rise more or less steadily as the bundling parameter is increase and cover a range of approximately 63 cents, more than five times the range obtained when a sharp boundary is utilised!

3. Frequency histogram for α = 0.50 premium estimator (based on 1,000 samples of 5,040 paths each)

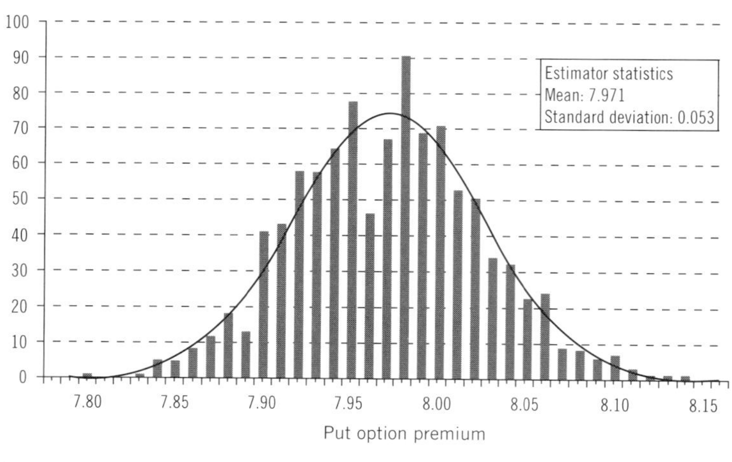

To study the efficiency of estimation, the "70 bundles by 72 paths per bundle" partition was used on 1,000 independent samples of 5,040 paths each. Each sample gives rise to an estimate of the put option premium. The frequency histogram of these 1,000 estimates is plotted in Figure 3. The mean of the estimates is 7.971 and the standard deviation of the estimates is 0.053. The solid line graph superimposed on the frequency histogram is that of a normal density function with the same mean and standard deviation as the option premium estimator. We can see that the algorithm produces premium estimates that are normally distributed. What seems surprising is that the premium estimator is *biased*. The mean estimate of $7.971 is 3 cents higher than the exact premium of $7.941, which is about 17.9 times the standard deviation of $5.3/\sqrt{1000}$ cents. Despite the bias, the algorithm can estimate the option premium quite tightly.

Estimator bias

In this section, we investigate the source of the bias in the option premium estimates that was discovered by means of the example presented in the last section. It turns out that the bias arises because the "optimisation" is done over a *finite* sample. The bias vanishes in the limit of infinite sample size. The description of the exercise-decision algorithm in the second section makes it evident that estimating the premium for an American option is equivalent to estimating the exercise-hold stock price boundary at each epoch at which the option can be exercised. Accordingly, we determined the "exact" boundary between holding and exercising the put option at each of the 12 exercise-decision epochs by using the Cox-Rubinstein binomial lattice that was described in the preceding section. With *full knowledge* of the exact exercise-hold boundaries, the American option premium was estimated again by simulation using the *same 1,000 samples* of 5,040 paths on which the results shown in Figure 3 were based. The resulting frequency histogram of the premium estimates is shown in Figure 4.

In Figure 4 the premium estimates are normally distributed. The standard deviation of the estimates is 5.3 cents, the same as in Figure 3. However, the mean of the estimates is $7.943, only 0.2 cents higher than the exact premium. This deviation is not statistically significant at a 5% level of confidence, since it is only about 1.2 times the standard deviation of $5.3/\sqrt{1000}$ cents. Thus, with full knowledge of the exact exercise-decision boundaries, the American option premium estimator is unbiased, even for finite samples of paths. We must conclude that the process of estimating the exercise-hold boundaries from a finite sample of paths introduces the bias. The following analysis demonstrates the truth of this assertion.

The exact price of an American option is the value given by the premium estimator equation in the second section when the *infinite sample space* of stock price paths and the *exact exercise-hold boundaries* are used. Determining the exact price of the option is equivalent to finding the exercise-hold boundaries at all exercise-decision epochs that maximise the value given by the premium estimator equation when the infinite sample space of stock price paths is used. An approximation to the exact price is obtained by finding the exercise-hold boundaries at all exercise-decision epochs that maximise the value given by the premium estimator equation when a *finite sample* of R stock price paths is used. A different approximation to the exact price is obtained by implementing the backward induction algorithm with eight substeps at each epoch that was described in the second section. This latter estimate of the exact option price is itself an approximation of the former estimate of the exact option price, by reason of the construction of the backward induction algorithm as an optimisation.

Let E_i denote the option premium estimate obtained when the ith sample of R paths is used together with some premium estimation method. The dependence of the estimate on the estimation method used is denoted by an appropriate superscript. The superscript ∞-optimal is used to represent the estimation method that utilises the exact boundaries determined from the infinite sample space of stock price paths. The superscript R-optimal is used to represent the estimation method that utilises the boundaries that optimise the value given by the premium estimator equation when the finite sample of R paths is used. Finally, the superscript R-algorithm is used to represent the estimation method that utilises the boundaries determined from the eight-substep backward induction algorithm applied to the finite sample of R paths As a consequence of the definitions of the various estimates and the

4. Frequency histogram for "best" premium estimator (based on 1,000 samples of 5,040 paths each)

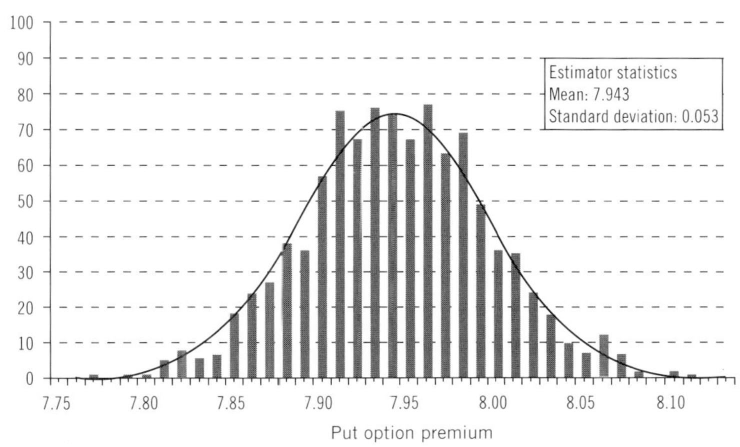

construction of the different estimation methods, the following inequalities hold for *any* sample I consisting of R paths:

$$E_i^{\infty\text{-optimal}} \leq E_i^{\text{R-optimal}} \text{ and } E_i^{\text{R-algorithm}} \leq E_i^{\text{R-optimal}}$$

Thus, the means of the various estimators computed over any finite number of samples of R paths each also satisfy the same inequalities. In practice, the strict inequality will hold "almost surely." When the sample size is infinite the inequalities become equalities. Because the ∞-optimal estimator is unbiased, the first inequality demonstrates that the R-optimal estimator must always have positive bias. The bias tends to zero as $R \to \infty$. Furthermore, the second inequality demonstrates that the R-optimal estimator must be positively biased relative to the R-algorithm estimator. The relative bias tends to zero as $R \to \infty$. It is indeterminable whether the R-algorithm estimator has positive or negative bias with respect to the ∞-optimal estimator. The two inequalities also show that we should not try too hard to "perfect" the R-algorithm estimator in the sense of making it better approximate the R-optimal estimator because the latter always has positive bias relative to the unbiased ∞-optimal estimator.

Table 2. Statistics for $\alpha = 0.50$ estimators of premiums for American put options (stock price volatility of 30%)

Stock price: 40
Option expiration: 3.00 years
Exercise interval: 0.25 years
Stock volatility: 30%
Annual interest rate: 7%

Strike price	"Exact" premium*	Estimator mean†	Estimator bias	Estimator standard deviation†
10	0.003	0.003	0.000	0.001
15	0.046	0.046	0.000	0.005
20	0.242	0.239	–0.003	0.012
25	0.744	0.744	0.000	0.020
30	1.689	1.694	0.005	0.027
35	3.172	3.185	0.013	0.038
40	5.247	5.268	0.021	0.044
45	7.941	7.968	0.027	0.055
50	11.255	11.289	0.034	0.063
55	15.136	15.161	0.025	0.059
60	19.469	19.485	0.016	0.054
65	24.100	24.109	0.009	0.044
70	28.894	28.899	0.005	0.034
75	33.764	33.763	–0.001	0.028
80	38.665	38.662	–0.003	0.024
85	43.576	43.574	–0.002	0.017
90	48.491	48.486	–0.005	0.015
95	53.407	53.400	–0.007	0.014
100	58.323	58.316	–0.007	0.012

*Calculated using the Cox–Rubinstein binomial model with 1,200 time intervals.
†Calculated using a simulation model with 100 samples of 5,040 paths and exercise boundary determined by first dominant string of 1s in the transition zone.

Example revisited

Now that we understand that the sign of the bias of the R-algorithm estimator is indeterminable but is likely to be positive if the R-algorithm estimates the R-optimal exercise-hold decision boundaries closely, we should conduct further empirical studies of the bias. Table 2 presents results obtained by using the R-algorithm estimator of the second section on 100 independent samples of 5,040 paths each by using a partition of 70 bundles by 72 paths per bundle. Results are shown for 3-year American put options with strike prices ranging from 10 to 100 in multiples of 5. All other assumptions are the same as in the earlier example. The "exact" premiums were calculated as before, by using the Cox–Rubinstein binomial lattice with 1,200 time intervals. The estimator bias ranges from a low of –0.7 cents to a high of +3.4 cents. The standard deviations of the estimates peak at 6.3 cents for a put option somewhat in the money. The premium estimates must be considered very accurate.

Table 3. Statistics for $\alpha = 0.73$ estimators of premiums for American put options (stock price volatility of 30%)

Stock price: 40
Option expiration: 3.00 years
Exercise interval: 0.25 years
Stock volatility: 30%
Annual interest rate: 7%

Strike price	"Exact" premium*	Estimator mean†	Estimator bias	Estimator standard deviation†
10	0.003	0.003	0.000	0.001
15	0.046	0.048	0.002	0.005
20	0.242	0.246	0.004	0.012
25	0.744	0.750	0.006	0.018
30	1.689	1.697	0.008	0.028
35	3.172	3.178	0.006	0.039
40	5.247	5.255	0.008	0.049
45	7.941	7.943	0.002	0.052
50	11.255	11.260	0.005	0.066
55	15.136	15.139	0.003	0.059
60	19.469	19.468	–0.001	0.058
65	24.100	24.094	–0.006	0.049
70	28.894	28.889	–0.005	0.037
75	33.764	33.752	–0.012	0.034
80	38.665	38.654	–0.011	0.032
85	43.576	43.566	–0.010	0.021
90	48.491	48.480	–0.011	0.021
95	53.407	53.398	–0.009	0.020
100	58.323	58.315	–0.008	0.016

*Calculated using the Cox–Rubinstein binomial model with 1,200 time intervals.
†Calculated using a simulation model with 100 samples of 5,040 paths and exercise boundary determined by dominant strings of both 0s and 1s in the transition zone.

Table 3 presents results similar to those in Table 2 but for a partition of 504 bundles by 10 paths per bundle. In this case, substep 6 of the exercise-decision algorithm was refined to take into account not only the first dominant string of 1s in the transition zone but also the last dominant string of 0s in the transition zone. As in substep 6, a boundary index is determined as the start of the first string of 1s the length of which exceeds the length of every subsequent string of

0s. Another boundary index is determined as the end of the last string of 0s whose length exceeds the length of every previous string of 1s. In many cases the two boundaries are identical, but, if not, the *dominant 0-string* boundary must occur before the *dominant 1-string* boundary. The boundary index actually used in the revised algorithm is the arithmetic mean of the two boundary indexes, rounded appropriately. The estimator bias shown in Table 3 ranges from a low of -1.2 cents to a high of +0.8 cents. The standard deviations of the estimates are generally a little higher than their counterparts in Table 2.

Table 4 presents results similar to those in Table 3 except that the stock price volatility has been doubled to 60%. Again, the estimator biases are small, ranging from -0.8 cents to +2.4 cents. The standard deviations of the estimates are much larger but are still very small when expressed as a percentage of the exact premiums.

Table 4. Statistics for $\alpha = 0.73$ estimators of premiums for American put options (stock price volatility of 60%)

Stock price: 40 Stock volatility: 60%
Option expiration: 3.00 years Annual interest rate: 7%
Exercise interval: 0.25 years

Strike price	"Exact" premium*	Estimator mean†	Estimator bias	Estimator standard deviation†
10	0.486	0.489	0.003	0.013
15	1.409	1.414	0.005	0.022
20	2.810	2.815	0.005	0.034
25	4.636	4.643	0.007	0.044
30	6.834	6.840	0.006	0.054
35	9.357	9.367	0.010	0.064
40	12.162	12.180	0.018	0.075
45	15.220	15.240	0.020	0.088
50	18.504	18.526	0.022	0.102
55	21.986	22.009	0.023	0.120
60	25.650	25.667	0.017	0.118
65	29.475	29.493	0.018	0.131
70	33.453	33.477	0.024	0.129
75	37.566	37.584	0.018	0.148
80	41.798	41.809	0.011	0.145
85	46.137	46.139	0.002	0.149
90	50.571	50.567	–0.004	0.149
95	55.086	55.078	–0.008	0.144
100	59.670	59.662	–0.008	0.144

*Calculated using the Cox–Rubinstein binomial model with 1,200 time intervals.
†Calculated using a simulation model with 100 samples of 5,040 paths and exercise boundary determined by dominant strings of both 0s and 1s in the transition zone.

Summary and conclusions

This paper has presented an algorithm for valuing American options in a path-simulation model and has demonstrated its accuracy by an example involving a put option on a non-dividend-paying stock for which the exact premium could be determined. The demonstration of the existence of a useful algorithm for valuing American options in a path-simulation model should remove what has been perceived as a major impediment to the use of simulation models in valuing a broker–dealer's derivatives book and in analysing the asset-liability condition of financial intermediaries.

In many situations involving the use of multifactor models to describe realistic market price behaviour, simulation is the only method that can handle the American option valuation problem satisfactorily. Furthermore, it is usually straightforward to apply a simulation technique, whereas solving complicated partial differential equations numerically generally requires great care as well as sophistication in applied mathematical methods. This chapter has not dealt with some of the complexities that arise in determining exercise–hold decision boundaries when multifactor stochastic models of asset price behaviour are utilised. Empirical studies I have conducted suggest that some modification to the algorithm presented here is required to handle those situations adequately. For example, the bundling must often be carried out in at least two dimensions rather than the single dimension presented in this work. Boundary points become boundary lines or surfaces.

BIBLIOGRAPHY

Black, F., E. Derman and W. Toy, 1990, "A One-Factor Model of Interest Rates and its Application to Treasury Bond Options", *Financial Analysts Journal* 46(1), pp. 33–39.

Bookstaber, R., D.P. Jacob and J.A. Langsam, 1986, "The Arbitrage-Free Pricing of Options on Interest-Sensitive Instruments", in J. Fabozzi (ed.), *Advances in Futures and Options Research*, Vol. 1, Part A, Greenwich, Conn.: JAI Press, pp. 1–23.

Boyle, P.P., 1977, "Options: A Monte Carlo Approach", *Journal of Financial Economics* 4, pp. 323–38 (Chapter 2 of this publication).

Clancy, R.P., 1985, "Options on Bonds and Applications to Product Pricing", *TSA* XXXVII, pp. 97–130.

Cox, J.C., S.A. Ross and M. Rubinstein, 1979, "Option Pricing: A Simplified Approach", *Journal of Financial Economics* 7, pp. 229–63.

Cox, J.C., and M. Rubinstein, 1985, *Options Markets*, Englewood Cliffs, N. J., Prentice-Hall.

Dyer, L.J., and D.P. Jacob, 1991, "An Overview of Fixed Income Option Models", in F.J. Fabozzi (ed.), *The Handbook of Fixed Income Securities*, Homewood, Ill., Business One Irwin, pp. 742–73.

Geske, R., and K. Shastri, 1985, "Valuation by Approximation: A Comparison of Alternative Option Valuation Techniques", *Journal of Financial and Quantitative Analysis* 20(1), pp. 45–71.

Hall, A.R., 1985, "Valuing the Mortgage Borrower's Prepayment Option", *Journal of the American Real Estate and Urban Economics Association* 13, pp. 229–47.

Ho, T.S.Y., and S.-B. Lee, 1986, "Term Structure Movements and Pricing Interest Rate Contingent Claims", *Journal of Finance* 41, pp. 1011–29.

Hull, J., 1989, *Options, Futures, and Other Derivative Securities*, Englewood Cliffs, N. J., Prentice-Hall.

Hull, J., and A. White, 1990, "Valuing Derivative Securities Using the Explicit Finite Difference Method", *Journal of Financial and Quantitative Analysis* 25, pp. 87–100.

Hull, J., and A. White, 1990, "Pricing Interest-Rate-Derivative Securities", *Review of Financial Studies* 3(4), pp. 573–92.

Hull, J., and A. White, 1990, "Valuing Interest Rate Options: Trinomial Versus Binomial Trees", *working paper*, University of Toronto.

Jacob, D.P., G. Lord and J.A. Tilley, 1987, "A Generalized Framework for Pricing Contingent Cash Flows", *Financial Management* 16(3), pp. 5–14.

Morgan, I.G., and E.H. Neave, 1990, "A Mean Reverting Process for Pricing Treasury Bills and Futures Contracts", *Proceedings of the 2nd AFIR Colloquium* 1, pp. 237–66.

Pedersen, H.W., and E.S.W. Shu, 1988, "Pricing of Options on Bonds by Binomial Lattices and by Diffusion Processes", *ARCH* 2, pp. 115–39.

Pedersen, H.W., E.S.W. Shu and A.E. Thorlacius, 1989, "Arbitrage-Free Pricing of Interest-Rate Contingent Claims", *TSA* XLI, pp. 231–65.

Pitts, M., 1985, "The Pricing of Options on Debt Securities", *Journal of Portfolio Management* 11, pp. 41–50.

Rendleman, R.J., Jr., and B.J. Bartter, 1980, "Two State Option Pricing", *Journal of Finance* 34, pp. 1093–110.

Rendleman, R.J., Jr., and B.J. Bartter, 1980, "The Pricing of Options on Debt Securities", *Journal of Financial and Quantitative Analysis* 15, pp. 11–24.

Ritchken, P., and K. Boenawan, 1990, "On Arbitrage-Free Pricing of Interest Rate Contingent Claims", *Journal of Finance* 45, pp. 259–64.

Stapleton, R.C., and M.G. Subrahmanyam, 1984, "The Valuation of Options When Asset Returns Are Generated by a Binomial Process", *Journal of Finance* 39, pp. 1525–39.

Tilley, J.A., 1992, "An Actuarial Layman's Guide to Building Stochastic Interest Rate Generators", *TSA* XLIV, pp. 509–64.

15

Numerical Valuation of High-Dimensional Multivariate American Securities

Jérôme Barraquand and Didier Martineau[1]

Salomon Brothers International and Long-Term Capital Management

We consider the problem of pricing an American contingent claim whose payoff depends on several sources of uncertainty. Several efficient numerical lattice-based techniques exist for pricing American securities depending on one or few (up to three) risk sources. However, these methods cannot be used for high-dimensional problems since their memory requirement is exponential in the number of risk sources.

There are several reasons motivating the development of efficient methods for multidimensional contingent claim pricing. In particular, applications exist in the pricing of over-the-counter (OTC) warrants, path-dependent instruments (Barraquand and Pudet, 1995), multidimensional interest rate term structure contingent claims (Heath, Jarrow and Morton, 1992), mortgage-backed securities (MBS), life insurance policies (Fabozzi and Pollack, 1987), and futures contracts with quality delivery options (Cheng, 1987; and Boyle, 1989).

For pricing purposes, financial assets can be divided into two major classes. The first class is that of assets whose future cashflows cannot be influenced by decisions from the holder taken after the purchase date. We will call European instruments all financial assets belonging to this first class. In particular, stocks, bonds, futures contracts, European options, swaps, caps, floors and MBS are European instruments. The second class is that of assets whose cashflows can be influenced *a posteriori* by the holder. American options belong to this class. We will call American instruments all financial assets belonging to this second class.

Following the general theory of arbitrage pricing, the theoretical price of a European contingent claim is the discounted expected value of its future cashflows under the so-called "risk-neutral" probability distribution of the underlying economic factors (Harrison and Kreps, 1979; Harrison and Pliska, 1981; Duffie, 1988; Karatzas and Shreve, 1988). Mathematically, computing the arbitrage price reduces to computing an integral (sum) over the space of the underlying economic factors. When the dimension of the space of the underlying economic factors is small, standard techniques for numerical integration can be used. In some cases, the integral can even be computed analytically (eg Black–Scholes formula). However, the computational complexity of evaluating the integral is clearly exponential in the dimension of the space. Efficient numerical techniques for pricing high-dimensional European claims are presented in Barraquand (1995).

The price of an American claim is the maximum over all possible cashflow monitoring strategies of the associated present values of cashflows. For example, the price of an American option is the maximum over all possible early exercise strategies of the corresponding present values. Since the space of cashflow monitoring strategies is generally huge, direct maximisation of the present value is rarely practical (see Bossaerts, 1989, for a discussion). However, when the underlying economy is modelled as a Markov process, one can use the Bellman principle of dynamic programming (Bellman, 1957) to compute the optimal monitoring strategy. American options are typically priced using a discrete approximation of the dynamic programming principle. This is the case in particular of the CRR model (Cox, Ross and Rubinstein, 1979) for American stock option pricing. This approach, however, becomes impractical when the underlying economic space has many dimensions since the dynamic programming algorithm requires a memory space that is exponential in the number of dimensions. This fact is known as the "curse of dimensionality" problem for dynamic programming.

Indeed, in general, the optimal monitoring strategy is a function associating to each economic state – ie to each value of the set of underlying securities – the optimal investment decision to be made. In particular, in the case of early exercise strategies for American options, the optimal early exercise strategy associates a date (the early exercise stopping time) to each value of the underlying assets. Hence, representing this exercise function requires in the worst case a memory space exponential in the number of underlying assets.

In this paper, we present a particular state space partitioning technique that attempts to circumvent the curse of dimensionality problem for American security pricing. More precisely, we partition the space of underlying assets (the state space) into a tractable number of cells, and we compute an approximate early exercise strategy that is constant over those cells. The hope is that, if the partition is appropriately chosen, the approximate strategy will be close to the actual optimal strategy. Such a partitioning technique is classically called a *state aggregation* technique.

Among the many possible ways of choosing a partition, one solution is to choose a real-valued function mapping the state (ie the prices of the underlying assets) that particularly influences the optimal strategy in the problem at hand. We call this function *stratification map*. Then, the partition chosen is a stratification of the state space into thin layers along this map. In other words, we limit our search to strategies that only depend on the stratification map, and not on the entire state itself. We call this technique *stratified state aggregation*.

In the case of American security pricing, an obvious candidate for the stratification map is the *payoff* of the security, ie the function representing the future cashflows associated with the security. When the stratification map chosen is the payoff of the American security, we call the technique *stratified state aggregation along the payoff* (SSAP).

After quantification of the payoff, the SSAP method can be combined with Monte Carlo simulation techniques in order to compute the set of conditional probabilities corresponding to changes in the payoff value over time. Using these conditional probabilities, an approximation of the American price can then be computed backwards in time using techniques reminiscent of the classical CRR integration method.

Since the true American price is the maximum over all possible early exercise strategies of the risk-neutral discounted cashflows, it is shown that, under suitable technical conditions, the approximate SSAP price is a lower bound of the true arbitrage price. Furthermore, this SSAP price is always greater than the price of the corresponding European security under the same technical conditions.

We implemented the SSAP method on American option pricing problems in dimensions ranging from 1 to 400. On all problems for which we could compare the SSAP method with known optimal solutions, the SSAP price was indistinguishable from the optimal theoretical price. In particular, in dimensions 1, 2, and 3 both put and call prices of options on the maximum of the underlying assets were computed accurately by the SSAP method. Also, the SSAP price of an American call on the maximum of n assets paying no dividends was indistinguishable from the European price for n ranging from 1 to 400, in accordance with a well-known theoretical result. (Indeed, since the discounted prices of assets paying no dividends are martingales under the risk-neutral measure, the discounted maximum of such n asset prices is a submartingale. Hence, the corresponding European call price is always higher than the immediate payoff.) In other cases, no other method exists to compare to our results. However, the SSAP price seems to constitute an accurate approximation of the American price in arbitrarily high dimensions.

To the best of our knowledge, the SSAP method is the first that is capable of computing American prices and exercise strategies in high-dimensional cases.

This paper is organised as follows: in the next section we relate our contribution to previous work in American security pricing, multidimensional asset pricing and Monte Carlo valuation. Following that, we recall the usual assumptions on the stochastic processes governing the evolution of securities prices, and the main results of the arbitrage pricing theory. Then we briefly review the current numerical methods used for American security pricing before presenting the method of stratified state aggregation. In the next section, we show how SSA prices can be computed through Monte Carlo simulation. Finally, we present numerical experiments illustrating the efficiency of the SSAP method.

Relation to other work

The theoretical analysis of optimal stopping times for early exercise of American options dates back to the work of McKean (1965). This theory was then further developed by several authors (Merton, 1973; Harrison and Kreps, 1979; Bensoussan, 1984; Karatzas, 1988; Jaillet, Lamberton and Lapeyre, 1988). Myneni (1992) surveys the theory of American option pricing.

The most widely used valuation technique for American options with one underlying asset is the binomial lattice approach of Cox, Ross and Rubinstein (1979). Cox and Rubinstein (1985) outline the principle of the multidimensional extension. Other numerical valuation techniques are presented in Geske and Johnson (1984), Barone-Adesi and Whaley (1987), and Barone-Adesi and Elliott (1991).

The valuation of options depending on several underlying assets has been extensively studied. Brennan and Schwartz (1979) address the problem of pricing options under two sources of risk by direct finite-difference approximation of the generalised Black–Scholes equation. In this example, the two sources of risk are the short-term and the long-term interest rates. The approach is clearly limited to a few assets since the memory space requirements and the computation time are both exponential in the number of underlying assets. Boyle, Evnine and Gibbs (1989) developed a multinomial lattice method for pricing multidimensional options, in the spirit of the approach outlined in Cox and Rubinstein (1985). According to the authors, the computation becomes very burdensome for more than two assets. In fact, the multinomial lattice approach can be viewed as a finite-difference approximation of the generalised Black–Scholes equation using an explicit Euler scheme and an appropriate change of variables (Brennan and Schwartz, 1978).

Stulz (1982) presents an analytical solution to the problem of pricing a European option on the maximum or minimum of two underlying assets. The analytical solution is generalised in Johnson (1987) to the case of an arbitrary number of assets, taking as given the cumulative multivariate normal distribution function. Boyle (1989) and Boyle and Tse (1990) developed an approximate method for the same problem. Although the problem is solved analytically in Johnson (1987), the approximate method does not require preliminary computation of the cumulative multivariate normal distribution function. To the best of our knowledge, no closed-form solutions have ever been obtained for American pricing problems.

The application of the Monte Carlo method to option pricing was first presented in Boyle (1977) – also Chapter 1 of this publication, in the context of claims contingent to a single underlying asset. It then was used by several authors for the valuation of path-dependent contingent claims. In particular, the method has been used for pricing MBS (see Schwartz and Torous, 1989; Hutchinson and Zenios, 1991). Barraquand (1995) presents the method of quadratic resampling for Monte Carlo valuation of European securities with many underlying assets. The method presented in this paper is an extension of this earlier work to the American pricing problem.

Arbitrage pricing theory

FUNDAMENTAL RESULTS

The arbitrage pricing theory is described in many textbooks. We refer the reader to Duffie (1992) for a comprehensive presentation.

We consider a complete market spanned by n securities with prices $X(t) = (x_1(t), \ldots, x_n(t))$, representing all the information available to investors at time t. We assume that X follows the Itô process

$$\forall i \in [1,n],$$
$$\frac{dx_i}{x_i} = \mu_{x_i}(x_1,\ldots,x_n,t)dt + \sum_{j=1}^{n} v_{ij}(x_1,\ldots,x_n,t)dw_j$$

The matrix $V = (v_{ij})_{(i,j)\in[1,n]^2}$ is called the volatility matrix, and the covariance of relative returns is the matrix $\mathcal{K} = (k_{ij})_{(i,j)\in[1,n]^2}$:

$$\mathcal{K} = VV^T$$

We consider the so-called *risk-neutral* information process $\tilde{X}$ for which all market prices for risk are zero:

$$\forall i \in [1,n],$$
$$\frac{d\tilde{x}_i}{\tilde{x}_i} = (r - d_{x_i})dt + \sum_{j=1}^{n} v_{ij}dw_j \qquad (1)$$

In the above equations, d_{x_i} is the dividend yield generated by the security x_i. A security is called a *European security* if future cashflows cannot be influenced by decisions from the holder taken after the purchase date (besides, of course, selling back the security). Since the market is assumed complete and X is assumed Markov, the cashflows are only functions of time and of the n

prices $x_1,...,x_n$. The cashflow generated by a European security C during the time dt can be written as

$$f_C(X,t)dt$$

A security is called an *American security* if it is not European, ie if future cashflows can be influenced by decisions from the holder taken after the purchase date. Then, the cashflows are not only functions of time and information but also functions of the decisions taken by the security's holder. A cashflow monitoring strategy u is a stochastic process associating with each time and information state a decision $u(X(t),t) \in U$, U being an appropriate *decision space*. We denote by CMS the space of the cashflow monitoring strategies. The cashflow generated by an American security C during the time dt can be written:

$$f_C(u,X,t)dt$$

The price, C, of an American contingent claim is the maximum over all possible cashflow monitoring strategies of the discounted expected value of future cashflows under the risk-neutral process $\tilde{X}$:

$$C(X(t),t) = \max_{u \in CMS} \tilde{E}_t\left(\int_t^{\infty} \frac{f_C(u(X(\tau),\tau), X(\tau),\tau)}{L(t,\tau)} d\tau\right)$$

where L(t,T) is the value at time T of one dollar invested at time t in the money market:

$$L(t,T) = \exp\left(\int_t^T r(X,s)ds\right)$$

and r(X,t) is the riskless short interest rate.

AMERICAN OPTIONS

By definition, an American call (put) option on an asset S with expiration date T and strike price K gives its owner the right to purchase (sell) *on or before time* T the asset S for the price K. The space of admissible CMS is composed of all adapted processes u taking the two possible values

$$\forall t \in [0,T], \; u(X,t) \in \{\text{exercise}, \text{no exercise}\}$$

and verifying at all times $t \leq T$

$$u(X(t),t) = \text{exercise} \\ => \forall \tau > t, \; u(X(\tau),\tau) = \text{no exercise}$$

That is, exercise cannot occur twice. For the sake of simplicity, we will assume that the payoff of the option at exercise time t^* is distributed during a small time interval $[t^*, t^* + \Delta t]$. We will also assume that exercise decisions u(X(t),t) are piecewise-constant, ie change only at the beginning of time intervals of duration Δt. For notational convenience, we will define the "spike" function $\delta(u)$ associating the value $1/\Delta t$ to the decision exercise and the value 0 to the decision no exercise.

The cashflow function of a call option is

$$f(u,X,t) = (S(X,t) - K)\delta(u)$$

The optimal early exercise strategy can be written:

$$u^*(X,t) = \begin{cases} \text{exercise} & \text{if } C(X,t) \leq S(X,t) - K \\ \text{no exercise} & \text{otherwise} \end{cases}$$

We see that the computation of the optimal early exercise strategy requires us to pre-compute the pricing relationship between the option and the underlying asset, which is what we were trying to compute in the first place.

The price of the American call (resp. put) option can be written as

$$C(X(t),t) = \max_{u(X(\tau),\tau),\tau \geq t, u \in CMS} \tilde{E}_t\left(\sum_{\tau=t}^{T} \frac{f(X(\tau),\tau)}{L(t,\tau)} \delta(u(X(\tau),\tau))\Delta t\right) \quad (2)$$

with

$$f(X,t) = \max(0, S(X,t) - K) \\ (\text{resp. } \max(0, K - S(X,t)))$$

More generally, any contingent claim entitling its holder to the single cashflow f on or before an expiration date T can be priced according to the above formula.

Numerical methods for American security pricing

STOCHASTIC DYNAMIC PROGRAMMING

The explicit numerical valuation of an American option using the above formula involves a maximisation over the set of all possible early exercise strategies. The strategy u can be any function associating to each current value of the underlying assets $X = (x_1,...,x_n)$ and each current time t the binary decision to exercise or not exercise. Since the number of such possible strategies is huge, direct maximisation is rarely practical (see Bossaerts, 1989, for a discussion of this).

The only practical technique to date consists in using the Bellman principle of dynamic programming. This principle can be applied since the information vector X is assumed to be a Markov process, and therefore the optimal early exercise strategy $u(X(t),t)$ only depends on time and the current vector $X(t)$.

Assuming that exercise decisions can only be taken at discrete time intervals of constant duration Δt, the maximisation problem (equation (2)) can be rewritten using the law of iterated expectations and the property $L(t,t^*) = L(t,t + \Delta t) \times L(t + \Delta t,t^*)$:

$$C(X(t),t) = \max_{u(X(t),t),u \in CMS} \tilde{E}_t \left[\frac{f(X(t))}{L(t,t + \Delta t)} \delta(u(X(t),t))\Delta t + \frac{1}{L(t,t + \Delta t)} \max_{u(X(\tau),\tau),\tau \geq t + \Delta t,u \in CMS} \tilde{E}_{t+\Delta t} \left(\sum_{\tau = t+\Delta t}^{T} \frac{f(X(\tau))}{L(t + \Delta t,\tau)} \delta(u(X(\tau),\tau))\Delta t \right) \right]$$

Examining successively the two cases $u(X(t),t) =$ exercise and $u(X(t),t) =$ no exercise, and using the expression of $C(X(t + \Delta t), t + \Delta t)$ from equation 2, we obtain

$$C(X(t),t) = \exp(-r\Delta t) \max(f(X(t)), \tilde{E}_t(C(X(t + \Delta t),t + \Delta t))) \tag{3}$$

assuming that the interest rate r is piecewise constant on intervals of duration Δt. The above recursive expression, called the Bellman equation, allows one to compute the price, C, of an American option by proceedings backwards in time from the expiration date, T.

LATTICE METHODS AND THE COX-ROSS-RUBINSTEIN APPROACH

The general method for solving the above Bellman equation is to quantify it using a finite difference or lattice method (see, for example, Duffie, 1992, Chapter 10). This is the paradigm underlying the original Cox-Ross-Rubinstein approach (Cox, Ross and Rubinstein 1979).

We will illustrate this approach using a number of simplifying assumptions. We will assume that the process of the underlying securities is jointly lognormal, ie that the mean and covariance matrix of relative returns are constant. We will also assume that the interest rate, r, and the dividend yields of the underlying securities are constant. Then, it can be shown that the solution of the Itô process (equation (1)) is:

$$\forall i \in [1,n], \quad x_i(t) = x_i(0) \exp\left(\left(r - d_{x_i} - \frac{1}{2} k_{ii}\right)t + \sum_{j=1}^{n} v_{ij}w_k(t)\right) \tag{4}$$

The Brownian motion, W, is approximated by an n-dimensional binomial process, $\bar{W}^{\Delta t}$ defined as follows:

$$\forall \epsilon = (\epsilon_1,\ldots,\epsilon_n) \in \{-1,1\}^n, \quad \text{Prob}\,(\bar{W}^{\Delta t}(t + \Delta t) = \bar{W}^{\Delta t}(t) + \epsilon\sqrt{\Delta t}) = \frac{1}{2^n}$$

The quantised Bellman equation writes:

$$C(W,t) = \exp(-r\Delta t) \times \max\left(f(X(t)), \frac{1}{2^n} \sum_{\epsilon \in \{-1,1\}^n} C(W + \epsilon\sqrt{\Delta t}, t + \Delta t)\right) \tag{5}$$

We implemented a simple variant of the above numerical scheme for an arbitrary number, n, of underlying assets. The method yields accurate results for $n \leq 3$, but its memory requirement is intrinsically exponential in n. Hence, it cannot be used for $n > 3$ with a quantification step, Δt, small enough to yield accurate results (see later section on experimental results).

State aggregation

STATE AGGREGATION PRICE

Classical numerical methods being unable to deal with high-dimensional American valuation problems, one must resort to alternative approximation schemes. State aggregation is a classical approximation technique for the numerical solution of stochastic optimal control problems (see, for example, Bertsekas, 1987; Kushner and Dupuis, 1992).

For the problem of American security pricing, the relevant state space in the Bellman equation (3) is the n-dimensional space of the underlying asset values $X = (x_1,\ldots,x_n)$. State aggregation consists in partitioning the state space into a tractable number of cells, and in computing an approximate early exercise strategy, $u(X,t)$, that is constant over those cells. The hope is that, if the partition is appropriately chosen, the approximate strategy will be close to the actual optimal strategy.

$\forall t \in \{0,\Delta t,\ldots,T\}$, let us consider a finite partition $P(t) = (P_1(t),\ldots,P_{k(t)}(t))$ of the state space R^n_+, ie a set of $k(t)$ subsets of R^n_+ verifying

$$\bigcup_{i\in[1,k(t)]} P_i(t) = R^n_+ \text{ and } \forall i \neq j,$$
$$P_i(t) \bigcap P_j(t) = \emptyset$$

We assume that the partition $P(0)$ has only two cells:

$$P_1(0) = \{X(0)\}, \text{ and } P_2(0) = R^n_+\backslash\{X(0)\}$$

Among the set CMS of all possible early exercise strategies, we consider the subset $\mathcal{U}(P)$ of piecewise constant strategies, ie of strategies $u(X,t)$ that are *constant* over each cell, $P_i(t)$, of the partition.

Then, we define the *state aggregation price*, $C^*(i,t)$, as the maximum over all possible piecewise constant strategies in $\mathcal{U}(P)$ of the expected risk-neutral discounted future cashflow conditionally to the event $X(t) \in P_i(t)$:

$$C^*(i,t) = \max_{u(X(\tau),\tau),\tau\geq t,\ u\in\mathcal{U}(P)} \tilde{E}_0\left(\sum_{\tau=t}^{T} \frac{f(X(\tau),\tau)}{L(t,\tau)}\right.$$
$$\left.\times\ \delta(u(X(\tau),\tau))\Delta t \,\middle|\, X(t) \in P_i(t)\right) \qquad (6)$$

Since $P_1(0) = \{X(0)\}$, and since $\mathcal{U}(P) \subset \text{CMS}$, the state aggregation price at initial time $C^*(1,0)$ is obviously upper-bounded by the true American price $C(X(0),0)$. Furthermore, since the strategy u_{Euro} corresponding to the European price consists in never exercising before the expiration date, it is clearly constant over the cells of any partition before expiration. Hence, by definition of the state aggregation price, the European price is upper-bounded by the state aggregation price at the initial time $C^*(1,0)$. We can state

> For any family of finite partitions P, the state aggregation price $C^* = C^*(1,0)$ is lower-bounded by the European price and upper-bounded by the American price:
>
> $$C^{Euro} \leq C^* \leq C^{Amer}$$

We will now derive a recursive backward valuation formula for the state aggregation price, under an additional Markovian assumption.

MARKOVIAN APPROXIMATION

We will now assume that the partition P is such that the process $I(t)$ defined by $X(t) \in P_{I(t)}(t)$ is approximately Markov under the risk-neutral measure:

$$\forall i,j,\phi,\ \tilde{E}_0(\phi(X(t+2\Delta t))\,|\,X(t) \in P_i(t),$$
$$X(t+\Delta t) \in P_j(t+\Delta t))$$
$$\approx \tilde{E}_0(\phi(X(t+2\Delta t))\,|\,X(t+\Delta t) \in P_j(t+\Delta t))$$

Applying again the law of iterated expectations to the definition of the state aggregation price in equation (6), we get:

$$C^*(I(t),t) = \max_{u(X(t),t),u\in\mathcal{U}(P)}$$
$$\tilde{E}_0\left[\frac{f(X(t))}{L(t,t+\Delta t)}\delta(u(X(t),t))\Delta t\right.$$
$$+\frac{1}{L(t,t+\Delta t)} \max_{u(X(\tau),\tau),\tau\geq t+\Delta t,u\in\mathcal{U}(P)}$$
$$\tilde{E}_0\left(\sum_{\tau=t+\Delta t}^{T}\frac{f(X(\tau))}{L(t+\Delta t,\tau)}\right.$$
$$\delta(u(X(\tau),\tau))\Delta t \mid X(t) \in P_{I(t)}(t),$$
$$\left.\left. X(t+\Delta t) \in P_{I(t+\Delta t)}(t+\Delta t)\right) \,\middle|\, X(t) \in P_{I(t)}(t)\right]$$

By examining successively the two cases $u(X(t),t) = \text{exercise}$ and $u(X(t),t) = \text{no exercise}$, and applying equation (6) at time $t + \Delta t$:

$$C^*(I(t),t) \approx$$
$$\max\left(\tilde{E}_0(e^{-r\Delta t}f(X(t)) \mid X(t) \in P_{I(t)}(t)),\right.$$
$$\left.\tilde{E}_0(e^{-r\Delta t}C^*(I(t+\Delta t),t+\Delta t) \mid X(t) \in P_{I(t)}(t))\right)$$

where the substitution of equation (6) at time $t + \Delta t$ is justified by the above Markovian approximation.

By definition of a partition, we have

$$1 = \sum_{j=1}^{k(t+\Delta t)} 1_{X(t+\Delta t)\in P_j(t+\Delta t)}$$

where 1_A is the indicator variable of the event A (ie takes the value 1 if A is true, and 0 otherwise).

Combining this with the above recursive equation, we get

$$\forall i \in [1,k(t)],$$

$$C^*(i,t) \approx \max\left(f_i(t), \sum_{j=1}^{k(t+\Delta t)} \pi_{ij}(t)C^*(j,t+\Delta t)\right)$$

where we have defined for notational convenience

$$f_i(t) = \tilde{E}_0(e^{-r\Delta t}f(X(t)) \mid X(t) \in P_i(t))$$

and

$$\pi_{ij}(t) = \tilde{E}_0\,(e^{-r\Delta t}1_{X(t+\Delta t) \in P_j(t+\Delta t)}|X(t) \in P_i(t))$$

Furthermore, the value at expiration date is determined by the terminal condition

$$C^*(i,T) = f_i(T)$$

RECURSIVE STATE AGGREGATION

We define the recursive state aggregation price, C_{SA}, as the solution of the following program:

$$C_{SA}(i,T) = f_i(T)$$

and

$$C_{SA}(i,t) = \max\left(f_i(t), \sum_{j=1}^{k(t+\Delta t)} \pi_{ij}(t)C_{SA}(j,t+\Delta t)\right)$$

When the partitions $P(t)$ are chosen in such a way that the process $I(t)$ is actually Markovian, the recursive state aggregation price, C_{SA}, is exactly the true state aggregation price, C^*. The implementation of a recursive state aggregation program proceeds in two steps: a) definition of an appropriate family of partitions; and b) the computation of the expected payoffs, $f_i(t)$, and the discounted conditional probability matrices, $\pi_{ij}(t)$. Then, the approximate price of an American contingent claim with terminal payoff f can be computed backwards in time using the above recursive equation.

Finite difference or lattice-based methods can be viewed as particular instances of the recursive state aggregation method. Indeed, each grid point or lattice point W^i can be viewed as the centre of the hypercube $\mathrm{Hyper}(W^i)$ defined by

$$\mathrm{Hyper}(W^i) = \{Z \in R^n, \forall j \in [1,n],$$
$$w_j^i - \tfrac{1}{2}\Delta w_j \le z_j < w_j^i + \tfrac{1}{2}\Delta w_j\}$$

Although the process $I(t)$ such that $W(t) \in \mathrm{Hyper}(W_{I(t)})$ is not actually Markovian, it can be approximated by a Markov process with reasonable accuracy for a fine enough lattice grid. In the lattice approach, this corresponds to approximating the Brownian motion W by the binomial process $W^{\Delta t}$. Then, the expected value over the hypercube $f_i(t)$ is approximated by the value in the centre, $f(X(W^i))$. Similarly, the conditional probability is simply taken from that of the binomial approximation, which yields the recursive formula, equation (5). As explained earlier, the limitation of this approach is the exponential growth of the number of lattice hypercubes in the number, n, of underlying assets.

STRATIFIED STATE AGGREGATION

One solution for choosing a partition with a tractable number of cells is to fix a real-valued function, mapping the state that particularly influences the optimal strategy in the problem at hand. We call such a function a *stratification map*. Then, the partition chosen is a stratification of the state space into thin layers along this map.

In other words, stratification consists in limiting the search to strategies that only depend on the stratification map, and not on the entire state itself. We call this technique *stratified state aggregation* (SSA).

In general, we can consider a vector-valued ℓ-dimensional stratification map $\ell < n$:

$$h: R^n \times R \to R^\ell$$
$$(X,t) \to h(X,t) \qquad (7)$$

and a family, Q, of partitions of R^ℓ. From the family Q and the map h, we can build the reciprocal image partition $P = h^{-1}(Q)$:

$$P_i(t) = \{X \in R^n_+, H(X,t) \in Q_i(t)\}$$

In the case of American security pricing, an obvious candidate for the stratification map is the *payoff* of the security. When the stratification map chosen is the payoff of the American security, we call the technique the *stratified state aggregation along the payoff* (SSAP).

Let us consider an American security with a single cashflow $f(X)$ on or before an expiration date T. In particular, in the case of an American call option, the cashflow is $f(X) = \max(0,S(X) - K)$, where K is the strike price. In order to illustrate the SSAP method, we set $\ell = 1$ and choose $h(X,t) = f(X)$ for the stratification map. We also take $k(t) = k$ constant for all times

$t \in [0,T]$. In the numerical examples developed in the later section on experimental results, we assume that $S(X) = \max_{i \in [1,n]} X_i$, and that the process X is jointly lognormal, of the form described earlier in our discussion of lattice methods and the CRR approach. The partitions, Q(t), of the image space $R^{\ell} = R$ are chosen logarithmic in all these examples, ie the interval $Q_i(t)$, $t > 0$ is of the form

$$\forall i \in [2,k-1],$$
$$Q_i(t) = (A(t)e^{B(t)(i-2)}, A(t)e^{B(t)(i-1)}]$$

and

$$Q_1(t) = (-\infty, A(t)],$$
$$Q_k(t) = (A(t)e^{B(t)(k-2)}, +\infty)$$

for adequate parameters A(t) and B(t). The cell $P_i(t)$ is then by definition

$$\forall i \in [2,k-1],$$
$$P_i(t) = \{X \in R^n_+,\ A(t)e^{B(t)(i-2)} < f(X) \le A(t)e^{B(t)(i-1)}\}$$

and

$$P_1(t) = \{X \in R^n_+,\ f(X) \le A(t)\},$$
$$P_k(t) = \{X \in R^n_+,\ f(X) > A(t)e^{B(t)(k-2)}\}$$

In our experiments, the number, k, of cells is set to 100. The numbers A(t) and B(t) are automatically adjusted so as to ensure

$$\text{Prob}\ (X(t) \in P_1(t)) \approx \text{Prob}\ (X(t) \in P_k(t)) \approx 0.1\%$$

The numerical results obtained with the SSAP method show that these empirical parameters are adequate for a broad range of American security pricing problems.

Monte Carlo estimation of American prices

GENERATION OF SAMPLE PATHS

Once the family of partitions, P, has been chosen, for example using the SSAP method, it remains to compute numerically the expected payoffs $f_i(t)$ and discounted conditional probabilities (Arrow–Debreu prices) $\pi_{ij}(t)$. These numbers can be expressed as integrals over the state space. In general, they must be computed numerically. The only general tractable method for computing such high-dimensional integrals is the Monte Carlo method.

It consists in generating a given number, M, of sample paths for the underlying assets price process X(t). In general, this can be done through direct numerical integration of the Itô equation (equation (1)). A simple explicit Euler scheme is given by

$$x_i(t+\Delta t) = x_i(t) \exp\left(\left(r - d_{x_i} - \frac{1}{2}k_{ii}\right) \times (X(t),t)\Delta t + \sum_{j=1}^{n} v_{ij}(X(t),t)\sqrt{\Delta t}\, z^t_j\right)$$

where z^t_j follow independent standard normal distributions for all j and t. $d = T/\Delta t$ being the number of time steps in $[0,T]$, we must draw a total of $M \times d \times n$ standard normal variates in order to generate M n-dimensional sample paths $X^1(t),\ldots, X^M(t)$ for all $t > 0$.

In general, Δt must be chosen small enough so as to reach a reasonable accuracy. In practice, a number of time steps $d = 100$ is sufficient in most asset pricing applications. However, when the joint process X(t) is assumed lognormal, as discussed earlier, d can be chosen much smaller. Indeed, the underlying assets price process X can then be obtained by equation (4) from a standard Brownian motion, W. In our experiments, we found that a number of time steps $d = 10$ is sufficient for American security pricing with lognormal underlying assets price processes.

CONDITIONAL PROBABILITIES AND PAYOFF EXPECTATIONS

Once the M sample paths $X^1(t),\ldots,X^M(t)$ have been computed, the number, $a_i(t)$, of samples crossing $P_i(t)$ and the sum, $b_{ij}(t)$, of discount factors over the samples moving from $P_i(t)$ to $P_j(t+\Delta t)$ are easily derived:

$$a_i(t) = \text{Card}\ \{p \in [1,M], X^p(t) \in P_i(t)\}$$
$$b_{ij}(t) = \sum_{\{p \in [1,M], X^p(t) \in P_i(t), X^p(t+\Delta t) \in P_j(t+\Delta t)\}} e^{-r\Delta t}$$

Similarly, the sum, $c_i(t)$, of payoff values over the samples crossing $P_i(t)$ is computed from

$$c_i(t) = \sum_{\{p \in [1,M], X^p(t) \in P_i(t)\}} e^{-r\Delta t} f(X^p(t))$$

By the law of large numbers we have the following identities:

$$\pi_{ij}(t) = \lim_{M\to\infty} \frac{b_{ij}(t)}{a_i(t)}, \quad f_i(t) = \lim_{M\to\infty} \frac{c_i(t)}{a_i(t)}$$

Table 1. Results of the SSAP method with one underlying asset

Call option prices: $x_1(0)$ = \$40, r = 5%

Parameters σ_1	T	K	European C_{BS}	European C_{SSAP}	European 4stdev	American C_{BS}	American C_{SSAP}	American 4stdev
20%	1	35	5.15	5.15	0.001	5.15	5.15	0.002
		40	1.00	1.00	0.006	1.00	1.00	0.008
		45	0.02	0.02	0.001	0.02	0.02	0.001
	4	35	5.76	5.76	0.003	5.76	5.76	0.005
		40	2.16	2.16	0.006	2.16	2.16	0.008
		45	0.50	0.50	0.004	0.50	0.50	0.004
	7	35	6.40	6.40	0.004	6.40	6.40	0.006
		40	3.00	2.99	0.010	3.00	3.00	0.010
		45	1.09	1.09	0.006	1.09	1.09	0.006
40%	1	35	5.38	5.38	0.003	5.38	5.40	0.007
		40	1.91	1.91	0.010	1.91	1.92	0.010
		45	0.41	0.41	0.003	0.41	0.41	0.003
	4	35	6.88	6.88	0.010	6.88	6.90	0.020
		40	3.96	3.96	0.020	3.96	3.97	0.020
		45	2.08	2.08	0.006	2.08	2.09	0.010
	7	35	8.07	8.08	0.010	8.07	8.10	0.020
		40	5.35	5.34	0.024	5.35	5.36	0.040
		45	3.40	3.40	0.012	3.40	3.42	0.020

Put option prices: $x_1(0)$ = \$40, r = 5%

Parameters σ_1	T	K	European P_{BS}	European P_{SSAP}	European 4stdev	American P_{PDE}	American P_{SSAP}	American 4stdev
20%	1	35	0.00	0.00	0.001	0.00	0.00	0.001
		40	0.83	0.83	0.008	0.84	0.84	0.008
		45	4.84	4.84	0.001	5.00	5.00	0.000
	4	35	0.19	0.19	0.003	0.19	0.19	0.004
		40	1.50	1.50	0.006	1.56	1.56	0.010
		45	4.77	4.77	0.004	5.06	5.07	0.010
	7	35	0.41	0.41	0.003	0.42	0.42	0.006
		40	1.86	1.86	0.010	1.96	1.96	0.012
		45	4.82	4.82	0.006	5.24	5.23	0.016
40%	1	35	0.24	0.24	0.003	0.24	0.24	0.003
		40	1.74	1.74	0.010	1.75	1.76	0.010
		45	5.23	5.23	0.003	5.27	5.29	0.015
	4	35	1.31	1.31	0.010	1.32	1.33	0.020
		40	3.31	3.31	0.020	3.36	3.37	0.020
		45	6.35	6.35	0.006	6.47	6.49	0.015
	7	35	2.08	2.09	0.012	2.12	2.13	0.015
		40	4.21	4.21	0.020	4.31	4.31	0.016
		45	7.12	7.13	0.012	7.36	7.34	0.030

BACKWARD INTEGRATION ALGORITHM

Using the above Monte Carlo estimates of the conditional probabilities and payoff expectations, an approximation of the American price can then be computed backwards in time using the simple algorithm described below.

❑ At time T, the approximate SSAP price is initialised at

$$C(i,T) = \frac{c_i(T)}{a_i(T)}$$

❑ At time $T = \Delta t$, we can compute for all $i \in [1,k]$:

$$C(i,T - \Delta t)$$

$$= \max\left(\frac{c_i(T - \Delta t)}{a_i(T - \Delta t)}, \sum_{j=1}^{k} C(j,T)\frac{b_{ij}(T - \Delta t)}{a_i(T - \Delta t)}\right)$$

❑ The above procedure is then applied recursively, backwards in time, to compute all the prices $C(i,T - 2\Delta t), C(i,T - 3\Delta t),\ldots,C(1,0) = C_{SSAP}$.

The memory required in the SSAP method is proportional to $k^2 \times d$, corresponding to the storage of the conditional probabilities $p_{ij}(t)$. The computation time is proportional to $M \times n^2 \times d + k^2 \times d$, the first term corresponding to the drawing of the M Monte Carlo sample paths and the second to the backwards integration. Hence, the memory and time complexities of the SSAP

Table 2. Prices for a call option with three underlying assets

Call option prices: $x_1(0) = x_2(0) = x_3(0) = \$40$
$\sigma_1 = 20\%$, $\sigma_2 = 30\%$, $\sigma_3 = 50\%$, r = 5%

Parameters			European			American		
ρ	T	K	C_{PDE}	C_{SSAP}	4stdev	C_{PDE}	C_{SSAP}	4stdev
0%	1	35	8.59	8.58	0.008	8.59	8.59	0.010
		40	3.84	3.83	0.010	3.84	3.84	0.012
		45	0.89	0.89	0.006	0.89	0.90	0.007
	4	35	12.55	12.53	0.020	12.55	12.55	0.016
		40	7.87	7.85	0.014	7.87	7.87	0.020
		45	4.26	4.25	0.014	4.26	4.27	0.014
	7	35	15.29	15.27	0.020	15.29	15.30	0.030
		40	10.72	10.70	0.016	10.72	10.73	0.035
		45	6.96	6.95	0.020	6.96	6.98	0.020
50%	1	35	7.78	7.77	0.010	7.78	7.78	0.012
		40	3.18	3.17	0.010	3.18	3.18	0.012
		45	0.82	0.82	0.004	0.82	0.83	0.004
	4	35	10.97	10.95	0.010	10.97	10.96	0.015
		40	6.69	6.67	0.016	6.69	6.69	0.020
		45	3.70	3.69	0.012	3.70	3.71	0.025
	7	35	13.23	13.21	0.016	13.23	13.24	0.030
		40	9.11	9.09	0.020	9.11	9.12	0.040
		45	5.98	5.98	0.016	5.98	5.99	0.020
100%	1	35	6.53	6.52	0.006	6.53	6.54	0.010
		40	2.38	2.37	0.010	2.38	2.38	0.012
		45	0.74	0.74	0.002	0.74	0.74	0.003
	4	35	8.51	8.50	0.008	8.51	8.53	0.016
		40	4.92	4.90	0.012	4.92	4.93	0.020
		45	2.97	2.96	0.008	2.97	2.99	0.020
	7	35	10.04	10.03	0.010	10.04	10.07	0.016
		40	6.64	6.63	0.016	6.64	6.67	0.030
		45	4.61	4.60	0.016	4.61	4.64	0.040

method are polynomial in n. This is to be contrasted with classical lattice methods, which are exponential in n.

Experimental results

A TEST CASE

We implemented the method of SSAP. We present below some numerical results for several European and American option pricing problems ranging from 1 to 400 underlying assets. In all the experiments, we assumed that the American options can only be exercised at $d = 10$ different dates during the life of the option. This corresponds to choosing a time step $\Delta t = T/d = T/10$. We experimented with several different payoff functions, in particular with payoffs corresponding to the maximum, the minimum, or the average of the n underlying assets. We obtained similar results for all these different payoff functions. We present below only the case of an option on the maximum of the underlying assets. Its payoff function is defined by:

$$f(x_1,\ldots,x_n) = \max(0, \max(x_1,\ldots,x_n) - K)$$

where K is the strike price of the option.

In all the experiments, we assumed that the underlying assets price process X(t) is lognormal, with a covariance matrix of relative returns $\mathcal{K}$ of the form

$$\forall i \in [1,n],\ k_{ii} = \sigma_i^2$$

and

$$\forall (i,j) \in [1,n]^2,\ i \neq j,\ k_{ij} = \rho\sigma_i\sigma_j$$

for $n + 1$ numbers $\sigma_1 > 0,\ldots,\sigma_n > 0$ and $-1/(n-1) \leq \rho \leq 1$.

Volatilities, σ_i, correlations, ρ, and interest rate, r, are counted in percent per year. The time to expiration, T, is counted in months, with the convention 1 month = 30 days. All asset and strike prices are counted in dollars.

Since SSAP uses Monte Carlo simulation, we report confidence intervals together with all results. These confidence intervals are computed from the central limit theorem, ie we assume that at a confidence level of 99.95% the error must be less than four times the observed standard deviation of the result. The confidence interval reported is four standard deviations.

Table 3. Prices for a put option with three underlying assets

Put option prices: $x_1(0) = x_2(0) = x_3(0) = \$40$
$\sigma_1 = 20\%$, $\sigma_2 = 30\%$, $\sigma_3 = 50\%$, $r = 5\%$

Parameters			European			American		
ρ	T	K	P_{PDE}	P_{SSAP}	4stdev	P_{PDE}	P_{SSAP}	4stdev
0%	1	35	8.59	8.58	0.008	8.59	8.59	0.010
		40	0.13	0.13	0.003	0.23	0.23	0.003
		45	2.26	2.27	0.012	5.00	5.00	0.000
	4	35	0.01	0.01	0.002	0.01	0.01	0.001
		40	0.25	0.25	0.006	0.44	0.45	0.006
		45	1.55	1.56	0.010	5.00	5.00	0.000
	7	35	0.03	0.03	0.004	0.04	0.04	0.002
		40	0.31	0.32	0.010	0.57	0.58	0.015
		45	1.41	1.42	0.020	5.00	5.00	0.000
50%	1	35	0.00	0.00	0.000	0.00	0.00	0.000
		40	0.38	0.39	0.003	0.48	0.49	0.006
		45	3.00	3.01	0.010	5.00	5.00	0.000
	4	35	0.07	0.08	0.006	0.09	0.09	0.004
		40	0.72	0.72	0.006	0.93	0.94	0.010
		45	2.65	2.66	0.010	5.00	5.00	0.000
	7	35	0.17	0.17	0.006	0.20	0.20	0.005
		40	0.91	0.91	0.012	1.19	1.21	0.010
		45	2.63	2.65	0.012	5.00	5.00	0.000
100%	1	35	0.01	0.01	0.001	0.01	0.01	0.001
		40	0.84	0.84	0.004	0.84	0.85	0.006
		45	4.18	4.18	0.003	5.00	5.00	0.000
	4	35	0.19	0.19	0.003	0.19	0.19	0.004
		40	1.51	1.51	0.006	1.56	1.57	0.008
		45	4.49	4.49	0.005	5.00	5.00	0.008
	7	35	0.41	0.41	0.004	0.42	0.42	0.004
		40	1.87	1.86	0.010	1.96	1.97	0.012
		45	4.70	4.70	0.008	5.20	5.20	0.012

All the simulations were run on a DEC 3000 model 500X workstation, with an Alpha AXP processor running at a clock rate of 200 MHz and 1 gigabyte of main memory.

ONE UNDERLYING ASSET

We first study the one-dimensional case. In this case, the SSAP price should converge toward the theoretical arbitrage price when both the number of time steps, d, and the number of cells, k, converge towards infinity. Both European calls and European puts can be priced according to the original Black-Scholes formula. These prices are reported in the columns European C_{BS} and European P_{BS} of Table 1. The American call can also be priced according to the same formula, since we assume the underlying asset pays no dividends. The price is reported in column American C_{BS}. For the American put, we computed the price using a variant of the lattice method presented earlier. We call this method PDE, since it consists in solving a partial differential equation. In dimension 1, it is essentially equivalent to the Cox-Ross-Rubinstein binomial lattice method. We used 120 time steps for T (time to expiration) ranging from one to four months, and 210 time steps for $T = 7$ months. The corresponding price is reported in column American C_{PDE}. The SSAP prices were computed using $M = 100{,}000$ samples and $k = 100$ buckets. The number of time steps was set to $d = 10$ in all the experiments.

The observed differences between the SSAP prices and the reference prices are below 0.7%. The confidence interval values are less than 1% of the reference prices. American put prices given by the SSAP method are very accurate even when the difference from the European put prices are important (up to 30 cents). The computation time for a price using the SSAP method is about 21 seconds, compared with less than one second with a classical integration method (PDE). In dimension 1, classical finite difference of binomial lattice methods should be preferred to the SSAP method.

THREE UNDERLYING ASSETS

In this case the SSAP method only finds an approximation of the optimal price. However, numerical experiments show that the SSAP price always remains within a few cents of the optimal theoretical price.

The European and American call and put option prices can be computed by the PDE method. This integration requires 120 time steps for T = 1 and T = 4 months and 210 time steps for T = 7 months. These results are reported in the columns European C_{PDE}, P_{PDE} and American C_{PDE}, P_{PDE} in Tables 2 and 3. The SSAP method was run using M = 100,000 samples and k = 100 buckets. The number of time steps was set to d = 10 in all the experiments. Results are presented in Tables 2 and 3.

The observed differences between the SSAP prices and the reference prices are below 1%. The confidence interval value is less than 1% of the reference prices except for very low prices, where it remains under 1 cent. American put prices given by the SSAP method are very accurate. The computation time of a price using the SSAP method is about 32 seconds, compared with 202 seconds for the classical integration method (PDE). In dimension 3, the SSAP method is as accurate and about six times faster than the classical integration method.

1. Linear behaviour of computation time for $0 < n \leq 60$

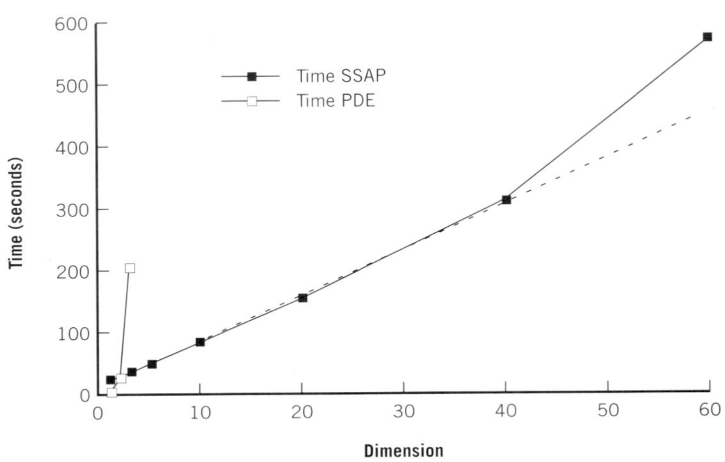

Table 4. Prices of a call option with 10 underlying assets

Call option prices: $x_1(0) = \ldots = x_{10}(0) = \$40$, $\sigma_1 = \ldots = \sigma_{10} = 40\%$, r = 5%

Parameters			European		American	
ρ	T	K	C_{SSAP}	4stdev	C_{SSAP}	4stdev
		35	12.66	0.010	12.66	0.010
	1	40	7.68	0.020	7.68	0.020
		45	2.98	0.020	2.98	0.020
		35	21.53	0.060	21.54	0.050
0%	4	40	16.62	0.030	16.62	0.030
		45	11.76	0.040	11.76	0.040
		35	27.91	0.060	27.92	0.060
	7	40	23.06	0.060	23.08	0.060
		45	18.24	0.040	18.25	0.040
		35	10.36	0.008	10.36	0.008
	1	40	5.54	0.010	5.54	0.010
		45	1.90	0.010	1.90	0.010
		35	16.52	0.020	16.53	0.020
50%	4	40	11.87	0.020	11.87	0.020
		45	7.81	0.040	7.81	0.040
		35	20.91	0.060	20.92	0.050
	7	40	16.38	0.040	16.38	0.040
		45	12.28	0.020	12.28	0.020
		35	5.41	0.003	5.42	0.010
	1	40	1.93	0.006	1.93	0.010
		45	0.42	0.004	0.42	0.004
		35	6.93	0.006	6.95	0.010
100%	4	40	4.00	0.012	4.02	0.020
		45	2.11	0.006	2.11	0.010
		35	8.14	0.010	8.16	0.020
	7	40	5.40	0.016	5.42	0.020
		45	3.44	0.012	3.45	0.020

TEN UNDERLYING ASSETS

We have previously compared the efficiency and accuracy of the SSAP method with that of the classical integration method (PDE). Here, we report results obtained with the SSAP method on American option pricing problems with 10 underlying assets (see Tables 4 and 5). Since no other method exists to compare to our results, and since the SSAP method only provides an approximation of the optimal price, we cannot guarantee the accuracy of the American premiums reported below. However, both the observed confidence intervals and the fact that the SSAP prices for the American calls without dividends equal the European prices lead us to believe that the SSAP method is reliable in general on 10-dimensional American pricing problems. The parameters of the SSAP method are again M = 100,000 and k = 100.

The differences between European and American call prices are less than 0.5%. Since the payoff is the maximum of n underlying assets prices without dividends, these two prices should indeed be identical. Confidence intervals are around 1% in the worst case. The computation time using the SSAP method is about 82 seconds.

EXPERIMENTAL TIME COMPLEXITY

We analyse in this subsection the computation time required by the SSAP method as a function of the number, n, of underlying assets. We compare these results when possible to those of classical integration methods (PDE). Observed

2. Quadratic behaviour of computation time for $0 < n \leq 400$

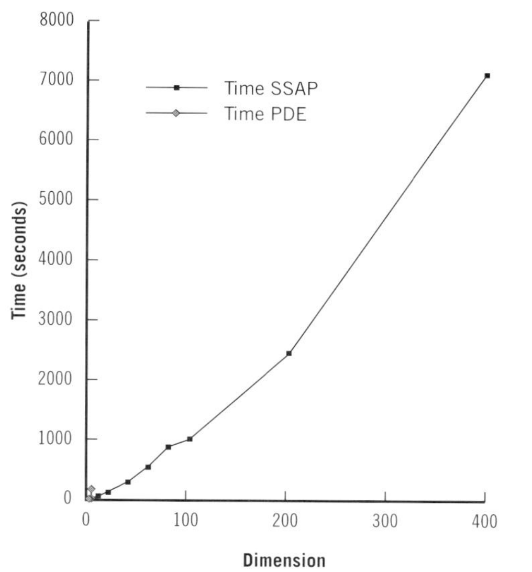

Table 5. Prices of a put option with 10 underlying assets

Put option prices: $x_1(0) = \ldots = x_{10}(0) = \$40$, $\sigma_1 = \ldots = \sigma_{10} = 40\%$, $r = 5\%$

Parameters			European		American	
ρ	T	K	P_{SSAP}	4stdev	P_{SSAP}	4stdev
		35	0.00	0.0000	0.00	0.0000
	1	40	0.00	0.0000	0.00	0.0000
		45	0.28	0.0060	5.00	0.0000
		35	0.00	0.0000	0.00	0.0000
0%	4	40	0.00	0.0008	0.01	0.0008
		45	0.06	0.0060	5.00	0.0000
		35	0.00	0.0000	0.00	0.0000
	7	40	0.00	0.0008	0.01	0.0010
		45	0.04	0.0030	5.00	0.0000
		35	0.00	0.0008	0.00	0.0006
	1	40	0.16	0.0030	0.26	0.0040
		45	1.49	0.0120	5.00	0.0000
		35	0.05	0.0040	0.07	0.0030
50%	4	40	0.32	0.0060	0.52	0.0060
		45	1.17	0.0200	5.00	0.0000
		35	0.10	0.0060	0.15	0.0060
	7	40	0.42	0.0120	0.69	0.0080
		45	1.18	0.0200	5.00	0.0000
		35	0.24	0.0020	0.24	0.0020
	1	40	1.73	0.0060	1.75	0.0080
		45	5.20	0.0040	5.27	0.0100
		35	1.30	0.0060	1.32	0.0100
100%	4	40	3.28	0.0160	3.34	0.0120
		45	6.31	0.0060	6.45	0.0100
		35	2.06	0.0100	2.11	0.0160
	7	40	4.18	0.0160	4.29	0.0160
		45	7.08	0.0150	7.30	0.0240

computation times are presented in Table 6. The number of Monte Carlo samples is $M = 100{,}000$. The number of cells is $k = 100$ as before.

Figure 1 shows that, for up to 40 underlying assets, the computation time is linear in n. The dominating term is the computation of the payoff function, which is linear in n $(M \times n \times d)$. Figure 2 shows that for larger n the complexity is quadratic in n $(M \times n^2 \times d)$, as expected.

For $n = 2$, the SSAP is comparable to the classical integral method. For $n = 3$, the SSAP method is faster by a factor of six. For $n > 3$, the SSAP method is the only one which can compute the price accurately. The integral (PDE) method is implemented using 60 time steps in the Cox-Rubinstein tree to obtain comparable precision. The exercise condition is also applied only 10 times during the life period of the option.

Conclusion

In this paper, we described a systematic numerical technique for pricing arbitrarily complex American contingent claims, ie generalised option contracts with possibilities of early exercise. Besides its obvious applications to trading and hedging in organised and OTC capital markets, American security pricing has many important applications in various areas of risk management, such as asset and liability management, and corporate investment decision-making. Using this technique, we were able to compute the prices of complex American instruments in a few tens of seconds on a workstation, and within a few seconds on a network of workstations.

Our approach essentially relies on appropriate state aggregation techniques that circumvent the intractability of the computation of the early exercise boundary, combined with a classical Monte Carlo simulation for the computation of the conditional probabilities in the backwards pricing formula. We call this method stratified

Table 6. Computation times as functions of the dimension

Dimension (N)	Time SSAP method (seconds)	Time PDE method (seconds)
1	21	0.12
2	25	23
3	32	202
5	46	Out of memory
10	81	Out of memory
20	151	Out of memory
40	309	Out of memory
60	567	Out of memory
80	894	Out of memory
100	1018	Out of memory
200	2438	Out of memory
400	7069	Out of memory

state aggregation along the payoff function (SSAP). We have successfully implemented the SSAP method for problems with up to 400 dimensions. To the best of our knowledge, no other method has ever been developed to date for pricing American contingent claims with many (more than three or four) underlying assets.

We feel that the method presented in this paper and the experimental results thus obtained make it possible to realistically envision the use of multidimensional stochastic models for practical real-world quantitative risk management problems. This capability of computing the joint influences of several tens of risk factors, such as interest rates of various terms in different currencies, equities and commodities of various kinds, and any other relevant economic variables, may dramatically increase the competitive advantage of quantitative methods over more traditional analytical techniques. An application of particular interest is the pricing and hedging of complex derivatives offered on international OTC markets. We plan to backtest on actual market data the performance of the SSAP method as compared to more classical delta-hedging techniques currently used on capital markets.

1 *This research benefited from discussions with Bruno Langlois. The authors wish to thank two anonymous referees for helpful comments.*

BIBLIOGRAPHY

Barone-Adesi, G., and R. Elliott, 1991, "Approximations for the Values of American Options", *Stochastic Analysis and Applications* 9, pp. 115-31.

Barone-Adesi, G., and R.E. Whaley, 1987, "Efficient Analytic Approximation of American Option Values", *Journal of Finance* 42, pp. 301-20.

Barraquand, J., 1995, "Numerical Valuation of High Dimensional Multivariate European Securities", *Management Science*.

Barraquand, J., and T. Pudet, 1995, "Pricing of American Path-Dependent Contingent Claims", *Mathematical Finance*.

Bellman, R., 1957, *Dynamic Programming*, Princeton University Press.

Bensoussan, A., 1984, "On the Theory of Option Pricing", *Acta Applicandae Mathematicae* 2, pp. 139-58.

Bertsekas, D.P., 1987, *Dynamic Programming, Deterministic and Stochastic Models*, Englewood Cliffs, N.J., Prentice Hall, Inc.

Bossaerts, P., 1989, "Simulation Estimators of Optimal Early Exercise", *unpublished manuscript*, Graduate School of Industrial Administration, Carnegie Mellon University.

Boyle, P.P., 1977, "Options: A Monte Carlo Approach", *Journal of Financial Economics* 4, pp. 323-38 (Chapter 1 of this publication).

Boyle, P.P., 1989, "The Quality Option and the Timing Option in Futures Contracts", *Journal of Financial Economics* 44, pp. 101-13.

Boyle, P.P., J. Evnine and S. Gibbs, 1989, "Numerical Evaluation of Multivariate Contingent Claims", *The Reviewournal of Financial Studies* 2, pp. 241-50.

Boyle, P.P., and Y. Tse, 1990, "An Algorithm for Computing Values of Options on the Maximum or Minimum of Several Assets", *Journal of Financial and Quantitative Analysis* 25(2), pp. 215-27.

Brennan, M., and E. Schwartz, 1978, "Finite-Difference Methods and Jump Processes Arising in the Pricing of Contingent Claims: A Synthesis", *Journal of Financial and Quantitative Analysis* 13, pp. 461-74.

Brennan, M., and E. Schwartz, 1979, "A Continuous Time Approach to the Pricing of Bonds", *Journal of Banking and Finance* 3, pp. 133-55.

Cheng, S., 1987, "Pricing Models for Multiple Currency Option Bonds", *unpublished manuscript*, Graduate School of Business, Stanford University.

Cox, J.C., and M. Rubinstein, 1985, *Options Markets*, Englewood Cliffs, N.J., Prentice Hall.

Cox, J.C., S.A. Ross and M. Rubinstein, 1979, "Option Pricing: A Simplified Approach", *Journal of Financial Economics* 7, pp. 229-63.

Duffie, D., 1988, *Security Markets: Stochastic Models*, Boston, Academic Press.

Duffie, D., 1992, *Dynamic Asset Pricing Theory*, Princeton University Press.

Fabozzi, F.J., and I.M. Pollack, 1987, *The Handbook of Fixed Income Securities*, second edition, Homewood, Illinois, Dow Jones-Irwin.

Geske, R., and H. Johnson, 1984, "The American Put Option Valued Analytically", *Journal of Finance* 39, pp. 1511-24.

Harrison, J.M., and D.M. Kreps, 1979, "Martingales and

Arbitrage in Multiperiod Securities Markets", *Journal of Economic Theory* 20, pp. 381–408.

Harrison, J.M., and S.R. Pliska, 1981, "Martingales and Stochastic Integrals in the Theory of Continuous Trading", *Stochastic Processes and their Applications* 11, pp. 261–71.

Heath, D.R., R. Jarrow and A. Morton, 1992, "Bond Pricing and the Term Structure of Interest Rates: A New Methodology for Contingent Claims Valuation", *Econometrica* 60(1), pp. 75–105.

Hutchinson, J.M., and S.A. Zenios, 1991, "Financial Simulations on a Massively Parallel Connection Machine", *International Journal of Supercomputer Applications* 5(2), pp. 27–45.

Jaillet, P., D. Lamberton and B. Lapeyre, 1988, "Inéquations Variationelles et Théorie des Options", *Comptes Rendus de l'Académie des Sciences de Paris* 307, pp. 961–65.

Johnson, H., 1987, "Options on the Maximum or the Minimum of Several Assets", *Journal of Financial and Quantitative Analysis* 22(3), pp. 277–82.

Karatzas, M., 1988, "On the Pricing of American Options", *Applied Mathematics and Optimization* 17, pp. 37–60.

Karatzas, I., and S.E. Shreve, 1988, *Brownian Motion and Stochastic Calculus*, Springer-Verlag.

Kushner, H.J., and P.G. Dupuis, 1992, *Numerical Methods for Stochastic Control Problems in Continuous Time*, Springer Verlag.

McKean, H., 1965, "Appendix: Free Boundary Problem for the Heat Equation Arising from a Problem in Mathematical Economics", *Industrial Management Review* 5, pp. 32–39.

Merton, R., 1973, "The Theory of Rational Option Pricing", *Bell Journal of Economics and Management Science* 4, pp. 141–83.

Myneni, R., 1992, "The Pricing of American Options", *Annals of Applied Probability* 2, pp. 1–23.

Schwartz, E.S., and W.N. Torous, 1989, "Prepayment and the Valuation of Mortgage-Backed Securities", *Journal of Finance* 44.

Stulz, R.M., 1982, "Options on the Minimum or the Maximum of Two Risky Assets: Analysis and Applications", *Journal of Financial Economics* 10, pp. 161–85.

Monte Carlo Methods for Pricing High-Dimensional American Options: An Overview

Mark Broadie and Paul Glasserman[1]
Graduate School of Business, Columbia University

American-style securities are those in which the holder or issuer can make decisions which influence the timing or magnitude of the cashflows. For example, standard American options allow the holder to choose when to exercise the option. Shout options allow the holder to choose when to "shout" and thereby lock in the intrinsic value of the option at the time of the shout. American swaptions give the holder the right to choose when to enter into a swap. Callable bonds give the issuer the right to buy back the bond at certain times and prices. Many considerations may influence the decision to exercise such options, but the arbitrage-free price is determined by the value-maximising strategy. Thus, pricing American-style securities requires the determination of optimal decision strategies.[2]

Closed-form expressions for the prices of American-style securities are rare, and an extensive literature on numerical pricing procedures has developed. Major approaches include lattice methods (for example, binomial and trinomial methods), techniques based on solving partial differential equations, integral equations, and variational inequalities. Finding optimal decision strategies is a dynamic programming problem, and these approaches share the feature that they work backwards in time in order to solve the optimal decision problem. They have been primarily applied to models where there is a single underlying asset or a single source of uncertainty, that is, the "one-dimensional" case. Advances in numerical procedures coupled with increases in computing power have led to extremely fast solution times for these one-dimensional problems. Indeed, several methods are fast enough to price large portfolios of American securities accurately in real time.

Many financial products have been introduced whose value depends on two or more underlying assets. Examples include basket options (options on the average of several underlying assets), outperformance options (options on the maximum of underlying assets), spread options (options on the difference of two assets), and quanto options (options whose payoff is adjusted by some quantity, typically a foreign exchange rate). Even when there is a single underlying asset, there has been a trend toward pricing models with multiple sources of uncertainty. Examples include models with stochastic volatility, credit or default risk, and stochastic convenience yields. The trend toward higher-dimensional models is especially evident in the interest rate area, where yield curve models with two to four factors are common and models with up to 10 factors are being tested. Multi-factor models also arise in the area of real options – options on real assets like land or technology or options to initiate or abandon a project.

One approach to pricing American securities with multiple assets or multiple state variables is to generalise the highly successful one-dimensional methods. Unfortunately, all of the major approaches mentioned before share the drawback that their computing requirements, both memory and computation time, are exponential in the problem dimension. In practice, this means that the methods are ineffective for more than about three state variables. In order to price high-dimensional European-style securities, ie those in which the holder does not have decisions to make, Monte Carlo simulation is often used. The main reason is that the convergence rate of Monte Carlo is typically independent of the problem dimension.[3] Simulation methods generate trajectories of assets or state variables forward in time. While the forward approach works well for European securities, it seems incompatible with the backward approach required to determine optimal decisions for pricing American securities.

The prevailing view was that simulation methods were not applicable to pricing American securities. For example, Hull (1997) writes: "A limitation of the Monte Carlo simulation approach is that it can only be used for European-style derivatives." In a similar vein, Campbell, Lo, and MacKinlay (1997) write that the "Monte Carlo approach may only be applied to European options, options that cannot be exercised early." This view has begun to change as several hybrid simulation-dynamic programming algorithms have been proposed to price American options.

Bossaerts (1989) and Tilley (1993) – also Chapter 14 of this publication, represent early attempts to extend the applicability of simulation to the pricing of American securities. Numerical results in both papers are limited to the case of a single underlying asset and extensions to higher-dimensional problems appear problematic at best. Bossaerts' method assumes a parametric form of the exercise decision. Tilley's procedure involves "bundling" paths into groups according to the underlying asset prices and then using these paths to estimate a single continuation value for all paths within a bundle. Further analysis of Tilley's method is given in Carriere (1996). Barraquand and Martineau (1995) – also Chapter 15 of this publication propose a stratification method which also involves partitioning paths, but they partition in the payoff space instead of the state space. The key point is that they reduce a high-dimensional problem to a single dimension since the exercise decision can only depend on the intrinsic option value, ie its payoff upon exercise. The resulting one-dimensional problem can be solved quickly by standard dynamic programming techniques. Their method is very fast at generating an approximate price. The quality of the approximation depends primarily on the extent to which the optimal exercise policy can be represented by a function of the single payoff statistic. For general multi-dimensional problems, the optimal exercise decision will depend on all of the state variables, and the option payoff is not a sufficient statistic for determining whether exercise is optimal. This implies that the stratification algorithm will not converge to the true American price. Nonetheless, for some types of problems it might provide a very good approximate solution. Raymar and Zwecher (1997a) extend the payoff stratification approach of Barraquand and Martineau by basing the exercise decision on a partition of two state variables, rather than one. The stratification approach is applied in an interest rate setting in Carr and Yang (1996, 1997) and Raymar and Zwecher (1997b).

A pricing method based on simulated trees is proposed in Broadie and Glasserman (1997a). The method generates random trees of underlying assets and state variables and then applies dynamic programming. This method not only generates an approximate price but also gives a valid confidence interval for the price of a high-dimensional, finitely exercisable American option. Furthermore, the method is convergent as the computational effort increases. Although the work is linear in the problem dimension, it is exponential in the number of exercise opportunities because the random trees do not recombine. There are some securities to which this method is ideally suited. For example, American swaptions typically have a limited number of exercise opportunities. Over-the-counter options often have "structured exercise opportunities," ie where exercise is allowed only at a finite number of prespecified dates. In some cases, the options might be continuously exercisable, but other considerations limit the optimal exercise policy to a small number of dates. This occurs with call options on stocks that pay discrete dividends. In order to apply the simulated tree method to American options with continuous exercise or a finite but large number of exercise opportunities, Broadie, Glasserman and Jain (1997) suggest using Richardson extrapolation and present some computational results.

In the simulated tree method, each branch emanating from a single node generates it own independent sub-tree. While this independence facilitates the convergence and analysis of the algorithm, it also leads to a computation time requirement that is exponential in the number of exercise opportunities. The method proposed in Broadie and Glasserman (1997b) is based on a different object termed a *stochastic mesh*. In this method, all of the information from one time step is used in estimating prices at the previous time step. The interlocking nature of the mesh introduces a dependence in the prices at different nodes at the same point in time. This considerably complicates the development and analysis of the method, but the main advantage is that the work is linear in the number of exercise opportunities. The stochastic mesh method gives confidence intervals and is convergent as the work increases. Numerical results given in Broadie and Glasserman (1997b) show that it is a viable method for pricing high-dimensional options when there is a finite, but possibly large, number of exercise dates.

Another approach to the problem is to specify a parametric family to represent possible exercise regions. Then simulation is combined with optimisation in order to find optimal decisions within the parametric family. The optimisation is carried out using techniques known as functional optimisation, sample path optimisation or stochastic approximation. Variations of this approach applied to option pricing are given in Bossaerts (1989), Fu and Hu (1995), Grant, Vora, and Weeks (1996, 1997), Li and Zhang (1996), and Gürkan, Özge and Robinson (1996, 1997). A related approach uses a parametric approximation to the dynamic programming value function, rather than the exercise region. Tsitsiklis and Van Roy (1997) develop such a method based on the idea of "neuro-dynamic programming" described in Bertsekas and Tsitsiklis (1996). The effectiveness of these approaches seems to be predicated on finding good parameterisations of exercise regions or value functions in higher-dimensional space. While this presents difficulties for applying these approaches to general models or payoff functions, it also offers opportunities to exploit the structure of specific instruments. Although some numerical tests have been done on single-asset problems, there has been relatively little investigation of these methods on multiple asset or multiple state variable problems, and so it is unclear how their complexity and effectiveness depend on the problem dimension. An alternative to parameterising the exercise region is to numerically solve for an exercise strategy in a low-dimensional model chosen to be similar to the higher-dimensional problem of interest. Clewlow and Strickland (1997) use a two-factor Hull–White model to generate an exercise strategy for a multifactor HJM model.

Although much of the interest in simulation-based approaches is driven by the desire to solve high-dimensional problems, there are many low-dimensional problems that are difficult to solve using standard techniques. Many of the new simulation-based approaches are potentially useful for low-dimensional problems because of the ease with which they can handle a wide range of models and payoff structures. One example is Dammon and Spatt (1992), who use a simulation-based algorithm to value dividend reinvestment and voluntary purchase plans.

While this paper is concerned specifically with the problem of pricing American-style securities that arise in financial theory and practice, large-scale dynamic programming models arise in many areas, including economics, industrial organisation, production planning and inventory control, reliability theory, and others. Many references to the literature in these areas can be found in the papers by Eckstein and Wolpin (1989) and Rust (1994) and the books by Bertsekas (1987) and Bertsekas and Tsitsiklis (1996). Overviews of numerical dynamic programming methods in economics are given in Judd (1997) and Rust (1996). A technique for solving discrete dynamic programming models using simulation and regression is proposed in Keane and Wolpin (1994). They give numerical results of their method applied to a model of occupational choice. Algorithms based on random successive approximation and random multigrids are proposed in Rust (1997). The theoretical analysis provided shows that the methods have provably good properties for a class of discrete dynamic programming problems. These methods require that the underlying state variables be restricted to a compact set, which limits the applicability to option pricing.

General references for Monte Carlo simulation are the book by Fishman (1996) and the paper by Geweke (1996). The first application of simulation to pricing European options was Boyle (1977). A recent survey of simulation methods applied to finance is given in Boyle, Broadie and Glasserman (1997) – see Chapter 2 of this publication.

In the next section we present descriptions of several methods for pricing American options.

Algorithm descriptions

In this section we give more detailed descriptions of four simulation-based methods for pricing American options. We begin with the bundling algorithm of Tilley (1993). Then we present the state stratification method of Barraquand and Martineau (1995), the simulated tree method of Broadie and Glasserman (1997a), and the stochastic mesh method of Broadie and Glasserman (1997b). The efficiency of all the simulation-based methods for pricing American options can be improved using standard variance reduction techniques. In the interest of brevity, we do not describe these important enhancements to the methods, but descriptions of many variance reduction techniques are given in Boyle, Broadie and Glasserman (1997) – also Chapter 2 of this publication, Fishman (1996), and Geweke (1996). All of the methods are used to price American options with a finite number of exercise opportunities. The section concludes by examining the convergence of the price of a discretely exercisable option to its continuously exercisable counterpart.

We frame the presentation of the methods in terms of the problem of computing the American option price Q, where

$$Q = \max_{\tau} E\left[h(\tau, S_\tau)\right] \tag{1}$$

and $S_t = (S_t^1, \ldots, S_t^n)$ is a vector-valued Markov process with a given initial state S_0. The components of S_t could represent asset prices, interest rates, exchange rates, or general state variables of the process. The payoff from exercise at time t in state $S_t = x$ is $h(t,x)$, and we assume $h(t,x) \geq 0$. For ease of notation we ignore discounting, but all of the methods we describe can take discounting into account in the obvious way. Exercise can occur at times $0 = t_0 < t_1 < \cdots < t_d = T$, which will often be written more simply as $t = 0,1,\ldots,T$. The stopping time, τ, is restricted to the finite set $t = 0,1,\ldots,T$.[4] It is common to have the transitions of the process S_t governed by risk-neutral probabilities.[5] Note that Q can also be written as

$$Q = E\left[h(\tau^*, S_{\tau^*})\right] \tag{2}$$

where τ^* is an optimal stopping time.

1. Illustration of Tilley's bundling method

The figure shows m = 9 sorted paths grouped into k = 3 bundles each at time t_4.

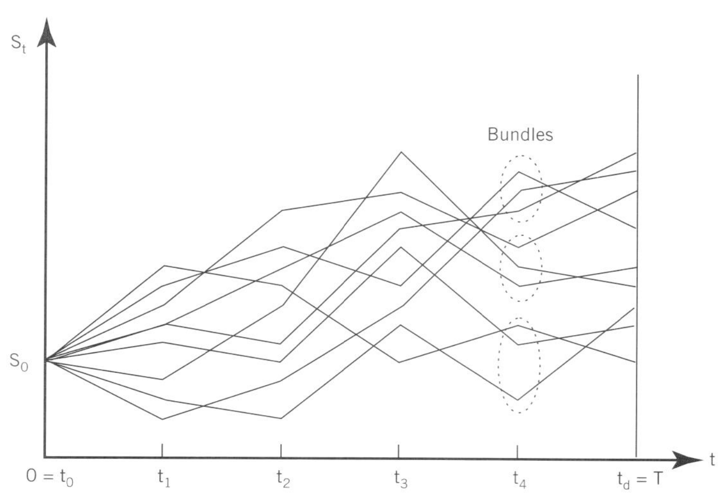

One idea for estimating the price Q is to simulate a path $S = (S_0, S_1, \ldots, S_T)$ and form the path estimate

$$\hat{P} = \max_{t=0,\ldots,T} h(t, S_t) \tag{3}$$

With this estimator, exercise occurs at the most favourable time for each path. It is called the *perfect foresight* estimator because the decision to exercise at some time t takes into account information from the future times $t + 1,\ldots,T$. The estimator is biased high because $\max_{t=0,\ldots,T} h(t,S_t) \geq h(\tau^*, S_{\tau^*})$ and hence $E\left[\max_{t=0,\ldots,T} h(t,S_t)\right] \geq E\left[h(\tau^*, S_{\tau^*})\right] = Q$. The perfect foresight estimator bias is typically severe, ie, $E\left[\hat{P}\right] \gg Q$.

Suppose that two paths are identical at time t and they separate thereafter. The estimator defined in equation (3) could very well set the decision to exercise at time t on one path but not on the other because of what happens at a later time. The policy defined by (3) is not a stopping time, and this leads to the bias problem. Tilley's idea is to force paths that are close to each other at time t to use the same estimate of the continuation value for determining the exercise decision. This will lead to a policy that is closer to a stopping time, and so it should reduce the "look-ahead" bias problem.

BUNDLING ALGORITHM

The description of the bundling algorithm in Tilley (1993) is restricted to the one-dimensional case, ie, $n = 1$. The method begins by simulating m paths, denoted $S(i) = (S_0, S_1(i), \ldots, S_T(i))$, for $i = 1,\ldots,m$. It then proceeds backwards from time T to time 0. The option value assigned by the algorithm to path i at time t will be denoted $\hat{Q}(t, S_t(i))$. At time T, the value of path i is assigned its payoff. Thus $\hat{Q}(T, S_T(i)) = h(T, S_T(i))$. At time $T - 1$, the paths are sorted in increasing order so that $S_{T-1}(i_1) \leq S_{T-1}(i_2) \leq \cdots \leq S_{T-1}(i_m)$. The sorted paths are then grouped into bundles, so that each bundle contains an equal number of paths. For example, bundle 1 contains paths $i_1,\ldots,i_k$, bundle 2 contains paths $i_{k+1},\ldots,i_{2k}$, etc (see Figure 1). To estimate the optimal action to take at time $T - 1$ for bundle 1, first estimate the continuation value as $C_1 = \Sigma_{j=1}^{k} \hat{Q}(T, S_T(i_j))/k$. Then for each path in bundle 1, assign the option value $\hat{Q}(T-1, S_T(i)) = \max(h(T-1, S_{T-1}(i)), C_1)$, for $i = i_1,\ldots,i_k$. Repeat for the remaining bundles at time $T - 1$. Then repeat this procedure for times $T - 2,\ldots,0$.[6]

The key idea is that the exercise policy will be closer to a stopping time since the same estimate of the continuation value is used for all paths within a bundle. If each bundle only contains a single path, then the algorithm coincides with the perfect foresight estimator. To get a good simulation estimate of the continuation value, each bundle should contain many paths. However, using the same continuation value for all of the paths within a bundle makes sense only if the paths are quite close to each other, and this argues for having fewer paths in each bundle. Thus there is a tradeoff between having too many or too few paths per bundle.

The bundling algorithm requires that the paths be stored so they can be sorted at each time step. The storage and sorting requirements can be significant with a large number of paths. A more serious issue is that the bundling idea based on a linear ordering does not easily generalise to higher dimensions. Even if this bundling could be accomplished, it is likely that either (i) each bundle will contain too few paths to give a good estimate of the continuation value, or (ii) the paths will be so far apart that the single estimated continuation value will be a poor estimate for many of the paths in the bundle.

PAYOFF STRATIFICATION ALGORITHM

In common with Tilley's method, Barraquand and Martineau (1995) also partition simulated paths into groups. However, they partition the one-dimensional option payoff space, not the n-dimensional state space. In effect, they reduce a high-dimensional problem to one dimension. As we will see, their approach also alleviates the need to store all of the paths simultaneously.

The main steps in the Barraquand and Martineau method are: partition the payoff space into K bins, then simulate m paths to estimate transition probabilities and path payoffs and, finally, use standard dynamic programming in the payoff space to estimate the option price. A more detailed description is given next.

In the first step, the payoff space is partitioned into K bins, indexed $k = 1,\ldots,K$. There are many ways to do this, and, for the standard geometric Brownian motion model for asset prices, Barraquand and Martineau recommend using bins that increase exponentially in size as the payoff increases. Next, simulate paths $S(i) = (S_0,S_1(i),\ldots,S_T(i))$, for $i = 1,\ldots,m$. Each element $S_t(i)$ is an n-vector, ie $S_t(i) = (S_t^1(i),\ldots,S_t^n(i))$. For each payoff bin k at time t, let $a_t(k)$ record the number of paths that fall into the bin. For each pair of bins k and ℓ at consecutive times t and $t + 1$, let $b_t(k,\ell)$ record the number of paths that fall into both bins. Also, for each bin k at time t, record the sum of the payoff values, $c_t(k) = \Sigma h(t,S_t(i))$, where the sum is over all paths i that fall into bin k at time t. As each path is generated, the records $a_t(k)$, $b_t(k,\ell)$, and $c_t(k)$ are incremented as indicated, and then the path is discarded.

The final step is to solve the resulting one-dimensional dynamic programming problem. Let $\hat{Q}(t,k)$ denote the estimated option value for bin k at time t. Starting at time T, set $\hat{Q}(T,k) = c_T(k)/a_T(k)$ for each bin $k = 1,\ldots,K$. Thus, the estimated value is the average payoff of the paths that finish at bin k at time T. Next work backwards from time $t = T - 1$ to time $t = 0$. For bin k at time t, set the estimated price to the maximum of the average payoff or the estimated continuation value. The average payoff is $c_t(k)/a_t(k)$. To estimate the continuation value, first approximate the transition probability from bin k at time t to bin ℓ at time $t + 1$ by $p_t(k,\ell) = b_t(k,\ell)/a_t(k)$ (see Figure 2). Then estimate the continuation value by $\Sigma_{\ell=1}^{K} p_t(k,\ell) \times \hat{Q}(t+1,\ell)$. In short, the recursive step is to set

$$\hat{Q}(t,k) = \max\left(\frac{c_t(k)}{a_t(k)}, \sum_{\ell=1}^{K} p_t(k,\ell)\hat{Q}(t+1,\ell)\right)$$

for bins $k = 1,\ldots,K$. This process is repeated backwards in time until time $t = 0$.

2. Illustration of Barraquand and Martineau's payoff stratification method

The figure shows $a_{t_2}(k) = 4$ paths passing through bin k at time t_2. Only one of these paths enters bin ℓ at time t_3, so the estimated transition probability from bin k to ℓ is $p_{t_2}(k,\ell) = 1/4$.

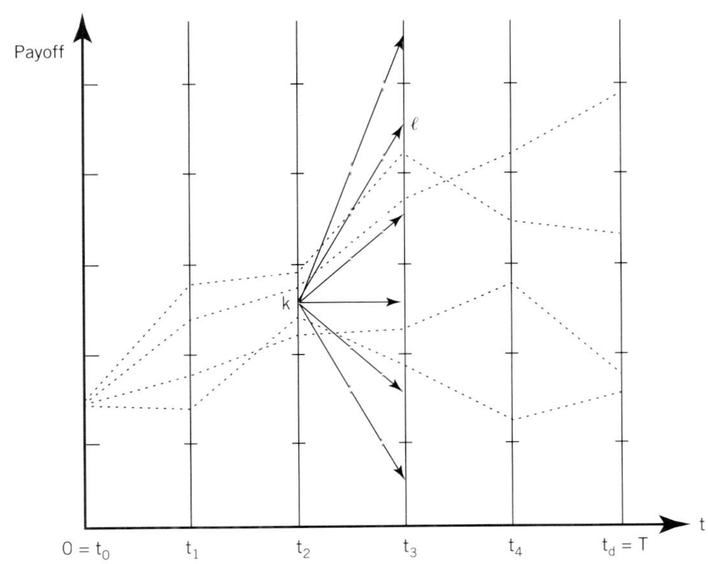

The main work in the method is in generating the paths. This effort is proportional to $n \times m \times d$, ie the work is linear in the problem dimension, the number of simulated paths, and the number of exercise opportunities. The work in the one-dimensional dynamic program is proportional to $K^2 \times d$, and this is typically negligible compared to the work in generating the paths. In short, it is very fast, in the sense that it does not require much more work to price American options than straightforward simulation would require for European options. The memory requirements of the method are modest, proportional to $K^2 \times d$ to store the transition probabilities.

The payoff stratification method is easy to implement, fast to run, and enjoys the flexibility typical of simulation-based methods. The main drawback of the method is that it does not converge to the correct value, even when the number of paths and the number of bins increase to infinity.

To illustrate the non-convergence result, consider using the method to price an American call option on the maximum of two dividend-paying underlying assets. This has become a standard test problem, and numerical results for this type of option are given in Barraquand and Martineau (1995). The payoff of this option is $h(t,S_t^1,S_t^2) = \max(\max(S_t^1,S_t^2) - K,0)$. At a fixed time $t < T$, the optimal exercise region is composed of two disjoint areas, as illustrated in Figure 3.[7] With payoff stratification only one decision is allowed along any "iso-payoff" curve, $h(t,S_t^1,S_t^2) = c$. But the optimal exercise policy involves both decisions along an iso-payoff curve, and so payoff stratification cannot converge to the optimal exercise policy. For general payoff functions of n state variables, the optimal exercise policy will typically depend on all n underlying state variables. Reducing the dimension of the problem to a single state variable, the payoff, leads to sub-optimal exercise decisions and implies that the stratification method will not converge to the correct value. Raymar and Zwecher (1997a) show that better results can be obtained by partitioning on two state variables rather than one.

3. Illustration of non-convergence with payoff stratification

The payoff of the max-option is $h(t,S_t^1,S_t^2) = \max(\max(S_t^1,S_t^2) - K,0)$, where the strike is $K = 100$. At a fixed time $t < T$, it is optimal to exercise in the two shaded areas. Along parts of the dashed iso-payoff curve labelled "Payoff = 30" it is optimal to exercise and on other parts it is optimal to continue. Stratifying on the payoff means that only one decision is ever possible along each iso-payoff curve.

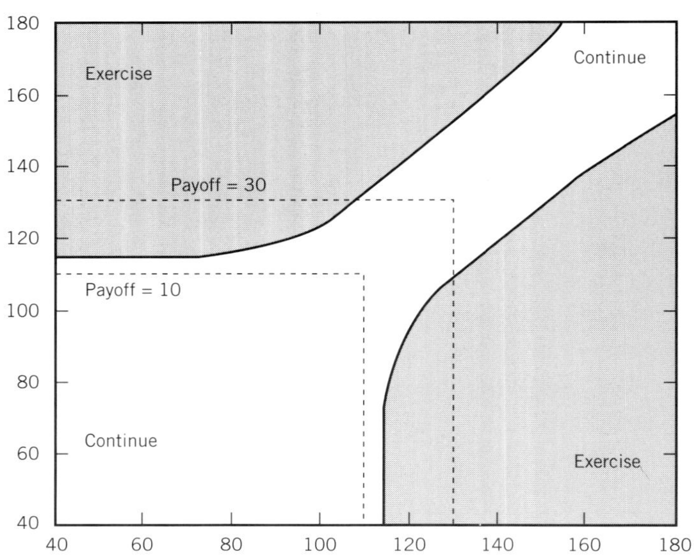

SIMULATION ESTIMATOR BIAS

The perfect foresight estimator defined in equation (3) is biased high because it takes into account information from the future. Estimators based on parametric representations of the exercise region are typically biased low. To see this, let τ denote the stopping time corresponding to *any* prespecified exercise policy. Since τ is not necessarily the optimal stopping time,

$$E[h(\tau,S_\tau)] \le \max_\tau E[h(\tau,S_\tau)] \equiv Q$$

and so the corresponding estimator is biased low. We call the estimator $\hat{Q}$ an *upper bound* if $E(\hat{Q}) \ge Q$, ie, if $\hat{Q}$ is biased high. Similarly, the estimator $\hat{q}$ is a *lower bound* if $E(\hat{q}) \le Q$, ie if $\hat{q}$ is biased low. In the absence of an unbiased estimator for the true price, Q, one can still obtain a valid confidence interval for Q using the following procedure proposed in Broadie and Glasserman (1997a). Compute N independent replications of the upper bound, $\hat{Q}$, and denote them $\hat{Q}^{(i)}$ for $i = 1,\ldots,N$. Combine them to form the estimate

$$\bar{Q}(N) = \frac{1}{N}\sum_{i=1}^{N} \hat{Q}^{(i)}$$

of the value $E(\hat{Q})$. Similarly, compute

$$\bar{q}(N) = \frac{1}{N}\sum_{i=1}^{N} \hat{q}^{(i)}$$

where $\hat{q}^{(i)}$, $i = 1,\ldots,N$, are independent replications of the lower bound, $\hat{q}$. Finally, form the

4. Illustration of confidence interval

Combining the upper end-point from the confidence interval for $E(\hat{Q})$ with the lower end-point from the confidence interval for $E(\hat{q})$ gives a valid confidence interval for **Q**.

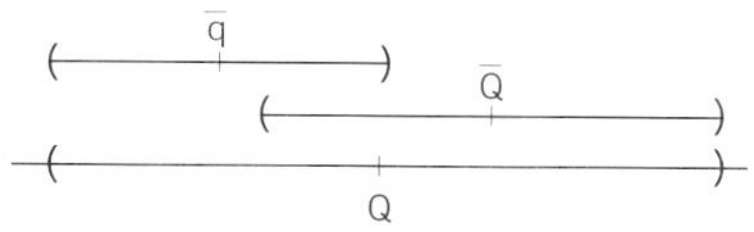

confidence interval

$$\left[\bar{q}(N) - z_{\alpha/2}\frac{s(\hat{q})}{\sqrt{N}}, \bar{Q}(N) + z_{\alpha/2}\frac{s(\hat{Q})}{\sqrt{N}}\right] \quad (4)$$

where $z_{\alpha/2}$ is the $1 - \alpha/2$ quantile of the standard normal distribution, and $s(\hat{q})$ and $s(\hat{Q})$ are the sample standard deviations of $\hat{q}$ and $\hat{Q}$, respectively.

By combining the upper confidence limit from the upper bound estimator and the lower confidence limit from the lower bound we obtain a valid confidence interval for the true price, Q, as illustrated in Figure 4. The resulting confidence interval will be conservative, ie the expected coverage will exceed the nominal coverage of $100(1 - \alpha)\%$, depending on the bias present in the estimators $\hat{Q}$ and $\hat{q}$.

Elements of high and low bias are present in the bundling and payoff stratification methods. High bias can arise because the same paths are used to estimate optimal exercise decisions and to compute option values ("look-ahead" bias). Low bias can occur because the estimated exercise strategy is not the optimal one. As the simulation effort is increased in the stratification method, the look-ahead bias will diminish, but since the estimated policy is not necessarily converging to the optimal policy, low bias will remain. If only a single estimator is used, it can be helpful to know the direction of the bias. In the previous two methods, an additional step can be added to accomplish this. As originally stated, both methods compute an approximation to the optimal exercise strategy. Once this policy is determined, another independent set of paths can be simulated and the option value under this policy can be estimated. This "additional pass" leads to an estimator that is biased low.

Biased estimators are often easy to define and compute. The difficulty lies in finding estimators that converge to the true value, Q, as the computational effort increases. The next two methods are similar in that they both generate lower and upper bounds that are asymptotically unbiased, ie converge to the correct value.

SIMULATED TREE ALGORITHM

In order to motivate this algorithm, consider the problem of estimating the American option price when there are only two possible exercise dates, at time $t = 0$ and $t = T$. In this case, the problem is to estimate

$$Q = \max(a,\mu) \quad (5)$$

where $a = h(0,S_0)$ is the *known* exercise value at $t = 0$, and $\mu = E(h(T,S_T))$ is the *unknown* continuation value.

Consider estimating Q using b independent and identically distributed samples $X_T(1),\ldots,X_T(b)$ from the distribution of S_T. Define

$$\Theta(X_T(1),\ldots,X_T(b),a) = \max(a,\bar{X}) \quad (6)$$

where $\bar{X} = \Sigma_{i=1}^{b} h(T,X_T(i))/b$. Then

$$\begin{aligned} E\left[\Theta(X_T(1),\ldots,X_T(b),a)\right] &= E\left[\max(a,\bar{X})\right] \\ &\geq \max(a,E\left[\bar{X}\right]) \\ &= \max(a,\mu) = Q \end{aligned}$$

The inequality on the second line is Jensen's. Hence Θ is biased high.

Although $\bar{X}$ is an unbiased estimate of μ, the max operator in the definition of Θ introduces bias into the estimation of Q. The high bias is the result of using the same samples $X_T(1),\ldots,X_T(b)$ to estimate the optimal decision and the continuation value. This bias can be eliminated by splitting the sample $X_T(1),\ldots,X_T(b)$ into two groups. The first group, $X_T(1),\ldots,X_T(b/2)$, is used to estimate the optimal decision and the second group, $X_T(b/2 + 1),\ldots,X_T(b)$, is used to estimate the continuation value. Formally, let $\tilde{X} = (2/b)\Sigma_{i=1}^{b/2} h(T,X_T(i))$ and $\check{X} = (2/b)\Sigma_{i=b/2+1}^{b} h(T,X_T(i))$ and define

$$\theta(X_T(1),\ldots,X_T(b),a) = \begin{cases} a & \text{if } \tilde{X} \leq a \\ \check{X} & \text{if } \tilde{X} > a \end{cases} \quad (7)$$

The expected value of this estimator can be written as

$$\begin{aligned} E\left[\theta(X_T(1),\ldots,X_T(b),a)\right] &= aP(\tilde{X} \leq a) + E(\check{X})P(\tilde{X} > a) \\ &= ap + \mu(1 - p) \\ &\leq \max(a,\mu) = Q \end{aligned}$$

In the second line p is defined as $P(\tilde{X} \leq a)$ and the third line follows since $0 \leq p \leq 1$. Hence θ is biased low. The low bias occurs because there is a positive probability that the estimated decision is not the optimal one. The estimators θ and Θ can be combined as in equation (4) to obtain a valid confidence interval for Q. Under mild restrictions on the distribution of S_T, both Θ and θ will be asymptotically unbiased and the confidence interval in equation (4) will shrink to the correct value as the branching parameter b increases.

When there are more than two exercise opportunities, a tree of the underlying assets and state variables is simulated. Starting from the initial state S_0 at time $t = 0$, b independent samples from the distribution of S_1 are generated. These are denoted $X_1(1),\ldots,X_1(b)$. From each node $X_1(i)$, b independent samples $X_2(i,j)$, $j = 1,\ldots,b$, from the conditional distribution of S_2 given $S_1 = X_1(i)$ are generated. This process is repeated until the last exercise opportunity (see Figure 5). The nodes in Figure 5 appear according to the order in which they are generated and not according to their node values, as would be the case in a lattice method. Also, each node represents an n-dimensional vector of underlying assets and state variables.

The estimators Θ and θ defined in equations (6) and (7), respectively, can be generalised to

5. Illustration of a simulated tree with b = 4 branches per node and exercise opportunities at times 0, t_1, and T

Connections between nodes illustrate dependence relations. For example, both $X_T(1,1)$ and $X_T(1,2)$ depend on $X_1(1)$ but neither depends on $X_1(2)$. The dashed box labelled 'Subtree 1' shows the subtree emanating from the node $X_1(1)$.

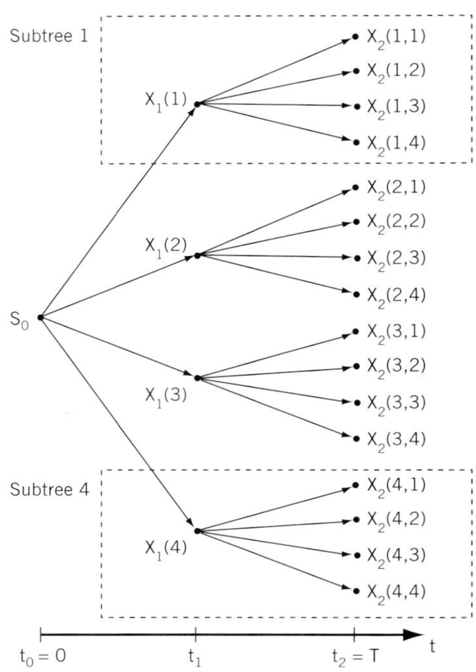

the simulated tree setting. To illustrate the definition of Θ, see subtree 1 emanating from node $X_1(1)$ in Figure 5. Viewed from this node, the pricing problem simplifies to the two exercise date problem in equation (5). Hence, $\Theta_1(1)$ can be defined using equation (6), and this estimate of the option value is assigned to node $X_1(1)$. Similarly, for $i = 2,\ldots,b$, $\Theta_1(i)$ can be computed corresponding to the estimated option values at nodes $X_1(i)$. Moving backwards to time 0, define $\Theta = \Theta_0$ using the analogue of equation (6). In particular, $\Theta_0 = \max(h(0,S_0),\bar{\Theta})$ with $\bar{\Theta} = \Sigma_{i=1}^{b}\Theta_1(i)/b$. The same recursive procedure can be used to define Θ for any simulated tree with a finite number of exercise opportunities.

We use Figure 5 to illustrate the definition of θ as well. Beginning at time t_1, define $\theta_1(i)$ at node $X_1(i)$, $i = 1,\ldots,b$, using equation (7). Moving backwards to time 0, define $\theta = \theta_0$ using the analogue of equation (7). In particular, define $\tilde{\theta}_1 = (2/b)\Sigma_{i=1}^{b/2}\theta_1(i)$ and $\breve{\theta}_1 = (2/b)\Sigma_{i=b/2+1}^{b}\theta_1(i)$. Then set $\theta = h(0,S_0)$ if $\tilde{\theta}_1 \le h(0,S_0)$. Otherwise, set $\theta = \breve{\theta}_1$. Thus, the first set of branches $(1, 2,\ldots,b/2)$ is used to estimate the optimal decision by checking whether $\tilde{\theta}_1 \le h(0,S_0)$ or $\tilde{\theta}_1 > h(0,S_0)$. If the estimated decision is to exercise, set θ to the exercise value. If the estimated decision is to continue, set θ to $\breve{\theta}_1$, the estimated continuation value. This recursive procedure can also be used to define θ when there are any finite number of exercise opportunities. More formal definitions of these estimators and variations are given in Broadie and Glasserman (1997a). They also prove that in the general setting with a finite number of exercise opportunities, Θ is an upper bound, θ is a lower bound, and both are asymptotically unbiased as the branching parameter b increases.

The non-recombining nature of the simulated trees leads to a work requirement that is exponential in the number of exercise opportunities. Stated differently, each branch emanating from a single node generates its own independent subtree. In Figure 5, the estimate of the option value at node $X_1(1)$ depends only on subtree 1. Potentially useful information from the other subtrees is ignored. In the next method, all of the information from one time step is used to estimate each of the option values at the previous time step.

STOCHASTIC MESH METHOD

As before, the problem is to compute the American option price

$$Q = \max_{\tau} E[h(\tau,S_\tau)]$$

where $S_t = (S_t^1,\ldots,S_t^n)$ is a vector-valued Markov process with a given initial state S_0 and τ is a stopping time taking values in the finite set $0 = t_0 < t_1 < \cdots < t_d = T$. Time will often be indexed more simply as $t = 0,1,\ldots,T$. Let $f(t,x,\cdot)$ denote the transition density of S_{t+1} given $S_t = x$ and let $f(t,\cdot)$ denote the marginal density of S_t (with S_0 fixed).

The stochastic mesh method generates random vectors $X_t(i)$ for $i = 1,\ldots,b$ and $t = 1,\ldots,T$. Since S_0 is given, set $X_0(1) = S_0$. The mesh points $X_t(i)$ are generated as independent and identically distributed samples from the *mesh density function* $g(t,\cdot)$. At this point the mesh density functions can be arbitrary, except for the technical requirement that $g(t,u) > 0$ if $f(t-1,x,u) > 0$ for some x. A specific choice is suggested later. An illustration of a stochastic mesh in dimension $n = 2$ is given in Figure 6. The arrows in Figure 6 illustrate how information is used in computing the mesh estimator. In particular, the information from all of the nodes at time t_2 is used to estimate each of the option values at time t_1.

The mesh estimator is defined in a recursive fashion. Starting at the option maturity, set

$$\hat{Q}(T,X_T(i)) = h(T,X_T(i)) \qquad (8)$$

for $i = 1,\ldots,b$. For times $t = T-1,\ldots,0$ and $i = 1,\ldots,b$, the mesh estimator is

$$\hat{Q}(t,X_t(i)) = \max\left(h(t,X_t(i)), \frac{1}{b}\sum_{j=1}^{b}\hat{Q}(t+1,X_{t+1}(j))w(t,X_t(i),X_{t+1}(j))\right) \qquad (9)$$

where $w(t,X_t(i),X_{t+1}(j)) \equiv f(t, X_t(i), X_{t+1}(j))/g(t+1, X_{t+1}(j))$ is a weight attached to the arc joining $X_t(i)$ at time t to $X_{t+1}(j)$ at time $t+1$. At time $t = 0$ only $i = 1$ is applicable in equation (9) and $\hat{Q} \equiv \hat{Q}(0,S_0)$ is the final mesh estimator of the true price Q.

Equation (9) says that the option value estimate $\hat{Q}(t,X_t(i))$ at node $X_t(i)$ is the maximum of the exercise value, $h(t,X_t(i))$ and the estimate of the continuation value. The continuation value is estimated by

$$\frac{1}{b}\sum_{j=1}^{b}\hat{Q}(t+1,X_{t+1}(j))\,\frac{f(t,X_t(i),X_{t+1}(j))}{g(t+1,X_{t+1}(j))}$$

The *likelihood ratio* $f(t,X_t(i),X_{t+1}(j))/g(t+1, X_{t+1}(j))$ adjusts for the fact that the mesh points $X_{t+1}(j)$ were generated by the density function $g(t+1,\cdot)$ and not from the natural density function $f(t, X_t(i), \cdot)$ of S_{t+1} given $S_t = X_t(i)$. The work in computing $\hat{Q}$ is proportional to $b^2 \times d$. That is, the work is quadratic in the mesh parameter, b, and linear in the number of exercise opportunities, d.

Unless one is careful about the choice of the mesh density function, the variance of the estimator $\hat{Q}$ can grow exponentially with the number of exercise opportunities. To avoid this problem, a choice recommended in Broadie and Glasserman (1997b) is

$$g(1,u) = f(0,S_0,u) \qquad (10)$$

and

$$g(t,u) = \frac{1}{b}\sum_{j=1}^{b} f(t-1,X_{t-1}(j),u) \text{ for } t = 2,\ldots,T \qquad (11)$$

Equation (11) says that the mesh density function $g(t,\cdot)$ is the average of the density functions $f(t-1, X_{t-1}(j),\cdot)$ of S_t given $S_{t-1} = X_{t-1}(j)$ for $j = 1,\ldots,b$. This choice also has the nice property that when applied to European options the recursive estimator defined in equation (9) simplifies to the estimator

$$\hat{Q} = \hat{Q}(0,S_0) = \frac{1}{b}\sum_{j=1}^{b} h(T,X_T(j)). \qquad (12)$$

6. Illustration of a stochastic mesh for n = 2, d = 2, and b = 3

For each time $t = t_1,t_2,\ldots,T$, random vectors $X_t(i)$ are generated for $i = 1,2,\ldots,b$. The arrows indicate how information is used. For example, the option value at node $X_1(2)$ is estimated using information from the nodes $X_2(i)$, $i = 1,\ldots,b$.

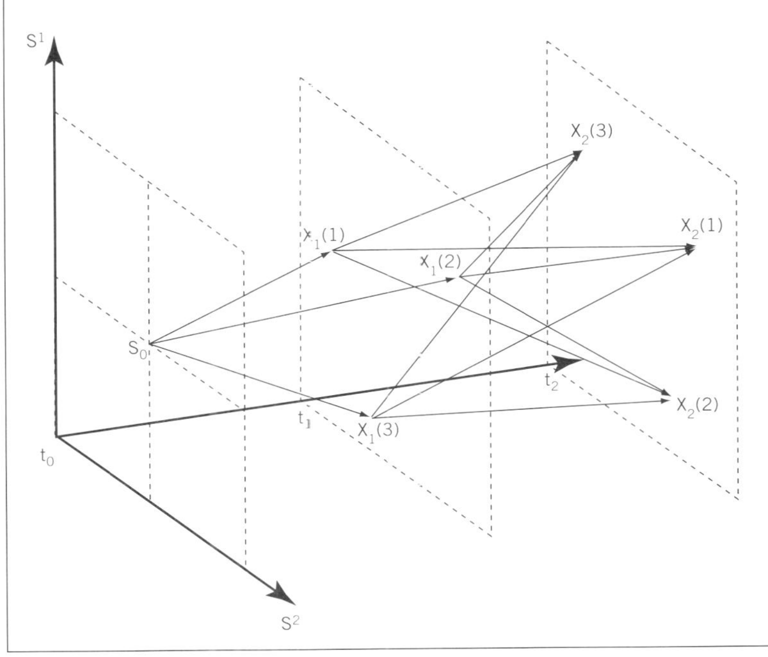

Equation (12) is the natural European option estimator that would result from using only the terminal mesh points $X_T(j)$ for $j = 1,\ldots,b$.

The mesh estimator $\hat{Q}$ is an upper bound and is asymptotically unbiased as the branching parameter b increases. In addition to this estimator, the stochastic mesh can also be used to define a related path estimator. This estimator is defined by simulating a trajectory $S = (S_0,S_1,\ldots,S_{\hat{\tau}})$ of the underlying process S_t until a stopping time $\hat{\tau} \le T$ is reached. The path S is simulated according to the natural density function of the process S_t. That is, the density of the simulated point S_{t+1} given $S_t = x$ is $f(t,x,\cdot)$. In order to estimate the optimal stopping time, at each point on the path the exercise value $h(t,S_t)$ is compared with the mesh estimate of the continuation value. The latter is estimated by $\hat{Q}(t,S_t)$ as defined in equation (9). If $h(t,S_t) < \hat{Q}(t,S_t)$, the decision is to continue and the next point, S_{t+1}, is simulated. If $h(t,S_t) \ge \hat{Q}(t,S_t)$, the decision is to exercise and the path estimator $\hat{q}$ is set to $h(t,S_t)$. In short, the path estimator is

$$\hat{q} = h(\hat{\tau},S_{\hat{\tau}}) \tag{13}$$

where approximate optimal policy determined by the mesh exercises at

$$\hat{\tau}(S) = \min\{t : h(t,S_t) \ge \hat{Q}(t,S_t)\} \tag{14}$$

Because $\hat{\tau}$ is not necessarily the optimal stopping time, the path estimator is a lower bound. It too is asymptotically unbiased as the branching parameter b increases because the exercise time under the policy specified by the mesh converges to that specified by the optimal policy. Hence multiple independent replications of $\hat{Q}$ and $\hat{q}$ can be used to form a valid confidence interval for Q as indicated in equation (4).

As with most simulation-based algorithms, substantial improvements are possible using variance reduction techniques. Several ideas tailored to the stochastic mesh method and crucial for the practical viability of the method are described in Broadie and Glasserman (1997b). Numerical results obtained with the method are also presented.

Table 1. Convergence of a discretely exercisable American option to one with continuous exercise

d	Discrete price	Difference (cont –discrete)	Ratio	2-point extrapolation	Difference (cont–extrapolation)
1	2.583	1.992	N/A	N/A	N/A
2	3.550	1.025	0.515	4.518	0.058
4	4.092	0.484	0.472	4.633	–0.057
8	4.342	0.234	0.483	4.592	–0.017
16	4.456	0.120	0.512	4.570	0.006
32	4.514	0.061	0.512	4.573	0.003
64	4.545	0.031	0.505	4.575	0.001
128	4.560	0.016	0.503	4.575	0.000
256	4.568	0.008	0.502	4.575	0.000
512	4.572	0.004	0.501	4.576	0.000
.	.	.	.	.	.
.	.	.	.	.	.
.	.	.	.	.	.
∞	4.576	0.000	0.500	4.576	0.000

Exercise dates are numbered 0,1,...,d. The "discrete price" column shows Q_d, the price with exercise at dates 0,1,...,d. The "difference" column shows the difference between the continuous and discrete prices. The "ratio" column shows the ratio of the current difference to the previous difference. The "2-point extrapolation" column shows $2Q_{2d} - Q_d$.

7. Convergence of a discretely exercisable American option to one with continuous exercise

Let Q_∞ denote the price of the continuously exercisable American option and Q_d the price with exercise allowed at times 0,1,...,d. The graph plots $Q_\infty - Q_d$ (on the vertical axis) versus d (on the horizontal axis). The slope of the line in the log–log plot converges to –1, illustrating the O(1,d) convergence of the discrete price to the continuous price.

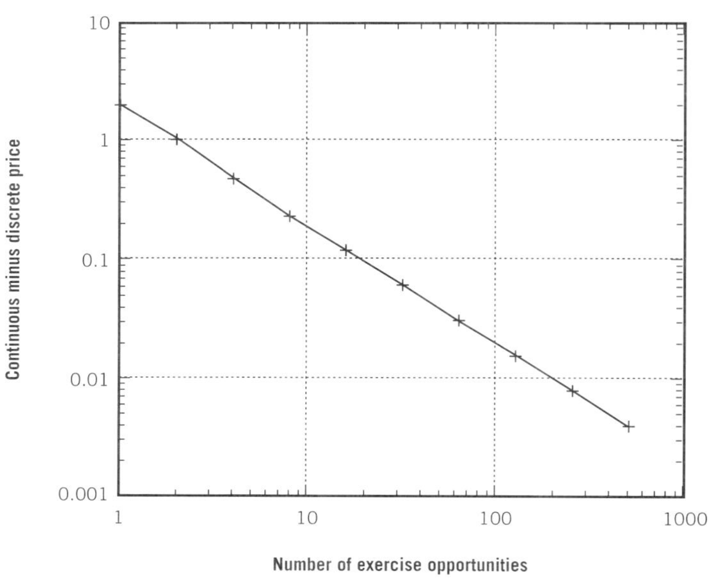

DISCRETE VERSUS CONTINUOUS EXERCISE

Until now we have assumed that the American option can be exercised at a finite set of dates. At some level, computer implementations of all numerical methods must work with a discretised problem. For example, suppose the standard binomial method of Cox, Ross and Rubinstein (1979) is used to price a single-asset American option with continuous exercise. If the binomial method is implemented with, say, 200 time steps, then there is an implicit assumption that the option can be exercised at only 201 dates. If the ultimate goal is to price a continuously exercisable option, it is important to understand how a discretely exercisable option converges to its continuously exercisable counterpart. Carverhill and Webber (1990) showed that for an option on a single underlying asset this convergence is $O(1/d)$, where d is the number of exercise opportunities.

To illustrate this property, consider an option on the geometric average of two assets. The payoff upon exercise is $h(t,S_t^1,S_t^2) = \max(\sqrt{S_t^1 S_t^2} - K, 0)$. We assume that S^1 and S^2 follow independent geometric Brownian motion processes with volatilities $\sigma_1 = \sigma_2 = 20\%$ and with dividend rates $\delta_1 = \delta_2 = 10\%$. There is a constant interest rate of $r = 5\%$. The strike is set to $K = 100$ and the option maturity is $T = 3$ years. The initial prices are $S_0^1 = S_0^2 = 100$.

For this option, the European value is \$2.583. This corresponds to $d = 1$ time period with two exercise opportunities at times $t = 0$ and $t = T$. The American value with continuous exercise is \$4.576, a difference of nearly \$2.[8] The convergence is illustrated in Table 1 and Figure 7. The table shows that the difference between the discrete and continuous prices is approximately halved for every doubling of the number of exercise opportunities. Richardson's extrapolation can be used to approximate the price of the continuously exercisable option using the prices of two or more discretely exercisable options. The application of this idea to American option pricing was first proposed in Geske and Johnson (1984). For this particular option, the discrete and continuous prices differ by 1.6 cents for $d = 128$. Nearly the same absolute difference is obtained using two-point Richardson extrapolation with $d = 4$ and $d = 8$. These results illustrate how a relatively small number of exercise dates can be used to accurately value an option with continuous exercise, ie one with an infinite number of exercise dates.

Conclusions

Securities whose value depends on multiple underlying assets and financial models with multiple stochastic factors are becoming increasingly common. This has led to a need for numerical algorithms to price high-dimensional American options. Simulation-based methods are attractive because their convergence rate is typically independent of the problem dimension. While the simulation of paths forwards in time is suitable for pricing European-style securities, it is not sufficient for solving the optimal exercise problem inherent in pricing American-style securities. In the past several years many hybrid simulation-dynamic programming algorithms have been proposed. This paper has surveyed the recent developments in this rapidly growing area.

1 *We thank an anonymous referee for several helpful suggestions.*

2 *Our use of the term "American" includes securities which permit decisions to occur at any point in time (eg continuous exercise) or only at a finite set of dates. Some authors prefer to use the term "Bermudan" for securities that have a finite number of decision times.*

3 *Quasi-Monte Carlo has become a popular tool for these problems as well. Brief overviews of the quasi-Monte Carlo method are given in Boyle, Broadie, and Glasserman (1997), Joy, Boyle and Tan (1996), and Press* et al. *(1992). For a detailed treatment see Niederreiter (1992).*

4 *Briefly, an exercise policy is a stopping time if the exercise decision depends only on the information available at the current time and not on the future behaviour of the underlying assets and state variables.*

5 *Alternatively, the transitions could be governed by real-world probabilities, with the market prices of risk included in the (discounted) payoff function.*

6 *Tilley (1993) introduces a "sharp boundary" variant of the method, but this idea is predicated on a specific form of the optimal exercise region. Furthermore, it is not obvious how to generalise the idea to higher dimensions.*

7 *Properties of the optimal exercise region are proved in Broadie and Detemple (1997).*

8 *Because the payoff is the geometric average, this two-dimensional pricing problem can be reduced to a problem in one dimension. The one-dimensional problem was solved using a standard binomial method with a large number of steps.*

BIBLIOGRAPHY

Barraquand, J., and D. Martineau, 1995, "Numerical Valuation of High Dimensional Multivariate American Securities", *Journal of Financial and Quantitative Analysis* 30(3), pp. 383–405 (Chapter 15 of this publication).

Bertsekas, D., 1987, *Dynamic Programming: Deterministic and Stochastic Models*, Englewood Cliffs, NJ, Prentice Hall.

Bertsekas, D., and J. Tsitsiklis, 1996, *Neuro-Dynamic Programming*, Belmont, MA, Athena Scientific.

Boessarts, P., 1989, "Simulation Estimators of Optimal Early Exercise", working paper, Graduate School of Industrial Administration, Carnegie Mellon University.

Boyle, P., 1977, "Options: A Monte Carlo Approach", *Journal of Financial Economics* 4, pp. 323–38 (Chapter 1 of this publication).

Boyle, P., M. Broadie and P. Glasserman, 1997, "Monte Carlo Methods for Security Pricing", *Journal of Economic Dynamics and Control* 21(8–9), pp. 1,267–322 (Chapter 2 of this publication).

Broadie, M., and J. Detemple, 1996, "American Option Valuation: New Bounds, Approximations, and a Comparison of Existing Methods", *Review of Financial Studies* 9(4), pp. 1,211–50.

Broadie, M., and J. Detemple, 1997, "The Valuation of American Options on Multiple Assets", *Mathematical Finance* 7(3), pp. 241–86.

Broadie, M., and P. Glasserman, 1997a, "Pricing American-Style Securities Using Simulation", *Journal of Economic Dynamics and Control* 21(8–9), pp. 1,323–52.

Broadie, M., and P. Glasserman, 1997b, "A Stochastic Mesh Method for Pricing High-Dimensional American Options", working paper, Columbia University.

Broadie, M., P. Glasserman and G. Jain, 1997, "Enhanced Monte Carlo Estimates for American Option Prices", *Journal of Derivatives* 5(1), pp. 25–44.

Campbell, J., A. Lo and A. MacKinlay, 1997, *The Econometrics of Financial Markets*, Princeton, NJ, Princeton University Press.

Carr, P., and G. Yang, 1996, "Simulating American Bond Options in an HJM Framework", working paper, Morgan Stanley, New York.

Carr, P., and G. Yang, 1997, "Valuation of Bermudan Derivatives by a General Multinomial Network", working paper, Morgan Stanley, New York.

Carriere, J.F., 1996, "Valuation of the Early-Exercise Price for Derivative Securities using Simulations and Splines", *Insurance: Mathematics and Economics* 19, pp. 19–30.

Carverhill, A., and N. Webber, 1990, "American Options: Theory and Numerical Analysis", in *Options: Recent Advances in Theory and Practice*, Manchester University Press.

Clewlow, L., and C. Strickland, 1997, "Pricing Interest Rate Exotics in a Multi-Factor Gaussian Interest Rate Model", presentation at the Tenth Annual Conference of the Financial Options Research Centre, University of Warwick.

Cox, J., S. Ross and M. Rubinstein, 1979, "Option Pricing: A Simplified Approach", *Journal of Financial Economics* 7, pp. 229–63.

Dammon, R.M., and C.S. Spatt, 1992, "An Option-Theoretic Approach to the Valuation of Dividend Reinvestment and Voluntary Purchase Plans", *Journal of Finance* 47, pp. 331–48.

Dixit, A., and R. Pindyck, 1994, *Investment Under Uncertainty*, Princeton University Press.

Eckstein, Z., and K. Wolpin, 1989, "The Specification and Estimation of Dynamic Stochastic Discrete Choice Models", *Journal of Human Resources* 24(4), pp. 562–98.

Fishman, G., 1996, *Monte Carlo: Concepts, Algorithms, and Applications*, New York, Springer-Verlag.

Fu, M., and J.Q. Hu, 1995, "Sensitivity Analysis for Monte Carlo Simulation of Option Pricing", *Probability in the Engineering and Information Sciences*, 9, pp. 417–46.

Geske, R., and H. Johnson, 1984, "The American Put Options Valued Analytically", *Journal of Finance* 39, pp. 1,511–24.

Geweke, J., 1996, "Monte Carlo Simulation and Numerical Integration", in H.M. Amman, D.A. Kendrick and J. Rust (eds), *Handbook of Computational Economics* 1, Elsevier Science, pp. 731–800.

Grant, D., G. Vora and D. Weeks, 1996, "Simulation and the Early-Exercise Option Problem", *Journal of Financial Engineering* 5(3), pp. 211–27.

Grant, D., G. Vora and D. Weeks, 1997, "Path Dependent Options: Extending the Monte Carlo Simulation Approach", *Management Science* 43, pp. 1,589–602.

Gürkan, G., A. Özge and S. Robinson, 1996, "Sample-Path Solutions of Stochastic Variational Inequalities, with Applications to Option Pricing", in J.M. Charnes, D.J. Morrice, D.T. Brunner, and J.J. Swain (eds), *Proceedings of the 1996 Winter Simulation Conference*, The Society for Computer Simulation, San Diego, CA, 337–44.

Gürkan, G., A. Özge and S. Robinson, 1997, "Sample-Path Solutions of Stochastic Variational Inequalities", working paper, Department of Industrial Engineering, University of Wisconsin-Madison.

Hull, J., 1997, *Options, Futures, and Other Derivatives*, 3rd edition, Upper Saddle River, NJ, Prentice Hall.

Joy C., P.P. Boyle and K.S. Tan, 1996, "Quasi-Monte Carlo Methods in Numerical Finance", *Management Science* 42, pp. 926–38 (Chapter 24 of this publication).

Judd, K., 1997, *Numerical Methods in Economics*, MIT Press (to appear).

Keane, M., and P. Wolpin, 1994, "The Solution and Estimation of Discrete Choice Dynamic Programming Models by Simulation and Interpolation: Monte Carlo Evidence", *Review of Economics and Statistics* 76(4), pp. 648–672.

Li, B., and G. Zhang, 1996, "Hurdles Removed", working paper, Merrill Lynch, New York.

Niederreiter, H., 1992, *Random Number Generation and Quasi-Monte Carlo Methods*, CBMS-NSF 63, SIAM, Philadelphia, Pa.

Press, W.H., S.A. Teukolsky, W.T. Vetterling and B.P. Flannery, 1992, *Numerical Recipes in C: The Art of Scientific Computing*, 2nd edn, Cambridge University Press.

Raymar, S., and M. Zwecher, 1997a, "A Monte Carlo Valuation of American Call Options on the Maximum of Several Stocks", *Journal of Derivatives* 5(1), pp. 7–23.

Raymar, S., and M. Zwecher, 1997b, "Monte Carlo Estimation of Multidimensional, Path-Dependent Equity and Fixed Income Options", working paper, Graduate School of Business, Fordham University.

Rust, J., 1994, "Structural Estimation of Markov Decision Process", in D. McFadden and R. Engle (eds), *Handbook of Econometrics* 4, North Holland, Amsterdam, pp. 3,082–143.

Rust, J., 1996, "Numerical Dynamic Programming in Economics", in H.M. Amman, D.A. Kendrick and J. Rust (eds), *Handbook of Computational Economics* 1, Elsevier Science, pp. 620–728.

Rust, J., 1997, "Using Randomization to Break the Curse of Dimensionality", *Econometrica* 65(3), pp. 487–516.

Tilley, J.A., 1993, "Valuing American Options in a Path Simulation Model", *Transactions of the Society of Actuaries* 45, pp. 83–104 (Chapter 16 of this publication).

Tsitsiklis, J., and B. Van Roy, 1997 "Optimal Stopping of Markov Processes: Hilbert Space Theory, Approximation Algorithms, and an Application to Pricing High-Dimensional Financial Derivatives", working paper, Laboratory for Information and Decision Sciences, MIT, Cambridge, MA.

Welch, R.L., and D.M. Chen, 1991, "Static Optimization of American Contingent Claims", *Advances in Futures and Options Research* 5, pp. 175–84.

FIXED INCOME

Introduction

Bruno Dupire

In banks, although at least half of the research effort on derivative pricing is on the fixed-income side, most of the theories and numerical methods are formulated in a stock world. Monte Carlo simulations do not escape this tendency, and very few papers deal with Monte Carlo pricing of fixed-income derivatives despite the fact that practitioners badly need them. This section gathers some of those rare papers.

Chapter 17, by Clewlow and Strickland, is set in a framework of Gaussian state variable yield curve models. In the case of Gaussian Heath–Jarrow–Morton (HJM) models, they perform a principal component analysis on the rates of the zero coupon associated with the instrument to be priced to reduce the dimension of the problem. For American options in high dimension, they suggest the use of an approximated exercise boundary estimated with a low (1 for instance!) dimension model.

The next chapter is on "Efficient and Flexible Bond Option Valuation in the Heath–Jarrow–Morton Framework" by Carverhill and Pang. They use the martingale variance reduction technique presented in Chapter 10 to price bond options, with zero coupons as control variates. In essence, this means that the bond option is projected on the zero coupons. The bond option minus the optimal portfolio of zero coupons has a reduced variance, which makes the simulation more accurate.

In the final chapter in this section, Oren Cheyette prices mortgage-backed securities with the Monte Carlo method. Actually, the complexity of the cashflows pretty much prevents the use of any other method. Cheyette develops a variance reduction method which makes use of the zero coupons as a control variate. More precisely, the sample paths are not generated by merely following the risk-neutral dynamics but are adjusted to price the zero coupons properly. Another way of interpreting this approach is in terms of calibration, which usually consists of computing the parameters of a model in order to correctly price (which means to match the market price of) a set of reference instruments. What is actually preferable is that the *implementation* of the model fits the market, which is what Cheyette's procedure ensures. His calibration depends on the precise set of Monte Carlo paths. He mentions another methodology, the mysterious "representative paths", which are not described but are probably low-discrepancy paths. The author develops (in the appendix) an important Markov specification of the Heath–Jarrow–Morton model. Although the derivative to be priced is path-dependent, the fact that the model is Markov (path-independent) eases the implementation considerably as only a few state variables need to be diffused.

In my view, the most interesting fixed-income Monte Carlo papers remain to be written (actually, some may have been, but I am unaware of them). The main issue for practitioners is the simulation of non-Markov HJM models, which cannot be implemented by other numerical methods. In this case, the whole yield curve needs to be kept at each time step and important issues such as optimal compact coding with finer representation of short maturities have to be addressed. Another application of paramount importance for practitioners is the Monte Carlo implementation of the Brace–Gatarek–Musiela model.

17

Pricing Interest Rate Exotics by Monte Carlo Simulation

Les Clewlow and Chris Strickland
Lacima Consultants Ltd and Warwick Business School; University of Technology, Sydney, Australia

For many interest rate exotic options – for example, options on the slope of the yield curve or American-featured options – a one-factor assumption for term structure evolution is inappropriate. These options derive their value from changes in the slope or curvature of the yield curve and hence are more realistically priced with multiple factor models. However, efficient construction of short-rate trees becomes computationally intractable as we increase the number of factors and, in particular, as we move to non-Markovian short-rate models.

In this chapter we describe a general framework for pricing a wide range of interest rate exotic options, including captions/floortions, yield spread options, average rate caps, barrier swaptions and American swaptions, under a very general family of multi-factor Gaussian interest rate models. The models belong to the multi-factor Gaussian class of Heath, Jarrow and Morton (HJM) (1992) models – their only required inputs are the initial yield curve and the volatility structures for pure discount bond price returns. The techniques developed do not require any restrictions on the form of the volatility functions. Our framework is based on a computationally efficient implementation of Monte Carlo integration utilising analytical approximations as control variates. These techniques extend the analysis of Clewlow, Pang and Strickland (1997) for pricing interest rate caps and swaptions.

In the next section we briefly review the general class of models with which we work, and the new general computational framework is described in the third section. Then follow the two main sections of this chapter. In the fourth section we show how the standard market instruments of caps and swaptions can be efficiently priced. The formula for caps, which has previously been described by Brace and Musiela (1995), is analytical in the sense of the Black–Scholes formula. We also show that swaption pricing in multi-factor Gaussian models reduces to an integration with dimension equal to the number of coupon payments due to the underlying bond. Although a similar result was obtained by Brace and Musiela (1995), our extension is the incorporation of an extremely efficient control variate, and the efficiency we achieve suggests that we can obtain realistic model calibration. In the fifth section we illustrate the general framework with examples of pricing a wide range of interest rate exotics, including path-dependent options, yield spread options, captions, compound swaptions, average rate caps, barrier swaptions and American swaptions.

As far as we are aware, the problems with implementing multi-factor models have, until now, restricted the application of the HJM model to standard instruments, and this exposition is the first to propose efficient implementation techniques for exotic options pricing.[1] Our approach is based, first, on a computationally efficient implementation of Monte Carlo integration that utilises eigen analysis of the covariance matrix of the pure discount bonds underlying the derivatives and, second, on the use of a general approach to obtaining analytical approximations which can be used as control variates. We also show how American-style interest rate derivatives can be efficiently handled in this framework. We obtain an approximate price by supplementing the Monte Carlo integration with Markovian short-rate trees in order to generate an approximate early exercise strategy. This method has the advantage that it handles high-dimensional problems very well. We compare this method to recent work on the application of Monte Carlo methods for pricing American options. In the sixth section we calibrate a two-factor model to a set of money market data and present computational evidence that indicates the viability of our methods, concentrating on accuracy and convergence. The final section provides a summary and conclusions.

The model

Heath, Jarrow and Morton (1992) extended the early term structure-consistent framework of Ho and Lee (1986) by proposing the following stochastic differential equation for the evolution of the instantaneous forward-rate curve:

$$df(t,T) = \alpha(t,T)dt + \sum_{i=1}^{n} \sigma_i(t,T,f(t,T))dz_i(t) \quad (1)$$

where $f(0,t)$ is the initially observed forward curve and dz_i are independent Wiener processes.[2] Equation (1) is the most general formulation of the HJM approach, with n sources of randomness and with the volatilities of forward rates allowed to be dependent on the level of the forward rates. As shown by HJM, the drift rate is determined by no arbitrage:

$$\alpha(t,T) = \sum_{i=1}^{n}\left\{\sigma_i(t,T,f(t,T))\left[\int_t^T \sigma_i(t,u,f(t,u))du\right]\right\} \quad (2)$$

and hence the forward-rate process is completely specified by the volatility functions. In the following analysis we concentrate on Gaussian versions of the approach, dropping the dependence of the volatility function on the rate levels. Although the original formulation of the HJM approach is in terms of forward rates, the model can be equivalently restated in terms of discount bond prices. Under this formulation (see, for example, Carverhill, 1995) bond price returns satisfy the stochastic differential equation

$$\frac{dP(t,T)}{P(t,T)} = r(t)dt + \sum_{i=1}^{n} \nu_i(t,T)dz_i(t) \quad (3)$$

where $P(0,T)$ is the initially observed discount function and represents the price of a bond which matures with value 1, with certainty, at time T. The T-maturity bond price return volatility is related to the forward-rate volatility function via

$$\nu_i(t,T) = -\int_t^T \sigma_i(t,u)du \quad (4)$$

The process for bond price returns is more intuitive than the forward-rate process: because the bonds are traded assets, the drift in a risk-neutral world is simply the instantaneous short rate. For the rest of the chapter we will work with this formulation as implementation is easier and more intuitive. Another reformulation of the HJM model is in terms of the implied process for the short rate. It can be shown (Carverhill, 1995) that the instantaneous spot interest rate, $r(t)$, given by $f(t,t)$, satisfies the stochastic differential equation

$$dr = \left[\frac{\partial f(0,t)}{\partial t} + \sum_{i=1}^{n}\left\{\int_0^t \nu_i(u,t)\frac{\partial^2 \nu_i(u,t)}{\partial t^2} + \frac{\partial \nu_i(u,t)^2}{\partial t}du + \int_0^t \frac{\partial^2 \nu_i(u,t)}{\partial t^2}dz_i(u)\right\}\right]dt + \sum_{i=1}^{n} \left.\frac{\partial \nu_i(u,t)}{\partial t}\right|_{u=t} dz_i(t) \quad (5)$$

The last component of the drift term, involving the integral with respect to the Brownian motions, indicates that the short rate is non-Markovian in general formulations of the HJM model.

The general framework

In our general framework we work in terms of the prices of pure discount bonds, which, under the risk-neutral measure, satisfy the stochastic differential equation (3) and can be written in integral form as

$$P(T,s) = P(t,s)\exp\left(\int_t^T r(u)du\right)Y(t,T,s) \quad (6)$$

where

$$Y(t,T,s) = \exp\left[\sum_{i=1}^{n}\left\{-\frac{1}{2}\int_t^T \nu_i(u,s)^2 du + \int_t^T \nu_i(u,s)dz_i(u)\right\}\right]$$

is an exponential martingale. Therefore, any set of pure discount bonds is jointly lognormally distributed at any set of future dates or, alternatively, the natural logarithms of pure discount bond prices are jointly normally distributed.

We consider any interest rate derivative to be a series of possibly contingent cashflows occurring on dates s_k, $k = 1,\ldots,m$, and, depending on discount bonds with maturity dates, also on the set of dates s_k, $k = 1,\ldots,m$, with face values c_k,

$k = 1,\ldots,m$. For example, the price of a European coupon bond call option at time t with strike K, maturity date T, on a bond with cash-flows c_k at dates s_k, $Call_{CB}(t,K,T,\{c_k\},s_k\})$, can be represented as

$$\begin{aligned} &Call_{CB}(t,K,T,\{c_k\},\{s_k\}) \\ &= E_t\left[P(t,T)Y(t,T,T)\max\left(0,\sum_{k=1}^{m} c_k P(T,s_k) - K\right)\right] \end{aligned} \quad (7)$$

where we have used $P(T,T) = 1 = P(t,T) \times \exp(\int_t^T r(u)du)Y(t,T,T)$. Note that although the dimensionality of the continuum of pure discount bond prices may be arbitrarily high, depending on the form of the volatility functions, the problem represented by equation (7) has a maximum dimensionality of m. That is, the expectation is taken with respect to the m-dimensional distribution of the m pure discount bonds underlying the coupon bond.

The natural logarithms of this set of pure discount bonds have an $m \times m$ covariance matrix, Σ, where the (k,j)-th element is represented by

$$\begin{aligned} \Sigma_{kj} &= Cov\left[\ln(P(T,s_k)),\ln(P(T,s_j))\right] \\ &= \sum_{i=1}^{n}\left\{\int_t^T \left[\nu_i(u,s_k)\,\nu_i(u,s_j)du\right.\right\} \end{aligned} \quad (8)$$

The covariance matrix Σ can be represented by an orthogonal set of m eigenvectors $\underline{w}_i$ and associated m eigenvalues λ_i such that

$$\Sigma = \Gamma\Lambda\Gamma' \quad (9)$$

where

$$\Gamma = \begin{vmatrix} w_{11} & w_{21} & \ldots & w_{m1} \\ w_{12} & w_{22} & \ldots & w_{m2} \\ \ldots & \ldots & \ldots & \ldots \\ w_{1m} & w_{2m} & \ldots & w_{mm} \end{vmatrix}$$

and

$$\Lambda = \begin{vmatrix} \lambda_1 & 0 & 0 & 0 \\ 0 & \lambda_2 & 0 & 0 \\ 0 & 0 & \ldots & 0 \\ 0 & 0 & 0 & \lambda_m \end{vmatrix}$$

and where the prime denotes transpose. The columns of Γ are the eigenvectors.

We now have a very natural Monte Carlo integration approximation of equation (7). Let M be the number of samples or simulations and ε_i, $i = 1,\ldots,m$, be independent standard normal random numbers. Therefore, we have

$$\begin{aligned} &Call_{CB}(t,K,T,\{c_k\},\{s_k\}) \\ &= \frac{1}{M}\sum_{j=1}^{M}\left[P(t,T)Y(t,T,T)\right. \\ &\quad \left.\times \max\left(0,\sum_{k=1}^{m} c_k P(t,s_k)Y_j(t,T,s_k) - K\right)\right] \end{aligned} \quad (10)$$

where

$$Y_j(t,T,s_k) = \exp\left[-\frac{1}{2}\sum_{i=1}^{m}\{w_{ik}^2\lambda_i\} + \sum_{i=1}^{m}\{w_{ik}\sqrt{\lambda_i}\varepsilon_i\}\right]$$

We also have a convenient way to trade off speed and accuracy by truncation of the eigenvector set to a sub-set of the more important (measured by the size of the eigenvalues) eigenvectors.

Finally, we have a natural, one-factor, Markovian approximation to the model corresponding to the Hull and White Extended Vasicek model (Hull and White, 1990). This will be particularly useful for the pricing of American-featured options.

Caps and swaptions pricing

In this section we briefly review the pricing of two of the market-standard interest rate derivatives – interest rate caps and European swaptions. The efficient pricing of these instruments was described in Clewlow, Pang and Strickland (1997).

Consider an s-year cap with reset period $\Delta\tau$ and a cap rate of r_c on an underlying principal of L, where the capped rate is the $\Delta\tau$-period spot rate (or Libor rate). The cap can be interpreted as a portfolio of "caplets" covering each of the $(s/\Delta\tau - 1)$ capping periods. If $R(t,\Delta\tau)$ is the $\Delta\tau$-period spot rate at time t for the period between t and $t + \Delta\tau$, then the payoff at time $t + (k + 1)\Delta\tau$ to the kth caplet $(k = 1,\ldots,s/\Delta\tau - 1)$ is given by

$$\max(0,R(t + k\Delta\tau,\Delta\tau) - r_c)\Delta\tau L \quad (11)$$

Now we can write the spot rate, $R(t,\Delta\tau)$, in terms of the price of a pure discount bond:

$$R(t,\Delta\tau) = \frac{1}{\Delta\tau}\left[\frac{1}{P(t,t + \Delta\tau)} - 1\right] \quad (12)$$

The payoff in equation (11) can be written as (see Hull, 1989, or Strickland, 1996)

$$\max\left(0,1 - \frac{1 + r_c\Delta\tau}{1 + R(t + k\Delta\tau,\Delta\tau)\Delta\tau}\right)L \quad (13)$$

or in terms of the pure discount bond price:

$$\max(0, 1 - (1 + r_c\Delta\tau)P(t + k\Delta\tau, t + (k+1)\Delta\tau))L \qquad (14)$$

Therefore a caplet is equivalent to a put option with a strike price of 1 and maturity date at the beginning of the capping period on a discount bond with face value $(1 + r_c\Delta\tau)$ and maturity date at the end of the capping period. A cap is therefore a portfolio of put options on discount bonds and a floor is a portfolio of call options on discount bonds.

A number of authors have shown (see, for example, Brace and Musiela, 1995, Clewlow, Pang and Strickland, 1997) that the price of a call option at date t with strike price K and maturity date T on a pure discount bond with maturity date s in this framework is given by

$$c(t,K,T,s) = P(t,s)N(h) - KP(t,T)N(h - w) \qquad (15)$$

where

$$h = \frac{\ln\left(\frac{P(t,s)}{P(t,T)K}\right) + \frac{1}{2}w}{\sqrt{w}}$$

and

$$w = \sum_{i=1}^{n}\left\{\int_t^T (\nu_i(u,s) - \nu_i(u,T))^2 du\right\}$$

The implication of equation (15) is that for multifactor Gaussian versions of the HJM model, evaluating European options on pure discount bonds is as straightforward as using the Black–Scholes equation. The calculation requires only univariate integrations involving the volatilities of the discount bonds maturing at the time of the option and the bond underlying the option. This integration will usually be analytical and, if not, standard numerical sub-routines can be used to perform the numerical integration efficiently (see, for example, Press *et al.*, 1992).

It is straightforward to show that European payer (receiver) swaptions can be priced as put (call) options on coupon bonds where the strike price of the option is set equal to the principal underlying the swap (see, for example, Brace and Musiela, 1995, or Strickland, 1996). Clewlow, Pang and Strickland (1997) showed that, since a coupon bond is a portfolio of discount bonds in which the cashflow associated with the longest-maturity discount bond is much larger than the other cashflows, the coupon bond is close to lognormally distributed and highly correlated with the longest-dated discount bond. They used this information to derive an analytical approximation for the coupon bond option in the form of a discount bond option with the same maturity date, T, strike price, K', and face value of the discount bond, L, where

$$L = \sqrt{\frac{\text{Var}[CB(T)]}{\text{Var}[P(T,S_M)]}} \qquad (16)$$

and

$$K' = K - \left(E_t[CB(T)] - L\frac{P(t,S_M)}{P(t,T)}\right) \qquad (17)$$

This discount bond option can also be used as a control variate for efficient and accurate Monte Carlo valuation of the coupon bond option.

Table 1 shows that the approximation is very accurate. The table gives prices of European at-the-money and 5% in- and out-of-the-money options with various maturities exercising into a new five-year swap. The maturities of the options are six months, 1.0 and 1.5 years. The first row for each maturity is the analytical approximation, the second and third rows giving the Monte Carlo value (with 1,000 simulations) and the standard error of the Monte Carlo estimate.

Carverhill and Pang (1995) - also Chapter 18 of this publication also look at the pricing of coupon bond options, showing how carefully chosen martingale control variates can be used to value the option. However, we can interpret their variance reduction scheme as a static delta hedge, whereas the methodology we employ here is equivalent to a continuously rebalanced delta hedge and the variance reduction we obtain is, therefore, superior.

Table 1. Swaption prices obtained by Monte Carlo simulation and the approximation

		Moneyness 0.95	1.00	1.05
Option maturity	0.5	0.0937	0.0077	0.0000
		0.0937	0.0077	0.0000
		0.0000	0.0000	0.0000
	1	0.0906	0.0108	0.0000
		0.0906	0.0107	0.0000
		0.0000	0.0000	0.0000
	1.5	0.0875	0.0133	0.0001
		0.0875	0.0131	0.0001
		0.0000	0.0000	0.0000

Pricing exotic interest rate derivatives

In this section we illustrate the efficiency and flexibility of our general framework by applying it to the valuation of yield spread options, captions, compound swaptions, average rate caps, barrier swaptions and American swaptions.

YIELD SPREAD OPTIONS

Let $yso(t,K,T,s_1,s_2)$ represent the price at time t of a yield spread option with strike price K that matures at time T. The payoff to the option is dependent on spot rates maturing at s_1 and s_2, with $t < T < s_1 < s_2$, is given by

$$
\begin{aligned}
& yso(T,K,T,s_1,s_2) \\
&= \max(0, R(T,s_1 - T) - R(T,s_2 - T) - K) \qquad (18)
\end{aligned}
$$

Thus, the payoff is the difference if positive between the spread between two yields and a fixed strike price, K.

The price of the option with payoff given by equation (18) under the risk-neutral measure is

$$
\begin{aligned}
& yso(t,K,T,s_1,s_2) \\
&= E_t\left[\exp\left(-\int_t^T r(u)du\right)\right. \\
&\left.\times \max(0, R(T,s_1 - T) - R(T,s_2 - T) - K)\right] \qquad (19)
\end{aligned}
$$

Writing the yields in terms of pure discount bond prices we obtain

$$
\begin{aligned}
& yso(t,K,T,s_1,s_2) \\
&= E_t\left[\begin{array}{l} \exp\left(-\int_t^T r(u)du\right) \\ \max\left(0, \left(\frac{1}{(s_1 - T)}\left(\frac{1}{P(T,s_1)} - 1\right)\right) - \left(\frac{1}{(s_2 - T)}\left(\frac{1}{P(T,s_2)} - 1\right)\right) - K\right) \end{array}\right] \qquad (20)
\end{aligned}
$$

By changing numeraire to the pure discount bond which matures at T (the forward measure), we obtain

$$
\begin{aligned}
& yso(t,K,T,s_1,s_2) \\
&= P(t,T)\, E_t^T\left[\max\left(0, \left(\frac{1}{(s_1 - T)}\left(\frac{1}{\frac{P(T,s_1)}{P(T,T)}} - 1\right)\right)\right.\right. \\
&\left.\left. - \left(\frac{1}{(s_2 - T)}\left(\frac{1}{\frac{P(T,s_2)}{P(T,T)}} - 1\right)\right) - K\right)\right] \qquad (21)
\end{aligned}
$$

where $E_t^T[\ ]$ indicates expectation with respect to the P(t,T)-numeraire and where

$$
\begin{aligned}
& \frac{P(T,s_k)}{P(T,T)} \\
&= \frac{P(t,s_k)}{P(t,T)} \exp\left[\sum_{i=1}^{n}\left\{-\frac{1}{2}\int_t^T (\nu_i(u,s_k) - \nu_i(u,T))^2 du\right.\right. \\
&\left.\left. + \int_t^T (\nu_i(u,s_k) - \nu_i(u,T)) dz_i(u)\right\}\right] \qquad (22)
\end{aligned}
$$

Rearranging equation (21) we obtain

$$
\begin{aligned}
& yso(t,K,T,s_1,s_2) \\
&= P(t,T)\, E_t^T\left[\max\left(0, \left(\frac{1}{s_1 - T}\left(\frac{P(T,T)}{P(T,s_1)}\right)\right)\right.\right. \\
&- \left(\frac{1}{s_2 - T}\left(\frac{P(T,T)}{P(T,s_2)}\right)\right) \\
&\left.\left. - \left(K + \frac{1}{s_1 - T} - \frac{1}{s_2 - T}\right)\right)\right] \qquad (23)
\end{aligned}
$$

It follows easily from equation (22) that

$$
\ln\left(\frac{1}{s - T}\frac{P(T,T)}{P(T,s)}\right) \sim N\left(\begin{array}{c} \ln\left(\frac{P(t,T)}{P(t,s)}\right) + \frac{1}{2}\left[\sum_{i=1}^{n} - \frac{1}{2}\int_t^T (\nu_i(u,s) - \nu_i(u,T))^2 du\right] - \ln(s - T); \\ \sum_{i=1}^{n}\int_t^T (\nu_i(u,s) - \nu_i(u,T))^2 du \end{array}\right) \qquad (24)
$$

Therefore the price of a yield spread option is equivalent to the price of a spread option on two lognormally distributed variables. This equivalence makes it possible to use the method of

Ravindran (1993) to derive a quasi-closed-form solution for the price of a yield spread option.[3] Let

$$X = \left(\frac{1}{s_1 - T}\left(\frac{P(T,T)}{P(T,s_1)}\right)\right)$$
$$Y = \left(\frac{1}{s_2 - T}\left(\frac{P(T,T)}{P(T,s_2)}\right)\right)$$
$$\hat{K} = \left(K + \frac{1}{s_1 - T} - \frac{1}{s_2 - T}\right)$$

and let E_x denote the expectation over X. Therefore we can write equation (23) as

$$\text{yso}(t,K,T,s_1,s_2) = P(t,T)E^T_{X,Y}\left[\max(0,X - Y - \hat{K})\right] \tag{25}$$

By using the following property of conditional expectations, we obtain

$$\text{yso}(t,K,T,s_1,s_2) = P(t,T)E^T_Y\left[E^T_X\left[\max(0,X - Y - \hat{K})|Y\right]\right] \tag{26}$$

It can be shown that the conditional distribution of X given Y is, again, lognormal; therefore, let

$$\ln X|\ln Y \sim N(\bar{\mu},\bar{\sigma}^2)$$

where

$$\bar{\mu} = \mu_1 + \frac{\rho\sigma_1}{\sigma_2}(\ln Y - \mu_2),\ \bar{\sigma}^2 = \sigma_1^2(1 - \rho^2)$$
$$\mu_i = \ln\left(\frac{P(t,T)}{P(t,s_i)}\right) + \frac{1}{2}\left[\sum_{i=1}^{n} - \frac{1}{2}\int_t^T (\nu_i(u,s_i) - \nu_i(u,T))^2 du\right] - \ln(s_i - T)$$
$$\sigma_i^2 = \sum_{i=1}^{n}\int_t^T (\nu_i(u,s) - \nu_i(u,T))^2 du$$

with

where

$$h = \frac{\bar{\mu} - \ln(\hat{K} + Y) + \bar{\sigma}^2}{\bar{\sigma}}$$

The price of the yield spread option can now be evaluated efficiently by a one-dimensional numerical integration of the expression in equation (27) over the distribution of Y. We present numerical results for this in the following section.

The interest rate spread options that are typically traded involve the spread between a short-maturity yield and a long-maturity swap rate. We define the price of a yield–swap-rate spread option under the risk-neutral measure as

$$\text{ySRso}(t,K,T,\{s_k\}) = E_t\left[\exp\left(-\int_t^T r(u)du\right) \times \max(0,R(T,s_1 - T) - SR(T,s_m) - K)\right] \tag{28}$$

where

$$SR(T,s_m) = \frac{1 - P(T,s_m)}{\sum_{k=1}^{m} P(T,s_k)}$$

This option fits into the general framework described in the third section and can therefore be efficiently priced by Monte Carlo integration. The analytical result for the yield spread option can be used as a control variate where necessary; however, the numerical results presented in the next section indicate that this is not usually necessary.

CAPTIONS

A caption is a European option on a cap. Let T be the maturity date of the European option with strike price K. Upon exercise the underlying will

$$\rho = \frac{\text{cov}[\ln X, \ln Y]}{\sigma_1\sigma_2} = \frac{\sum_{i=1}^{n}\left\{\int_t^T (\nu_i(u,s_1) - \nu_i(u,T))(\nu_i(u,s_2) - \nu_i(u,T))du\right\}}{\sqrt{\sum_{i=1}^{n}\int_t^T (\nu_i(u,s_1) - \nu_i(u,T))^2 du}\sqrt{\sum_{i=1}^{n}\int_t^T (\nu_i(u,s_2) - \nu_i(u,T))^2 du}}$$

The inner expectation of equation (26) can be evaluated analytically, yielding

$$E_X\left[\max(0,X - (Y + \hat{K})|Y\right] = \exp\left(\bar{\mu} + \frac{\bar{\sigma}^2}{2}\right)N(h) - (Y + \hat{K})N(h - \bar{\sigma}) \tag{27}$$

be an S-year cap with reset period $\Delta\tau$ and cap rate r_c. Using the results of the third (general framework) section, we interpret the underlying cap as portfolio of European discount bond put options and, hence, the caption as an option on a portfolio of discount bond options. Therefore, the price of a caption at time t is given within the general framework by

$$\text{caption}(t,K,T,s,r_c,\Delta\tau) = E_t\left[P(t,T)Y(t,T,T)\max\left(0,(1+r_c\Delta\tau)\sum_{k=1}^{s/\Delta\tau-1} p\left(T,\frac{1}{1+r_c\Delta\tau},T+k\Delta\tau,T+(k+1)\Delta\tau\right)-K\right)\right] \quad (29)$$

Since the underlying cap can be priced analytically as a portfolio of pure discount bond options, the price of the caption can be obtained by Monte Carlo integration of the payoff of the caption at date T over the distribution of the underlying pure discount bonds at this date. To apply a good control variate for this option, we use a one-factor approximation to the covariance structure and apply Gaussian quadrature directly on the payoff.

COMPOUND SWAPTIONS

A European compound swaption is a European option on a European swaption where, as before, we interpret the swaption as a European coupon bond option. Let T represent the maturity date of the compound option having strike price K. On exercise the underlying will be an option maturing at time T_u and strike price K_u on a coupon bond with cashflows $\{c_k\}$ at times $\{s_k\}$. If the option is a call option on a receiver swaption, then the time t price in our general framework is given by

$$\begin{aligned}&\text{compound\_swaption}(t,K,T,K_u,T_u,\{c_k\},\{s_k\})\\ &= E_t\left[P(t,T)Y(t,T,T)\times\max(0,\text{Call}_{CB}(T,K_u,T_u,\{c_k\},\{s_k\})-K)\right] \quad (30)\end{aligned}$$

where $\text{Call}_{CB}(T,K_u,T_u,\{c_k\},\{s_k\})$ is defined by equation (7). Instead of using Monte Carlo integration to evaluate the underlying option price, we approximate this option value with the analytical approximation of the previous section. To apply a good control variate for this option, we again use a one-factor approximation and apply Gaussian quadrature directly to the payoff of the compound option.

AVERAGE RATE CAPS

We consider only fixed-strike Asian call options, but the method generalises easily for floating-strike versions and put options. We define the payoff of an average rate caplet to be determined by the average level of the rate to be capped (eg three-month Libor) observed at a set of predetermined dates over the capping period. Specifically, the payoff of the caplet for the period from T to $T + \Delta\tau$ is set by the average level of the $\Delta\tau$-period yield observed at dates $s_k = T - \Delta\tau + k\Delta t$, $k = 1,\ldots,m$, between $T - \Delta\tau$ and T, where $\Delta t = \frac{\Delta\tau}{m}$. The payoff at time $T + \Delta\tau$ is therefore

$$\text{ARcaplet}(T+\Delta\tau) = \max(0,A(T)-r_c)\Delta\tau L \quad (31)$$

where $A(T) = \frac{1}{m}\sum_{k=1}^{m} R(s_k,\Delta\tau)$ and L is the principal amount of the caplet.

Therefore the price of the average rate caplet at time t in terms of the prices of pure discount bonds is given by

$$\begin{aligned}&\text{ARcaplet}(t)\\ &= E_t\left[P(t,T+\Delta\tau)Y(t,T+\Delta\tau,T+\Delta\tau)\times\max\left(0,\frac{1}{m}\sum_{k=1}^{m}\frac{1}{\Delta\tau}\left(\frac{1}{P(s_k\ s_k+\Delta\text{-})}-1\right)-r_c\right)\Delta\tau L\right] \quad (32)\end{aligned}$$

If the averaging only takes into account the last observed Libor rate (ie at date T), then the average rate caplet is the same as the standard caplet, and so this becomes the natural control variate.

BARRIER SWAPTIONS

Without loss of generality we concentrate on a down-and-out receiver swaption. We define a down-and-out swaption as a standard swaption except that if the underlying swap rate is below the predetermined barrier level, H, on any of the swap reset dates, $s_k = t + k\Delta\tau$, $k = 1,\ldots,m$, up to the maturity date of the swaption, $T = s_{m'} < s_m$, the swaption expires worthless. The payoff to the down-and-out receiver swaption is therefore

$$\text{DAORswaption}(T) = \max\left(0, \sum_{k=1}^{m} SR(t)\Delta\tau P(T,s_k) - 1\right) 1_{SR(s_k)>H;\ k=1,\ldots,m'} \tag{33}$$

where $SR(s_k)$ is the par swap rate at the reset date s_k and $1_{\text{condition}}$ is the indicator function which takes the value 1 if condition is true and zero otherwise. Therefore the value of the barrier swaption is

$$\text{DAORswaption}(t) = E_t\left[P(t,T)Y(t,T,T)\max\left(0, \sum_{k=m'+1}^{m} SR(t)\Delta\tau P(T,s_k) - 1\right) 1_{SR(s_k)>H;\ k=1,\ldots,m'}\right] \tag{34}$$

The natural control variate for this instrument is the standard European swaption for which we have the analytical approximation given in the previous section.

AMERICAN SWAPTIONS

We conclude this section by proposing a method for obtaining an accurate approximation for the value of American-style options. It has long been considered that one of the drawbacks of using Monte Carlo simulation for derivative pricing is that it is difficult to value American options (see, for example, the comments of Hull, 1997, p. 364, and Campbell, Lo and MacKinley, 1997, p. 390). The problem arises because simulation methods generate trajectories of state variables forward in time, whereas a backward dynamic programming approach is required to efficiently determine optimal decisions for pricing American options. We show in this section that by supplementing the Monte Carlo integration of previous sections with Markovian short-rate trees, in order to generate an early exercise strategy, we can obtain approximations for prices of American-style derivatives efficiently. This methodology is an alternative to recent work in the area and one that can very easily be applied to multi-factor problems. Tilley (1993) – also Chapter 14 of this publication represents one of the earliest attempts to extend simulation for pricing American options, bundling paths into groups according to the underlying asset price and applying dynamic programming using these paths to estimate a single continuation value for all paths within a bundle.[4] His method, however, is difficult to extend to multi-factors.[5] Broadie and Glasserman (1997b) propose a method based on simulated trees that generates both a lower and upper bound, both convergent and asymptotically unbiased as the computational effort increases. The main drawbacks from the perspective of this chapter are that the work is exponential in the number of early exercise opportunities (resulting from the fact that each node generates its own independent sub-tree) and that it is, again, problematic to extend to high dimensions. Broadie and Glasserman (1997a) develop a "stochastic mesh" whereby all the information from one time step is used to estimate the prices at the previous time step. Although this interlocking nature complicates the development of the algorithm, the advantage is that the work is linear in the number of early exercise opportunities. Although the authors claim that the method is viable, results from the tables in their paper suggest that, with the dimensions typical in the analysis of this chapter, computation times can be prohibitive (Broadie and Glasserman, 1997a, comments on Table 4).

In order to obtain real-time approximations we pursue an alternative strategy that involves numerically solving for an early exercise strategy in a low-dimension model similar to the higher-dimensional model of interest. Perhaps the published material that is closest in spirit to ours can be illustrated by the papers of Fu and Hu (1995) and Li and Zhang (1996), who specify a parametric family to represent possible exercise regions.[6]

To illustrate our techniques we derive the price of a Bermudan swaption that exercises into an existing swap. This instrument, which is currently popular with practitioners, can be interpreted as an American (or Bermudan) coupon bond option. Consider an option with strike price K and maturity date T on a coupon bond with cashflows c_k on dates s_k, $k = 1,\ldots,m$, which can be exercised on any of the dates $s_k \le s_{m'} = T$ (so-called Bermudan-style exercise). As before, in the case of a swaption the dates would be separated by the swap reset period, $\Delta\tau$, and the cashflows would be equal to the product of the swap rate and the reset period.

Let $K^*(t)$ be the value of the coupon bond $CB(t)$ above which it is optimal to exercise the call option early. The value of the Bermudan coupon bond option is given by

$$C_{CB}(t,K,T,\{c_k\},\{s_k\}) = E_t\left[P(t,s^*)Y(t,T,T)\max(0,CB(s^*)-K)\,\middle|\,CB(s^*) > K^*(s^*),s^* \in [s_1,\ldots,s_m]\right] \quad (35)$$

We obtain an approximate early exercise boundary by using a single-factor Gaussian Markovian short-rate approximation to our multi-factor non-Markovian model. Our approximating model is essentially the Hull and White Extended Vasicek model (Hull and White, 1990). For models of this class we can obtain the early exercise boundary by building a recombining tree for the short rate and determining the critical coupon bond prices at the early exercise dates by backwards induction in the tree. Our reason for choosing a one-factor rather than a two- or higher-factor Markovian short-rate approximation is that we can obtain a very accurate early exercise boundary by increasing the number of time steps in the tree. This allows us to study the error introduced into the option value by the approximation without having to consider the discretisation error introduced by using a tree.

The stochastic differential equation for the short rate in our approximating model is

$$dr = [\theta(t) - \alpha r]dt + \sigma dz \quad (36)$$

The time-dependent function in the drift, $\theta(t)$, allows the model to fit exactly the initial term structure of interest rates. The parameters α and σ describe the speed of mean reversion and volatility of the short rate, respectively, and determine the volatilities of forward rates, yields or pure discount bond prices. The volatility function of instantaneous forward rates is given by

$$\sigma(t,s) = \sigma e^{-\alpha(s-t)} \quad (37)$$

We "best fit" in a least-squares sense at the standard market maturities 0.25, 0.5, 1, 2, 3, 4, 5, 7 and 10 years the Hull–White volatility function to the total instantaneous forward-rate volatility of our multi-factor model. A detailed numerical example is given in the following section, where we also investigate the accuracy of this approximation technique.

Since a version of this work was first presented we have become aware of two authors who have suggested the same general approach.[7] Hull (1998, p. 8.18) suggests that when valuing American options "one idea is to estimate the early exercise boundary using a Markov model and then use HJM in conjunction with Monte Carlo simulation". Rebonato (1996, p. 332) says that "... it might be very useful to use 'in parallel' a tree-based model which displays distributional assumptions as close as possible to the chosen HJM model. This 'germane' model mapped onto a recombining tree could provide an (almost) internally consistent approach to pricing and risk managing a complex options book."

Calibration and numerical results

In this section we calibrate a two-factor version of our multi-factor model to market data and evaluate numerical results for the instruments discussed in the preceding section. The only other published attempts of which we are aware to calibrate a model of this type to market prices of standard interest rate derivatives are those of Brace and Musiela (1995) and Brace, Gatarek and Musiela (1997).

CALIBRATION TO STANDARD INSTRUMENTS

Table 2 gives the market data used in the example. The data consist of Eurodollar interest rates with maturities from one month out to 12 months, interest rate swap levels from two years out to 30 years, with the volatility information consisting of standard Black volatilities for interest rate caps with maturities from one year out to 10 years and interest rate floors with the same maturities. The reset frequency for both the swaps and caps is semi-annual.

Figure 1 shows the resulting yield curve. The yield curve is upward-sloping from a little over

Table 2. Money market data for rates and volatilities for June 21, 1995

Eurodollar rates (actual/360)

Today	Day	1 month	3 months	6 months	12 months
June 21, 1995	4	5.595	5.581	5.621	5.677

Interest rate swaps (annual actual/360)

2 years	3 years	5 years	7 years	10 years	15 years	20 years	30 years
5.808	5.9535	6.188	6.3245	5.5355	6.7955	6.8925	6.9665

Interest rate caps

1 year	2 years	3 years	4 years	5 years	7 years	10 years
25.5	27.7	27.5	26.7	25.85	23.5	22

Interest rate floors

1 year	2 years	3 years	4 years	5 years	7 years	10 years
26.5	28.75	28.45	27.75	27.25	24	22.5

5.5% for the short end to just over 6.5% for a maturity of 10 years. To construct the yield curve we used the Eurodollar rates out to 12 months and then bootstrapped into the swap rates.

The volatility functions are calibrated to the caps data of Table 2 and for these examples are assumed to be time-homogeneous and piecewise flat. We assume that there are nine segments determining the volatility structure and that the maturities of these segments match the maturity structure of the instruments underlying the yield and volatility curves – ie 0 to three months, three to six months, six months to one year, one to two years, two to three years, three to four years, four to five years, five to seven years, and seven years onwards. The volatility functions are determined by minimising the sum of the squared differences between the market prices of the caps and the prices of the caps under the two-factor model.

1. Spot rate yield curve for June 21, 1995

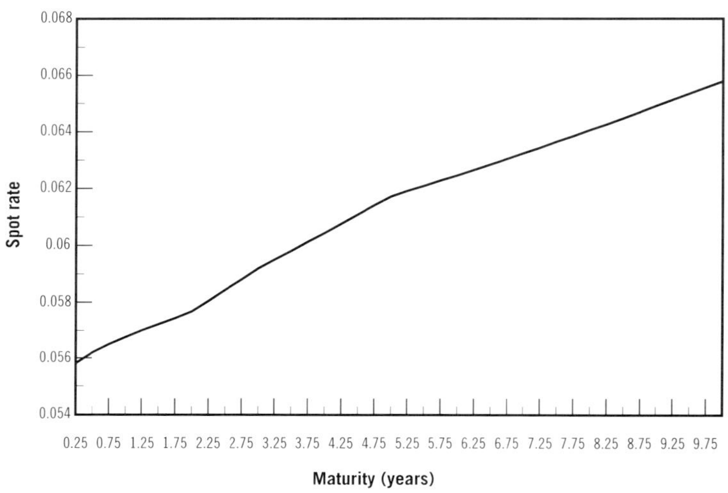

2. Calibrated time-homogeneous piecewise flat volatility function for a two-factor version of the general model

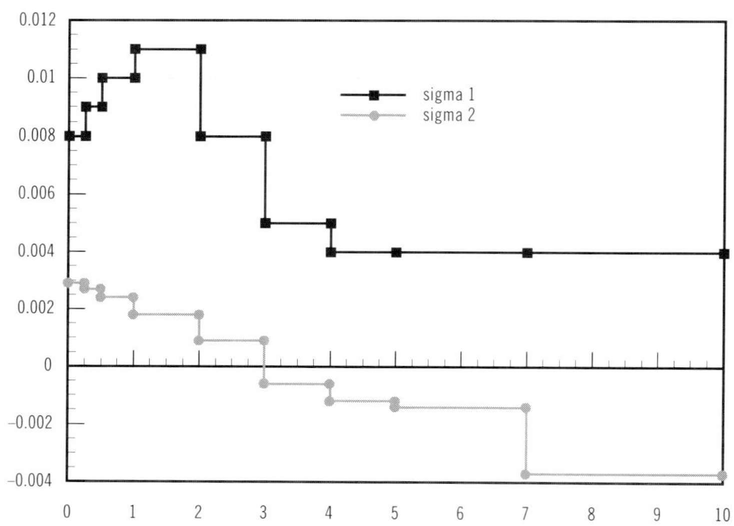

Table 3. Prices of yield spread options

Strike (K)	yso	t	ySRso	se	t
–0.010	0.00005	0	0.00172	0.00003	0
–0.005	0.00000	0	0.00014	0.00001	0
0.000	0.00000	0	0.00000	0.00000	0
0.005	0.00000	0	0.00000	0.00000	0
0.010	0.00000	0	0.00000	0.00000	0

Figure 2 shows the resulting volatility functions. The first volatility function (sigma 1) exhibits the peak at around two years that is commonly observed and then declines in an approximately exponential fashion before flattening out at longer maturities. The second volatility function (sigma 2) shows a roughly linearly decreasing level of absolute volatility with increasing maturity. These volatility structures broadly match the parallel shifts and twists in the yield curve which are typically found with principal components analysis.

NUMERICAL RESULTS

In this section we price the interest rate exotics described in the previous section and illustrate the efficiency of our methodology. All option prices in this section are evaluated with antithetic control variates and the specific control variates described in the previous section for each instrument.

Yield spread options Table 3 shows the prices of yield spread options with one year to maturity where the first rate has a maturity at the option maturity date of three months and the second rate has a maturity of five years. A range of typical strike prices is shown. The number of simulations is 1,000. Prices are first shown under the assumption that the two rates are simple spot rates (the column labelled "yso") using the quasi-analytical formula represented by equations (26) and (27). The prices are then shown under the assumption that the short rate is a simple spot rate and the long rate is a swap rate using the Monte Carlo simulation of equation (28) with the simple spot rate spread option as a control variate (the column labelled "ySRso"). Column "se" is the standard error for the Monte Carlo simulation, and the times taken in seconds to evaluate the prices are in the columns headed "t". We see that for this instrument the standard error is very small even though we are only performing 1,000 simulations, indicating that the control variate works well. As we increase the strike price to 1%

Table 4. Convergence of yield spread option prices

nr	ySRso	se	t
100	0.00175	0.00009	0
200	0.00180	0.00006	0
300	0.00178	0.00005	0
400	0.00175	0.00004	0
500	0.00171	0.00004	0
600	0.00174	0.00004	0
700	0.00173	0.00003	0
800	0.00172	0.00003	0
900	0.00173	0.00003	0
1,000	0.00172	0.00003	0

Table 5. Prices of a one-year option on a five-year cap

Strike (K)	Approx	t	MC price	se	t
0.04854	0.01213	0	0.01198	0.00007	3
0.05393	0.00948	0	0.00934	0.00008	3
0.05933	0.00724	0	0.00711	0.00007	3

Table 6. Convergence of caption prices

nr	MC price	se	t
1,000	0.00947	0.00024	0
2,000	0.00945	0.00017	0
3,000	0.00935	0.00014	1
4,000	0.00938	0.00012	1
5,000	0.00940	0.00011	1
6,000	0.00939	0.00010	2
7,000	0.00936	0.00009	2
8,000	0.00935	0.00008	3
9,000	0.00932	0.00008	3
10,000	0.00934	0.00008	4

the value of the option decreases to zero, indicating that the probability of the spread between the rates exceeding 1% is negligibly small.

Table 4 shows the convergence of the price of the yield-swap rate spread option as a function of the number of simulations, "nr", for a strike of -1%. This shows that with 1,000 simulations we are indeed getting a price accurate to the fourth decimal place.

Captions Table 5 gives the prices of a one-year option on a five-year cap. The reset frequency for the underlying cap is assumed to be six months. The strike levels are chosen to be at-the-money forward and 10% in- and out-of-the-money. The second column, "Approx", is the analytical approximation using Gaussian quadrature. The Monte Carlo price, "MC price", is evaluated using equation (29) with 10,000 simulations and the analytical approximation as a control variate.

Table 6 shows the convergence of the prices of the "at-the-money" one-year option into a five-year cap with semi-annual resets. The standard error is reduced from 0.00024 for 1,000 simulations to 0.00008 for 10,000 simulations.

Table 7. Prices of compound swaptions

Strike (K)	Approx	t	MC price	se	t
0.00062	0.00046	0	0.00046	0.00004	32
0.00069	0.00044	0	0.00045	0.00004	32
0.00075	0.00042	0	0.00043	0.00004	32

Compound swaptions Table 7 shows the prices of compound swaptions evaluated using equation (30) where the first option has six months to maturity and is written on a swaption which itself has six months to maturity and has as its underlying a swap with an original maturity of five years. The strike prices are again chosen to reflect "at the money" and 10% "in the money" and "out of the money". The column headed "Approx" is the analytical approximation derived by reducing to a one-factor model and applying Gaussian quadrature. The column "MC price" is the Monte Carlo simulated price, with the standard errors and calculation time also shown for 10,000 simulations.

Table 8 shows the convergence of compound swaption prices for the at-the-money option as the number of simulations is increased from 100 to 1,000. The standard error is seen to reduce from 0.00013, taking 3 seconds, to 0.00004, taking 32 seconds.

Average rate caps The average rate cap provides a good illustration of the efficiency gains we obtain by determination of the effective dimensionality of the problem via the eigen analysis of the pure discount bond covariance matrix. The size of the largest three eigenvalues for this case are 0.002508, 0.000004 and 0.000001 with the remaining eigenvalues being less than 10^{-7}. This implies that we can reduce the dimension of the simulation and still achieve good results. Table 9 gives the prices of fixed-strike average rate caplets ("AR caplet") for the period between two and two and a half years for a decreasing number

Table 8. Convergence of compound swaption prices

nr	MC price	se	t
100	0.00058	0.00013	3
200	0.00045	0.00008	6
300	0.00045	0.00007	9
400	0.00045	0.00007	13
500	0.00046	0.00006	16
600	0.00047	0.00005	19
700	0.00047	0.00005	22
800	0.00047	0.00004	26
900	0.00045	0.00004	29
1,000	0.00045	0.00004	32

Table 9. Prices of average rate caplets for numbers of factors

NOE	AR caplet	se	t
10	0.002179	0.000007	7
9	0.002176	0.000007	7
8	0.002173	0.000007	7
7	0.002189	0.000007	6
6	0.002182	0.000007	5
5	0.002179	0.000007	5
4	0.002177	0.000007	4
3	0.002189	0.000007	3
2	0.002207	0.000006	4
1	0.002241	0.000002	2

Table 10. Prices of average rate caplets

NOE	AR caplet	se	t
0.04	0.00927	0.00000	1
0.05	0.00530	0.00000	1
0.06	0.00219	0.00001	1

of retained eigenvectors ("NOE") from the full set down to the primary eigenvector only. We assume that there are five equally spaced fixings for the average during this period, which results in a full set of ten eigenvectors. The control variate is the standard caplet with two years to maturity.

Table 9 confirms the results of the eigenvalue analysis – significant time savings can be made by considering only the first three or four factors with only a small loss of accuracy. Table 10 gives a set of average rate caplet prices. Prices are based on 10,000 simulations, and the table gives the prices of caplets under various strike prices from 4% to 6%. The standard errors are again seen to be very small, with the simulations taking only one second.

3. Overall two-factor model instantaneous forward-rate volatility function and fitted Hull–White volatility function

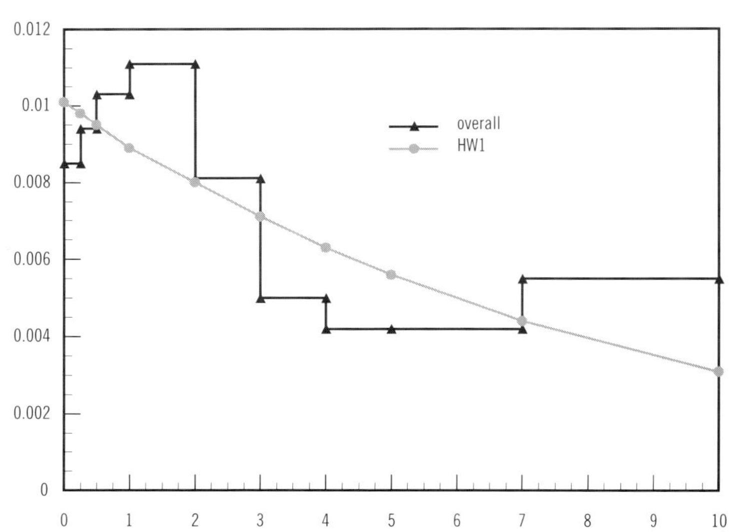

Table 11. Convergence of average rate caplet prices

nr	AR caplet	se	t
1000	0.00222	0.00002	0
2000	0.00219	0.00002	0
3000	0.00218	0.00001	0
4000	0.00218	0.00001	0
5000	0.00217	0.00001	0
6000	0.00218	0.00001	0
7000	0.00218	0.00001	0
8000	0.00218	0.00001	1
9000	0.00218	0.00001	1
10,000	0.00218	0.00001	1

Table 12. Convergence of barrier swaption prices

NOE	DAOR swaption	se	t
1,000	0.00025	0.00002	4
2,000	0.00023	0.00002	7
3,000	0.00022	0.00001	10
4,000	0.00022	0.00001	14
5,000	0.00022	0.00001	17
6,000	0.00022	0.00001	20
7,000	0.00021	0.00001	24
8,000	0.00022	0.00001	27
9,000	0.00022	0.00001	31
10,000	0.00022	0.00001	33

Table 11 shows the convergence of average rate caplet prices as the number of simulations increases from 1,000 to 10,000 for the two-year maturing caplet with a strike price of 6%. Again, the results indicate that we can achieve real-time pricing.

Barrier swaptions Table 12 shows the convergence of the down-and-out receiver swaption price, "DAOR swaption", where the swaption has three years to maturity on an underlying swap with five years to maturity. The strike price of the swaption is 5% and the barrier is set at a level of 4.5%.

Bermudan swaption The methodology for pricing options with early exercise features involves approximating the general n-factor non-Markovian Gaussian model with a single-factor Markovian short-rate model which corresponds to the Hull–White extended Vasicek model. Figure 3 shows the overall instantaneous forward-rate volatility function of our two-factor model and the least-squared-fitted Hull–White volatility function. The best-fit values for the parameters describing the volatility function in the approximating model are $\alpha = 0.117552$ and $\sigma = 0.010059$.

We price a 1.5-year Bermudan option on a three-year semi-annual 5% coupon bond with early exercise dates on the coupon dates (ie early exercise opportunities at 0.5 and 1.0 years). The second stage of the process is to determine the early exercise boundary points in the restricted model. The early exercise boundary points, in terms of the level of the underlying coupon bond above which it is optimal to exercise the option, are found to be 0.986202 at 0.5 years and 0.982956 at one year. The value of the European and Bermudan coupon bond option values derived from the short-rate tree are 0.151789 and 0.163046, respectively. The corresponding values obtained via our two-factor model using the European analytical approximation as a control variate and with 10,000 simulations are 0.151789 and 0.157342 with standard errors of 0.000000 and 0.000054, respectively.

The robustness of the procedure we have outlined depends on the accuracy of the early exercise strategy derived from the single-factor Markovian short-rate model. We can illustrate the accuracy of the strategy by evaluating the Bermudan option price for a range of values for the two points determining the early exercise boundary centred on the values obtained from the single-factor short-rate model. Figure 4 shows the results of this procedure. The y- and x-axis are centred on the critical coupon bond prices for the first (0.986202) and second (0.982956) early exercise dates identified from the short-rate tree, respectively. The true value for the Bermudan option should be the maximum value possible as a function of the early exercise boundary points. Figure 4 indicates that the approximate values obtained from the single-factor Markovian short-rate model are very close to optimal because the maximum value for the option lies approximately at the centre of the plot.

Summary and conclusions

In this chapter we have described a framework for evaluating interest rate exotics under a very general class of multi-factor Gaussian non-Markovian interest rate models. We began by showing that, within our framework, and with the use of an innovative control variate, we can calibrate the model realistically to caps and swaption prices.

The second main part of this chapter represents the first published attempt to answer the concerns of a number of authors as to the ability of multi-factor versions of HJM to price path-dependent claims efficiently. We have developed a general methodology for pricing interest rate exotics that is very flexible and, as examples, we looked at yield spread options (for which we have developed an analytical solution if the two rates are

4. Bermudan swaption prices for a range of values determining the early exercise boundary

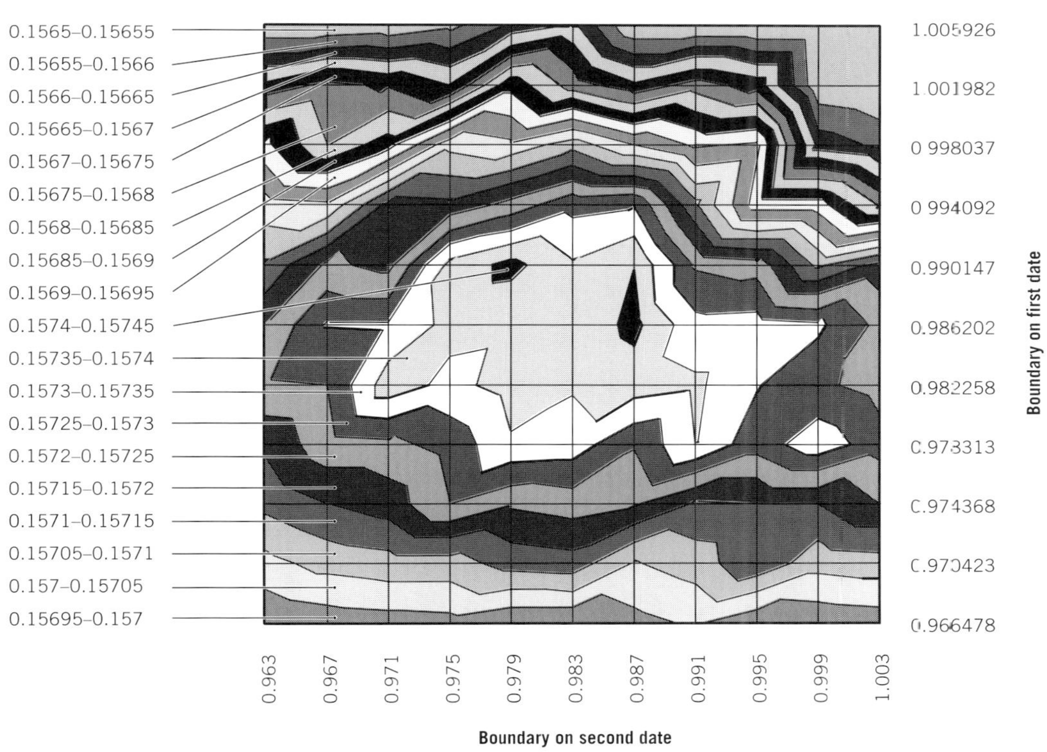

Libor rates), captions/floortions, average rate caps, barrier caps, compound swaptions and barrier swaptions. The nature of the formulation means that we do not have to model every element of the observed term structure, only the pure discount bonds which correspond to underlying cashflows. Applying this to interest rate exotics yields valuation formulas in terms of expectations that are efficient to compute. A further contribution of this paper is that our general framework lends itself in a natural way to efficient sampling of the state space by truncation of the eigenvector representation of the covariance structure to a subset of the important factors. Finally, this representation leads to a general method of obtaining analytical approximations that can also be used as control variates for the simulation.

Although until recently it has been considered that one of the drawbacks of using Monte Carlo simulation for derivative pricing is that it is difficult to value American options, we showed that by supplementing the Monte Carlo integration with Markovian short-rate trees to generate an early exercise strategy, we are able to obtain approximations to the prices of American-style derivatives. We compared our approximation technique with recent research on Monte Carlo methods for pricing American options.

We completed the chapter by calibrating a general two-factor HJM model to a set of money market data – making this one of the first studies to look at multi-factor implied calibration to market data and the first to look at Eurodollar data – and producing numerical results that demonstrate the viability of the Monte Carlo method and the computational efficiency for all the instruments considered. For all options the methodology is computationally extremely efficient, with many path-dependent option prices calculated to within reasonable accuracy in a few seconds. The results obtained when the state space is sampled by truncating the eigenvector set are also impressive. We find that the first three or four factors account for nearly all of the total variation, and so by reducing the full dimensionality of the continuum of discount bond prices to these important factors we obtain an efficient approximation of the option value. Our results show that reduction from 10 factors to three reduces the accuracy of the approximation only slightly while more than halving the computation time.

The results we obtain for pricing a Bermudan swaption by simulation are similarly encouraging. We tentatively conclude that for typical shapes of volatility functions implied by market data, the non-Markovian character of the general model does not rule out approximating the early exercise strategy by using a Markovian restriction.

The authors would like to thank Martin Cooper (Tokai Bank Europe), Stewart Hodges (Financial Options Research Centre, University of Warwick) and Harry Tsivanidis (SBC Warburg) for many stimulating discussions. This work was partially supported by sponsors of the Financial Options Research Centre: Deutsche Morgan Grenfell, Foreign and Colonial Management, Kleinwort Benson Securities, Mitsui Bussan Commodities and SBC Warburg.

1 *Brace and Musiela (1995) deal with caps and swaption pricing, Carverhill and Pang (1995) look at pricing discount bond and coupon bond options, and Amin and Morton (1994) price Eurodollar futures options.*

2 *Note that throughout this paper we work entirely in the standard risk-neutral measure (see, for example, Carverhill, 1995).*

3 *We would like to thank Palle Broman for assistance in verifying this result.*

4 *Barraquand and Martineau (1995) also propose a method that involves partitioning paths – but in the payoff space instead of the state space.*

5 *Raymar and Zwecher (1997) extend Barraquand and Martineau (1995) by basing the early exercise decision on a partition of two state variables.*

6 *Although some numerical tests have been performed on single-asset problems, there has been little investigation of the application of these methods to multiple state variable problems.*

7 *The idea behind pricing American options in this way was first presented at the September 1996 Financial Options Research Centre Annual Conference.*

BIBLIOGRAPHY

Amin, K.I., and A.J. Morton, 1994, "Implied Volatility Functions in Arbitrage Free Term Structure Models", *Journal of Financial Economics* 35, pp. 141-80.

Barraquand, J., and B. Martineau, 1995, "Numerical Valuation of High-Dimensional Multivariate American Securities", *Journal of Financial and Quantitative Analysis* 30 (3), pp. 383-405 (Chapter 15 of this publication).

Black, F., 1976, "The Pricing of Commodity Contracts", *Journal of Financial Economics* 3, pp. 167-79.

Brace, A., and M. Musiela, 1995, "Multi-Factor Gaussian Heath-Jarrow-Morton", *Mathematical Finance* 2, pp. 254-83.

Brace, A., D. Gatarek and M. Musiela, 1997, "The Market Model of Interest Rate Dynamics", *Mathematical Finance* 7, pp. 127-55.

Broadie, M., and P. Glasserman, 1997a, "A Stochastic Mesh Method for Pricing High Dimensional American Options", working paper, Graduate School of Business, Columbia University, New York.

Broadie, M., and P. Glasserman, 1997b, "Pricing American Style Securities Using Simulation", *Journal of Economic Dynamics and Control* 21(8/9), pp. 1323-52.

Campbell, J., A. Lo and A. MacKinley, 1997, *The Econometrics of Financial Markets*, Princeton University Press.

Carverhill, A.P., 1995, "A Simplified Exposition of the Heath-Jarrow-Morton Model", *Stochastics and Stochastic Reports* 53, pp. 227-40.

Carverhill, A.P., and K. Pang, 1995, "Efficient and Flexible Bond Option Valuation in the Heath, Jarrow, and Morton Framework", *Journal of Fixed Income* 5(2), pp. 70-7 (Chapter 18 of this publication).

Clewlow, L., K. Pang and C.R. Strickland, 1997, "Implementing Multi-Factor Gaussian HJM for Caps and Swaptions Pricing", preprint 97/76, Financial Options Research Centre, Warwick Business School, University of Warwick.

Fu, M., and J.Q. Hu, 1995, "Sensitivity Analysis for Monte Carlo Simulation of Option Pricing", *Probability in the Engineering and Information Sciences* 9, pp. 417-46.

Heath, D., R. Jarrow and A. Morton, 1992, 'Bond Pricing and the Term Structure of Interest Rates: A New Methodology for Contingent Claim Valuation", *Econometrica* 60(1), pp. 77-105.

Ho, T.S.Y., and S.-B. Lee, 1986, "Term Structure Movements and Pricing Interest Rate Contingent Claims", *Journal of Finance* 41(5), pp. 1011-29.

Hull, J., 1997, *Options, Futures and Other Derivative Securities*, 3rd edition, Prentice Hall.

Hull, J., 1998, "Understanding and Implementing Important New Developments In Single and Multi-Factor Modelling for Interest Rate Derivatives", presentation at Global Derivatives '98.

Hull, J., and A. White, 1990, "Pricing Interest Rate Derivative Securities", *The Review of Financial Studies* 3(4), pp. 573-92.

Li, B., and G. Zhang, 1996, "Hurdles Removed", working paper, Merrill Lynch.

Press, W.H., S.A. Teukolsky, W.T. Vetterling and B.P. Flannery, 1992, *Numerical Recipes in C: The Art of Scientific Computing*, second edition, Cambridge University Press.

Ravindran, K., 1993, "Exotic Options: The Building Blocks and Their Applications", in I. Nelken (ed.), *The Handbook of Exotic Options*, Chicago, Irwin Professional Publishing.

Raymar, S., and M. Zwecher, 1997, "A Monte Carlo Valuation of American Call Options on the Maximum of Several Stocks", *Journal of Derivatives* 5(1), pp. 7-23.

Rebonato, R., 1996, *Interest-Rate Options Models*, London, John Wiley.

Strickland, C.R., 1996, "A Comparison of Models for Pricing Interest Rate Derivatives", *The European Journal of Finance* 2(3), pp. 261-87.

Tilley, J.A., 1993, "Valuing American Options in a Path Simulation Model", *Transactions of the Society of Actuaries* 45, pp. 83-104 (Chapter 14 of this publication).

18

Efficient and Flexible Bond Option Valuation in the Heath, Jarrow and Morton Framework

Andrew Carverhill and Kin Pang
University of Science and Technology, Hong Kong; Morgan Stanley Dean Witter & Co

Term structure models have contrasting strengths and weaknesses when applied to valuing options on the term structure. The Heath, Jarrow and Morton (HJM) (1992) model is more flexible, and takes as its inputs the initial term structure and the volatility of the term structure.[1] This is convenient for the user, who can directly observe the initial term structure and can estimate, or take a view on, the volatility. Moreover, the HJM model gives the option value elegantly as an expectation involving these inputs, and does not require the risk premium of the market for this. The model is thus reminiscent of the classical Black–Scholes model for equity options.

The equilibrium (or factor) models, which begin with the models of Vasicek (1977) and Cox, Ingersoll and Ross (CIR) (1985), by contrast, start from a collection of randomly evolving state variables (or factors, which usually include the short rate). These, together with a risk premium associated with each one, are taken to characterise the entire term structure at any time. Thus, the term structure and its volatility structure are outputs, and not inputs, in the equilibrium model. This can be inconvenient to the user because these outputs may not perfectly match what is observed or desired. Because they are appropriate to the model and not what is empirically observed or desired, this can lead to the model returning values that are clearly inaccurate.

At the same time, option valuation is often easier in the equilibrium models than in more general versions of the HJM model. The former require solving a diffusion equation similar to the Black–Scholes equation, while the latter most generally require a Monte Carlo simulation or a related numerical integration technique.

We attempt to combine the advantages of the HJM model and the equilibrium models. Our approach is to retain the HJM framework and Monte Carlo procedure, and to make it more efficient by using carefully chosen martingale variance reduction (MVR) variates. The MVR technique is standard in the field of stochastic simulation, and we summarise it in a panel. The technique has been used by Carverhill and Clewlow (1994) (see also Clewlow and Carverhill, 1994) for equity-based exotic options.[2]

Our approach contrasts with that in much of the related literature. Usually the approach is to extend the equilibrium model to be more general, as in Hull and White (1990, 1993), Fong and Vasicek (1991), Longstaff and Schwartz (1992), and Duffie and Kan (1996), or to specialise the HJM model so that the Monte Carlo technique is unnecessary, as in Ho and Lee (1986) and Heath, Jarrow and Morton (1990, 1992).[3] Extensions of the equilibrium model tend to be incomplete, however, and specialisations of the HJM model tend to detract severely from its generality.

The model of Duffie and Kan is perhaps the closest counterpart to our approach. Duffie and Kan take an exogenously given initial term structure and volatility structure and then derive state variables (or factors) that will provide these structures through the equilibrium model. They have to contend with the highly non-linear character of this procedure, and to get a close match they may need many factors.

We concentrate here on the European call (or put) with strike price X and maturity date s, written on a bond with coupon payments $c_1, \ldots, c_n$ at times $q_1,\dots,q_n$ beyond time s. The time 0 value of this option is given in the HJM model by the formula

$$E_0[\{\pm\{(c_1 P_0^{q_1} R_0^{s,q_1} + \dots + c_n P_0^{q_n} R_0^{s,q_n}) - X P_0^{s} R_0^{s,s}\}\}^+] \tag{1}$$

where E_0 denotes the risk-neutral expectation at time 0, and

$$R_0^{s,q} = \exp\left\{\sum_{j=1}^{m}\int_{\rho=0}^{s}\left[\nu_\rho^{j,q}\, dW_\rho^j - \tfrac{1}{2}(\nu_\rho^{j,q})^2\, d\rho\right]\right\} \tag{2}$$

in which $\{\nu_t^{j,q}\}_{j=1,\dots,m;t\le q}$ is the volatility structure, and we take "+" for a call and "-" for a put. See Heath, Jarrow and Morton (1992) or Carverhill (1995b).

The volatility structure is defined by the Itô equation

$$dP_t^q / P_t^q = r_t\, dt + \sum_{j=1}^{m} \nu_t^{j,q}\, dW_t^j \tag{3}$$

in which P_t^q is the time t price of the pure discount bond to mature at time q; r_t is the short rate at time t; and $dW_t^1,\dots,dW_t^m$ are the differential increments of independent standard Brownian motions.

We first show how to improve the Monte Carlo simulation to solve (1) by using martingale variance reduction variates. We implement our procedure in the context of the familiar Fong and Vasicek (1991) model.

Ensuring that the interest rates stay positive in the model is an important consideration, and not just from a purist point of view. Rogers (1994) points out that even a small probability of negative interest rates can have a large effect on prices for long terms to maturity. In fact, in implementations of the HJM model, the volatility is often assumed to be non-random, ie independent of the term structure itself, although this leads directly to the possibility of negative interest rates. We show how to ensure that interest rates stay positive in our procedure, and we implement it in the context of the Longstaff and Schwartz (1992) model.

MVR variates for Monte Carlo simulation of the HJM model

The Monte Carlo technique is to simulate, for many independent trials, the vector $(R_0^{s,s}, R_0^{s,q_1},\dots,R_0^{s,q_n})$ as given in (2), and thence the magnitude

$$\{\pm\{(c_1 P_0^{q_1} R_0^{s,q_1} + \dots + c_n P_0^{q_n} R_0^{s,q_n}) - X P_0^{s} R_0^{s,s}\}\}^+ \tag{4}$$

This magnitude is the discounted payoff of the option, and each trial involves taking a sample of the m-dimensional Brownian trajectory $\{(W_t^1,\dots,W_t^m)\}_{t\in[0,s]}$, in a suitable time discretisation. The "naïve" Monte Carlo estimate of the option value is then just the mean of (4) over the many trials.

In general, the MVR variates associated with a Monte Carlo simulation are a vector of martingales starting from zero, which are generated simultaneously with the trials themselves. Our choice for these variates is the vector

$$\{(R_0^{t,s} - 1, R_0^{t,q_1} - 1,\dots,R_0^{t,q_n} - 1)\}_{t\in[0,s]} \tag{5}$$

(Note that the $R_0^{t,q}$s are exponential martingales so long as the volatility $v_t^{j,q}$ is bounded over t, j, q - see Carverhill (1995b).)

The variance-reduced trial is then

$$\begin{aligned}&\{\pm\{(c_1 P_0^{q_1} R_0^{s,q_1} + \dots + c_n P_0^{q_n} R_0^{s,q_n}) - X P_0^{s} R_0^{s,s}\}\}^+\\ &- \{\beta_{(n+1)}(R_0^{s,s} - 1)\\ &+ \beta_1(R_0^{s,q_1} - 1) + \dots + \beta_n(R_0^{s,q_n} - 1)\}\end{aligned} \tag{6}$$

The β_is here are chosen to minimise the variance of (6) (see Panel 1). The MVR technique works because (6) must have the same expectation as (4), since $E_0(R_0^{s,q_i} - 1) = 0$ for each i.

Our choice of MVR variate is efficient because the R_0^{s,q_i}s have already been calculated for (4). Also, with these R_0^{s,q_i}s, the procedure will automatically give the correct value (without any random error) if it is applied to valuing the bond itself, which corresponds to X = 0 in the call valuation. In fact, in this case the variance is clearly reduced to zero by taking $\beta_i = c_i P_0^{q_i}$ in (6).

This variance-reduced Monte Carlo procedure can be implemented directly when the volatility structures $\{v_t^{j,q}\}_{t\le q,j=1,\dots,m}$ are non-random, ie they do not depend on the term structure itself. Also, this procedure works provided only that the volatility structures are bounded, and it will be valuable for volatility structures for which the model cannot easily be reduced to a finite-dimensional factor (ie equilibrium) model.[4]

A case in point would be with volatility of the shape

$$v_t^{1,t+\tau} \equiv v^1(\tau) = (\rho - \eta)\exp\{-\alpha\tau\} + \eta \tag{7}$$

$$v_t^{2,t+\tau} \equiv v^2(\tau) = \eta/2 + \eta(1 - \exp\{-\alpha\tau\}) \tag{8}$$

which we might expect to obtain from a principal components analysis of the term structure of volatility.[5]

PANEL 1

THE MARTINGALE VARIANCE REDUCTION TECHNIQUE

In its "naïve" form, the Monte Carlo technique may be expressed as: suppose we want to estimate the mean, μ, of a probability distribution, and we are able to generate independent random samples (trials) $\{y_1, y_2, \ldots\}$ from the distribution. Then the simple Monte Carlo estimator of μ, based on M samples, is just the sample mean:

$$\hat{\mu} = \frac{1}{M} \sum_{i=1}^{M} y_i \tag{A1}$$

If we denote the variance of each y_i (and hence of the distribution) by Var(y), then by the central limit theorem the variance of $\hat{\mu}$ (ie its standard error squared, as an estimator of μ) is given by Var(y)/M. Also, if we can achieve a certain variance with, say, M samples, to reduce this by a factor of, say, X, will require us to generate XM samples.

Now, suppose that in parallel with generating $\{y_1, y_2, \ldots\}$ we also generate the antithetic variates $\{y_1^*, y_2^*, \ldots\}$, which are also independent and of the same distribution as the y_is, but negatively correlated with them. Then the variates $\bar{y}_i \equiv \frac{1}{2}(y_i + y_i^*)$ will again have mean μ, but they will have variance $\mathrm{Var}(\bar{y}) \equiv \frac{1}{2}\mathrm{Var}(y)[1 + \mathrm{Corr}(y,y^*)] \leq \frac{1}{2}\mathrm{Var}(y)$. Moreover, the sample mean

$$\hat{\bar{\mu}} = \frac{1}{M} \sum_{i=1}^{M} \bar{y}_i \tag{A2}$$

will have variance less than half that of $\hat{\mu}$. If the pair y_i, y_i^* is less than twice as expensive to generate than y_i, calculating $\hat{\bar{\mu}}$ will be a more efficient way to estimate μ than calculating $\hat{\mu}$. This is the antithetic variate technique.[1]

The martingale variance reduction technique is to generate simultaneously with the samples $\{y_1, y_2, \ldots\}$ the i.i.d. variates $\{x_1, x_2, \ldots\}$, where each x_i is an m-vector $(x_i^1, \ldots, x_i^m)^T$ (the symbol T denoting "transpose"), whose increments have zero mean, ie they are martingales. Thus, the (m + 1)-vectors (y_i, x_i) are independent for different values of i, but their components are correlated among themselves in a constant way. Then, for any m-vector ξ, the quantity

$$\hat{\mu}_c = \frac{1}{M} \sum_{i=1}^{M} \left(y_i - \sum_{j=1}^{m} x_i^j \xi_j \right) \tag{A3}$$

will also be an estimator of μ, and the variance of this is minimised if we choose ξ so that

$$\Sigma_{xx}\, \xi = \Sigma_{xy} \tag{A4}$$

where Σ_{xx} is the covariance matrix of x, and Σ_{xy} is the m-vector of covariances between x and y.

If we know the matrix Σ_{xx} and vector Σ_{xy}, then equations (A3) and (A4) let us use the control variates x_i to reduce the variance in our simple Monte Carlo estimation (A2) of μ. If we do not know Σ_{xx} and Σ_{xy}, the procedure to obtain the variance-reduced estimator $\hat{\mu}_c$ is: find $\hat{\beta} \equiv (\hat{\beta}_0, \ldots, \hat{\beta}_m)^T$ to fit, in a least-squares sense, the system of equations

$$X_i\, \beta = y_i, \quad i = 1, \ldots, M \tag{A5}$$

where $X_i = (1, x_i^1, \ldots, x_i^m)$. Then $(\hat{\beta}_1, \ldots, \hat{\beta}_m)$ is the choice of $(\xi_1, \ldots, \xi_m)$ that will minimise the variance of the estimator (A3), and $\hat{\beta}_0$ is the corresponding variance-reduced estimator $\hat{\mu}_c$ of μ.

In fact, $\hat{\beta}$ can be calculated simply as

$$\hat{\beta} = (X^T X)^{-1} X^T y \tag{A6}$$

provided none of the control variates is redundant. If some of the control variates are redundant, or highly collinear, then $X^T X$ will be singular, or nearly singular.[2] In this case, however, because $X^T X$ is symmetric, we can work with its pseudo-inverse, which is obtained by projecting along the directions given by the eigenvectors with zero eigenvalue.

The m × m matrix $X^T X$ and m-vector $X^T y$ can be calculated conveniently as the simulation proceeds using the equations

$$(X_{n+1}^T X_{n+1})_{j,k} = (X_n^T X_n)_{j,k} + x_j^{n+1} x_k^{n+1} \tag{A7}$$

$$(X_{n+1}^T y_{n+1})_j = (X_n^T y_n)_j + x_j^{n+1} y_{n+1} \tag{A8}$$

where $x_0^{n+1} = 1$.

1 We have implemented this technique in the HJM framework, but we do not report the results here because the variance reduction is quite feeble.

2 This tends to be the case when applying the technique because the bonds corresponding to the coupons are highly correlated.

Table 1. Simulation results – Fong and Vasicek model

Three-year European call on six-year pure discount bond

20 steps/year, 50,000 simulations				100 steps/year, 50,000 simulations			
Strike	0.738147	0.745603	0.753059	Strike	0.738147	0.745603	0.753059
Moneyness	99%	100%	101%	Moneyness	99%	100%	101%
Simple MC	0.010815	0.007504	0.004889	Simple MC	0.010872	0.007599	0.004998
(std. error)	0.000058	0.000049	0.000040	(std error)	0.000059	0.000050	0.000042
MC with MVR	0.010766	0.007463	0.004855	MC with MVR	0.010953	0.007665	0.005050
(std. error)	0.000027	0.000027	0.000026	(std error)	0.000028	0.000028	0.000027

Three-year European call on six-year 10% semi-annual coupon bond

20 steps/year, 50,000 simulations				100 steps/year semi-annual coupon bond			
Strike	0.989276	0.999269	1.009262	Strike	0.989276	0.999269	1.009262
Moneyness	99%	100%	101%	Moneyness	99%	100%	101%
Simple MC	0.014142	0.009687	0.006197	Simple MC	0.014215	0.009809	0.006338
(std error)	0.000075	0.000063	0.000052	(std error)	0.000076	0.000065	0.000053
MC with MVR	0.014090	0.009644	0.006164	MC with MVR	0.014359	0.009933	0.006439
(std error)	0.000014	0.000016	0.000017	(std error)	0.000015	0.000016	0.000017

A simple and reasonable way to allow for stochastic volatility is simply to replace (3) by

$$dP^q_t/P^q_t = r_t\,dt + \sigma_t \sum_{j=1}^{m} v^{j,q}_t\,dW^j_t \qquad (9)$$

where σ_t represents the general level of volatility and satisfies an Itô equation

$$d\sigma_t = \xi(\sigma_t)\,dt + \eta(\sigma_t)\,dWt \qquad (10)$$

where dWt is the increment of standard Brownian motion, with constant correlation coefficients – say, $dW_t.dW^i_t = \rho_i dt$ – with the basic factors.

It would also be easy, but more complicated, to assign different σ_ts to each random factor. Our MVR procedure can accommodate this extra volatility factor if we replace (2) by

$$R^{s,q}_0 = \exp\left\{\sum_{j=1}^{m}\int_{\rho=0}^{s}\left[\sigma_\rho \nu^{j,q}_\rho\,dW\rho^j - \tfrac{1}{2}\sigma^2_\rho(\nu^{j,q}_\rho)^2 d\rho\right]\right\} \qquad (11)$$

(See Heath, Jarrow and Morton (1992) and Carverhill (1995b).)

The Fong and Vasicek (FV) (1991) model can be cast in this form, as in fact can all the Gaussian models, and we present numerical results for our variance-reduced Monte Carlo procedure applied to this model. The FV model starts from the equations

$$dr_t = \alpha(\bar{r} - r_t)\,dt + \sqrt{\nu_t}\,dx_t \qquad (12)$$

$$d\nu_t = (\bar{\nu} - \nu_t)\,dt + \xi\sqrt{\nu_t}\,dy_t \qquad (13)$$

for the short rate r_t and variance $v_t \equiv \sigma^2_t$, in which dx_t and dy_t are standard Brownian motions with correlation ρ, ie $dx_t dy_t = \rho dt$. Fong and Vasicek then solve this model in the form $P^q_t(r, v) = \exp\{-rD(q - t) + vF(q - t) + G(q - t)\}$ for certain functions D, F, and G, and from this we can write

$$dP^q_t/P^q_t = r_t\,dt - \sqrt{\nu_t}\,D(q - t)\,dx_t + \xi\sqrt{\nu_t}\,F(q - t)\,dy_t \qquad (14)$$

corresponding to (9).[6] (Also, (13) corresponds to (10) if we translate from ν to $\sigma \equiv \sqrt{\nu}$.)

Table 1 gives results for this model, with parameters $\alpha = 1.5$, $\gamma = 1.0$, $\xi = 0.1$, $\rho = 0.5$, $r_0 = 0.1$, $\bar{r} = 0.1$, $v_0 = 0.01$, and $\bar{v} = 0.01$. From these results we see that the variance reduction works better for the coupon bond than for the pure discount bond; the reduction in standard error is on the order of four, rather than two. These factors four and two indicate savings on the order of about 16 and four, respectively, in the number of iterations required to achieve a given standard error.

The reason that the variance reduction is more effective for the coupon bond is simply that there are more control variates because there are more coupons. Recall, however, that using these variates incurs hardly any extra cost because they have already been calculated in the Monte Carlo procedure itself.

We have chosen to implement the FV model simply because it is familiar. The strength of our procedure is its flexibility, however; it will work for a very general volatility structure, which can be chosen or estimated directly, and when there is no analytic solution and no easy reduction of the model (such as Duffie and Kan seek, and which the Fong and Vasicek model provides) to a set of factors such as (r_t, v_t) of equations (12) and (13).

Ensuring non-negative interest rates

Perhaps the simplest way to ensure that interest rates cannot go negative is to throw a factor $\sqrt{r_t}$ into the volatility, replacing equation (3) by

$$dP_t^q / P_t^q = r_t\,dt + \sqrt{r_t}\sum_{j=1}^{m} \nu_t^{j,q}\,dW_t^j \qquad (15)$$

and keeping $\nu_t^{j,q}$ non-random. We could also throw in a factor σ_t, but we omit this for the sake of clarity. Then, following Heath, Jarrow and Morton (1992), we have

$$df_t^q = r_t\,\alpha_t^q\,dt + \sqrt{r_t}\sum_{j=1}^{m} \sigma_t^{j,q}\,dW_t^j \qquad (16)$$

where

$$\sigma_t^{j,q} = -\frac{\partial}{\partial q}\nu_t^{j,q}, \quad \alpha_t^q = -\sum_{j=1}^{m} \nu_t^{j,q}\,\sigma_t^{j,q} \qquad (17)$$

so

$$r_t\,(\equiv f_t^t) = f_0^t + \int_{\rho=0}^{t} r_\rho\,\alpha_\rho^t\,dt + \sum_{j=1}^{m}\int_{\rho=0}^{t} \sqrt{r_t}\,\sigma_\rho^{j,t}\,dW_\rho^j \qquad (18)$$

Also, we have

$$R_0^{s,q} = \exp\left\{\sum_{j=1}^{m}\int_{\rho=0}^{s}\left[\sqrt{r_\rho}\,\nu_\rho^{j,q}\,dW_\rho^j - \tfrac{1}{2}r_\rho(\nu_\rho^{j,q})^2\,d\rho\right]\right\} \qquad (19)$$

The idea is now to simulate the $R_0^{t,q}$s and r_t simultaneously as t evolves over (0,s), using (18) and (19), and continue the MVR procedure.

From (18) it is impossible for r_t to go negative as long as the initial forward rates $\{f_0^q\}_{q\geq 0}$ are never negative. Also, if r_t is never negative, then neither is f_t^q for any t, q; this can be seen since the RHS always decreases in the equation

$$\exp\left\{-\int_{\tau=t}^{q} f_t^\tau\,d\tau\right\}(\equiv P_t^q) = E\left[\exp\left\{-\int_{\tau=t}^{q} r_\tau\,d\tau\right\}\right] \qquad (20)$$

The Longstaff and Schwartz model fits indirectly into this framework, and we present numerical results for this model. Again, we have chosen to implement our procedure for this model simply because it is familiar, but the strength of our procedure is its generality. The formulation above, using the square root of r_t, will work as long as the volatility structures $\{\nu_t^{j,q}\}_{t\leq q}$ in (15) are all bounded.[7]

The Longstaff and Schwartz model starts from the equations

$$dx_t = (\gamma - \delta x_t)dt + \sqrt{x_t}\,dW_t^0 \qquad (21)$$

$$dy_t = (\eta - \varepsilon y_t)dt + \sqrt{y_t}\,dW_t^1 \qquad (22)$$

where (x_t, y_t) are related linearly to (r_t, v_t) via the equations $r = \alpha x + \beta y$ and $v = \alpha^2 x + \beta^2 y$, and W_t^0, W_t^1 are orthogonal Brownian motions. Longstaff and Schwartz obtain functions A, B, C and D such that

$$P_t^q(r, v) = A^{2\gamma}(q-t)\,B^{2\eta}(q-t) \times \exp\{\kappa(q-t) + C(q-t)\,r + D(q-t)\,v\}$$

in which $\kappa = \gamma(\delta + \phi) + \eta(\nu + \varphi)$, $\phi = \sqrt{2\alpha + \delta^2}$, $\varphi = \sqrt{2\beta + \nu^2}$, and ν is a constant.

Table 2. Simulation results – Longstaff and Schwartz model

Three-year European call on six-year pure discount bond

20 steps/year, 50,000 simulations				100 steps/year, 50,000 simulations			
Strike	0.646075	0.652601	0.659127	Strike	0.646075	0.652601	0.659127
Moneyness	99%	100%	101%	Moneyness	99%	100%	101%
Simple MC	0.020339	0.017744	0.015321	Simple MC	0.020581	0.017977	0.015545
(std error)	0.000113	0.000104	0.000096	(std error)	0.000113	0.000105	0.000097
MC with MVR	0.020390	0.017791	0.015363	MC with MVR	0.020552	0.017951	0.015522
(std error)	0.000057	0.000056	0.000055	(std error)	0.000057	0.000056	0.000055

Three-year European call on six-year 10% semi-annual coupon bond

20 steps/year, 50,000 simulations				100 steps/year 50,000 simulations			
Strike	0.879675	0.888560	0.897446	Strike	0.879675	0.888560	0.897446
Moneyness	99%	100%	101%	Moneyness	99%	100%	101%
Simple MC	0.025770	0.022224	0.018932	Simple MC	0.026076	0.025518	0.019213
(std error)	0.000142	0.000131	0.000120	(std error)	0.000143	0.000132	0.000120
MC with MVR	0.025858	0.022305	0.019005	MC with MVR	0.026056	0.025501	0.019199
(std error)	0.000046	0.000047	0.000047	(std error)	0.000047	0.000048	0.000048

From this we see that

$$dP_t^q / P_t^q = r_t dt + \sqrt{x}\,\alpha \times [C(q-t) + \alpha D(q-t)]\, dW_t^0 + \sqrt{y}\,\beta\,[C(q-t) + \beta D(q-t)]\, dW_t^1 \quad (23)$$

With parameters $\alpha = 0.05$, $\beta = 0.06$, $\gamma = 0.8$, $\delta = 0.8$, $\varepsilon = 0.5$, $\eta = 0.9$, and $\nu = 0.5$, the results are given in Table 2. Our conclusions for the LS implementation are essentially the same as for the FV implementation. The variance reduction is better for the coupon bond and requires fewer iterations for a given standard error, by a factor of about nine.

Summary and commentary

Our implementation of the HJM model for valuing options on the term structure is efficient and flexible, and can avoid the possibility of negative interest rates. To ensure flexibility, our implementation is basically Monte Carlo simulation; we have made this more efficient by using martingale variance reduction variates.

The MVR technique is well established in the field of stochastic simulation, and our contribution has been to make an appropriate choice of MVR variates. When valuing a coupon bond option, our MVR variates are associated with the pure discount bonds that make up the coupons. These must be calculated when simulating the option payoff itself, so using these MVR variates essentially imposes no extra cost in calculation. To ensure that interest rates never become negative, we insert the factor $\sqrt{r_t}$ in a way that mimics the CIR model.

When we have cast the models of Fong and Vasicek (1991) and Longstaff and Schwartz (1992) in our framework and apply our variance-reduced Monte Carlo procedure, we get more variance reduction when there are more coupons. The reason is that there are then more MVR variates. For a three-year option on a six-year bond that pays semi-annual coupons (so that there are seven MVR variates, corresponding to the number of coupons plus the option payment itself), the standard error for a given number of simulations is reduced by a factor of about four. This translates to a reduction in the number of simulations by a factor of about 16 to achieve a given standard error.

The motivation for applying our procedure to these models is simply that they are familiar. The strength of our procedure is that it will work very generally – we require only some regularity conditions on the volatility of the term structure.

Is our HJM implementation preferable to its rivals, the equilibrium (factor) models? This depends on the priorities of the user. If flexibility and accuracy in fitting market data are needed, our approach is valuable. Factor models such as Duffie and Kan's (1996) can achieve this flexibility by including more factors, although this may be difficult because of the basically non-linear character of this procedure. The time taken to solve the factor model will grow geometrically with the number of factors, while the time taken by the Monte Carlo procedure will grow less dramatically as the number of factors increases.

1 *Technically, the HJM model is not separate from the equilibrium models, but rather includes them as special cases. The HJM model takes the initial term structure and volatility structures, but does not attempt to explain them in terms of state variables.*

2 *In these articles the approach to the design of the MVR variates is different; the design is based on the delta-hedge.*

3 *This is a retrospective interpretation, since Ho and Lee (1986) predates Heath, Jarrow and Morton (1992). The most common implementations of the HJM model use the volatility of the Vasicek (1977) model or the Cox, Ingersoll, and Ross (1985) model. See Heath, Jarrow and Morton (1990, 1992). Carverhill (1994) shows that if the volatility is time-homogeneous and non-random, and if the short rate is Markovian (which is necessary for compatibility with a one-factor model), then the volatility must be as in Vasicek. Rogers (1994) makes the point that practicable implementations of the HJM model often push the model into formulation that would more conveniently start from a factor model.*

4 *See the appendix to Carverhill (1995b).*

5 *The significant aspect of this volatility structure is that (8) represents a tilting of the term structure. Such a tilt seems to be difficult to incorporate into a factor structure in a time-homogeneous way.*

6 *Fong and Vasicek obtain the functions D, F, and G from a series of transformations. Selby and Strickland (1993) obtain them as series expansions. To implement our procedure, we must transform to orthogonal Brownian motions.*

7 *See the appendix to Carverhill (1995b). Note, however, that we cannot replace* $\sqrt{r_t}$ *with, say,* r_t *in (15) because we would not then be able to ensure that* r_t*, given as a solution corresponding to (18), and* P_t^q *itself are well -defined and non-explosive.*

BIBLIOGRAPHY

Carverhill, A.P., 1994, "When is the Short Rate Markovian?", *Mathematical Finance* 4(4), pp. 305–12.

Carverhill, A.P., 1995a, "A Note on the Models of Hull and White for Pricing Options on the Term Structure", *Journal of Fixed Income* 5(2), pp. 89–96.

Carverhill, A.P., 1995b, "A Simplified Exposition of the Heath, Jarrow and Morton Model", *Stochastics* 53, pp. 227–40.

Carverhill, A.P., and L.C. Clewlow, 1994, "Quicker on the Curves", *Risk* 7(5), pp. 70–1 (Chapter 10 of this publication).

Clewlow, L.C., and A.P. Carverhill, 1994, "On the Simulation of Contingent Claims", *Journal of Derivatives* 2(2), pp. 66–74.

Cox, J.C., J.E. Ingersoll and S.A. Ross, 1985, "A Theory of the Term Structure of Interest Rates", *Econometrica* 53, pp. 385–407.

Duffie, D., and R. Kan, 1996, "A Yield-Factor Model of Interest Rates", *Mathematical Finance* 6, pp. 379–406.

Fong, H.G., and O. Vasicek, 1991, "Interest Rate Volatility as a Stochastic Factor", Working paper, Gifford Fong Associates.

Heath, D., R. Jarrow and A. Morton, 1990, "Bond Pricing and the Term Structure of Interest Rates: A Discrete Time Approximation", *Journal of Financial and Quantitative Analysis* 25(4), pp. 419–40.

Heath, D., R. Jarrow and A. Morton, 1992, "Bond Pricing and the Term Structure of Interest Rates: A New Methodology for Contingent Claims Valuation", *Econometrica* 60(1), pp. 77–105.

Ho, T.S.Y., and S.B. Lee, 1986, "Term Structure Movements and Pricing Interest Rate Contingent Claims", *Journal of Finance* XLI, pp. 1,011–29.

Hull, J., and A. White, 1990, "Pricing Interest Rate Derivative Securities", *Review of Financial Studies* 3(4).

Hull, J., and A. White, 1993, "One Factor Interest Rate Models and the Valuation of Interest Rate Derivative Securities", *Journal of Financial and Quantitative Analysis* 28(2), pp. 235–54.

Longstaff, F.A., and E.S. Schwartz, 1992, "Interest Rate Derivative Securities", *Journal of Financial Studies* XLII(4), pp. 1259–1281.

Rogers, L.C.G., 1994, "Which Model of the Term-Structure of Interest Rates Should One Use?", *Proceedings of the IMA Workshop on Mathematical Finance*, Springer Series in Applied Mathematics 65, pp. 63–116.

Selby, M.J.P., and C. Strickland, 1993, "Computing the Fong and Vasicek Pure Discount Bond Pricing Formula", *Journal of Fixed Income*.

Vasicek, O., 1977, "An Equilibrium Characterization of the Term Structure", *Journal of Financial Economics* 5, pp. 177–88.

19

Term Structure Dynamics and Mortgage Valuation

Oren Cheyette
BARRA Inc

The past decade has seen an enormous expansion of both the quantity of securitised mortgages in the market and the variety of collateralised mortgate obligation (CMO) structures. Mortgage-backed securities (MBS) offer attractive yields with little credit risk, and the newer CMOs include tranches with relatively low prepayment risk. Because of the implicit government guarantee of the principal, the main risk to an MBS pass-through or CMO investor lies in uncertainty as to the timing of the repayment in different interest rate scenarios. Evaluation of the cost of this uncertainty is the primary subject of this chapter.

The task of mortgage valuation has two major components. The first is to model the cashflows of MBS collateral and, for a CMO, of the deal's tranches. This requires a prepayment model for the collateral and a model for distributing cashflows to the CMO tranches. Prepayments are generally described by an empirical model based on statistical studies of actual prepayment experience as a function of market conditions. CMO cashflow modelling is primarily an algorithmic logic problem. While the problem is potentially quite difficult for some of the recent deals, the CMO rules themselves are given in the prospectus, and the difficulty is in translating these into a computer program. This and the prepayment modelling problem, while interesting and crucial to mortgage valuation, are not discussed here.

My subject is the second part of the mortgage modelling problem, namely the problem of computing the present value of uncertain cashflows. Because prepayments are largely driven by economic incentives, they give rise to the familiar negative convexity of MBS. To a varying extent depending on their structure, CMO tranches are also subject to cashflow variability due to prepayment risk. CMOs often include capped and floored floaters and inverse floaters whose rate-dependent cashflows can also be valued by the techniques described here.

There now exists a fairly well developed theory of the pricing of interest rate-dependent cashflows. Various models based on these ideas are in widespread use for modelling corporate bonds and mortgages. In this chapter I will review both the theory and its application to mortgage valuation. For completeness I have included some of the mathematical detail in an appendix.

The components of a valuation algorithm for interest rate-sensitive cashflows consist of a theoretical valuation formula giving the present value of uncertain cashflows as a function of the term structure dynamics, a model of the term structure dynamics, and an algorithm for evaluating the formula.

Valuation of interest rate-sensitive securities

The fundamental valuation formula for securities subject to interest rate risk is surprisingly simple considering the effort expended on developing models for bonds, mortgages and derivative securities. The formula incorporates most of the popular option pricing models – and all of the arbitrage-free ones. Examples are the Ho–Lee (1986) model, with one random factor, constant volatility and parallel yield curve shifts, the Black–Derman–Toy (BDT) (1990) model, with one factor and proportional volatility, and the Brennan–Schwartz (1979) model, with two factors and proportional volatility.

The valuation formula is easily stated the present value of a security is the average over all possible future *risk-neutral* interest rate paths of the discounted present values of its cashflows, where the discount factor is given by the cumulative one-period rates along the future path.

Symbolically,

$$\text{PV} = \text{Average of the sum over periods of} \left\{ \frac{\text{Cashflow at period i for this rate path}}{(1 + r_1)(1 + r_2)\cdots(1 + r_i)} \right\}$$

where $r_1,\ldots,r_i$ are the successive one-period rates along a sample rate path. This equation can be written more formally as

$$\text{PV} = \lim_{N\to\infty} \frac{1}{N} \sum_{\text{(paths r)}} \sum_{\text{(periods i)}} \frac{\text{CF}_i(\text{path})}{(1 + r_1)(1 + r_2)\cdots(1 + r_i)}$$

where N is the number of sample rate paths.

In the absence of interest rate uncertainty (zero volatility), the average over paths collapses because all paths are identical, and the formula reduces to ordinary static cashflow discounting. The formula also reduces to the static cashflow discount case when the cashflows are certain (as for an option-free bond). As will be discussed, this happens because the distribution of interest rate paths is chosen to make it happen.

The purpose of this section is primarily to explain exactly what is meant by the term "risk neutral" in the specification of the valuation formula. Once this has been clarified, it is a simple matter to derive the present value formula.

The derivation is based in part on the work of Vasicek (1977), Jamshidian (1989a) and Heath, Jarrow and Morton (1989). Readers uninterested in the derivation are invited to skip to the closing paragraphs of this section, where the meaning and use of the formula are discussed.

For simplicity, let us assume that there is just a single source of price risk for fixed income assets. We may identify this risk as uncertainty of the short rate (the one-period rate), and assume that longer rates are perfectly correlated with the short rate. This is an obviously unrealistic assumption, but including multiple sources of risk complicates the argument without affecting the basic ideas.

Because the short rate may have a systematic drift component (due, for example, to the slope of the yield curve) as well as a random component, it is simpler to think of the source of risk as some uncertain economic variable on which rates depend, with no built-in drift. We may denote this variable by z, and by assumption its expected change is 0. As a convention, we also assume that its one-period variance is 1. A coin flip with payoff 1 for heads and –1 for tails is such a random variable. The change in, say, the 10-year rate over one period is then the 10-year rate volatility times the result of the coin flip, plus any systematic drift.

The first step in the argument is to determine the relationship between the riskiness of an asset and its expected rate of return. The riskiness of the asset is simply the sensitivity of the asset price to the random variable z multiplied by the volatility of z.

A simple argument now shows that the expected one-period return on any asset is given by the short rate plus a premium proportional to the riskiness of the asset. Consider a portfolio of two assets, A and B. As there is only one source of risk, their price movements are perfectly correlated, and the portfolio can be made riskless by choosing appropriate amounts of the two. If the portfolio contains \$1 of asset A, we must be short $\frac{\text{riskiness of A}}{\text{riskiness of B}} \times$ \$(1 of B) to cancel the risk. We may scale down the resulting portfolio by the riskiness of A; the new portfolio looks like

$$\left(\frac{\$1}{\text{Riskiness of A}} \text{ of A}, \frac{-\$1}{\text{Riskiness of B}} \text{ of B} \right)$$

Because the portfolio is riskless, absence of arbitrage requires that it earn exactly the one-period rate during the next period – that is, the risk premium of A is cancelled by that of B. For this to happen, the risk premium of A must be proportional to its riskiness, and likewise for B, with the same constant of proportionality. The symbol λ is used for this ratio. The value of λ and the term structure of interest rates incorporate all the effects of the market's risk aversion.

Denoting A's riskiness by σ_A, we may write for the one-period change in the price, P_A, of A:

$$\frac{\Delta P_A}{P_A} = r + \lambda\sigma_A + \sigma_A \Delta z$$

This formula incorporates the expected return on A attributable to the short rate, the risk premium, and the random fluctuation. Let us define a new random variable, $\tilde{z}$, such that $\Delta\tilde{z} = \Delta z + \lambda$. In terms of this new variable, the price change of A is

$$\frac{\Delta P_A}{P_A} = r + \sigma_A \Delta\tilde{z}$$

The risk premium no longer appears explicitly in the equation for the price change, but we have changed nothing because we have built a drift into $\tilde{z}$ to account for the risk premium (we have biased the payoffs of the coin flip). The key observation is that there is nothing special about asset A. Because λ does not depend on which asset we are talking about, the same substitution of $\tilde{z}$ for z removes the explicit risk premium from the price change equation of *all* assets.

If $\tilde{z}$ had no drift, all assets would be expected to appreciate at the same rate independent of their riskiness. We therefore define the *risk-neutral* distribution of rate paths to be the distribution generated by assuming $\tilde{z}$ to be an unbiased random variable. With this choice of probability for possible future rate paths, all assets appreciate on average period-by-period at the short rate.

This is not to say that these rate paths are predictions of future rates. They are simply a set of possible rate paths chosen so that the expected rate of return on all assets is the short rate, when averaged over these paths. That it is possible to find such a probability distribution for the paths at all is a non-trivial consequence of the no-arbitrage argument given above.

If we had started out without any risk premium, that is, with $\lambda = 0$, we would have obtained the the same formulas for the price movements. In this case the biased coinflip $\tilde{z}$ is the same as the unbiased one, z. It is for this reason that the interest rate paths are referred to as risk-neutral.

For the risk-neutral distribution of future rate paths, a "money market" portfolio consisting of \$1 invested in a series of one-period bonds (call this portfolio M) is expected to earn the same rate of return as any other investment, say, asset A. Therefore, on average, the ratio of the value of A to the value of M will not change. Over any particular rate path the ratio will change because A is exposed to market risk, while M is riskless period-by-period; but when averaged over the risk-neutral set of possible future rate paths the ratio will be constant.

This ratio is $P_A/\$1$ today. Along some particular rate path, the ratio evolves to[1]

$$\frac{P_A \text{ at period } i}{\$1 \times (1+r_1)(1+r_2)\cdots(1+r_i)}$$

Averaging over the paths, we then have

$$P_A = \text{Average over risk-neutral paths of} \left\{\frac{P_A \text{ at period } i}{(1+r_1)(1+r_2)\cdots(1+r_i)}\right\}$$

To obtain the formula in terms of A's cashflows, we use the notion of a security that pays \$1 at period n only if a particular rate scenario is realised up to that period (called an Arrow–Debreu primitive security). We can duplicate the cashflows of any security with a portfolio of Arrow–Debreu securities.

For example, a zero-coupon bond maturing at period n is replicated by an equally weighted portfolio of all Arrow–Debreu securities that mature at period n – one for each possible rate path. A one-period cap for the period n to n+1 with strike R is replicated by a portfolio of those Arrow–Debreu securities that pay off when the rate at n is above R, weighted in proportion to the cap's payoff. According to the last formula, the present value of an Arrow–Debreu security maturing at period n is

$$P_{A-D} = \text{Average over risk-neutral paths of} \left\{\frac{\$1 \text{ if it is the right path, otherwise } \$0}{(1+r_1)(1+r_2)\cdots(1+r_i)}\right\}$$

which may be more precisely expressed as

$$P_{A-D} = \text{Average over risk-neutral paths of the sum over periods of} \left\{\frac{\text{Cashflow of this A–D security at period } i}{(1+r_1)(1+r_2)\cdots(1+r_i)}\right\}$$

Because the present value of a portfolio of securities is equal to the sum of the present values of the individual securities, and since the denominator of the valuation formula above does not depend on which security we are valuing, we obtain the valuation formula given at the beginning of this section.

Although it is intuitively appealing, this formula is not so simple to use in practice. To begin with, determining the ith cashflow is not generally a simple problem. For bonds with optional sinking funds or American call provisions, one needs a full-blown option model just to determine the cashflows (ie the cashflow at period i depends on whether the option is exercised then, determination of which in turn requires an option model that depends on the later cashflows). Similarly, determining a mortgage's cashflows requires a prepayment model, which typically produces cashflows dependent on the specific rate history experienced along the path.

More significantly, this discussion has made no mention of the detailed relationship between the random source of risk z and the interest rate. Before we can use the valuation formula, we need to know how to construct the risk-neutral rate paths. It is here that the specification of interest rate volatility enters, and one finds the relationship between typical rate paths and the current forward rate.

Finally, though simple to state, the valuation formula is not easily translated into an efficient computer model. While it lends itself to a direct translation into a Monte Carlo algorithm, such a scheme generally requires vast amounts of computer time to price even the simplest securities because of the large number of interest rate paths that must be sampled to obtain an accurate answer.[2]

A shortcut is available for securities whose cashflows at period n depend only on interest rates at that period (such as callable bonds without optional sinking funds), provided that the interest rate evolution is of a special type known as a Markov process. In this case, one can show that the valuation formula satisfies a certain partial differential equation similar to that behind the Black–Scholes formula and use standard techniques for solving that equation. This is the basis for the bond option models mentioned earlier.

Unfortunately, mortgage cashflows are dependent on the entire history of rates, so this shortcut is not useful. Nevertheless, by applying so-called variance reduction methods to the Monte Carlo approach one can obtain quite accurate results in a reasonable amount of time, as is discussed later.

Term structure dynamics

The time evolution of the term structure of interest rates in an arbitrage-free model is governed by the current term structure and by a specified rate- and maturity-dependent volatility function. While calculating the details of the rate dynamics is a fairly technical problem (an outline is given in the appendix), the main features of the dynamics can be explained by fairly simple financial arguments.

If there were no uncertainty about the movement of rates, the short rate would evolve according to the current forward rate curve. This is because otherwise there would be an arbitrage between a long zero and a series of reinvested one-period bonds. The returns on both portfolios over any period are certain by assumption, so they must be the same. This is not a consequence of risk-neutrality, as in this case there is no risk.

If we "switch on" volatility, two things happen. First, if the risk premium λ is positive, a downward drift is added to the evolution along the forward curve. This gives the holders of longer, riskier bonds an added expected return proportional to their exposure to rate risk, as required.[3]

A more subtle consequence, whose details depend on the assumed term structure of volatility (the dependence of volatility on maturity), is that the forward curve tends to steepen. For simplicity, assume in the argument below that the fluctuations are independent of maturity (the parallel-shift model). The steepening then follows from a simple arbitrage argument.

Consider the price behaviour of two straight bonds: a zero-coupon bond with a modified duration of seven years, and a longer coupon bond with the same modified duration (with a maturity of around 10 years). The coupon bond has the same risk exposure as the zero but a higher convexity. (A zero-coupon bond has the lowest convexity of any straight bond with the same duration.) Because of its greater convexity, the coupon bond will outperform the zero over short horizons whether yields fluctuate up or down.

This appears to represent an arbitrage opportunity. One could buy the coupon bond with proceeds from shorting the zero and have guaranteed profits. This gain is offset, on average, by a systematic steepening of the term structure. The steepening has more of a negative effect on the higher-convexity coupon bond than on the zero, and *in expectation* balances the outperformance of the coupon bond from the yield fluctuations.

Because the excess return of the coupon bond over the zero is of second order in the size of the yield fluctuation (ie proportional to the square of the size of the fluctuation), the rate of steepening is proportional to the square of the volatility. For a typical volatility of 12% in an 8% yield environment the required steepening is quite modest: an increase of about 15 basis points annually in the spread between the overnight rate and the 30-year spot rate. While this effect is smaller than the typical fluctuations we are ignoring (the yield curve twists and flexes that would be included in a multi-factor model), it must be included to obtain arbitrage-free pricing.[4]

While this gradual steepening is modest over the span of a year, it adds up over longer periods. Over 30 years it would produce a steepening of 450bp between the 30-year spot rate and the short rate. Clearly, this makes the parallel-shift model somewhat unrealistic. Another problem with the parallel-shift model is that it cannot account for steepening over the short run, nor for inversion of the yield curve.

These seemingly contradictory difficulties can both be resolved by introducing mean reversion of the short rate. Mean reversion is the name given to the tendency of rates to return towards a long-run level if they get too high or too low. Mean reversion can be incorporated into the model in a natural way by making the yield curve fluctuations non-parallel.

Specifically, a term structure of volatility that declines with maturity (long bond yield less volatile than short rate) implies mean reversion. The reason is simply that if very long forward rates do not fluctuate (which still permits long yields to fluctuate), then the tendency of the short rate to drift along the forward curve will always bring it back towards the long forward rate.

Mean reversion is described mathematically as a rate that tells how fast the short rate tends back towards its long-run level. A mean reversion strength of 5% per year, corresponding to a mean reversion timescale of 20 years, means that a 100bp deviation produces a 5bp per year drift towards the mean. The mean reversion timescale is the amount of time over which mean reversion has a significant effect.

The distribution of rates along different random paths is initially unaffected by mean reversion. Only after enough time has elapsed does the effect of the mean reversion on the width of the rate distribution become apparent. Mean reversion can also be specified by giving the short and long rate volatilities. With a flat 8% yield curve, specifying a short rate volatility of 15% and a long bond yield volatility of 10% is equivalent to a mean reversion time of about 20 years. This amount of mean reversion is difficult to see directly in the motion of the short rate. The best evidence for mean reversion is the observed decreasing term structure of volatility.

Mean reversion moderates the tendency of the yield curve to steepen because very long term forward rates are effectively "anchored". Starting from an initially flat term structure, the short end of the forward curve steepens, while the long end remains flat. (Because there are no fluctuations of very long forward rates, that part of the curve does not steepen.) On average this results in a humped forward curve, which gives a natural-looking par bond yield curve rising steeply at the short end and flattening out at the long end.

Deviations of the short rate from equilibrium cause the yield curve to invert or become steeper as the short rate deviates above or below equilibrium. This is consistent with the observed behaviour of the yield curve.

A family of zero-coupon spot rate curves with different departures of the short rate from equilibrium that results from incorporating mean reversion is shown in Figure 1. The curves assume a 20-year mean reversion timescale and about 15% annual volatility of the short rate. These are long-run equilibrium spot curve shapes towards which the initial term structure evolves.

The same spot curve shapes result no matter what market risk parameter is assumed Changing the risk premium simply changes the long-term mean rate level. One sees in these curves the competition between:

- the steepening attributable to the no-arbitrage requirement; and
- the tendency of mean reversion to bring the short and long rates to the same level.

The dynamics of the term structure also depend on how one assumes interest rate volatility depends on rates. The Ho–Lee model assumes that volatility is a constant (eg 100bp per year), independent of the interest rate, maturity and current time. The model suffers the defects that rates can become negative, that there is no mean reversion, and that volatility does not increase with rates.

While these problems make the model somewhat unrealistic, it still yields reasonably good security valuations. As its creators themselves say, it is primarily useful as the simplest arbitrage-free model, against which other models can be

1. Long-term zero-coupon rates in a mean-reverting term structure model, with 20-year mean reversion and 14% annual volatility of short rates

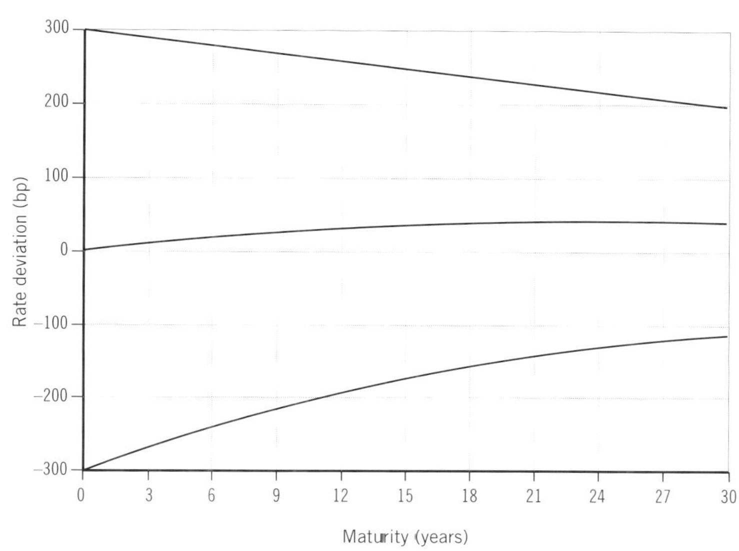

measured. It lends itself to fairly straightforward analytical methods, and one can find closed-form solutions for European puts and calls on bonds. These solutions look very much like the Black–Scholes formula for stock options: as rate fluctuations are normally distributed, zero-coupon bond price fluctuations are lognormal (although with decreasing volatility as they approach maturity), just as are stock price fluctuations in the Black–Scholes model.

When one incorporates both mean reversion and the dependence of volatility on interest rates, it becomes difficult to find analytic solutions (except for the mean-reverting extension of the Ho–Lee model). In general, the evolution of the term structure from a given time forward depends on how one got there, ie the rate process is history-dependent (not a Markov process). This is not too unreasonable because, as we have noted, the term structure's shape changes in a way that depends on its volatility, and the volatility in turn depends on the term structure.

The BDT model is an example of a model that incorporates time-dependent volatility and dependence of volatility on rates, and for which the rate dynamics are path-independent. Because of the way the model is specified, however, it is hard to disentangle the effects of mean reversion and the dependence of volatility on maturity and on time.

In this model, in fact, one does not explicitly specify the dependence of volatility on maturity. Rather, one inputs a "forecast" of short rate volatilities; to obtain mean reversion one must assume that in the future short rates will be less volatile than today. This is of course a very unrealistic assumption. One would much prefer to be able to specify a realistic term structure of volatility that incorporates the declining volatility of rates with maturity without implicitly forecasting that overall volatilities will decline in the future.

A model with just such features is described briefly in the next section and in some detail in the appendix. The model permits the volatility to be almost any desired function of the short rate, while the term structure of volatility has a fairly general form that accommodates mean reversion. The model has the advantage that the relationship between mean reversion of short rates and the volatility term structure is transparent.

Evaluating the pricing formula

We have developed two approaches to the problem of efficiently evaluating the average over paths in the pricing formula. One is a reduced-variance Monte Carlo algorithm, in which a random number generator is used to generate risk-neutral interest rate paths. The second is a modified scheme in which we choose a set of representative interest rate paths without the "noise" introduced by the use of random numbers. This "representative path" scheme is generally faster than the Monte Carlo method by as much as a factor of 50 for a given level of accuracy. Because the Monte Carlo method is fairly intuitive as well as being a commonly accepted methodology, I have compared the two methods in the data given below.

A direct transcription of the pricing formula into a Monte Carlo algorithm would be the following. Generate a risk-neutral interest rate path using a random number generator for the variable $\tilde{z}$ defined earlier. Along this path, compute the security's cashflows period by period as a function of the current interest rate. Compute the progressive one-period discount factors, and add up the discounted cashflows. Do this for many paths and take the average.

Unfortunately, this method is extremely slow to achieve reasonable accuracy. There are two main reasons for this. The most important is that the use of random numbers to generate simulation paths introduces statistical noise into the method. A simple argument based on the law of large numbers shows that this statistical error goes away very slowly as the number of paths is increased. For example, to reduce the error by a factor of 10 requires a hundredfold increase in the number of paths.

A second reason is that the simulation algorithm has to be calibrated to the initial spot curve so that option-free securities will be priced as though simply discounted off the initial term structure. Specifically, one must find the period-by-period average drift of interest rates that properly reproduces the initial spot curve. This leads to two problems.

The first is that, for many interest rate models, the required drift cannot be calculated analytically. This problem is addressed by the interest rate model described mathematically in the appendix. The basic idea behind this model is that one can describe a large class of interest rate processes (including, for example, a lognormal rate process with mean reversion) as *two-component* processes — that is, processes that can be described in terms of two "state variables" — where the first component is the short rate and the second component is a kind of cumulative volatility. The entire term structure is determined at any future date in terms of the initial term structure and the two state variables. While the short-rate process alone is history-dependent, the combined short-rate and cumulative volatility process is history-independent and is automatically consistent with the initial term structure. That is, one can analytically calibrate the interest rate model to an arbitrary initial term structure.[5]

The fact that for this family of interest rate models one can describe the entire term structure analytically in terms of the two state variables is especially important for the mortgage valuation problem. The reason is that mortgage prepayments are determined by mortgage rates, which are in turn dependent on long-term yields. So one needs to know the entire spot curve – not just the short rate – to determine a mortgage security's cashflow at each point along a simulation path. This is in contrast to many one-factor models (such as the BDT model), where one knows only the short rate at each simulation point.

The second issue arises from the statistical error discussed previously. Very simply, the option values that we are trying to obtain by simulation (eg the value of the prepayment option in a particular MBS) are typically a small fraction (a few per cent) of the overall value of the security. So the statistical errors on the overall security value – even if it had no embedded option – are likely to swamp the option value, which is the "interesting" part. This problem can be overcome by means of a so-called variance reduction method. This entails valuing a zero-coupon bond maturing at the same time as each cashflow jointly with the security. One then adjusts the discount factor to each cashflow date so that each zero-coupon bond is priced correctly according to the initial term structure. By this adjustment, any security with deterministic cashflows will be priced with no statistical error. The only remaining source of random error is variability of the cashflows due to options.

With the two-state interest rate model and variance reduction method described here, moderately accurate values (of order ±8 tics for a current coupon pass-through) can be obtained with 128 sample paths at a computation speed of about seven paths per second on a 386 PC. The algorithm requires no numerical calibration to the interest rate term structure.

Figure 2 shows a random set of 20 sample rate paths with no mean reversion. Note that, because volatility increases with the interest rate, interest rates can easily get very big. The upward drift of the risk-neutral rates is also apparent, as is the increasing speed of drift when rates (and therefore volatility) are high.

Figure 3 shows a set of 20 sample rate paths with 20-year mean reversion, using the same set of random numbers. The effect of mean reversion on the widening of the distribution and on the tendency of the risk-neutral rates to grow is obvious – both are significantly reduced.

A significant improvement over the Monte Carlo method can be had by, in effect, eliminating the use of random numbers. A method for doing this is to choose "representative" sample paths using a systematic procedure such that the

2. Sample paths with 11% short rate volatility and no mean reversion

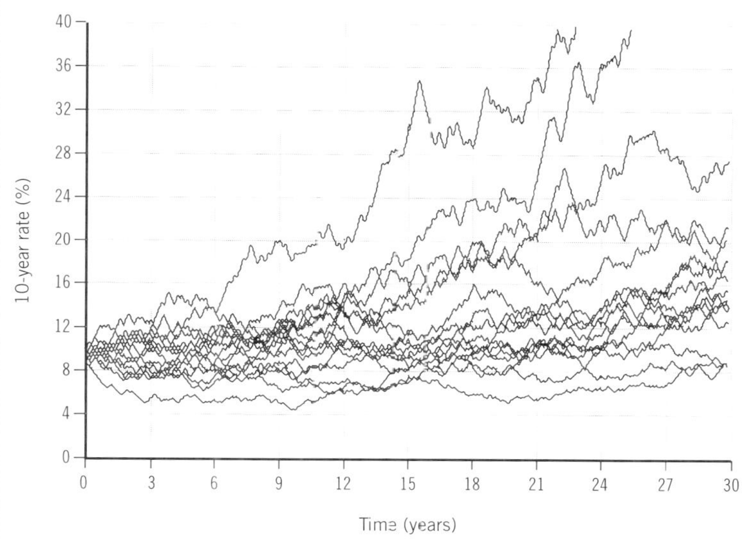

3. Sample paths with 11% short rate volatility and 20-year mean reversion

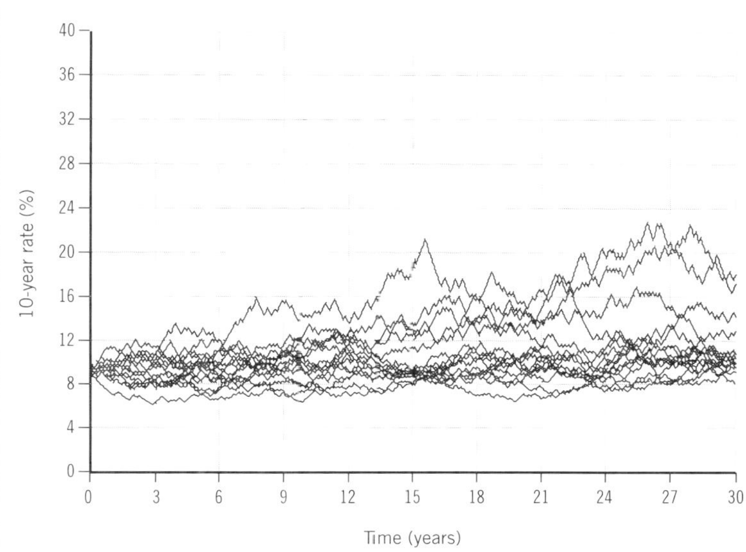

paths accurately sample the path distribution of the interest rate model. This "representative path" method has two advantages over the reduced-variance Monte Carlo algorithm. One is that it samples the set of possible paths widely and uniformly with relatively few paths. The second is that it eliminates the statistical uncertainty of the result by avoiding the use of random numbers. Instead of a large statistical uncertainty, there is a small systematic bias (of unknown sign and magnitude because the methodology is not exact if we use only a finite number of paths).

4. Valuation of a 9% sequential tranche (receiving the last 25% of the principal) using the Monte Carlo and representative path algorithms. The rate process was lognormal with no mean reversion

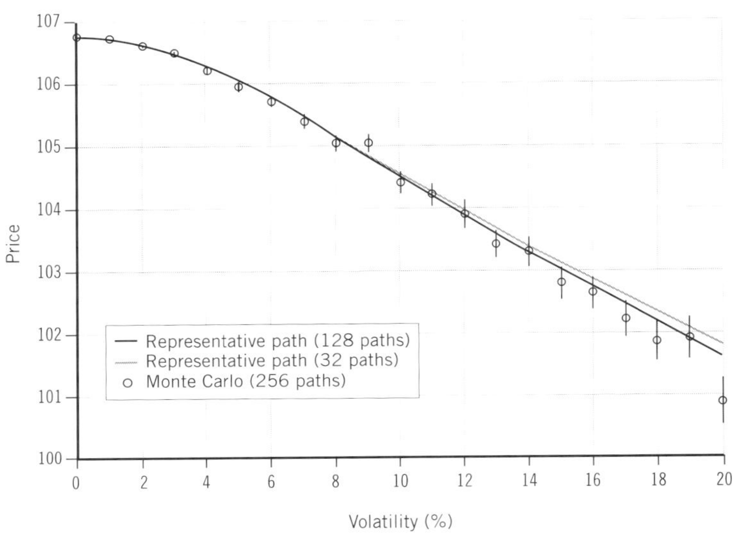

5. Valuation of sequential tranche using the Monte Carlo and representative path algorithms with 10-year, 25-year and no mean reversion

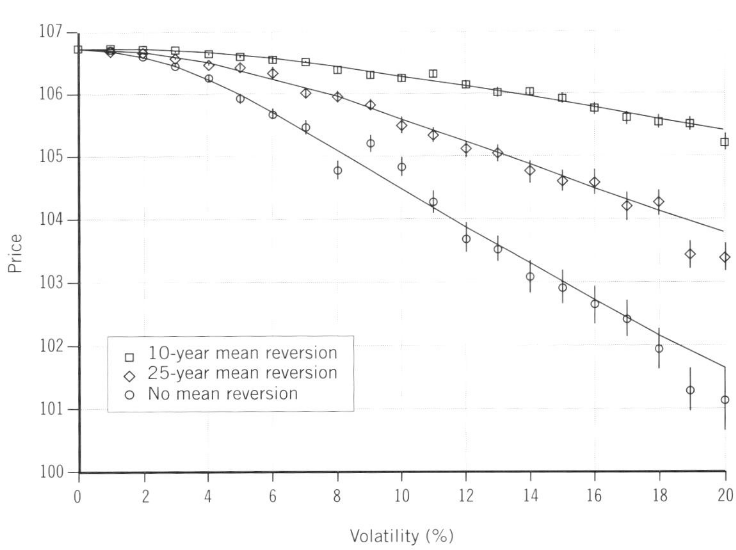

Random sequences produce a clumpy distribution of paths, which smooths out fairly slowly as the number of paths is increased. This clumpiness gives rise to the $1/\sqrt{N}$ error in Monte Carlo evaluation. Non-random methods can produce significantly faster convergence, though the systematic bias can be hard to analyse and control.

Figure 4 shows a comparison of this methodology with the Monte Carlo method for a sequential CMO tranche receiving the final quarter of the principal of 9% collateral. The prepayment model was statistically estimated from 12 months of Federal National Mortgage Association (FNMA) data and incorporates aging, seasonality and refinancing incentive. The input treasury term structure is that of early December 1991. The solid curve is made from 21 points, each computed with the representative path method using 128 paths. Each calculation takes approximately nine seconds.

The graph shows the computed tranche value as a function of volatility. Higher volatility poses more of a challenge for any pricing scheme, because it must sample a wider range of likely scenarios. The dashed curve is the same method with only 32 paths (for which each point takes 3.5 seconds). The circles with error bars are obtained using the Monte Carlo algorithm with 256 paths. Each calculation took about 42 seconds of computer time. The error bars represent estimates of the one standard deviation uncertainty of the result (based on the variance of subsample means) – the exact answer should be within the error bar range roughly two-thirds of the time.

It is apparent that the two algorithms are converging to the same result. It is also clear that even at fairly high volatility the 32-path run of the representative path algorithm gives results whose systematic errors are as small as the random uncertainties of the Monte Carlo with 256 paths.

Figure 5 shows the results when mean reversion is included. Three sets of calculations were done for the same tranche as above, with no mean reversion (the same as Figure 4) and with 25- and 10-year mean reversion timescales. Again, the curves were made with the representative path method (128 paths per point) and the symbols with Monte Carlo (256 paths per point). As one would expect, mean reversion reduces the effect of volatility. Increasing mean reversion and increasing volatility have opposite effects on the time value of the prepayment option.

Table 1 presents a comparison of calculation time and error size (statistical for the Monte Carlo algorithm, systematic for the representative path method).

To compute the error in the representative path method, the correct result was estimated by running the Monte Carlo with over 300,000 paths (which took 15 hours), for which the residual statistical pricing error was less than 0.01%. The representative path method is seen to give a better result in 10 seconds than does the Monte Carlo after almost six minutes.

Table 1. Comparison of calculation time and error size for the Monte Carlo and representative path methods

	Representative path			Monte Carlo		
Number of paths	Time (s)	Price err. (%)	OAS err. (bp)	Time (s)	Price. err. (%)	OAS err. (%)
32	3.6	0.36	5.0	6.15	1.0	14
128	9.5	0.11	1.5	22	0.5	6.9
1024	54	0.09	1.2	170	0.18	2.5
2048	98	0.08	1.1	340	0.12	1.7
8192	341	0.07	1.0	1356	0.06	0.9
32768	1211	0.057	0.8	5423	0.03	0.4

Comparison is for the same tranche as shown in Figures 4–6, with 12% volatility and no mean reversion. Calculations were done on a 386 computer running at 25 MHz with a math coprocessor.

Conclusions

I have attempted in this chapter to clarify the option theoretical issues underlying the mortgage pricing problem. The key valuation formula, which states the intuitive result that an asset's value is the expected present value of its cashflows (with respect to a risk-neutral distribution of paths) was justified using arbitrage arguments.

The dynamics of the risk-neutral rate paths were shown to include both the systematic drift implied by the current forward curve and a competing upward bias due to volatility. Volatility produces a steepening of the term structure because of the convexity arbitrage; this steepening in turn gives an upward bias to the rate drift.

The probability distribution of real-world future rate paths must also account for the market price of risk, which lends a downward drift bias to the term structure. Fortunately, the valuation formula does not depend on knowledge of this parameter – which is difficult to estimate empirically – just as the Black–Scholes equity option formula does not depend on the actual expected growth rate of the stock price.

The valuation formula was shown to lend itself to direct calculation on a computer using the Monte Carlo method, although the calculations are time-consuming. A more efficient approach to valuation was also briefly described and a number of sample comparisons presented. For practical purposes, the "representative path" scheme seems to give good results with much less computational effort and no exposure to the random errors of the Monte Carlo method.

Appendix

I give here a formal derivation of the valuation formula for interest rate-sensitive assets and a mathematical discussion of arbitrage-free rate processes. The main new result contained in the following discussion is the representation of a general class of interest rate models as two-state Markov processes, and the associated two-dimensional diffusion equation (20) for asset values, with an arbitrary short rate volatility function and no parameter fitting of coefficients required to match the initial forward rate curve.[6]

THE EXPECTATION FORMULA FOR PRICES

Parts of the following derivation can be found in papers by Vasicek (1977), Jamshidian (1989a and b) and Heath, Jarrow and Morton (1989). The derivation starts with the arbitrage argument for a single market price of risk that is independent of the specific asset. Let $P_T(t)$ denote the price of a zero-coupon bond of maturity T at time t. The bond price is related to the forward curve $r_s(t)$ by

$$P_T(t) = e^{-\int_t^T ds\, r_s(t)} \qquad (1)$$

The instantaneous short rate at time t is $r(t) \equiv r_t(t)$. Assume the following general form (with a one-dimensional stochastic process) for the bond price process:

$$dP_T(t) = P_T(t)[\mu_T dt + \sigma_T cz] \qquad (2)$$

where μ_T is the drift rate, σ_T is the volatility (which in general depends on rates and time), and z is a Brownian motion. Consideration of a portfolio of two bonds with prices P_S and P_T leads to a specific form for μ_T. Let the portfolio of cost Q contain weights 1 of the first bond and α

of the second. Its price process is then

$$\begin{aligned} dQ(t) &= dP_S(t) + \alpha dP_T(t) \\ &= [P_S(t)\mu_S + \alpha P_T(t)\mu_T]dt \\ &\quad + [P_S(t)\sigma_S + \alpha P_T(t)\sigma_T]dz \end{aligned} \tag{3}$$

Set $\alpha = -(P_S(t)\sigma_S/P_T(t)\sigma_T)$. Then the cost is $Q = P_S(t)(1 - \sigma_S/\sigma_T)$, and its process is $dQ(t) = P_S(t)(\mu_S - (\sigma_S/\sigma_T)\mu_T)dt$, which represents a certain return. Because its instantaneous return is riskless, arbitrage freedom requires that it be the short rate, $dQ(t) = r(t)Q(t)dt$. Equating these two gives

$$\frac{\mu_T - r}{\sigma_T} = \frac{\mu_S - r}{\sigma_S} \tag{4}$$

Because the left side of this equation is independent of S and the right side is independent of T, each side must be independent of both S and T. Denote the value of each side by λ. This is the market risk premium, as is seen by writing $\mu_T = r(t) + \lambda\sigma_T$. λ is the payoff in expected return for bearing risk; the excess expected return is linearly proportional to the riskiness of the portfolio. Note that, while λ is independent of the bond maturity, it may depend on t and r.

Observe also that nothing in this argument requires the securities in the portfolio to be zero-coupon bonds. Equation (2) describes the price fluctuations of any asset if T is taken simply to be a label. The argument then allows us to conclude that all assets can be expected to appreciate at a rate given by the sum of the short rate and a global constant times the asset's volatility. The argument can be extended to the case of multiple risk factors; one finds that each risk factor has its own λ, and the return premium of an asset is the sum of the products of the λs and the corresponding risk exposures.

The risk-neutral process is obtained by defining a new stochastic variable $\tilde{z}$, putting $d\tilde{z}(t) = dz(t) + \lambda dt$. Written in terms of $\tilde{z}$, λ disappears from (2); using the equation for μ in terms of λ and σ, the price process is

$$dP_T(t) = P_T(t)[r(t)dt + \sigma_T d\tilde{z}] \tag{5}$$

Taking $\tilde{z}$ to be a Brownian motion defines the risk-neutral weighting of paths.

Define a money market portfolio, M, which is an initial investment of $1 continually reinvested at the short rate. The value of this riskless portfolio follows the process $dP_M = r(t)P_M dt$, which has solution $P_M(t) = e^{\int_0^t ds\, r(s)}$. For any asset A, the ratio P_A/P_M follows the process[7]

$$\begin{aligned} d\left(\frac{P_A}{P_M}\right) &= \left(\frac{P_A}{P_M}\right)\sigma_A(\lambda(t)dt + dz(t)) \\ &= \left(\frac{P_A}{P_M}\right)\sigma_A d\tilde{z}(t) \end{aligned} \tag{6}$$

Therefore, the value ratio is a martingale with respect to the risk-neutral probability measure on the stochastic process $\tilde{z}$, for which $E[d\tilde{z}] = 0$, ie

$$E\left[P_A(r(t),t)e^{-\int_0^t ds\, r(s)} \middle| r(0)\right] = P_A(0) \tag{7}$$

where the expectation is with respect to the risk-neutral weighting of paths starting at the initial value r(0). To convert this to an expression in terms of the cashflows of A, one can use the argument in terms of Arrow–Debreu securities as in the first section of the main text. Alternatively, one can approximate the process r as a step function (a series of constant values), and write

$$\begin{aligned} &P_A(0) \\ &= E\left[P_A(r(t),t)e^{-\int_0^t ds\, r(s)} \middle| r(0)\right] \\ &= E\Bigg[\int_0^{\epsilon_1} ds C_A(s;r)e^{-r(0)s} \\ &\quad + e^{-r(0)\epsilon_1} E\left[P_{A_{\epsilon_1}}(r(t),t)e^{-\int_{\epsilon_1}^t ds\, r(s)} \middle| r(\epsilon_1)\right] \Bigg| r(0)\Bigg] \\ &= E\Bigg[\int_0^{\epsilon_1} ds\, C_A(s;r)e^{-r(0)s} \\ &\quad + \int_{\epsilon_1}^{\epsilon_2} ds\, C_A(s;r)e^{-r(0)\epsilon_1 - r(\epsilon_1)s} \\ &\quad + e^{-r(0)\epsilon_1 - r(\epsilon_1)(\epsilon_2 - \epsilon_1)} \\ &\quad \times E\left[P_{A_{\epsilon_2}}(r(t),t)e^{-\int_{\epsilon_2}^t ds\, r(s)} \middle| r(\epsilon_2)\right] \Bigg| r(0)\Bigg] \\ &\;\vdots \\ &\to E\left[\int_0^t ds\, C_A(s;r)e^{-\int_0^s du\, r(u)} + e^{-\int_0^t ds\, r(s)} P_{A_t}(r(t),t) \middle| r(0)\right] \end{aligned} \tag{8}$$

where $C_A(s;r)$ denotes the cashflow rate of A at time s along the rate path r, and A_s denotes asset A without the cashflows to time s. We thus obtain the valuation formula by means of a limiting process as the size of the time steps goes to 0, the stepped rate process converges to a continuous one, and we let time t get larger than any of the cashflows of A.

THE EVOLUTION OF THE TERM STRUCTURE OF INTEREST RATES

Using the relation (1), and applying Itô's lemma, we obtain for the forward rate fluctuations

$$dr_T(t) = \sigma_T'(t)(\sigma_T(t) - \lambda(t))\,dt - \sigma_T'(t)\,dz(t) \qquad (9)$$

where $\sigma_T'(t) = \frac{d\sigma_T(t)}{dT}$. This can be formally integrated to give

$$r_T(t) = r_T(0) + \int_0^t ds\, \sigma_T'(s)(\sigma_T(s) - \lambda(s)) - \int_0^t dz(s)\,\sigma_T'(s) \qquad (10)$$

which gives the forward rate in terms of its initial value at time 0, the cumulative drift, and the cumulative stochastic fluctuations. Putting $T = t$ and differentiating, we obtain the short rate process

$$dr(t) = \left[r_t'(0) - \sigma_t'(t)\lambda(t) + \int_0^t ds \frac{d}{dt}\,\sigma_t'(s)(\sigma_t(s) - \lambda(s)) - \int_0^t dz(s)\,\sigma_t''(s)\right]dt - \sigma_t'(t)dz(t) \qquad (11)$$

where $r_t'(0)$ denotes the maturity derivative of the initial forward rate curve. The term in brackets is in general path-dependent since it depends on the volatility history which, in turn, depends on the path. The risk-neutral rate process is obtained by the substitution of $\tilde{z}$ for z, which eliminates both appearances of λ in (11).

For an arbitrary forward rate volatility function $\sigma_T'(t)$ it is difficult to proceed any farther in characterising the rate process. Unless $\sigma_T'(t)$ has a special form, there is no simple relationship between forward rates of different maturities at future times. If, however, $\sigma_T'(t)$ is *separable* into a time- and rate-dependent term and a maturity-dependent term, the description of the process can be considerably simplified. Let $\sigma(r,t)$ be the short rate volatility, and assume that the volatility term structure can be written as

$$\sigma_T'(t) = \frac{\sigma(r,t)}{f(t)}\,f(T) \qquad (12)$$

which implies that $\sigma_T(t) = \frac{\sigma(r,t)}{f(t)}(F(T) - F(t))$, where $\frac{dF(T)}{dT} = f(T)$. Substituting into equation (10), we obtain

$$r_T(t) = r_T(0) + f(T)\left(\int_0^t ds\,\frac{\sigma^2(r,s)}{f^2(s)}\,(F(T) - F(s)) - \int_0^t d\tilde{z}(s)\,\frac{\sigma(r,s)}{f(s)}\right) \qquad (13)$$

Note that the second integral is independent of the maturity, T. Taking $T = t$ and solving for the stochastic integral we have

$$-\int_0^t d\tilde{z}(s)\,\frac{\sigma(r,s)}{f(s)} = \frac{r(t) - r_t(0)}{f(t)} - \int_0^t ds\,\frac{\sigma^2(r,s)}{f^2(s)}(F(t) - F(s)) \qquad (14)$$

which can be substituted into equation (13) to give

$$r_T(t) = r_T(0) + \frac{f(T)}{f(t)}\,(r(t) - r_t(0)) + f(T)(F(T) - F(t))\int_0^t ds\,\frac{\sigma^2(r,s)}{f^2(s)} \qquad (15)$$

The short rate process (equation 11) can now be written

$$d(r(t) - r_t(0)) = \left[v^2(t) + \frac{d\ln f(t)}{dt}\,(r(t) - r_t(0))\right]dt - \sigma(r,t)d\tilde{z}(t) \qquad (16)$$

where $v^2(t)$ is the cumulative volatility,[8] defined by

$$v^2(t) = f^2(t)\int_0^t ds\,\frac{\sigma^2(r,s)}{f^2(s)} \qquad (17)$$

One sees that, provided the volatility is a decreasing function of maturity, the short rate is mean-reverting.

The forward rate curve (equation 15) is now entirely determined by the deviation of the short rate from its value implied by the initial forward curve and by the cumulative volatility along the path.[9] The prices of bonds are therefore determined by these two quantities. The yield of a zero-coupon bond, defined by the relation $P_T(t) = \exp(-(T - t)y_T(t))$, is easily computed from equation (15):

$$y_T(t) = \frac{1}{T - t}\left[\ln\frac{P_t(0)}{P_T(0)} + \frac{F(T) - F(t)}{f(t)} \times \left(r(t) - r_t(0) + \frac{F(T) - F(t)}{2f(t)}\,v^2(t)\right)\right] \qquad (18)$$

Differentiating equation (17), we find

$$\frac{dv^2(t)}{dt} = \sigma^2(r,t) + 2\frac{d \ln f(t)}{dt} v^2(t) \tag{19}$$

It follows that the pair (r,v^2) is a two-dimensional Markov process. The Feynman–Kac formula applied to the valuation equation (8) (see, eg, Friedman, 1975, chapter 6) then implies that the price $P_A(r,v^2,t)$ of a security, with cashflows $C_A(r,v^2,t)$ dependent only on the forward curve at t, obeys the two-dimensional diffusion equation

$$\frac{\partial P_A}{\partial t} + \frac{\sigma^2(r,t)}{2}\frac{\partial^2 P_A}{\partial r^2} + \left(r_t'(0) + v^2 - \frac{r - r_t(0)}{\tau(t)}\right)\frac{\partial P_A}{\partial r} + \left(\sigma^2(r,t) - \frac{2v^2}{\tau(t)}\right)\frac{\partial P_A}{\partial (v^2)} - rP_A + C_A(r,v^2,t) = 0 \tag{20}$$

where

$$\frac{1}{\tau(t)} \equiv -\frac{d \ln f(t)}{dt} \tag{21}$$

This is the analogue of equations (15) of Vasicek (1977) and (32) of Jamshidian (1989a), the difference being that this two dimensional equation allows for an arbitrary short rate volatility function and a fairly general term structure of volatility within the constraints of the separability condition. Note that, in distinction to the one-dimensional diffusion equation obtained by a number of authors, this equation contains no undetermined coefficients to be fitted to the forward rate. All the coefficients are determined explicitly by the initial forward rate curve and the chosen volatility structure. One can verify that zero-coupon bond prices as given by equation (18) and the definition above satisfy equation (20). The equation is two-dimensional even though the stochastic process is one-dimensional because r(t) alone is not a Markov process.[10]

A natural assumption to introduce here is that the volatility structure is given by the product of an arbitrary short rate volatility function and a *relative* maturity function: $\sigma'_T(t) = \sigma(r,t)B(T-t)$, with the normalisation $B(0) = 1$. Together with the separability assumption, this implies a very simple form for the volatility. From equation (12) we have $B(T-t) = f(T)/f(t)$. Using $\partial B(T-t)/\partial T = -\partial B(T-t)/\partial t$ we have $\partial \ln f(T)/\partial T = \partial \ln f(t)/\partial t$, which implies that $\tau(t)$ must be constant; τ is the mean-reversion timescale, and we have $f(t) = e^{-t/\tau}$. With this form for f(t), equation (15) becomes

$$r_T(t) = r_T(0) + e^{-(T-t)/\tau}[(r(t) - r_t(0)) + \tau(1 - e^{-(T-t)/\tau})v^2(t)] \tag{22}$$

The short rate process (equation 16) now takes the form

$$d(r - r_t(0)) = \left(v^2(t) - \frac{1}{\tau}(r - r_t(0))\right)dt - \sigma(r,t)d\tilde{z}(t) \tag{23}$$

Note both the upward drift due to the forward curve steepening induced by the volatility and the tendency of the short rate to revert towards the forward curve, with timescale τ. The zero-coupon yield (equation 18) is now given by

$$y_T(t) = \frac{1}{T-t}\ln\frac{P_t(0)}{P_T(0)} + \frac{1 - e^{-(T-t)/\tau}}{(T-t)/\tau}(r(t) - r_t(0)) + \frac{T-t}{2}\left(\frac{1 - e^{-(T-t)/\tau}}{(T-t)/\tau}\right)^2 v^2(t) \tag{24}$$

This form, which is a generalisation of the one found by Vasicek, is shown in Figure 1 (assuming a flat long-term forward rate structure) for different values of $r(t) - r_t(0)$ and with $v^2(t) = 0.1$, which is a typical long-run value with the quoted volatility and mean reversion. Increasing v^2 increases the steepness of the yield curve. Note that v^2 does not tend to grow without limit: from equation (19) one sees that the growth of v^2 is damped. For example, if σ is constant, v^2 approaches a limiting value of $\frac{\tau}{2}\sigma^2$. When $\sigma(r,t)$ is non-constant the coupled equations (16) and (19) may not be so easy to solve, but in general the mean reversion constrains the growth of the cumulative volatility.[11]

COMPARISON WITH OTHER MODELS

The continuum limit of the Ho–Lee model is a special case of the model described here, with $1/\tau \rightarrow C$ and $\sigma(r,t)$ constant. In this case $v^2(t) = \sigma^2 t$, which is non-stochastic, and r is a Markov process. Explicit solutions for the values of European options on zero-coupon bonds have been given by Jamshidian (1989a) and Heath, Jarrow and Morton (1989) for this model. As mentioned earlier, the model allows negative interest rates and has no mean reversion. Because r is Markov, the process can be simulated on a binomial lattice of short rates (more generally, one has a one-dimensional diffusion equation for asset values in terms of the short rate, for which numerous numerical solution methods exist). A simpler form of this model – with mean reversion included, but not calibrated to an arbitrary initial term structure – was originally proposed by Vasicek (1977).

There are also a number of models, typically specified as binomial lattice approximations, that are lognormal in the short rate and in which, by construction, the short rate follows a Markov process.[12] In their simplest continuum form they can be represented by

$$dr = \theta(r,t)dt + \sigma(t)r\,dz \qquad (25)$$

The drift rate $\theta(r,t)$ is partly determined by the requirement that the valuation formula (8) reproduce observed treasury bond prices. This condition alone is not enough to completely determine θ, so one must assume a particular form, eg drift proportional to short rate, or independent of short rate.

One difference between these lognormal rate models and the models described in the previous section is evidently this indeterminacy of the form of θ. In general one could put $\theta = a(r,t)b(t)$ with almost any function $a(r,t)$ and still fit $b(t)$ to the observed term structure. This appears to pose a substantial problem for parameter estimation from real data.

Another difference between these models is that the forward rate volatility structure is given only implicitly and may depend in a non-trivial fashion on the shape of the forward yield curve. If one wishes to fit a term structure of volatility from, say, market prices for liquid options, the only way to do so is to impose a forecast of future short rate volatility, $\sigma(t)$.

To obtain the typically observed declining volatility of rates with maturity, one must forecast that short rate volatility will become very small in the future. This is clearly unrealistic. For valuation of path-independent securities such as caps, swaptions on fixed principal, or callable bonds, the advantage of these methods is that binomial lattice algorithms can be very fast and are well suited to pricing American options. For path-dependent securities this advantage disappears.

The approach described in this chapter then has the advantages that: the initial rate curve appears explicitly; and one can impose a realistic term structure of forward rate volatility. (It is possible to extend the lattice models by introducing variable "up" and "down" probabilities in a rate-dependent fashion so as to obtain mean reversion and a term structure of volatility without the requirement of declining short rate volatility. This extension eliminates some of the simplicity of these models and still leaves the relationship between the volatility term structure and the short rate process rather opaque.)

Arbitrage-free models with multiple stochastic factors have been proposed by Brennan and Schwartz (1979), Jamshidian (1989a) and Heath, Jarrow and Morton (1989). The expectation formula (7) is the same regardless of how many factors are included. There is a market risk premium for each source of uncertainty and the expected rate of return on an asset can be written generally as

$$\mu_A(t) = r(t) + \sum_{i=1}^{n} \lambda_i \sigma_{Ai}(t) \qquad (26)$$

where σ_{Ai} is the contribution of the ith uncertainty to the volatility of A. The forward rate evolution equation for the difference $r_T(t) - r_T(0)$ can then be written as a sum of n terms of the same form as those in equation (10). The evolution of the term structure is then determined by n rate fluctuation terms and n cumulative volatilities; the present value of path-independent cashflows satisfies a $2n$-dimensional diffusion equation.

One can, of course, fit a more complicated volatility structure with multiple factors. In the case of a Monte Carlo implementation, the extra work involved is fairly minor, while lattice methods become much more cumbersome. However, inclusion of multiple factors requires estimation of additional parameters from market information, which may be difficult.

1 *We can assume that cashflows from* A *prior to period* i *are reinvested in either* A *or* M *– it doesn't matter which one – and the accumulated value included in* P_A.

2 *The Monte Carlo algorithm entails the use of a random number generator to create a set of random rate paths. The cashflows along each rate path are then obtained (from a prepayment model, for example), discounted, and the results averaged over the paths.*

3 *The expected total return for longer bonds is higher than for shorter ones even if the yield curve is inverted, as long as there is a positive market risk premium.*

4 *This apparent arbitrage has led some to conclude that one must pay for convexity. As this argument shows, one does not pay directly: rather, one "pays" through exposure to the steepening of the yield curve, which reduces the expected total return.*

5 *This two-component model also leads to a valuation equation for bonds. This is a two-dimensional diffusion equation in which all coefficients of the drift terms are given explicitly in terms of the initial term structure of rates and term structure of volatility (unlike, for example, the BDT model, which contains an unknown function that must be fitted to the term structure). The equation is derived in the appendix.*

6 *After this work was first published in 1992, the same model was independently discovered by Ritchken and Sankarasubramanian (1995).*

7 *Without altering the present value, we can augment the description of asset* A *by the prescription that its cashflows are to be reinvested in* M *until maturity.*

8 *The cumulative volatility is a measure of how much the short rate is likely to deviate from its expected value. If* $\sigma(r,s)$ *is constant,* $v^2(t)$ *is simply the variance of the distribution of* $r(t)$.

9 *Equation 15 implies that the shape of the yield curve is independent of the market risk premium for a given short rate and cumulative volatility. The risk premium will affect only the typical values of these quantities.*

10 *Because of the restriction on the cashflows this partial differential equation unfortunately cannot be used for mortgage pricing.*

11 *If the volatility grows too quickly with the interest rate level then, with non-zero probability, there are "exploding" sample paths that run away to infinity in finite time. These can be avoided by capping the volatility, as suggested by Heath, Jarrow and Morton (1989).*

12 *Generally, any model defined by specifying a process for the short rate together with the pricing formula (7) and a prescription for fitting the short rate process to the current term structure can be used to define an arbitrage-free model where the short rate is a Markov process.*

BIBLIOGRAPHY

Black, F., E. Derman, and W. Toy, 1990, "A One Factor Model of Interest Rates and Its Application to Treasury Bond Options", *Financial Analysts Journal* January, pp. 33–9.

Brennan, M.J., and E.S. Schwartz, 1979, "A Continuous Time Approach to the Pricing of Bonds", *Journal of Banking and Finance* 3, pp. 133–55.

Cox, J.C., J.E. Ingersoll and S.A. Ross, 1985, "A Theory of the Term Structure of Interest Rates", *Econometrica* 53, March, pp. 385–407.

Friedman, A., 1975, *Stochastic Differential Equations and Applications*, Academic Press.

Heath, D., R. Jarrow and A. Morton, 1992, "Bond Pricing and the Term Structure of Interest Rates: A New Methodology for Contingent Claims Valuation", *Econometrica* 60(1), pp. 77–105.

Ho, T.S.Y., and Lee, S.-B., 1986, "Term Structure Movements and Pricing Interest Rate Contingent Claims", *Journal of Finance* 56, December, pp. 1,011–29.

Jamshidian, F., 1989a, "The One-Factor Gaussian Interest Rate Model: Theory and Implementation", Merrill Lynch working paper, January.

Jamshidian, F., 1989b, "The Multifactor Gaussian Interest Rate Model and Implementation", Merrill Lynch working paper, September.

Ritchken, P., and L. Sankarasubramanian, 1995, "Volatility Structure of the Forward Rates and the Dynamics of the Term Structure", *Mathematical Finance* 5(1), pp. 55–72.

Vasicek, O., 1977, "An Equilibrium Characterization of the Term Structure", *Journal of Financial Economics* 5, pp. 177–88.

VAR

Introduction

Bruno Dupire

Over the last few years, value-at-risk (VAR) has reached the status of a full-blown field, competing with pricing in terms of complexity as much as in terms of the resources committed by banks. It now challenges the prominence of reward in the classical risk/reward trade-off. It is used to understand the risk of big portfolios, to allocate capital between different trading units, and to compute regulatory capital along the BIS guidelines.

VAR is a measure of the potential loss incurred by a portfolio under a certain set of assumptions. It is commonly defined as a quantile (typically 99%) of the P&L distribution of the portfolio over a certain time horizon (typically 10 days). To compute properly, it would require the accurate valuation of the whole portfolio under a collection of scenarios. This means that we need to identify the risk factors which affect the position and make assumptions about their dynamics. The computational burden of these operations has, for a long time, spurred the need for simplification – either of the portfolio valuation (linear or quadratic approximations) or of the assumptions (Gaussian returns). The increase in computer power and the advent of better algorithms have made Monte Carlo methods affordable, even for large portfolios of thousands of deals on hundreds of assets.

Chapter 20, by Evan Picoult, provides an overview of the methods used to generate market factor scenarios and to revalue the portfolio. He presents a method that combines parametric (as opposed to historical) simulation of the market factors with what he calls parametric (as opposed to exact) portfolio valuation. This is either a Taylor expansion of the portfolio value function or a discrete version of it (grid of factor sensitivities).

The following chapter, by Julian Shaw, warns against non-normality of returns and non-linearity of value function and correlation, which capture only linear causality. He advocates the use of Latin hypercube techniques (or extensions, such as Stein's algorithm) to sample the space of market variables properly on each component. This fits the structure of the value function, which is usually close to additive, in the sense that many deals of the portfolio depend mainly on one variable. To perform a principal component analysis would destroy this structure.

Chapter 22, by Gerald Quinlan, also suggests the use of Stein's algorithm. He fits the empirical data with a four-parameter density function and estimates correlation among risk factors. He shows the method's superior consistency in estimating the 97.5% and 99% confidence interval over the standard Monte Carlo method.

Owen and Tavella, in Chapter 23, discuss the application of scrambled nets, which are a hybrid of low-discrepancy sequences and Monte Carlo. Digits of the binary representation of the original deterministic sequence are shuffled to respect the initial low discrepancy. The introduction of randomness increases the convergence order by 0.5.

Pricing entails estimating an expectation, VAR a quantile or other statistics. The nature of VAR, which focuses on unlikely events, makes it specially sensitive to the model assumption and the precise shape of the tails. VAR is at a confluence of complexities, in terms of products, dynamics, dimensionality, portfolio size and histogram statistics. Because of this, it is a wonderful domain for the application of Monte Carlo simulations. However, the accuracy requirement is substantially lower than in the case of pricing, which directly affects P&L.

Calculating Value-at-Risk with Monte Carlo Simulation

Evan Picoult
Citibank

Value-at-risk (VAR) has become a widely used form of measuring price risk. Methods for calculating VAR differ in their broad assumptions and in the details of their implementation. No method has yet become an industry standard. This chapter categorises and compares different VAR models in terms of their two primary components: (1) their method of simulating changes in market factors and (2) their method of transforming a simulated change in market factors into a simulated change of portfolio value. The primary focus of this chapter is on the calculation of VAR by Monte Carlo simulation. We describe techniques for the statistical simulation of market factors that incorporate correlations and "specific" risk (which is a form of correlation risk). We also describe an efficient technique for transforming a change in market factors into a change of portfolio value that incorporates complex non-linearities and cross-terms of a Taylor-series expansion.

Basic concepts of risk measurement

GENERAL DEFINITION OF VAR

The most prominent risk in trading is price risk – the risk of an economic loss caused by the decrease in the market value of a portfolio of contracts. VAR is a form of measuring price risk that has become widely used. It is a statistical estimate, at a specified confidence level, of the potential loss in the economic value of a portfolio of contracts over a specified time horizon. As a statistical estimate, VAR is derived from the probability distribution of the portfolio's potential change in economic value. That probability distribution will be based (implicitly or explicitly) on some assumptions about and method of simulating the joint probability distribution of changes in market rates and some method for transforming simulated changes in market rates into simulated changes of the portfolio's value.

VAR is used by many firms to quantify and limit their price risk. Some firms also use it as the basis for the allocation of economic capital to trading desks for the price risk they take. Economic capital for price risk can serve as a standard for measuring business performance (return on risk) and as a common unit for comparing and aggregating the various risks (credit, price, liquidity, operational, etc) taken in the course of doing business.

Another use of VAR is in the assessment of regulatory capital for market risk. In the Bank of International Settlements (BIS) internal models approach (Panel 1), regulatory capital for market risk is proportional to the VAR that is calculated with a firm's internal model under certain general parameters which are specified by regulation.

Models for calculating VAR can and do differ in their broad methodology, in their assumptions and in the details of their implementation. At the broadest level, VAR models can be differentiated with regard to the method and detailed assumptions underlying their simulation of changes in market rates and their transformation of changes in market rates into changes in portfolio value. Issues regarding the method of simulation include the assumed shapes of the probability distributions of changes in market rates, the assumed nature of correlations of changes in market rates, the general means by which correlations are incorporated into simulations and the particular method that is used to model the correlations underlying specific risk. Issues regarding the method for transforming changes of market rates into changes of portfolio value include the means by which and the extent to which non-linearities are incorporated, the extent to which cross-terms of a Taylor-series expansion of change

PANEL 1

BIS INTERNAL MODELS APPROACH

The Basle committee on banking supervision was set up in 1975 by the Group of Ten (G10) countries to co-ordinate and standardise regulatory capital requirements for the risks of banking. The first regulatory capital requirements were established for credit risk. As of 1998, two basic approaches have been established for ascertaining regulatory capital for market risk.

The standard model approach

The first version of the "standard model" approach was proposed by BIS in 1993. It was later updated. The standard model is a rule-based approach, whose essential feature is that transactions are classified into different categories and total regulatory capital is calculated by applying arithmetic rules to the notional principal in each category. This type of approach was criticised by many individual banks and by the International Swaps and Derivatives Association (ISDA) for its simplistic and cumbersome method for ascertaining market risk.

Internal model approach

The essential feature of the "internal model" approach is that regulatory capital is a multiple of the value-at-risk (VAR) calculated by a firm with its own internal model. To use an internal model a firm must meet certain qualitative standards. Regulations do not specify the details of the internal model, but they do specify some quantitative standards the internal model must adhere to. These standards are:

- volatilities and correlations must be derived from a sample period of at least one year;
- assume a static portfolio;
- calculate VAR at the 99% confidence level assuming an instantaneous 10-day change in market rates. If one assumes that changes in market rates are normally distributed, that day-to-day changes in rates are independent and that a portfolio is linear, then this market shock is equivalent to a 7.36 sd loss; and
- the BIS regulations make an important distinction between general market risk and specific risk similar to that discussed in the main text.

A bank's internal model for calculating VAR is complete if it can calculate VAR for both general market risk and specific risk. According to regulations, a complete calculation of VAR for specific market risk would have two features:

1) it would take into account that the price change of an individual equity is not necessarily perfectly correlated with the price change of an equity index of that country. It would take into account that a change in the yield of a debt security with specific risk is not necessarily perfectly correlated with a change in a general market yield of that currency. (See the main text for a description of the types of debt securities with specific risk); and

2) it would also take into account, in some unspecified way, event risk – that is, the potential discontinuous change in a debt or equity security arising from an issuer-specific event.

in value need to be and are incorporated, the suitability of the transformation given the confidence level at which VAR is measured and some aspects of specific risk.

Therefore, VAR models vary by the extent to which and the means by which they capture different aspects of market risk. Because of the increased reliance on VAR as a measurement of market risk, it is important to understand the strengths and weaknesses of different methods of measuring VAR. The significance of the strengths and weaknesses of any given VAR model must be evaluated in the context of the types of contracts whose risk is being measured, the purpose of the risk measurement and the confidence level at which potential loss is estimated.

More generally, the limitations of VAR models as such should always be kept in mind. It would be inappropriate to use only a single form of risk measurement to monitor and control the market risks of trading. For example, liquidity risk and event risk are not well incorporated into VAR measurements. Consequently, there is a need to have separate measures of and limits for liquidity and event risk.

This chapter begins with a discussion of revaluation models, market factors and market risk. Descriptions are then given of different methods for simulating changes in market factors, with a strong focus on techniques for statistical simulation. Finally, an efficient method for transforming simulated changes in market factors into simulated changes in portfolio value is presented.

The amount of regulatory capital under the internal model approach will depend on whether a bank can meet only standard 1) or can meet both standard 1) and standard 2). In order to describe the calculations needed by the BIS standards, several terms must be defined. Let us designate:

- $VAR_{BIS\_general}$ The VAR calculated under BIS standards for general market risk;
- $VAR_{BIS\_specific}$ The VAR calculated under BIS standards for specific market risk;
- VAR_{BIS_total} The total VAR calculated under BIS standards; and
- "Reg Capital" The amount of regulatory capital needed for market risk.

The required calculations are set out in the box below.

Hybrid: internal model plus standard model

A hybrid approach allows banks to use an internal model to calculate the regulatory capital for "general" market risk while using the standard model to calculate regulatory capital for specific risk.

If a bank can meet standards 1) and 2), the relationship between VAR_{BIS} and regulatory capital will be:

$$\text{Reg. Capital} = \text{Maximum } [VAR_{BIS\_total}\text{ (today)}, K \times VAR_{BIS\_total}\text{ (60-day average)}]$$

where: VAR_{BIS_total} (today) = Today's total VAR measured under BIS parameters;
VAR_{BIS_total} (60-day average) = VAR_{BIS_total} averaged for the last 60 days;
K = Multiplication factor. Normally set to 3.0.

The actual multiplication factor can vary from 3.0 to 4.0 depending on the outcome of backtesting.

If a bank can only meet standard 1), the relationship between VAR_{BIS} and regulatory capital will be:

$$\text{Reg. Capital} = \text{Maximum } [VAR_{BIS\_total}\text{ (today)}, K \times VAR_{BIS\_total}\text{ (60-day average)}] + VAR_{BIS\_specific}$$

where: VAR_{BIS_total} (today) = Today's total VAR measured under BIS parameters;
VAR_{BIS_total} (60-day average) = VAR_{BIS_total} averaged for the last 60 days;
K = Multiplication factor. Normally set to 3.0;
$VAR_{BIS\_specific}$ = VAR due to specific risk.

The actual multiplication factor can vary from 3.0 to 4.0 depending on the outcome of backtesting.

REVALUATION MODELS AND MARKET FACTORS; GENERAL AND SPECIFIC MARKET RISK

The economic value (market value) of a portfolio of traded contracts is measured by marking them to market. Simple cash-like instruments in liquid markets are valued directly by means of their price in the market – the market value of the position is simply obtained by multiplying the number of units held by the market price per unit. Examples of such instruments are spot foreign exchange, spot equities or spot treasury securities. In contrast, forward and derivative contracts are generally valued indirectly by means of discounted cashflow revaluation models. Such models are used to calculate the expected present value of the future cashflows of a contract.

Discounted cashflow models represent the expected present value of a contract as a function of several elements: the form of the contract and its terms and conditions; the current date; the current magnitude of market factors; and (particularly for path-dependent contracts) the magnitude of market factors on previous dates.

Market factors are the basic market prices and rates in terms of which contracts are valued. Market factors are often classified as general or specific; some examples are given in Table 1. General market factors reflect the current and the market's expectations and uncertainties about future macroeconomic conditions (eg level of economic growth, inflation, etc). Specific market factors reflect the current and the market's expectations about both general macroeconomic

PANEL 2

GRIDS OF FACTOR SENSITIVITIES

A grid of factor sensitivities is a set of changes in a contract's (or portfolio's) market value that corresponds to a set of specified changes in a single market factor – eg the changes in a portfolio's market value corresponding to 10 changes in the spot US dollar/Deutschmark rate from −5% to +5% in steps of 1%. As we shall see in the main text, grids of factor sensitivities are a key component of an efficient calculation of VAR by Monte Carlo simulation.

More generally, we can define grids of factor sensitivities of different dimensionality. A one-dimensional grid of factor sensitivities is the set of changes in the market value of a contract (or portfolio) that corresponds to a set of changes in only one market factor, all other market factors held constant. A two-dimensional grid of factor sensitivities is the set of changes in a contract's (or portfolio's) market value corresponding to a set of concurrent changes in two market factors, all others held constant. Higher-dimensional grids can also be defined. As will be explained, the dimensionality of a grid is related to the dimensionality of cross-terms in a Taylor-series expansion of the change in portfolio value.

conditions and the financial conditions and risks of specific issuers.

A revaluation model uses the current value of market factors in several ways: to estimate the expected future floating or contingent cashflows; to discount all fixed, expected future floating and expected future contingent cashflows to present value; and to convert the present values as measured in different currencies into a common denominator.

Although the market value of a debt security should be assessed directly from its market price, its price risk is usually analysed by means of a discounted cashflow model. This enables one to ascertain the sensitivity of the market value of the debt security to changes in the term structure of interest rates (and interest rate spreads). The benefit of this analysis is that it facilitates the integration of the price risk of debt securities with the interest rate price risk of other types of contract.

Table 1. Examples of general and specific market factors

General market factors	Specific market factors
The term structure of treasury or Libor interest rates	Spot price of equities
Spot foreign exchange rates	Spot price of corporate securities or, equivalently, their yields (ie their spreads to a base yield curve)
Spot equity indices	Spot price of a sovereign debt security issued in another country's currency – eg a South American government bond denominated in US dollars
Spot and forward commodity prices	The term structure of the implied volatilities of options on each of the above
The term structure of the implied volatilities of options on each of the above	

STATISTICAL AND NON-STATISTICAL MEASURES OF RISK

The two basic forms of measuring price risk are a non-statistical or scenario measurement and a statistical measurement.

A non-statistical or scenario measurement of price risk is the calculation of the change in the market value of a contract (or a portfolio of contracts) that would occur for a particular scenario (ie for a particular change in each market factor) independently of the likelihood of that scenario occurring.

An example of a scenario measurement is a factor sensitivity, the change in the market value of a contract (or portfolio) that would occur for a specific change in a single market factor, all other market factors held constant – eg the change in the value of a portfolio of contracts to a +6bp change in three-month US dollar Libor, all other market factors unchanged. A factor sensitivity is defined in terms of the particular market factor that is changed (eg six-month US dollar Libor, spot US dollar/yen, etc) and the magnitude and direction of its change. A factor sensitivity is a broader concept than and not identical to an instantaneous delta.

Another form of risk measurement is historical scenario analysis. This assesses the change in portfolio value that would occur if market factors were to change in the near term as they did during some historically stressful event – eg as they did during the interval from immediately before to immediately after the stockmarket crash of 1987, or before and after Iraq's invasion of Kuwait, etc.

In contrast to a scenario analysis, a statistical measurement of price risk is the potential loss of a portfolio's market value at a specified confidence level rather than for a specific scenario. In effect, one measures the probability distribution of the potential changes in the portfolio's value corresponding to thousands of scenarios of potential changes in market rates, with each scenario having a particular probability of occurring. VAR is an example of a statistical measurement of price risk. It is an order statistic in which the potential loss is represented by a single number, its value at some confidence level.

Most measurements of VAR assume a static portfolio. That is, they measure the potential loss in the value of the portfolio over the very near term, given a set of potential changes in market rates, under the assumption that the composition of the portfolio remains unchanged. This should be contrasted with the assumptions and processes that underlie the accurate calculation of pre-settlement counterparty credit exposure,[1] in which it is necessary to model and simulate changes in market rates over long periods of time and concurrently simulate the changing structure of the portfolio over time as floating rates are set and cashflows settle.

Stress testing is another form of measuring price risk. It is a measure of potential loss under extreme market conditions. A stress test can be a non-statistical or a statistical measurement. In a non-statistical stress test the potential loss of economic value is calculated for a specified scenario of very large market changes. The stressed scenario can be either historical (corresponding to the changes in market factors that occurred during some historical stressful event) or imagined (ie a hypothetical scenario of large changes in market rates that may not have yet occurred). A statistical stress test would calculate VAR under non-standard conditions – eg at a confidence level that was much higher than usual or with non-standard volatility or correlation assumptions. As discussed below, a meaningful calculation of VAR at a very high confidence level is not trivial.

THE BASICS OF MEASURING VAR

Holding period and assumption of static portfolio

VAR is usually measured as the potential loss in value of a static portfolio due to changes in market rates. For many, if not most, portfolios of traded products, an accurate or realistic measurement of price risk over a time period longer than 24 hours would require more complex modelling than is typically done for a VAR calculation. To calculate the potential loss in portfolio value over a period longer than one day, it would be necessary to model:

- ❑ The structural changes in the portfolio that could occur contractually during the period. For example, to compute VAR over a two-week period, it would be necessary to simulate for each day of that period: the potential daily changes in all market factors; the concurrent setting each day of floating rates for some of the forwards, swaps and options, if required by the contract's terms and conditions; and the concurrent daily settlement of fixed, floating or contingent cashflows, as contractually specified. As floating rates are set and cashflows settle, the portfolio's remaining value and its factor sensitivities will change.
- ❑ Depending on the purpose of the VAR calculation, it might be necessary to simulate changes in the portfolio's hedges during a long period of risk, given some rules for hedging, the simulated changes in market rates and assumptions about or models of market liquidity. The most prudent VAR calculation would assume that no hedging would occur (or that hedging could not occur because of market illiquidity).

Even the BIS internal models approach effectively assumes a static portfolio, although it nominally refers to a "10-day holding period". In actuality, the internal models approach requires a firm to calculate the potential change in portfolio value assuming a static portfolio and a market rate shock equal to a 10-day move at the 99% confidence level.[2]

VAR versus the counterparty exposure profile

The need to model both the potential changes in market rates over time and the concurrent structural changes in the portfolio over time is one of the factors that normally differentiates the calculation of a pre-settlement counterparty exposure profile from the calculation of VAR. Another factor that distinguishes the calculation of a counterparty exposure profile from the calculation of VAR is that the former critically rests on legal and credit assumptions about the enforceability of netting and other risk mitigants which are irrelevant to the calculation of VAR.

Confidence level and non-linear portfolios

Firms differ in the confidence level at which they normally calculate VAR. The adequacy of any particular VAR model has to be judged in the context

1. Steps in calculating VAR

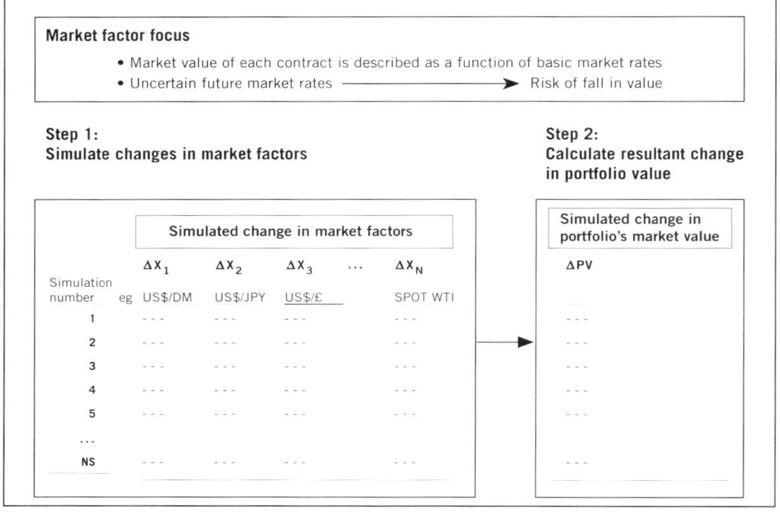

of the type of contract whose risk is being measured, the confidence level at which VAR is measured and the purpose of the VAR measurement. This is most important for a non-linear portfolio:

❑ The higher the confidence level, the more important is the need to model precisely the changes in portfolio value that would occur for large changes in market rates. For example, at high confidence levels it is almost never sufficient to approximate the change in an option portfolio by taking into account only the delta and gamma term of a Taylor-series expansion.

❑ A related point is that the width of the grid of factor sensitivities used in parametric simulation (see below) needs to be commensurate with the confidence level at which VAR is measured.

❑ The higher the confidence level, the more important is the need to incorporate fat tails in the assumed distribution of changes in market rates.

2. Methods for calculating VAR

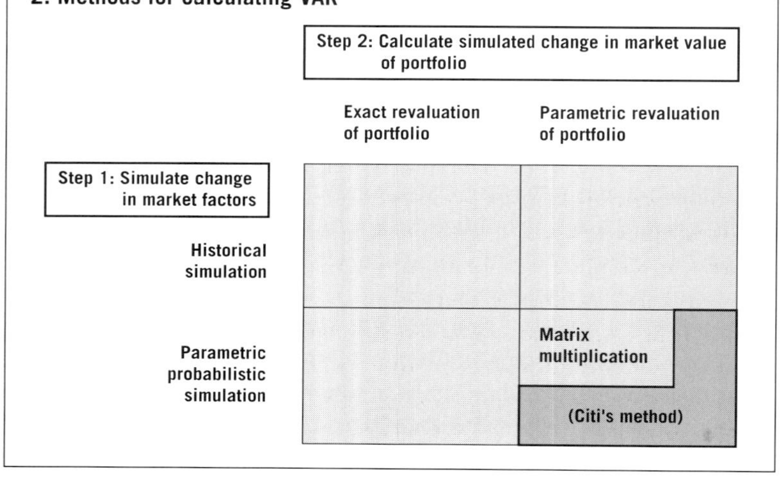

Two basic components of VAR calculation The two basic steps in the calculation of VAR are the simulation of changes in market rates and the calculation of the resulting change in portfolio value. In Figure 1 the simulation of market factors is represented by a two-dimensional array of numbers. Each row of numbers corresponds to a particular scenario of simulated changes in market rates. Each column of numbers corresponds to the simulated changes in a particular market factor. There should be a column for each market factor on which the value of the contracts in the portfolio depends.

VAR methodologies differ with respect both to how market rates are simulated and to how changes in market rates are transformed into changes in portfolio value. The basic alternatives are described in Figure 2. The two basic methods for simulating changes in market factors are historical simulation and parametric statistical simulation. The two basic ways of transforming changes in market rates into changes in portfolio value are exact contract revaluation and parametric portfolio revaluation. As we shall see, matrix multiplication is a special, limited case of combining statistical parametric simulation with parametric portfolio revaluation.

I shall first describe methods for simulating changes in market rates and then look at methods for transforming these simulated changes into changes of portfolio value.

Simulating changes in market factors

MARKET FACTOR SIMULATION

The different forms of simulation can be most easily understood as different methods for populating the rows of Figure 1 – ie different methods of generating a scenario of simulated changes in market rates.

In historical simulation the rows are populated by the actual changes in market factors that occurred between specified pairs of dates in the past. For example, under this method one scenario for a simulated change in each market factor, between today and tomorrow, would be the actual change in those market factors that occurred between a consecutive pair of dates in the past. By this method a thousand simulations of the potential changes in market factors between today and tomorrow could be obtained from a thousand pairs of past dates. Historical simulation has some strengths and some weaknesses (see Panel 3).

The essence of parametric statistical simulation is the use of statistical parameters (eg standard deviation, correlation, etc) derived from time

PANEL 3

STRENGTHS AND WEAKNESSES OF HISTORICAL SIMULATION

The strengths of historical simulation are:

❑ It does not make any explicit assumptions about the shapes of the probability distributions of changes in market rates.

❑ It does not assume that correlations are stable and in fact makes no explicit assumptions about correlations (although it assumes implicitly that past changes in market rates, whatever they may have been, are useful for estimating the potential loss from future changes in market rates).

❑ Therefore, historical simulation can incorporate fat tails, skewness and dynamic correlations (correlations that are a function of the magnitude of the change in market rates) if the historical sample period had such features.

Its weaknesses are:

❑ A problem with obtaining a sufficient number of historical simulations – eg to generate 1,000 independent simulations of a one-day change in rates, it would be necessary to obtain a time series of four years (roughly 250 days per year times four). To generate 1,000 independent historical simulations of an N-day change in rates, it would be necessary to obtain a time series of N thousand days. If N were four, it would be necessary to have a time series for 16 years! This would be impossible to obtain for many market factors and would be meaningless for most, even if the data could be obtained, because of the structural changes in markets that would have occurred over that time period.

❑ A problem with trends for short sample periods. For example, assume that historical simulation is done on the basis of the last 100 business days. Assume that during that period the Japanese yen consistently fell against the US dollar. Consequently, a large spot foreign exchange position that was short yen against the US dollar would gain in value for almost all the historical simulations based on the last 100 days. Under this condition the VAR would be small. This would tend to significantly understate the actual risk – the potential loss that would occur should the trend in the market (a falling yen) reverse itself.

series of past changes in market rates to simulate future changes in market rates. In other words, in parametric statistical simulation the rows of Figure 1 are populated by generating random changes in market factors based on historically derived statistical parameters. Parametric statistical simulation can take several forms.

The simplest form of parametric simulation (implicitly) assumes that near-term changes in market factors are normally distributed and that correlations of changes in market rates are stable. On these assumptions, future changes in market rates are simulated in terms of the standard deviations and correlations of past changes in market rates. The statistical parameters are calculated for some historical sampling period. Even in its simplest form, parametric statistical simulation entails many choices, such as: the length of the sample period (eg three months, three years); and whether the volatilities and correlations should be derived by giving the terms of the historical time series equal or unequal weights (eg exponential weights).

In a more complex form, statistical simulation would depend on more complex assumptions than normality and stable correlations. It would incorporate "fat tails" and skewness in the probability distributions of changes in market rates and dynamic correlations of changes in market rates (eg the correlation could depend on the magnitude of changes in market rates). Incorporating fat tails, skewness and dynamic correlations is particularly important if the VAR measurement is to represent accurately the potential loss that could occur at a very high confidence level. However, it is non-trivial to incorporate such features into statistical simulation.

Monte Carlo simulation is a general method of modelling stochastic processes (ie processes involving human choice or processes for which we have incomplete information). It simulates such a process by means of random numbers drawn from probability distributions that are assumed to describe accurately the uncertain components of the process being modelled. Monte Carlo simulation is extensively used in physics and engineering as well as finance.

PARAMETRIC STATISTICAL SIMULATION OF GENERAL MARKET FACTORS

Overview: different means of incorporating correlations into simulation The means by which correlations can be incorporated into a parametric statistical simulation of changes in market rates is not self-evident. A statistical simulation will be driven by random numbers, and a good random number generator will (and should) generate only independent random numbers.

The calculation of VAR at a very large global financial institution may require the simulation of between one and two thousand general market factors (taking into account spot exchange rates, the term structures of yield curves, the term structures of implied volatilities and so forth) plus hundreds or more specific market factors. The remainder of this section will be given to a discussion of the simulation of general market factors. A later section will describe the simulation of specific market factors.

Under the basic form of parametric statistical simulation, a VAR that depended on N general market factors would require N volatilities and $N(N - 1)/2$ correlations. One simplification concerns the materiality of correlations. For example, it may be possible to identify sets of market factors that have material correlations among themselves but which have zero, or near-zero correlation with other sets of market factors. The assumption of zero correlation between different classes of market factors results in a block diagonal correlation matrix. Even when such a simplification is employed, there remains the challenge of generating simulations within each set of correlated market factors.

A full parametric statistical simulation of N market factors, X_j, with no simplifying assumption of zero correlation, begins with a variance–covariance matrix constructed from N volatilities and $N(N - 1)/2$ correlations. Several paths can be followed from that starting point depending on the coherence of the covariance matrix. The simplest situation occurs when the covariance matrix is positive definite (see discussion below).

When a matrix is positive definite, statistical simulation can be done by means of either: (1) a square-root matrix obtained by Cholesky decomposition of the variance–covariance matrix; or (2) principal component analysis.

The essential steps of simulating changes in N market factors, X_j, by means of Cholesky decomposition are as follows:

1. Construct matrix **V**, the $N \times N$ variance–covariance matrix.
2. Use Cholesky decomposition to construct matrix **A**, the $N \times N$ lower-left corner square-root matrix of matrix **V** – that is, matrix **A** is defined as $\mathbf{AA}^T = \mathbf{V}$, where $\mathbf{A}^T$ is the transpose of **A** (ie $\mathbf{A}^T_{j,i} = \mathbf{A}_{i,j}$) and where $A_{i,j} = 0$ for $j > i$.
3. Use matrix **A** to linear transform a set of N independent draws from a uniform normal distribution into a set of correlated changes of the N market factors, X_j. Cholesky decomposition is described more fully below.

The essential steps in simulating N market factors, X_j, by means of a principal component analysis are as follows:

1. Construct matrix **V**, the $N \times N$ variance–covariance matrix.
2. Find matrix **C**, the $N \times N$ matrix for which $\mathbf{CVC}^{-1}$ is a diagonal matrix with elements $\lambda^2_{k,k}$.
3. Use matrix **C** to linear transform the initial set of N correlated market factors, X_j, into a set of N independent market factors, U_k. The set of market factors, U_k, have zero correlations and have standard deviations equal to $\lambda_{k,k}$. In other words, the set U_k are the eigenvectors of the variance–covariance matrix and the set $\lambda^2_{k,k}$ are its eigenvalues.
4. The simulated change in the set of N market factors, X_j, is equal to a linear transformation of the simulated change in the independent market factors, U_k.

Described less mathematically: under a principal component analysis the change of each of N *correlated* market factors, X_j, is represented as a linear combination of N *independent* market factors, U_k. Therefore, a simulated change of each X_j can be obtained from a linear combination of the simulated changes of the N independent factors, U_k. The simulation of changes in U_k is straightforward because of their independence.

To contrast the two methods, let us consider the simulation of a single yield curve. Let the yield curve be represented by the yield at each of 13 tenors, from one month to 30 years. The first step for either method is to construct the 13×13 variance–covariance matrix from the historical volatilities at each tenor and the correlation of changes of the yields at different tenors. A simulation by Cholesky decomposition would use the square-root matrix to simulate the changes in the yield at each of the 13 tenors. A principal component analysis would represent changes in the yield curve at each of 13 tenors as a linear sum of 13 independent market factors (the eigenvectors). For example, changes in the yield curve could be represented as the sum of: (1) a "parallel" change, in which changes in yield at each tenor are perfectly correlated, always either increasing or decreasing, with a tenor-specific volatility; (2) a greater or lesser "tilting" (more or less steepness) of the yield curve around a pivot point; (3) a greater or lesser "bending" of the yield curve; and so forth.

The principle component decomposition of a yield curve is analogous to a Fourier decomposition of a vibrating string, in which the vibrating string is represented as the sum of simple periodic vibrations whose vibrational frequencies are multiples of a base frequency (ie overtones). Just as most of the string's vibration can usually be described by just a few overtones, most of the change in the yield curve can be described by only a few of the principal components. For example, under the principal component analysis described above, 13 points on the yield curve are simulated as the linear sum of 13 independent factors. However, to a very good approximation, most of the simulated change in the yield curve can be accounted for by the first three factors: a parallel change, a tilting and a bending of the curve. Using this approximation, changes in each of the 13 market factors (ie the yield curve at 13 tenors) would be simulated as a sum of the simulated changes in three independent factors (ie the magnitude and sign of the parallel change, the tilting and the bending).

Cholesky decomposition

This section describes in more detail the use of Cholesky decomposition to simulate changes in market factors. Cholesky decomposition is described in three steps: the simulation of one market factor; the simulation of two market factors; and the simulation of N market factors. To begin with, several terms are defined that will be used throughout the remaining discussion.

- ❑ X_k is market factor k.
- ❑ ΔX_k is the daily change in market factor k (historical or simulated). This is defined for most market factors as the percentage change. For some market factors, such as interest rates, it is often better described as an absolute change.
- ❑ σ_k is the historical daily volatility of market factor k – ie the historical standard deviation of ΔX_k.
- ❑ $\rho_{k,j}$ is the historical correlation between daily changes in market factors k and j.

For simplicity, it is sometimes assumed that correlations between different classes of market factors are zero – eg that the correlation between changes in spot foreign exchange and equity indices are zero. It is not necessary to make this assumption and the remaining discussion does not assume that the simplification has been made.

Simulating changes in one market factor by statistical simulation

Under the assumption of a normal distribution, the daily change in single market factor k is simulated as

$$\Delta X_k = \text{Mean}_k + \sigma_k \times Z \qquad (1)$$

where Mean_k is the historical average change in market factor k, which is usually taken as zero for a one-day period of risk (in what follows I will assume a mean of zero) and Z is a random number drawn from a uniform normal distribution. Z is the simulated number of standard deviations, up or down, that market factor k has changed.

Simulating changes in two market factors by statistical simulation A method needs to be defined by which two or more independent random numbers are transformed into the simulated correlated changes of two or more market rates. We will show how this is done for the simulation of two market factors, k and j

Independent change Assume that changes in market factors k and j are independent. Under the assumption of a normal distribution, the daily change in these market factors can be simulated as follows (assuming that the mean changes are zero):

$$\Delta X_k = \sigma_k \times Z_1 \qquad (2)$$

$$\Delta X_j = \sigma_j \times Z_2 \qquad (3)$$

That is, changes in each market factor are simulated by separate random draws, Z_1 and Z_2. This can be rewritten more elegantly as a matrix multiplication:

$$\begin{bmatrix} \Delta X_k \\ \Delta X_j \end{bmatrix} = \begin{bmatrix} \sigma_k & 0 \\ 0 & \sigma_j \end{bmatrix} \begin{bmatrix} Z_1 \\ Z_2 \end{bmatrix} \qquad (4)$$

Perfectly correlated changes When market factors k and j are perfectly correlated ($\rho = +1$), changes in their values can be simulated as

$$\Delta X_k = \sigma_k \times Z_1 \qquad (5)$$

$$\Delta X_j = \sigma_j \times Z_1 \qquad (6)$$

That is, they are simulated by the same random number, Z_1. This can be rewritten in matrix form as

$$\begin{bmatrix} \Delta X_k \\ \Delta X_j \end{bmatrix} = \begin{bmatrix} \sigma_k & 0 \\ \sigma_j & 0 \end{bmatrix} \begin{bmatrix} Z_1 \\ Z_2 \end{bmatrix} \quad (7)$$

General case The most general case will be represented by the following matrix multiplication:

$$\begin{bmatrix} \Delta X_k \\ \Delta X_j \end{bmatrix} = \begin{bmatrix} A_{1,1} & A_{1,2} \\ A_{2,1} & A_{2,2} \end{bmatrix} \begin{bmatrix} Z_1 \\ Z_2 \end{bmatrix} \quad (8)$$

where the matrix **A** transforms the two independent random numbers Z_1 and Z_2 (drawn from a uniform normal distribution) into correlated changes of market factors k and j. Matrix **A** should only depend on some combination of the volatilities and correlations of market factors k and j. It can be shown that one way to solve for **A** is by means of Cholesky decomposition of the covariance matrix, in which matrix **A** is set equal to the lower-left square-root of the variance–covariance matrix of market factors k and j. That is, matrix **A** is defined as the matrix for which

$$\mathbf{A} \times \mathbf{A}^T = \begin{bmatrix} A_{1,1} & 0 \\ A_{2,1} & A_{2,2} \end{bmatrix} \times \begin{bmatrix} A_{1,1} & A_{2,1} \\ 0 & A_{2,2} \end{bmatrix} = \mathbf{V} \quad (9)$$

$$\text{where } \mathbf{V} \equiv \begin{bmatrix} \sigma_k^2 & \sigma_k \sigma_j \rho_{k,j} \\ \sigma_k \sigma_j \rho_{k,j} & \sigma_j^2 \end{bmatrix}$$

is the variance–covariance matrix. It is straightforward to solve for the components of **A**:

$$\mathbf{A} = \begin{bmatrix} \sigma_k & 0 \\ \sigma_j \rho_{k,j} & \sigma_j \sqrt{1-\rho_{k,j}^2} \end{bmatrix} \quad (10)$$

To see that this makes sense, we substitute this value of **A** into the above equation for simulating changes in market factors k and j and obtain

$$\Delta X_k = \sigma_k \times Z_1 \quad (11)$$

$$\Delta X_j = \sigma_j \times \rho_{j,k} \times Z_1 + \sigma_j \times \sqrt{1-\rho_{1,2}^2} \times Z_2 \quad (12)$$

This has a simple interpretation. The first market factor, k, is simulated by one component, random number Z_1, as if it were a stand-alone market rate. The second market factor, j, is simulated by two components. One component (dependent on Z_1) is perfectly correlated with changes in market factor k, while the second component (dependent on Z_2) is independent of changes in market factor k. The above process does not depend on the order in which the market factors are arrayed. That is, the same distribution of simulated changes would have been obtained if the order of market factors k and j were transposed.

This process can be generalised to N market factors. The simulation of N market factors by Cholesky decomposition is given by

$$\begin{bmatrix} \Delta X_1 \\ \Delta X_2 \\ \cdots \\ \Delta X_N \end{bmatrix} = \begin{bmatrix} A_{1,1} & 0 & \cdots & 0 \\ A_{2,1} & A_{2,2} & \cdots & 0 \\ \cdots & \cdots & \cdots & \cdots \\ A_{N,1} & A_{N,2} & \cdots & A_{N,N} \end{bmatrix} \begin{bmatrix} Z_1 \\ Z_2 \\ \cdots \\ Z_N \end{bmatrix} \quad (13)$$

Again, the order in which the market factors are arrayed is irrelevant. This type of analysis can be applied to a variance–covariance matrix of any size that is positive definite.

There is a connection between principal component analysis and Cholesky decomposition. The $N \times N$ variance–covariance matrix can only be decomposed by Cholesky decomposition if it is positive definite. A matrix is positive definite if all of its eigenvalues are greater than zero. In other words, an $N \times N$ variance–covariance matrix **V** can only be decomposed by Cholesky decomposition if an $N \times N$ matrix **C** can be found for which $\mathbf{CVC}^{-1}$ is an $N \times N$ diagonal matrix with positive diagonal elements – ie $\lambda_{kk}^2 > 0$.

There are several reasons why the $N \times N$ matrix might not be decomposable (ie might not be positive definite):

❑ *It has eigenvalues of zero* This will occur if a market factor is a linear combination of other market factors. One reason this could happen would be if a market rate (such as an exchange rate) were defined as a weighted basket of other market rates, with unvarying weights. It could also happen if the time series of some market factor had been obtained by linearly interpolating two other market factors – eg if seven-year yields had been obtained by interpolation of five- and 10-year yields.

❑ *It has negative eigenvalues* Negative eigenvalues correspond to eigenvectors (ie principal components) with imaginary standard deviations (ie sd < 0)! This could occur for at least two reasons. The first is related to the fact that all volatilities and correlations will be expressed by a value that is rounded off to some number of significant figures. The rounding off is equivalent to introducing noise, which can, in some cases, result in very small negative eigenvalues.

A second reason is that volatilities and correlations may have been derived from time series of different lengths. For example, assume that the default condition is to measure volatilities and correlations for a sample period of the most recent three years. For some market factors it may not be possible to obtain three years' worth of data. Assume we have three years of data for market factors A and B but only six months of data for market factor C. Assume that changes in market factors A and C were highly correlated during the six months for which data were available for market factor C. Similarly, assume that changes in market factors B and C were highly correlated during the six-month period. In contrast, assume that during the last three years changes in market factors A and B were only moderately correlated. Because we have calculated correlations over time periods of different lengths we have an inconsistency: the correlations of (A,C) and (B,C) will be high, whereas the correlation of (A,B) will be moderate! This inconsistency can cause negative eigenvalues, which will make it impossible to obtain the square-root matrix by Cholesky decomposition.

If the Cholesky decomposition fails, a singular value decomposition (SVD)[3] can be used to identify the immediate cause of the failure. If the SVD indicates a very small, immaterial negative eigenvalue, procedures can be followed to fix the problem. It would be incorrect to use a principal components analysis mechanically, with no regard to the sign and magnitude of the eigenvalues. Simulating market rates when the matrix has material, negative eigenvalues is not a meaningful exercise.

This also illustrates the risk of a mechanical use of matrix multiplication even for a linear portfolio. If the covariance matrix has material, negative eigenvalues, it is inconsistent and corrupt. Even when the portfolio is linear and the value-at-risk is to be calculated by simple matrix multiplication, it is important to test the coherence of the variance-covariance matrix. This can be tested either by a principal components analysis (with a test of the eigenvalues), by a Cholesky decomposition (to test if the matrix is even decomposable) or, more generally, with a singular value decomposition.

PARAMETRIC STATISTICAL SIMULATION OF SPECIFIC MARKET FACTORS

Specific market risk can only be understood by distinguishing it from general market risk.

"General" market risk is the risk that the economic value of a contract or portfolio of contracts could fall because of changes in general market factors. General market factors are market rates and prices that are independent of the credit risk of a particular issuer. They are determined by both the current and the market's view of the potential future macroeconomic conditions of an economy (eg inflation, general level of economic activity), the supply and demand for certain classes of investment and the supply and demand for particular commodities.

Examples of general market factors are: Libor and/or treasury yield curves, spot foreign exchange rates, equity indexes commodity prices and their respective term structures of implied volatilities.

"Specific" market risk is the risk that the economic value of a contract or a portfolio of contracts could decrease because of changes in specific market factors, apart from changes in general market factors. Specific market factors are market prices and rates that partly reflect the issuer-specific risk of debt or equity. Examples of specific market risk are:

❑ The price of an equity could change even if the price of a broad equity index within that country remained unchanged. Or, more generally, the percentage change in the price of an equity may not be perfectly correlated to the percentage change in the price of a broad equity index within that country.

❑ The price of a corporate bond could change even if the price of a treasury security of the same duration were unchanged. More generally, the change in a corporate yield curve (at some tenor) may not be perfectly correlated to changes in the base yield curve (eg treasury or Libor) of that country (at the same tenor).

❑ The price of a Eurobond issued by a government A but denominated in the currency of country B could change even if the base yield curve in country B were unchanged. More generally, changes in the yield of a Eurobond (of a specific tenor) issued by government A and denominated in the currency of country B may not be perfectly correlated to changes in the base yield curve of country B.

These three examples illustrate that specific risk is a form of correlation risk – the risk that the change in a market factor associated with a particular issuer may not be perfectly correlated to the change of a similar general market factor. The types of contract with specific risk include:

❑ Individual equities and derivatives on individual equities.

❑ Individual corporate debt securities and derivatives on corporate debt securities.

❑ Credit derivatives.

The two broadest categories of instruments with specific risk are equities/equity derivatives and debt securities/debt security derivatives. A brief description follows of how the price risk of each can be incorporated into a VAR calculation.

Correlations and event risk: for general market risk and specific risk

General market risk and specific market risk are similar: to incorporate either into a VAR calculation requires that correlations be taken into account; and each is subject to event risk.

For both types of risk, a portfolio could lose economic value because of three types of changes in market rates:

❑ Normal day-to-day fluctuations in market rates. The potential loss at some confidence level can be described by a VAR calculation.

❑ Large market jumps that are rare but predictable. The potential loss for this type of market move can, in principle, also be incorporated into a VAR calculation if the VAR model is sufficiently robust – that is, if its simulation of changes in rates and its transformation of the change in market rates into the change in portfolio value are properly modelled for large changes in rates.

❑ Very large, unpredictable changes in a particular market rate – ie event risk. These types of event are difficult to incorporate into a VAR calculation because of the rarity of their occurrence.

Event risk for general market rates includes: a large sudden currency devaluation; a large dramatic decrease in equity indices (eg the October 1987 crash); a large sudden change in some commodity prices (eg during the Iraqi invasion of Kuwait).

Event risk for specific market rates includes: a large unexpected change in the market's perception of the issuer's financial health; unexpected news about a financial restructuring, merger or failed merger of the corporate issuer.

VAR should never be the sole means of measuring, monitoring and limiting market risk. Event risk (and liquidity risk) are not well incorporated into VAR models. Such risk should be measured, monitored and limited by controls other than VAR – such as concentration limits, issuer limits, cross-border limits, etc.

Simulating specific risk of individual equities

There are two broad techniques for simulating the price changes of individual equities:

❑ Simulate each equity price as a basic market factor. For L equities, one would need L individual volatilities and $L(L-1)/2$ correlations of changes in the price of all pairs of equities. However, to incorporate equity risk into the VAR of a general trading portfolio (which will have factor sensitivity to many types of market factor), it would also be necessary to calculate the correlation of changes in each equity's price to changes in every other market factor.

❑ Simulate changes in the price of each equity by means of factor analysis. In this approach the simulated change in the price of each individual equity is set equal to the sum of independently simulated components. At a minimum there must be two components: one component should have perfect correlation with the change in a general equity index; and the second (the residual component) should be independent of changes in the index and also be independent of changes in the price of other equities. With these two basic components one will be able to account for the fact that some of the change in the price of an individual equity is correlated with changes in the general market while part of the change is specific to the firm that issued the equity.

A two-factor model will simulate changes in each equity's price with only two components: one that is perfectly correlated with the equity index; and the residual component. One type of three-factor model will simulate changes in each equity's price with three components: one that is perfectly correlated with the equity index; a component that is independent of changes in the equity index but is partially correlated with the changes in the price of other equities in its industry; and a residual component that is independent of changes in the equity index and independent of price changes of any other equity, including equities within the same industry. Equity factor models with a much greater number of factors have been developed.

Mathematical expression of two-factor model

Under a two-factor model, the simulated change in the equity index is

$$\Delta X_{index} = \sigma_{index} \times Z_{index} \tag{14}$$

and the simulated change in each equity, j, is

$$\Delta X_j = \sigma_j \times \rho_{j,index} \times Z_{index} + \sigma_j \times \sqrt{1-\rho^2_{j,index}} \times Z_j \tag{15}$$

where σ_{index} is the daily volatility of the equity index; σ_j is the daily volatility of equity j; $\rho_{j,index}$ is the correlation of changes in equity j and the equity index; Z_{index} is the random number used to simulate change in the equity index; and Z_j is the random number used to simulate change in the residual component of equity j. Again, let ΔX be the daily percentage change and assume that the equity and the index are within the same country.

The first term of equation (15) is the component of the simulated change in the price of equity j that is perfectly correlated with the change in the index. The second term of equation (15) is the component of the simulated change of equity j that is independent of the change in the index and independent of the change in the price of other equities – ie the random number Z_j is specific to the change in the price of equity j.

This formula is identical in form to the formula we obtained above using the Cholesky decomposition for two market factors. However, as we shall see, the simulation of an index and two or more equities does not have the same form as the Cholesky decomposition of the equivalent number of market factors.

By defining a beta of the equity as

$$\beta_{j,index} = \left(\frac{\sigma_j}{\sigma_{index}} \right) \times \rho_{j,index} \quad (16)$$

equation (15) can be rewritten in a form that may be more familiar, so it is left to the reader to rewrite the formula.

"Diversifiable" and "undiversifiable" risk In a two-factor model, the first term of equation (15) – ie the change in the price of equity j that is perfectly correlated with the change in the index – is called the "undiversifiable" risk component. The second component of equation (15) – ie the change in the price of equity j that is independent of the change in the index and is specific to that firm, is called the "diversifiable" risk component. The labelling of the components as diversifiable or undiversifiable arose in the context of portfolio theory for investments. Consider an investment portfolio that consists of bought positions in individual equities. That portfolio will be linear and its VAR can be calculated by matrix multiplication, as described below. It can be shown that the relative contribution of the diversifiable risk to the total VAR can be made as small as one wishes simply by diversifying the constituents of the portfolio. For a sufficiently large and diversified portfolio, its total VAR will predominantly be due to the undiversifiable risk – ie the change in the value of the portfolio will be highly correlated to changes in the general equity index.

However, it would be a mistake to assume that the diversifiable risk is necessarily immaterial in a trading portfolio. A large trading portfolio will consist of long and short positions in the underlying equities as well as in derivative contracts on individual equities and equity indices. As a consequence, depending on the structure of the portfolio, the contribution of the undiversifiable risk to total VAR may actually be relatively small, while the contribution of the diversifiable risk to total VAR may be relatively very large. This could occur, for example, if traders tended to hedge derivatives on an individual stock with an offsetting derivative on an equity index.

How many factors are there in a two-factor equity model? One should not confuse the use of the phrase "two-factor" or "three-factor" equity model with the term "three-factor" yield curve model. They have very different meanings. In the three-factor yield curve model, described above, the simulated change in 13 market factors, X_j (the change in yield at each of 13 tenors), is a function of three random numbers, one for each independent component of the simulation:

$$\begin{bmatrix} \Delta X_1 \\ \Delta X_2 \\ \dots \\ \Delta X_{13} \end{bmatrix} = \begin{bmatrix} A_{1,1} & A_{1,2} & A_{1,3} \\ A_{2,1} & A_{2,2} & A_{2,3} \\ \dots & & \\ A_{13,1} & A_{13,2} & A_{13,3} \end{bmatrix} \begin{bmatrix} Z_1 \\ Z_2 \\ Z_3 \end{bmatrix} \quad (17)$$

In contrast, consider 100 equities within a country, plus a general equity index for that country. In a two-factor simulation model, we would need 101, not two, random numbers to simulate the equity index and the 100 equities.

For the equity index:

$$\Delta X_{index} = \sigma_{index} \times Z_{index} \quad (18)$$

For each of the 100 individual equities:

$$\begin{bmatrix} \Delta X_1 \\ \Delta X_2 \\ \dots \\ \Delta X_{100} \end{bmatrix} = \begin{bmatrix} A_{1,index} & A_{1,1} & 0 & 0 \\ A_{2,index} & 0 & A_{2,2} & 0 \\ \dots & \dots & \dots & \dots \\ A_{100,index} & 0 & 0 & A_{100,100} \end{bmatrix} \begin{bmatrix} Z_{index} \\ Z_1 \\ \dots \\ Z_{100} \end{bmatrix} \quad (19)$$

In the two-factor model changes in each equity are simulated as the sum of two components, one that is common to all the equities in that country and the other idiosyncratic to each equity.

Consider a three-component model applied to 150 equities, which we will assume fall into 10 industry groups of 15 equities each. The simulation of the portfolio of 150 equities plus the equity index would require 161 random numbers in a three-factor model: one random number for the general equity index; 10 random numbers for each of the 10 industry factors; plus 150 random numbers for each of the 150 residual factors, one per equity. The 161 random numbers would be used to simulate changes in the three components for each equity.

Simulating specific risk of individual debt securities

Debt securities can also be simulated by factor analysis. In the liquid markets, the specific risk of debt securities can be simulated quite well by a three-factor specific-risk yield model. In illiquid, emerging markets there may not be a sufficient number of debt securities to specify a yield curve. In such markets, debt securities tend to trade by price, not yield. A simple two-factor price model, similar to a two-factor equity model, can be used in such markets to simulate changes in the value of debt securities.

I use the convoluted phrase "three-factor specific-risk yield model" to distinguish this method for incorporating specific risk (with three factors per debt security) from a three-factor yield curve model, in which an entire yield curve is simulated with three components.

A three-factor specific-risk yield model will represent the yield of a debt security, of some specified tenor, as the sum of three components:

❑ A base yield curve within the country, such as a Libor or treasury yield curve.

❑ A general market spread within the country that would be specified, in principle, as a function of: the type of issuer (eg municipality, corporate), which can affect taxes for the investor; the risk rating of the issuer; and the tenor of the security.

❑ A residual spread per issuer. The residual spread could be positive or negative.

The yield of each debt security could then be simulated as the sum of three terms: (1) the simulated change in the base yield; (2) the simulated change in the general market spread; and (3) the simulated change in the residual spread. In many markets it will be difficult to obtain the time series needed to derive the historical volatilities of each component.

Debt security specific risk can be analysed with more complex factor analysis than that described here. For example, one could include industry components and so forth.

Simulating changes in portfolio value

THE LARGER PICTURE

The next step in the calculation of VAR is the transformation of simulated changes of market rates into simulated changes in portfolio value. We shall examine three such transformations:

❑ Matrix multiplication – which is limited in its applicability to linear portfolios.

❑ Exact contract revaluation.

❑ Parametric portfolio revaluation.

Parametric portfolio revaluation is a more efficient method for calculating VAR than exact revaluation and in many contexts, when appropriately implemented, can be as precise. Each method for simulating changes in portfolio value will be analysed from several perspectives:

❑ The required information that must be fed to the VAR calculator.

❑ The cost of obtaining and feeding that information.

❑ The potential weaknesses or inapplicability of the method.

MATRIX MULTIPLICATION AS A LIMITED FORM OF PARAMETRIC REVALUATION

The standard calculation of VAR by matrix multiplication is a particular form of combining the parametric statistical simulation of changes in market factors and the parametric representation of the sensitivity of the portfolio to changes in market factors. Calculating VAR by means of matrix multiplication is attractive because it entails using a simple closed-form equation and does not require simulation. However, matrix multiplication is only appropriate for a linear portfolio.

A portfolio is linear if the change in its market value is directly proportional to the change in each market factor on which its value depends:

$$\Delta PV = \sum_{k=1}^{N} FS_k \times \Delta X_k \qquad (20)$$

where $FS_k = \delta PV/\delta X_k$ is the factor sensitivity of the portfolio to a unit change in market factor k. FS_k equals the sum of the factor sensitivity of each contract in the portfolio to a unit change in market factor k.

This linear equation is the simplest example of a parametric representation of the change in a portfolio's value as a function of changes in market factors. The usefulness of this representation can be seen by contrasting the number of revaluations required for two alternative ways of simulating a change in the portfolio's value. Assume that the portfolio has 1,000 contracts and that each contract's value depends on only one market factor, X. Assume that we have generated 10,000 simulated changes in market factor X and want to ascertain the corresponding distribution of changes in portfolio value.

❑ Under an exact revaluation method, each of the 1,000 contracts would have to be revalued 10,001 times. One revaluation would be done at the current market rate to ascertain the contract's current market value. Each contract would have to be revalued an additional 10,000 times to ascertain its market value for each of the 10,000 simulated changes in the market factor X. Consequently, this method would require a total of 10,001,000 (1,000 × 10,001) exact contract revaluations.

❑ Under the assumption of linearity, each contract would need to be revalued only twice to determine its factor sensitivity: once at the current market value of market factor X and once for a unit change in X. (For most contracts, the factor sensitivity can be determined analytically and would not require two revaluations.) The portfolio's factor sensitivity to X is the sum of each contract's factor sensitivity. The change in the value of a linear portfolio to each of the 10,000 simulations of changes in X would be equal to the product of the portfolio's factor sensitivity and the simulated change in X. Therefore, instead of 10,001,000 exact contract revaluations, this method would require: 2,000 exact contract revaluations (two per contract times 1,000 contracts) to determine the factor sensitivity; and 10,000 multiplications of the portfolio's factor sensitivity by the simulated change in market factor X.

In fact, under the condition of linearity, it is not even necessary to perform these simulations to calculate the portfolio's VAR. Instead, VAR can be calculated by a closed-form equation.

Under the assumption of linearity, the variance of ΔPV is given by

$$\text{Variance}(\Delta PV) = \sum_{k=1}^{N} FS_k^2 \times \sigma_k^2 + \sum_{j \neq k} FS_j \times FS_k \times \rho_{j,k} \times \sigma_j \times \sigma_k \qquad (21)$$

This can be rewritten as a multiplication of matrices:

$$\text{Variance}(\Delta PV) = [FS_1 \;\; FS_2 \;\; \cdots \;\; FS_N] \times \begin{bmatrix} \sigma_1^2 & \sigma_1\sigma_2\rho_{1,2} & \cdots & \sigma_1\sigma_N\rho_{1,N} \\ \sigma_2\sigma_1\rho_{1,2} & \sigma_2^2 & \cdots & \sigma_2\sigma_N\rho_{2,N} \\ \cdots & \cdots & \cdots & \cdots \\ \sigma_N\sigma_1\rho_{1,N} & \sigma_N\sigma_2\rho_{2,N} & \cdots & \sigma_N^2 \end{bmatrix} \times \begin{bmatrix} FS_1 \\ FS_2 \\ \cdots \\ FS_N \end{bmatrix} \qquad (22)$$

or

$$\text{Variance}(\Delta PV) = FS^T \times \mathbf{V} \times FS \qquad (23)$$

where FS is an N-dimensional vector of factor sensitivities and **V** is the $N \times N$ variance-covariance matrix.

So far, we only have an equation for the variance in ΔPV. We must now connect this to VAR. The standard deviation of ΔPV is equal to the square root of Variance(ΔPV). VAR is normally specified as the potential loss at some confidence level. The relationship between ΔPV at some confidence level and the standard deviation of ΔPV depends on the shape of the probability distribution of ΔPV.

If we assume that changes in market factors are normally distributed and that correlations of changes in market factors are stable, the probability distribution of ΔPV will be normal. Consequently, the relationship between ΔPV at some confidence level and the number of standard deviations of ΔPV will be known. The relationship can be described as:

$$\begin{aligned} VAR_{portfolio} &= \text{Scale factor} \times \text{Standard deviation}(\Delta PV) \\ &= \text{Scale factor} \times \text{Square root}[\text{Variance}(\Delta PV)] \end{aligned} \qquad (24)$$

where the scale factor translates a confidence level into its corresponding number of standard deviations. For example, a scale factor of two corresponds to a VAR at the 97.7% confidence level.

The matrix multiplication calculation of VAR can be rewritten in an equivalent form. Define a "market factor" VAR, to market factor k as

$$VAR_k = \text{Scale factor} \times FS_k \times \sigma_k \qquad (25)$$

Therefore

$$\left(VAR_{portfolio}\right)^2 = \left[VAR_1 \quad VAR_2 \quad \cdots \quad VAR_N\right] \times \begin{bmatrix} 1 & \rho_{1,2} & \cdots & \rho_{1,N} \\ \rho_{1,2} & 1 & \cdots & \rho_{2,N} \\ \cdots & \cdots & \cdots & \cdots \\ \rho_{1,N} & \rho_{2,N} & \cdots & 1 \end{bmatrix} \times \begin{bmatrix} VAR_1 \\ VAR_2 \\ \cdots \\ VAR_N \end{bmatrix} \quad (26)$$

In general, VAR_k is not the incremental amount of the portfolio's VAR due to market factor k. It is the VAR corresponding to market factor k in isolation, assuming no other market factors have any volatility. The sum of all the VAR_ks for all market factors k will be greater than $VAR_{portfolio}$ (unless changes in all the market factors are perfectly correlated, in which case the sum will equal $VAR_{portfolio}$). Methods have been developed to measure the incremental change in $VAR_{portfolio}$ that would occur for an incremental change in a position or cashflow.[4]

The matrix multiplication method is only exact for a linear portfolio. For non-linear portfolios simulation is necessary. There have been some attempts to take non-linearity into account by incorporating the effect of a gamma term in the matrix multiplication. In general this is a futile exercise for a large portfolio with many options. The incorporation of non-linearities in the calculation of VAR by means of a Taylor-series expansion will generally require many terms beyond the first two. In general, for an option portfolio, the higher the confidence level of the VAR calculation, the more Taylor-series terms are needed. The explanation for this is given below.

EXACT REVALUATION OF EACH CONTRACT FOR EACH SIMULATION

In this method each contract is revalued for each of the N scenarios of simulated changes in market rates. There are two basic ways this could be done.

In the first, the set of N simulated scenarios (eg the rows of Figure 1) could be fed to each back-office product processor, in which each contract would be revalued N + 1 times – once using the actual end-of-day market rates, and once for each of the N scenarios of simulated changes in market rates. The N + 1 revaluations could then be fed from each back-office system into a central price risk database. By this method, N scenarios of simulated changes in market rates will be transformed into N simulated changes in portfolio value.

Alternatively, a central transaction database could be fed a detailed description of each contract's terms and conditions from each back-office product processor. The transaction database could then feed the detailed description of each contract to the VAR calculator. The VAR calculator would need to revalue each contract N + 1 times, based on: (1) the appropriate revaluation formula or process for each contract; (2) the set of current market rates; and (3) the set of N scenarios of simulated changes in market rates. With this information, the VAR calculator could transform N scenarios of simulated changes in market rates into N simulated changes of portfolio value.

This is both the most precise method of transforming changes of market factors into changes of portfolio value and the most costly method to implement. It can be used for all types of portfolio – in contrast to the limited use of matrix multiplication. The primary cost of this method is the resources required to revalue each contact thousands of times. If a firm implements this method by creating a global transaction database, it will also incur costs in the building and maintenance of the centralised database and in the maintenance of the VAR calculator, which would have to contain or have access to every revaluation formula/process needed to value every form of contract in the portfolio.

We have so far only considered two methods – one that is relatively inexpensive to implement but is inapplicable to non-linear portfolios (matrix multiplication), and an exact method which can be very costly to implement (exact revaluation). Parametric portfolio revaluation is an alternative approach that combines efficiency and accuracy, including the ability to take convexity into account.

PARAMETRIC REVALUATION OF PORTFOLIO

Parametric portfolio revaluation is another way of transforming simulated changes in market factors into simulated changes in portfolio value. In this method, the functional relationship between the change in portfolio value and changes in market factors is represented in terms of a finite number of parameters. The parameters are derived from a small number of exact revaluations of each contract in the portfolio.[5] Two types of parametric portfolio revaluation are the familiar Taylor-series expansion and the less familiar grids of factor sensitivity.

Taylor-series expansion

The concept of a Taylor-series expansion should be familiar to everyone who has taken first-year calculus. In a Taylor-series expansion the change in the value of a function of several independent variables is represented as an infinite polynomial series whose parameters are the partial derivatives of the function with respect to changes in the independent variables.

The total market value of a portfolio, $PV_{portfolio}$, is the sum of the values of each of its component contracts. Given the revaluation formula, PV_m, of each contract, m, we have

$$PV_{portfolio} = \Sigma PV_m(T\&C_m, X_j(t), t) \quad (27)$$

where $T\&C_m$ are the terms and conditions of each contract and $X_j(t)$ are the values of each of the market factors on which the contract's value depends. For the Taylor-series expansion, $X_j(t)$ are the independent variables. Therefore we have

$$\begin{aligned}\Delta PV_{portfolio} &= \sum_j \left(\frac{\delta PV}{\delta X_j}\right)\Delta X_j \\ &+ \frac{1}{2}\sum_j \left(\frac{\delta^2 PV}{\delta X_j^2}\right)\Delta X_j^2 \\ &+ \frac{1}{2}\sum_{j\neq k} \left(\frac{\delta^2 PV}{\delta X_j \delta X_k}\right)\Delta X_j \Delta X_k \\ &+ \frac{1}{6}\sum_j \left(\frac{\delta^3 PV}{\delta X_j^3}\right)\Delta X_j^3 \\ &+ \frac{1}{6}\sum_{j\neq k} \left(\frac{\delta^3 PV}{\delta X_j^2 \delta X_k}\right)\Delta X_j^2 \Delta X_k \\ &+ \frac{1}{6}\sum_{j\neq k\neq l} \left(\frac{\delta^3 PV}{\delta X_j \delta X_k \delta X_l}\right)\Delta X_j \Delta X_k \Delta X_l + \ldots \end{aligned} \quad (28)$$

where j, k, l, etc, are summed over all market factors. The portfolio's partial derivatives are summed over each contract's partial derivatives.

This polynomial is one form of parametric portfolio revaluation. The parameters in the polynomial are the instantaneous partial derivatives of the change of the portfolio's value with respect to changes in its market factors. These parameters are derived for each contract from the contract's exact revaluation formula and then summed across all contracts in the portfolio. The first set of parameters are the instantaneous deltas, the second set are the instantaneous gammas, the third set are the instantaneous cross-gammas, and so forth. Potentially, an unlimited set of parameters, representing higher-order derivatives, could be included.

The number of sets of parameters that need to be included in a Taylor-series expansion to achieve a specified degree of accuracy in the VAR calculation depends on the composition of the portfolio, the confidence level at which VAR is to be measured and the relationship between the set of market factors used for simulation and the market factors on which the contract's value depends. This latter point is not obvious and will be explained below when we discuss cross-terms in a Taylor-series expansion.

One-, two-, three- and N-dimensional terms

For reasons that will be clearer below, we can reorganise the terms of the Taylor-series expansion into the following subset of terms:

$$\begin{aligned}PV_{portfolio} &= \text{One-dimensional terms} \\ &+ \text{Two-dimensional terms} \\ &+ \text{Three-dimensional terms} + \ldots\end{aligned}$$

Let us now define each of the terms and explain what we mean by dimensionality. We will begin with the "one-dimensional" terms.

$$\begin{aligned}&\text{One-dimensional terms} \\ &= \sum_j \left(\frac{\delta PV}{\delta X_j}\right)\Delta X_j \\ &\quad + \frac{1}{2}\sum_j \left(\frac{\delta^2 PV}{\delta X_j^2}\right)\Delta X_j^2 + \frac{1}{6}\sum_j \left(\frac{\delta^3 PV}{\delta X_j^3}\right)\Delta X_j^3 + \ldots \\ &= \sum_j \text{polynomial function of} \left(\Delta X_j\right)\end{aligned} \quad (29)$$

The one-dimensional terms are the sum of polynomial functions of the change of each single market factor. The Taylor-series expansion will consist of only one-dimensional terms if the revaluation formula of each contract can be expressed as a separable function of each market factor – that is, if the sensitivity of the value of a contract to changes in each market factor is independent of the magnitude of the change of any other market factor.

Two-dimensional terms

$$= \frac{1}{2}\sum_{j\neq k}\left(\frac{\delta^2 PV}{\delta X_j\,\delta X_k}\right)\Delta X_j\,\Delta X_k + \frac{1}{6}\sum_{j\neq k}\left(\frac{\delta^3 PV}{\delta X_j^2\,\delta X_k}\right)\Delta X_j^2\,\Delta X_k + \ldots$$

$$= \sum_{j\neq k} \text{polynomial function of} (\Delta X_j, \Delta X_k) \quad (30)$$

The two-dimensional terms are the sum of polynomials of pairs of market factors. The Taylor-series expansion will depend on two-dimensional terms if the sensitivity of the market value of a contract to changes in a market factor is conditional on the magnitude of the change in one other market factor. An example would be a spot exchange rate and the corresponding implied volatility of a foreign exchange option. Under certain conditions, the change in the value of an option corresponding to a change in the spot exchange rate (particularly for a large change) can depend on the magnitude of the change in the implied volatility. Other examples will be discussed below.

Three-dimensional terms =

$$\frac{1}{6}\sum_{j\neq k\neq l}\left(\frac{\delta^3 PV}{\delta X_j\,\delta X_k\,\delta X_l}\right)\Delta X_j\,\Delta X_k\,\Delta X_l + \ldots$$

$$= \sum_{j\neq k\neq l} \text{polynomial function of} (\Delta X_j, \Delta X_k, \Delta X_l) \quad (31)$$

3. One-dimensional grid of factor sensitivities: changes in portfolio value for set of specific changes in market factor

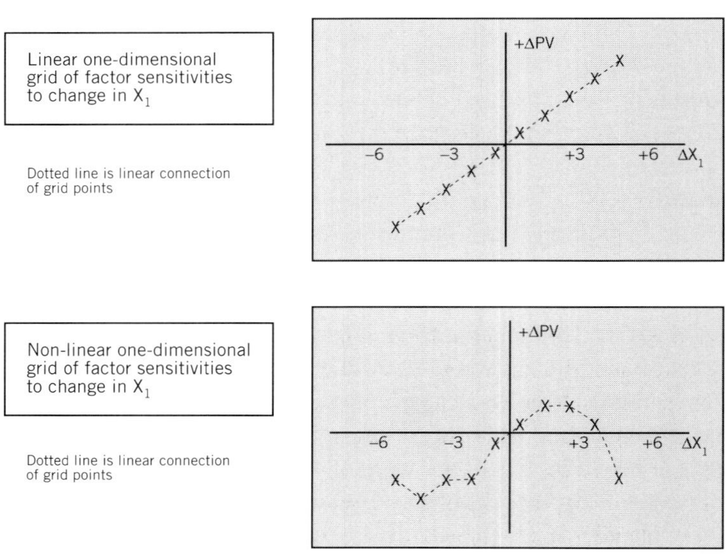

The three-dimensional terms will be important if the sensitivity of a contract to a change in a market factor is conditional on the magnitude of changes in two other market factors. An example of when this can be important will be given below.

Consequently, two factors must be considered when deciding which terms of a Taylor-series expansion to include in a simulation: the order of derivatives (ie delta, gamma, third derivative, etc) that are required; and the dimensionality of the cross-terms that are needed.

A linear portfolio needs only the one term per market factor – the delta. The linear portfolio can therefore be completely described with only the first term of the one-dimensional polynomial.

For small changes in the market, a single in-the-money European option with a long time to expiration can be reasonably described by three one-dimensional polynomials – ie as a separable function of changes in the underlying spot rate, changes in the implied volatility and changes in the interest rate. The change in the value of the option in response to small changes in the spot rate can be reasonably well described with only the first two terms (ie the delta and the gamma) of the one-dimensional polynomial.

However, if VAR is to be measured at a very high confidence level and if the portfolio contains many bought and sold options of sufficiently different strike prices and with different times to expiration, then, at a minimum, many terms of the one-dimensional polynomial with respect to changes in the spot rate will have to be included. Consider a graph of the change in the portfolio's value as a function of a large range of changes in the underlying spot rate (eg see Figure 3). The function could have many bending points depending on the constituent options. As a simple example consider a portfolio with a small position in bought-at-the-money puts and bought-at-the-money calls plus a very large position in sold-out-of-the-money puts and sold-out-of-the-money calls. Assume that for small changes in the spot price, up or down, the portfolio would gain in value. Assume that for sufficiently large changes in the spot price, up or down, the portfolio would experience a very large loss in value. Consequently, calculating VAR simply on the basis of instantaneous delta and gamma will result in an erroneous measurement of the portfolio's risk. In this example, the higher the confidence level at which VAR is measured, the larger the error.

In addition, depending on the context and the confidence level at which VAR is being measured, it may be necessary to include some terms of the two-dimensional polynomial with respect to concurrent changes in the spot price and the implied volatility. This will be needed because the sensitivity of the portfolio to changes in the underlying rate may depend on the magnitude of the concurrent change in implied volatility. There is no simple rule to determine the minimum number of parameters needed to achieve a specified degree of accuracy in the VAR calculation – it is very much a contextual issue that depends on the precise composition of the portfolio and the confidence level at which VAR is to be measured.

Multi-dimensional terms and yield curves
Earlier I made the statement that the number of terms of the Taylor-series expansion required for simulation will depend in part on the relationship between the market factors used for simulation and the market factors on which the market value of the contracts in the portfolio directly depend. To illustrate this point, let us consider a single interest rate cap option, ie a caplet – an option on a forward interest rate. Let us assume that the caplet's payout depends on the forward Libor rate in the interval between three and six months from today. Let us further assume that we are simulating changes in the yield curve in terms of zero-coupon rates at tenors one month, three months, six months, nine months, and so forth. The change in the forward Libor rate for the interval between three and six months forward is nearly a linear function of the changes in three-month Libor and the six-month Libor. However, because the market value of the caplet is a non-linear function of the change in the forward rate, the value of the option cannot be written as a separable function of three-month Libor and six-month Libor. The market value of the caplet depends on the joint changes of three- and six-month Libor. Consequently, a Taylor-series expansion of the caplet would have to include some two-dimensional cross-terms – ie some terms of the two-dimensional polynomial function of the change in three-month Libor ($\Delta Rate_{3\text{-month Libor}}$) and six-month Libor ($\Delta Rate_{6\text{-month Libor}}$).

Now consider a caplet whose payout depends on the forward Libor rate in the interval between 4.5 months and 7.5 months from today. Changes in the forward Libor rate in this interval will be nearly a linear function of changes in three-month, six-month and nine-month Libor. Changes in the market value of the caplet will therefore depend on the joint changes of three-, six- and nine-month Libor. Consequently, the Taylor-series expansion should include a three-dimensional polynomial with respect to the joint changes in three-, six- and nine-month Libor.

The materiality of including the two- and three-dimensional polynomials will depend on the context, including the confidence level. If a model were to accurately calculate VAR at the extreme confidence level required by the BIS (effectively, a 7.36 standard deviation confidence level – see footnote 2), these higher-dimensional polynomials would have to be included in the calculation for interest rate options.

Grids of factor sensitivities
A related form of parametric portfolio revaluation are grids of factor sensitivities. As explained above, a single-factor sensitivity is the change in the portfolio's value for a specified change in a particular market factor. A grid of factor sensitivities is a set of factor sensitivities corresponding to a set of discrete changes in some market factor(s).

The efficiency of doing Monte Carlo simulation with grids of factor sensitivities is that they can be created from a limited number of exact revaluations of each contract. Once the grids have been created, they can be used to transform an unlimited number of simulated scenarios of changes in market factors into the corresponding simulated change in portfolio value. For a simulated change in each market factor, ΔX_j, the corresponding change in portfolio value, ΔPV, can be obtained by interpolation by using the grids of factor sensitivities as look-up tables to ascertain the change in PV corresponding to each change in X_j.

A one-dimensional grid of factor sensitivities is a set of changes in portfolio value corresponding to a discrete set of changes of a single market factor. It can most easily be represented as a graph – eg Figure 3, which depicts two one-dimensional grids of factor sensitivities, one for a linear portfolio, the other for a non-linear portfolio. For each, the change in portfolio value is shown for a set of 12 changes in a single market factor: up six units and down six units. The 12 factor sensitivities are calculated by revaluing each contract in the portfolio 13 times – once at the current market rate and once for each shift in the market factor.

The one-dimensional grid of factor sensitivities to changes in market factor X_j corresponds to the one-dimensional terms of the Taylor-series expansion – ie a one-dimensional grid of factor sensitivities to changes in market factor X_j corresponds to a polynomial function of ΔX_j.

A two-dimensional grid of factor sensitivities is a set of changes in portfolio value corresponding to a set of concurrent changes in two market

4. Two-dimensional grid of factor sensitivities: ΔPV as a surface over a two-dimensional plane

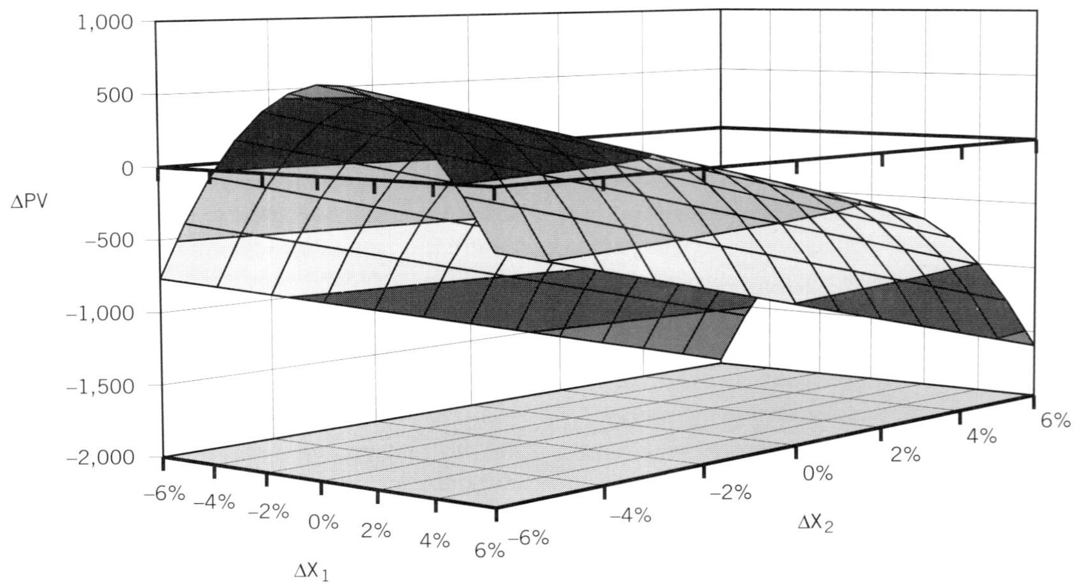

factors (eg X_i and X_j). Figure 4 shows an example for concurrent changes in X_1 (eg changes in a spot exchange rate) and X_2 (eg changes in the implied foreign exchange volatility for a parallel shift in the term structure of implied volatility). The change in portfolio value for a two-dimensional grid is a surface over a two-dimensional plane. The grids of sensitivities along each axis, keeping the other market factor unchanged, are identical to one-dimensional grids.

The two-dimensional grid of factor sensitivities to market factors X_j and X_k corresponds to a combination of the one-dimensional and two-dimensional terms of the Taylor-series expansion terms – ie it corresponds to the polynomial function of ΔX_j, the polynomial function of ΔX_k and the polynomial function of $(\Delta X_j, \Delta X_k)$.

Creating grids of factor sensitivities Grids of factor sensitivities should be created from each back-office product processor for each of the market factors, X_j, that affect the value of the contracts on that product processor. Some market factors (eg US dollar Libor) are ubiquitous in that they are needed for the revaluation of many types of product. Consequently, grids of factor sensitivities to a common market factor may be created by many back-office product processors. It is not necessary to add grids for the same market factor into a single grid. The computer can use each grid as a look-up table and then add up the total changes in portfolio value corresponding to each simulated scenario.

The width of the grids of factor sensitivities effectively sets boundaries on the confidence level at which potential loss can be calculated. This is because a grid of factor sensitivities can only be used as a look-up table if the magnitude of the simulated change in the market factor lies within the range of changes in market factor that were used in constructing of the grid.

Summary

VAR is a useful measure of market risk when properly calculated. The calculation of VAR rests on two components: (1) the simulation of changes in market factors; and (2) the transformation of simulated changes of market factors into the simulated change of portfolio value. Different methods of implementing each component have been described, but most attention was focused on statistical parametric simulation and parametric portfolio revaluation.

The strengths and weaknesses of different methods of measuring VAR were considered and an explanation given of how the suitability of any VAR method must be judged primarily in the context of the types of contract whose risk is being measured and the confidence level at which the risk is being measured.

I would like to acknowledge and thank my colleagues at Citibank, with whom many of the ideas presented here were developed: David E. Lawrence, Ron Hylton, Pratap Sondhi, Gary Lachmund, Greg Chu and Nenad Marinovich (a former colleague now at Salomon Brothers). Any errors are, of course, my responsibility.

1 *The portfolio method of calculating pre-settlement credit exposure and the representation of exposure as a portfolio exposure profile were developed in 1991 by David Lawrence and myself with critical technical assistance from Byron Nicas. See: D. Lawrence and E. Picoult, 1991, "A New Method For Calculating Pre-Settlement Risk",* The Tactician *4, no. 3 (internal Citibank publication); D. Lawrence, 1995, in* Derivative Credit Risk, Advances in Measurement and Management, *Risk Publications; E. Picoult, 1996, "Measuring Pre-Settlement Credit Risk on a Portfolio Basis", in* Risk Measurement and Systemic Risk, Proceedings of a Joint Central Bank Research Conference, November 1995, *Board of Governors of the Federal Reserve System.*

2 *If one assumes that market rates are lognormally distributed and one also assumes that daily changes in market rates are independent of one another, then the 10-day 99% shock in market rates is equivalent to a 7.36 standard deviation measure of VAR.*

3 *For a description of singular value decomposition (SVD) and Cholesky decomposition see, for example, Press, W., S. Teukolsky, W. Vettering and B. Flannery, 1992,* Numerical Recipes in C, The Art of Scientific Computing, Second Edition *(Cambridge: Cambridge University Press).*

4 *Mark B. Garman, 1996, "Improving on VAR",* Risk *9 (5), pp. 61–63. Garman describes a technique for analysing the incremental change in the VAR (calculated by matrix multiplication) that would result from adding a transaction to a portfolio. The general principle, applicable to any portfolio, is that the incremental change in total VAR resulting from adding a new transaction to the portfolio could be positive, nearly zero or negative, depending on context. It depends on what is already in the portfolio and it depends on the assumed volatilities and correlations of changes in market factors.*

5 *Ron Hylton, of Citibank, developed this method for measuring VAR in 1994. While Taylor-series expansion is quite familiar, Hylton is the first person I know of to use grids of factor sensitivities to measure VAR.*

21

Beyond VAR and Stress Testing

Julian Shaw
NatWest Markets

Value-at-risk (VAR) is the z in the following proposition: Given a holding period of x days and a confidence level of y%, find the VAR z which satisfies the following condition: With our current portfolio, over the next x trading days with probability at least y% we will lose no more than \$z, assuming no change in portfolio composition during that period.

Symptomatic of the confusion reigning in the risk management world is the fact that x ranges from 1 to 250 and y ranges from 95 to 98 and beyond in different firms.

It seems obvious (to me at least) that risk management departments should focus on the risk of *extreme* events. By definition, a 95% confidence level will be exceeded at least once a month on average – it is forecasting what will happen *on those days* that is the point of risk management. If there is to be a single confidence level, it should be set much higher, certainly to the equivalent of no more than one trading day per year, ie 99.6%. In any case we really want to approximate the *entire* P&L distribution, not just one confidence level.

Conversely, I favour holding periods which approximate the time it would take to neutralise the position, albeit under adverse conditions. This depends on the specific market and the portfolio, but one day seems reasonable for many markets, 10 trading days seems sufficient for all but the most illiquid markets and one year seems absurdly long. We really want a methodology that allows us to *specify* any holding period so that we can adjust it to suit the portfolio we are evaluating.

What is wrong with VAR?

The essential problem with conventional VAR is neither the confidence level nor the holding period, both of which can be adjusted, but rather the method by which it is typically computed. This relies on the following proposition:

Assume that the changes in the risk factors are multivariate normal, where σ_i is the standard deviation of F_i and ρ_{ij} is the correlation coefficient for changes in F_i versus changes in F_j (Assumption 1)

Assume that portfolio value is driven by K factors F_i, and that we can approximate the change in portfolio value, $\Delta PFVal$, as the sum of the sensitivity to each factor times the change in the factor, ie:

$$\Delta PFVal = \sum(\delta_{F_i} \times \Delta F_i) \qquad \text{(Assumption 2)}$$

Then $\Delta PFVal$ is normally distributed too with standard deviation σ given (in matrix form) by

$$\sigma^2 = [\delta_{F_i}] \times [COV_{ij}] \times [\delta_{F_i}]^1$$

where $COV_{ij} = \rho_{ij} \times \sigma_i \times \sigma_j$.

Thus, assuming a mean return of approximately zero, y-percentile VAR can be estimated as $NORMSINV(y) \times \sigma$.[1] For example, 95-percentile VAR is $1.64 \times \sigma$ and 98-percentile VAR is $2.05 \times \sigma$. Indeed any confidence level VAR can be estimated from σ.

The only justification for computing VAR with this algorithm is that it is *easy*. It does *not* provide a good estimate of what we really need, which is realistic estimates of the probabilities of extreme losses. The proposition is correct, but the assumptions are wrong when applied to typical financial markets.

Assumption 1 is wrong

It is well known that changes in financial time series have much fatter tails than would be the case if they were normally distributed. As has been pointed out many times, assuming a normal distribution for equity returns, the crash of October 1987 was a greater than 20σ event and hence would only be expected once every many millions of years!

There is a lot of faith in the risk management world in the applicability of the central limit theorem. This says roughly that the sum of a number of distributions tends toward normality, even if the individual distributions are not normal.[2] Hence, it is hoped that portfolio returns will tend to normality even though the individual components are not normal, and that returns over longer periods will be normal even though daily returns are not normal.

But neither is necessarily true. We examined the returns of a test portfolio from Cox (1995). This portfolio is long 5% Nikkei 225 index, short 10% S&P 500, long 15% DAX, long 20% FTSE-100 and short 30% CAC40. It has net zero exposure, the absolute values of the exposures sum to 80% and it is continually readjusted to maintain them. Whereas Cox considered three years of bi-weekly data, we consider 10 years of daily data for the same portfolio. The returns have much fatter tails than those of a normal distribution with the same mean and standard deviation, as shown in Figure 1.

Nor is it true that returns over longer periods are necessarily closer to normality than returns over shorter periods. As Ray (1993) points out, three-month changes in US three-month bills have been even less normal than one-day changes in the same instruments.

1. Long and short equity portfolio: all data

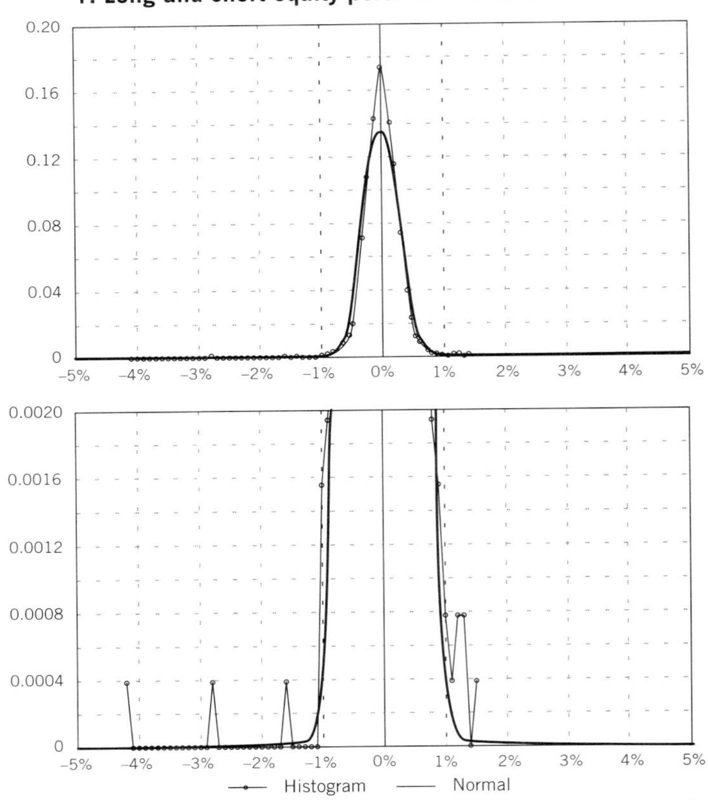

Two views of the same chart. Notice how the assumption of normality would lead us to grossly underestimate the probability of extreme returns.

One reason for the disappointing lack of normality is that portfolio components do not have fixed correlations and there is serial correlation between returns in different periods. Because the normal distribution is familiar, this leads to the hypothesis that returns are normal but with *changing* standard deviation, so returns are only *conditionally* normal. Various volatility estimation techniques are used, ranging from simple sample standard deviation models to exponential weighting (such as RiskMetrics) or more sophisticated Garch techniques[3] to estimate the variance of this hypothesised continually changing normal distribution. Volatility changes *do* account for some of the fatness in the tails of return distributions but not enough – especially at the extremes, as the following argument shows.[4]

We seem to see a 20%+ fall in the S&P500 about once every 60 years, or every 15,000 trading days. This implies that *if* the annualised volatility were constant, it would have to be approximately $\sqrt{250} \times -20\%/(\text{NORMSINV}(1/(60 \times 250)))$, ie 83%. If volatility is not constant, this implies that volatility would have to be even higher than 83% on some days – but predicted volatilities never get that high (typically they are around 20%) except ... the day *after* a crash!

As Jackwerth and Rubinstein (1996) observe, options traders seem to have learned a lesson from the crash of '87. Although traders have always continually revised their at-the-money option implied volatilities, since the crash they have also charged a hefty implied volatility premium for out-of-the-money options, giving rise to the volatility "smile". This means that the risk-neutral probabilities of 3 and 4 standard deviation declines are, respectively, 10 and 100 times more likely than under the assumption of lognormality. So options traders do not believe that (log) returns are conditionally normal, let alone unconditionally normal. Some day risk analysts may learn the same lesson.

Actually, most people in risk management *do* know that returns have fat tails, but they do not know what to do about it. This has led to some feeble suggestions for coping with the problem. The regulators' knee-jerk reaction is to load VAR estimates by some *ad hoc* multiple, but this is just fudging.

One authority on risk management justifies the use of 98-percentile VAR with the following argument: if there are two unimodal distributions and one has fatter tails, then it must also have a thinner peak and the density functions must cross somewhere. Empirically, for financial time series that place is located around the 98th percentile, so the 98-percentile VAR should be computed and reported assuming normality[5] *(even though this misleads naive management into underestimating the probability of the more extreme events they are really interested in!)*

At the conclusion of a technical paper which assumes normality, another author approvingly quotes the Federal Reserve Board as follows:

"Assuming a normal distribution, the probability of experiencing a four standard deviation event is approximately ... once in 130 years. In practice, however, such unusual market movements are seen in most major markets on average almost every year."

She concludes: "Clearly, there is much more to risk management than calculating VAR" (Bagg, 1996), but she does not explain what that "more" is. If VAR (computed as specified above) cannot give reliable estimates of the consequences of financial events that occur as often as *once a year*, then what *is* it good for?

Assumption 2 is wrong

Assumption 2 says that the effects of changes in the risk factors are *linear* in each variable and *orthogonal*, ie the total effect of a number of changes is just the sum of the individual effects.

For portfolios that incorporate derivatives (which means the global portfolios of every major investment bank) this is wrong. The value of such portfolios is not linear in the market risk factors (that is the point of derivatives!) and may not even be monotonic in them. The maximum loss could occur at any point, not necessarily at some combination of extreme changes in the risk factors.

A more subtle point is that the effects of changes are not orthogonal – the effect of changing the value of the underlying security and changing the implied volatility is not simply the sum of effects of the two individual changes. Note that the commonly used "greek" approximation

$$
\begin{aligned}
\Delta \text{ Portfolio value} \;\sim\; & \delta \times \Delta S \\
+ \; & \gamma \times (\Delta S)^2 \\
+ \; & \theta \times \Delta T \\
+ \; & \kappa \times \Delta\sigma \\
+ \; & \rho \times \Delta r
\end{aligned}
$$

which expresses changes in derivative portfolio value in terms of delta, gamma, theta, kappa and rho is non-linear in changes in the value of the underlying only (because of the gamma term) and it assumes orthogonality. (It is a poor global approximation anyway for many derivative portfolios.)

Some claim that in practice the linear approximation to derivatives risk is good enough. But this is not true when the effects of large moves are considered. Figure 2 shows the P&L of a simple derivatives portfolio as a function of the price of the underlying security. Clearly no linear or, for that matter, quadratic function will be a good global approximation to a P&L like this.

Is this example unreasonable? I do not believe it is. Real derivative portfolios are not random collections of derivative positions – they are *managed* by traders who keep their portfolios locally hedged. Delta-neutral (and delta-gamma-neutral) portfolios are the rule, not the exception. However, contrary to the implicit assumption of those who feed delta equivalents into their VAR computations, these portfolios are not riskless!

Also, it is perfectly natural to compare the risk of a derivative business with the risks of vanilla businesses – indeed, the point of VAR as compared to the old *ad hoc* greek and other market-specific risk measures is that it can express the risks of diverse businesses *commensurately*. If risk management cannot realistically estimate the risks of derivative businesses, then it is not doing its job.

Correlation is bunk!

A more subtle problem with the assumption of multivariate normality concerns the use of correlation as the measure of association between risk factor changes.

Implicit in the use of correlation is the assumption that the relationship between a pair of risk factors F_i and F_j can be summed up by one

2. A simple derivatives portfolio and its "greek" approximations

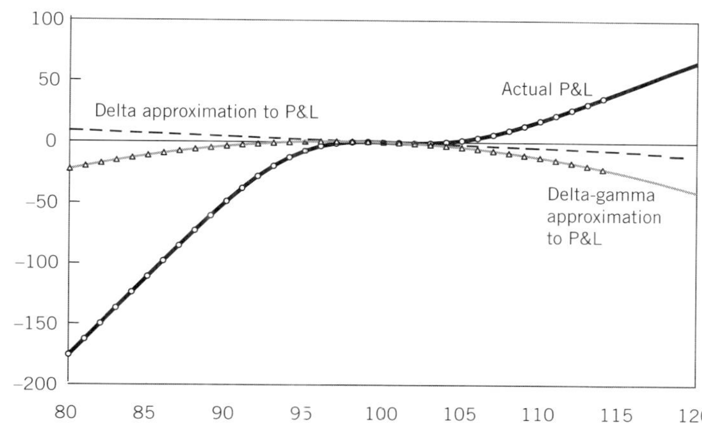

number, the (linear) correlation coefficient ρ_{ij}. An approximate characterisation of ρ_{ij} is that if the $\langle F_i, F_j \rangle$ observations are plotted on a scatter diagram and an ellipse is drawn around them, then ρ_{ij} is approximately

$$\pm[1.0 - (\text{minimum width of the ellipse}) / (\text{maximum width of the ellipse})]$$

where the sign is plus if the ellipse leans to the right and minus if it leans to the left. Thus the skinnier the ellipse, the higher the absolute value of the correlation.

A precise characterisation of ρ_{ij} is that it is the square root of the fraction of the variation in F_j that is explained by the *straight line* of best fit drawn through the graph of F_j, versus F_i where "best" means minimising the sum of the squared error terms.

3.

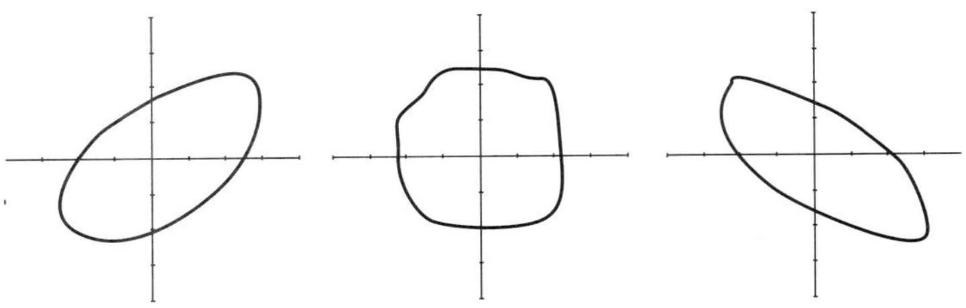

a) Positive correlation   b) Low correlation   c) Negative correlation

4.

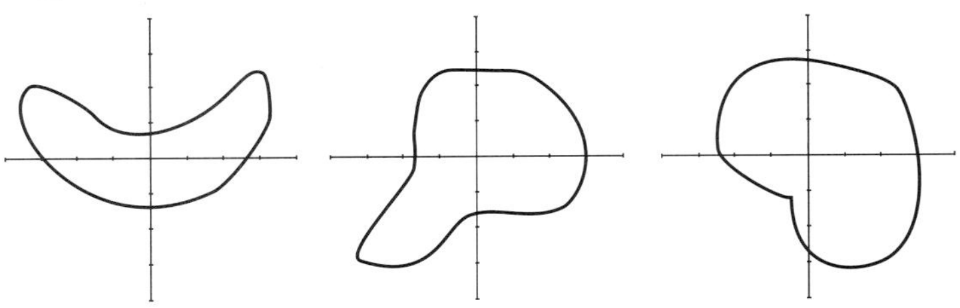

a) Implied volatility vs change in the underlying   b) Factors with low correlation in normal times that crash together   c) Factors with low correlation in normal times that do not crash together

5.

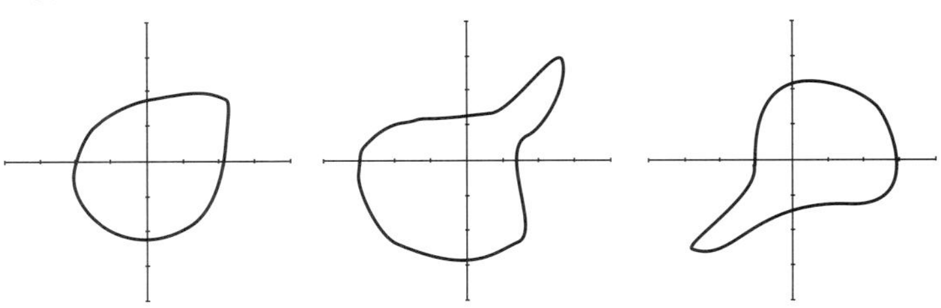

a) Stocks vs treasuries   b) Treasuries vs junk bonds   c) Stocks vs junk bonds

Our fundamental objection to correlation of *any type* is that it assumes that a relationship between two factors can be summed up by one number. A (linear) correlation coefficient will only capture relationships of the types indicated in Figure 3.

A little thought about financial factors reveals that this does not include all the relationships of interest in financial simulation. For example, we know that there is a relationship between changes in implied volatility and changes in the corresponding underlying security. In general, a big move up or down in the underlying security tends to be associated with a rise in implied volatility. This non-linear relationship might be represented as in Figure 4a.

For another example, consider equity indexes around the globe. Under normal market conditions the correlation is only moderate, but given a very large move the correlation is extreme. While the crash of the S&P500 alone in October 1987 was improbable enough under the hypothesis of normally distributed changes, the simultaneous crashes of all the major equity markets was even more astronomically improbable, given the hypothesis of multivariate normality and the typical correlations between equity markets.

The true relationship is better represented by Figure 4b. Again, there is a relationship, but it is non-linear. Ordinary linear correlation estimated over a typical time period will overestimate the actual effect of small moves and greatly underestimate the effect for big moves.

Finally, there might be financial variables which have low conventional correlation but rarely crash together. Figure 4c illustrates this situation.

For yet another example, consider Bookstaber's example of stocks versus treasuries versus junk bonds (Chew, 1994). In normal times junk bond prices are primarily driven by rates and, hence, correlate with treasuries. But when equities crash, junk debt is driven by equities. These relationships are indicated in Figure 5.

Implementation problems of VAR

The fundamental problems with conventional VAR computation discussed above are compounded by the shortcuts that are required to make conventional VAR "practical".

Typically, the variance/covariance methodology uses historical data (possibly with additional data such as implied volatilities) to estimate an underlying multivariate normal distribution. This is not easy, since there are thousands of risk factors to consider in computing firmwide VAR but typically only about 2,500 days or 10 years of data. So even if the world were multivariate normal, the error in estimating all the correlations would be huge – especially if they are time-varying.

Variance/covariance VAR modellers have two ways of coping with the multiplicity of risk factors. The first is to reduce the number of factors by using factor models. For example, changes in the yields of all bonds are modelled as combinations of shifts in a few key rates. For another example, implied volatilities of all maturities are assumed to move by some deterministic function of changes in three-month implied volatility (for example, the square root of time).

These cures can be worse than the disease because many factor models fail to pick up the risks of *typical* trading strategies, which can be the greatest risks run by an investment bank. According to naive yield factor models, huge spread positions between on-the-run bonds and off-the-run bonds are riskless! According to naive volatility factor models, hedging one-year (or longer dated) implied volatility with three-month implied volatility is riskless, provided it is done in the "right" proportions – ie the proportions built into the factor model! It is the *rule*, not the exception, for traders to put on spread trades which defeat factor models *since they use factor-type models to identify richness and cheapness!*

The second method used by variance/covariance VAR modellers to cope with the multiplicity of factors is explicitly or implicitly to set huge numbers of the correlations between factors to zero – for example, by estimating the VAR of a firm's total portfolio as the square root of the sum of the squares of sub-portfolio VARs. The standard deviation of the firm's total P&L is therefore estimated to be the square root of the sum of the squares of the firm's fixed-income risk, equity risk, FX risk and so on.

On average, these correlations are close to zero, but in extremes risk factors tend to move together. If there is a huge move in equities, there is likely to be a large move in interest rates too. They might go down (as in 1987) or they might go up (as in the mini-crash of 1989), but it is very likely there will be some kind of extreme move in rates. This fact is contradicted by VAR implementations that explicitly or implicitly set such correlations to zero.

What is wrong with stress testing?

Stress testing is often suggested as a "complementary" methodology to value-at-risk, which fails to provide reasonable estimates of the probability of extreme losses because of the assumptions incorporated in its method of computation. In practice, stress testing is used as a band-aid for poor VAR. Despite the methodologies of VAR proponents, VARs *are* the right numbers to ask for because, if correctly computed, they take into account the joint probability distribution of changes in risk factors and the diversification of exposures to these factors in portfolios.

There are various ways of performing stress testing, but in general hypothetical extreme scenarios are created and corresponding hypothetical P&Ls are computed. The two obvious questions are:

- Where do the scenarios come from? and
- What should we do with the results?

One method is to dream up global scenarios. But dreaming up multiple *global* scenarios which include all the changes in the thousands of risk factors to which a typical investment bank is exposed is hard, since there are so many changes to consider. If a hypothesis is that the Deutschmark sharply appreciates against the dollar, what is the hypothesis about, say, the Austrian Schilling and the Dutch guilder, or the yen or the Dax, or interest rates, and so on and so on?

Another method is to generate many *local* scenarios and so consider a few risk factors at a time. For example, given an FX option portfolio, we might compute the hypothetical P&L for each currency pair under a variety of exchange rate and implied volatility scenarios. The problem then is how to amalgamate the results. One way is simply to add the worst-case results for each of the sub-portfolios for the portfolio as a whole. But this obviously ignores the portfolio effect and fails to reward cross-hedging.

Does this matter? *Yes it does!* Both these factors can be critical to risk-adjusted return and traders' whole approach to trading. Cross-hedging, spreading and diversification are the *modus operandi* of traders. Cross-hedging Austrian Schilling or Dutch guilder risk with positions in Deutschmarks is certainly not riskless, but it is certainly not as risky as the sum of individual naked positions, which is a vast and useless overestimate. Risk management methodologies which cannot provide realistic estimates of the risks of such commonplace strategies are not adding value.

Artificial scenarios in general have the problem that, since there is no probability measure associated with them, they are not very useful in controlling risk as opposed to just limiting the loss under those scenarios we happen to think up. This is the problem with Taleb's alternative to conventional VAR (Taleb, 1997). Taleb says that risk managers should prepare "a list of associated crisis scenarios without unduly attaching probabilities to the array of events, until such time as we can show a better grasp of probability of large deviations ...". He does not say when that time will come, nor does he explain what one should do with these scenarios. In any case their usefulness is contradicted by his dictum that "nothing predictable can be truly harmful".

An alternative approach is to use global scenarios from history to create a time series of hypothetical P&Ls. This would let us calculate the amounts that we would have made or lost daily over, say, the last 10 years if we had held our current portfolio. This immediately provides some useful information, such as the worst losses we would have experienced. As we shall see, with some analysis we can get the VAR estimates that we really needed all along.

Obviously a *purely* historical approach cannot encompass risks that we know exist but which have not yet manifested themselves and are hard to quantify – the prospect that an exchange rate which is now pegged will be allowed to float, for example. What we want is a method for incorporating this extra information into a realistic VAR system to get *conditional* VAR.

Large-scale market risk simulation

We have attacked conventional VAR and stress testing, but what do we propose to put in their place? The simple but hardly straightforward answer is *simulation*. In theory, we can simply take a large set of scenarios which we think characterises the joint distribution of risk factor changes, compute the P&L under each scenario and then sort the P&Ls to constitute our estimate of the entire P&L distribution. We can then simply read off or interpolate any confidence level VAR we wish.

If the changes in risk factors happen to be multivariate normal and our exposures to them are linear, we would get the same answer for portfolio VAR as we do when we use the matrix multiplication short cut, apart from simulation error. However, the exposures to the risk factors no longer have to be linear and orthogonal. We now have the freedom to chose any multivariate distribution we like.

Another way to compute VAR is to create a time series of the simulated P&L of today's portfolio under the historical changes and then apply econometric analysis to this time series to estimate tomorrow's P&L distribution. Again, we would expect that if the joint distribution is normal and our exposures are linear, our VAR estimates should approximate VAR computed by the standard short cut – but we are not forced to assume either multivariate normality or linear exposures to the risk factors.

These freedoms give rise to two big problems:

PROBLEM 1

Computational and software feasibility In many cases revaluing a portfolio thousands of times is not computationally feasible, either because it contains too many instruments or, as with path-dependent options, because the valuation algorithms are too slow. Also, building software capable of revaluing every instrument in the total portfolio of an investment bank is an enormous task. Fortunately, as we will see, it is also an unnecessary one.

PROBLEM 2

Methodology If we are going to use an alternative joint distribution of risk factor changes, where are we going to get it from? Alternatively, if we are going to use raw history and then analyse the simulated P&L time series, what precisely are we going to do?

Computational and software complexity of market risk management

Before we start proposing many thousands of revaluations to compute VAR by simulation, let us consider the work required to produce the one revaluation that every investment bank does daily.

Typically there are hundreds, if not thousands, of portfolios of thousands of securities organised by currency, product type, bank division and trader. They are managed by a plethora of computer systems, ranging from large commercial or in-house systems to spreadsheets.

Each of these systems contains static data defining the securities, plus descriptions of the portfolios of the positions in these instruments held by the bank. At the end of the day (or more frequently) each system takes a subset of the world's market data and uses it to mark its books to market. The profit or loss of each book, trader, division and the entire firm is then computed.

The simple-minded approach to simulation is to pull all the transactions into a central database and replicate all the various mark-to-market systems in the risk management system. The cost of building such a system, reconciling it with the various front office systems and continually updating it to cope with new products would be monstrous.

It is true that we do not require great accuracy in the VAR number itself, but this is no great saving since small errors in the values of underlying instruments can lead to large errors in the P&L and, therefore, large errors in VAR estimates. And any discrepancy in valuation between the front office system and the risk management system will lead to loss of confidence in risk estimates.[6]

Typical commercial VAR systems which use simulation either only attempt to value restricted classes of instruments or try to do the whole lot and fail. Note that these "implementation" problems are usually left out of risk management texts, but *this is where global risk management projects get stuck* – they take so long replicating existing systems that they never get around to risk management for more than vanilla products. Even if the software feasibility problem is solved, the systems run into computational feasibility problems if more than conventional VAR computation is attempted. Their software is too slow to revalue portfolios the many times that simulation-based VAR requires.

Another dream is that all businesses within a firm will use the *same* front office system and that the risk management system can be built on top of this. But no single system as yet in existence adequately handles all markets; each business quite rightly wants the system that best serves its own trading needs.

Constructing the global P&L approximation function

Our solution to the computational and software feasibility problems is to construct a global P&L approximation function (GPLAF) from the outputs of the firm's existing front office systems. The two requirements of a GPLAF are that it is fast to compute (so that we can recompute it many thousands of times in our P&L simulations) and that it provides a good approximation *globally*, ie it gives a reasonable approximation to the firm's P&L (even ±20% would be acceptable) over a wide range of risk factor changes. Note that this behaviour is the opposite of typical "greek" approximations, which are unnecessarily good for small changes in risk factors but can be poor approximations for large changes.

Most front office systems already implicitly specify a P&L approximation function in the form of sensitivities (deltas, PVBPs, key rate durations, option greeks and so on), from which a P&L approximation function can be strung together. So why not use this? This is appropriate for some well-behaved portfolios, even though it was only designed to apply for small changes of risk factors.

But generally the classic greek approximation for derivative portfolio value is not good enough to use as a global approximation because of the inherent properties of derivative portfolios and because of the way traders use the greeks. As we remarked earlier, using classic dynamic hedging techniques for hedging derivative portfolios, option traders typically keep their greeks as flat as possible – which means they are well hedged for small moves but not necessarily for large ones.

The first step in constructing the GPLAF is to identify the factors that drive each of the subportfolio's P&Ls. This is a much harder task than might be imagined, and where risk managers who actually understand the strategies that the traders are employing are invaluable.

The next step is to construct P&L approximations for each of the subportfolios. The GPLAF will simply be the sum of the approximation functions for the P&Ls of the individual portfolios. Again, this is a task best performed in conjunction with individual risk managers who know what the front-end systems under their supervision can deliver.

Given that any decent front-end system can generate "what-if" P&L scenarios, we can construct a GPLAF which interpolates and extrapolates from these scenarios. This is a generalisation of the implicit P&L approximation used in conventional VAR, where the effects of individual risk factors are extrapolated from the effects of small moves, eg 1bp.

6. P&L of portfolio of exotic options

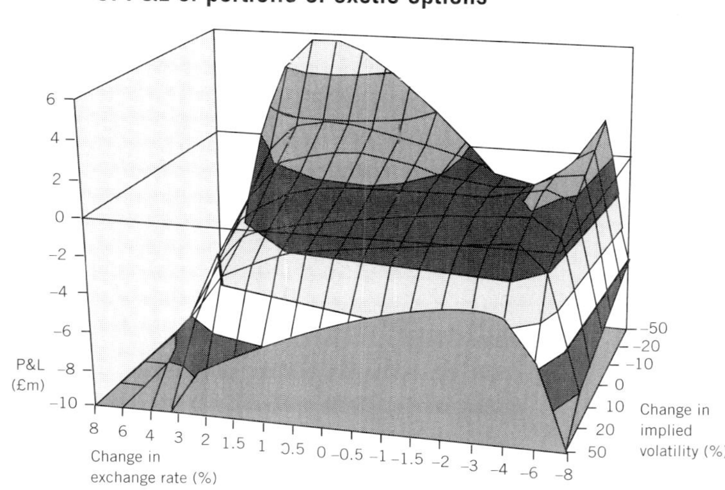

The P&L of a real non-linear and non-monotonic portfolio (of exotic FX options) which cannot be globally approximated by greeks. We will approximate this P&L function the same way the chart is created – by interpolation.

7. Two-dimensional linear interpolation and extrapolation

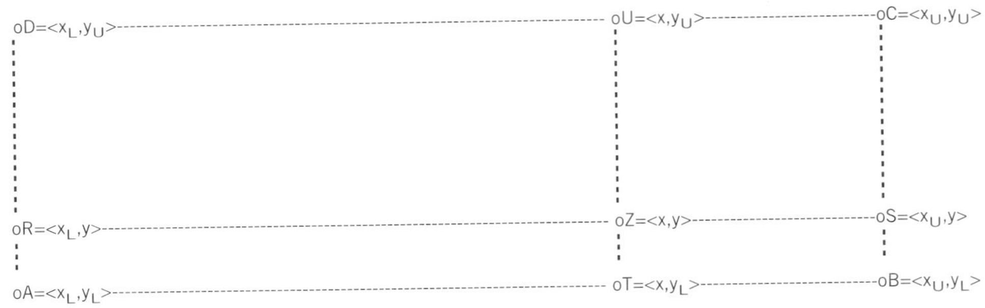

Let Z be the point at which we want to interpolate the value of the function F and let A, B, C and D be the four gridpoints at which we have computed F. Let $T = (x_U-x)/(x_U-x_L)$ and let $u = (y_U-y)/y_U-y_L)$. Then the bilinear interpolation approximation is

$$F(Z) \sim F(A) \times tu + F(B) \times (1-t)u + F(C) \times (1-t)(1-u) + F(D) \times t(1-u)$$

Our generalisation is to approximate total P&L as a sum of interpolations from vectors of scenarios (where there is little interaction between risk factors) or matrices (where there is interaction between two factors, such as changes in volatility and the value of the underlying security). Computing these vectors and matrices is overhead (performed by the individual front office systems) and the simulation itself just interpolates and sums these values. Interpolation is an extremely fast process and huge numbers of *approximate* simulations are feasible.

What about interactions between three or more factors? We could extend our method to allow interpolation and extrapolation from three- or higher dimensional arrays, but the number of grid points in an n-dimensional space with k points per dimension is k^n, so we want to avoid this since the overhead threatens to become infeasible. In the case of higher-dimensional complex interactions between risk factors or uncertainty over what the important risk factors really are, a more automated approach can be taken.

Elucidating and approximating the effects of large numbers of variables on some output variable from examples is a classic problem of artificial intelligence. A front-end system can be required to revalue portfolios under a number of scenarios of changes in the valuation inputs. These scenarios can be chosen using Latin hypercube techniques or their generalisations (see below). Then artificial intelligence techniques can be used to create a quick approximation function that has "learned" the relationships between the inputs and the output – the P&L. For example, abductive inference techniques generate networks of polynomials, which approximate the output as a function of the input (Drake and Yim, 1997).

Of course, no matter what approximation scheme we use, it is always possible to come up with a pathological P&L function that will defeat it. Suppose there are only, say, 60 risk factors affecting a P&L and there is one combination of up moves and down moves that will bankrupt the firm. The number of combinations of up and down moves that would have to be checked to be certain of finding the disaster combination is 2^{60}, that is, $\sim 10^{36}$, which is clearly infeasible. Even in the Black–Scholes formula there are interactions between all the input variables.

Fortunately, however, serious interactions between risk factors tend to be limited (in particular, to interactions between implied volatility and the price of the underlying security) because more complex interactions tend to make the securities unprofitable to hedge.

The details of how to implement this scheme may vary between sub-portfolios, but the goal is always the same – a fast-to-compute function which is a good global approximation to the P&L. These functions can be programmed on spreadsheets and provide a real-time approximation to P&L with live data feeds, which can then be verified against actual P&L, especially on days with moves of a reasonable size.

This scheme has several advantages. It is software efficient – we do not try to replicate existing software. It is also computationally efficient – we do not end up demanding impossible numbers of complete revaluations. And, crucially for any global risk management project, it is also *organisationally* efficient – we make maximal use of our talent in the frontlines of risk management, we can incrementally reconcile each piece of our global approximation function with results from the traders' own software, and we can incrementally improve our approximation as computing power and algorithms in the various front-office systems improve.

Once we have a function that gives a good approximation of the firm's P&L with any combination of risk factor changes, how are we going to use it? As we saw earlier, the classic method is to create a huge sample which approximately characterises the joint distribution of tomorrow's risk factor changes, apply the GPLAF to this sample and sort the results. An alternative approach is to simulate the P&L that we would have observed over time and perform analysis on this simulated P&L time series to predict the distribution of tomorrow's P&L.[7] Let us consider each strategy in turn.

Estimating and simulating the joint distribution of risk factor changes

How does one estimate the joint distribution of hundreds, if not thousands, of risk factors? Our short cut is to assume that the correlation "structure" remains constant. Estimating time-varying correlations is an attempt (fraught with estimation risk) to compensate for the inaccuracy of the initial linear correlation hypothesis, a hypothesis we have discarded. We will avoid explicitly estimating the correlation structure and instead use a sampling technique which ensures that our sample has a structure similar to that of the historical data, but we will allow the individual risk factor distributions to vary over time.

ESTIMATING INDIVIDUAL RISK DISTRIBUTIONS

Using historical data, we want to parameterise the individual distributions in the sense that we want to fit simple, smooth probability distributions that are more realistic than the classic normal or lognormal distributions. Details of several families of such distributions have been published. In particular, Ramberg *et al.* (1979) have created a parametric fitting family that is particularly simple and fast to sample from. Given any specified mean, μ, variance, σ^2, skew and kurtosis (within bounds), they show how to construct a distribution from their family that has almost exactly the same moments.

Of course, there are alternatives to RTDM. Rachev *et al.* in a number of papers report excellent results fitting Weibull and double-Weibull (two Weibull distributions pasted together in the middle) distributions to many financial time series including very fat tailed unconditional distributions, such as the dollar/Bulgarian Lev exchange rate. (Chobanov *et al.*, 1996).

An obvious strategy is to use one of these distributions fitted to our historical data, but in doing so we would throw away any temporal information in the data. Although we have criticised techniques such as Garch because they assume conditional normality, they do seem to have some power. On average, when they predict a high volatility the returns are, in fact, more dispersed.

We would like to make use of such predictions without swallowing the normality assumption. We can use Garch or (even better) implied volatility if it is available or any other volatility predictor in a non-parametric way by plotting actual historical moves versus the volatility predicted for that day. We should see the historic distribution widening out as the predicted volatility increases. In that case a sensible strategy is to fit a function not to the entire history of historical changes but instead to those changes which occurred when predicted volatility was near its current predicted level.

As an example of our method, in Figures 8–11 overleaf we consider the dollar/sterling exchange rate, to which we have fitted double-Weibull distributions conditional on either implied volatility or Garch-predicted volatility.

So far we have modelled each of the individual risk factor distributions. How can we use them for multivariate simulation?

An efficient simulation method which preserves correlation structure

It is well known that the variance of an estimate of an average provided by Monte Carlo sampling decreases in proportion to the inverse of the square root of the number of trials. To improve the accuracy of the estimate by a factor of 10, the number of trials must be increased by a factor of 100. A simple technique that greatly speeds up convergence is stratified sampling. In this technique, if N simulations are required the cumulative probability range interval $[0,1]$ is sliced into M equal-size buckets and N/M simulations are performed in each bucket. This avoids the clustering which occurs with naive sampling.

In a naive generalisation of stratified sampling to multivariate simulation, if we wanted each of the K variables to be individually stratified M ways we would end up with M^K buckets, which is infeasible since we have a lot of variables. A feasible alternative is Latin hypercube sampling, in which samples are drawn in such a manner that each variable is *individually* stratified, as

8. $/£ January 1988–April 1996 daily log returns

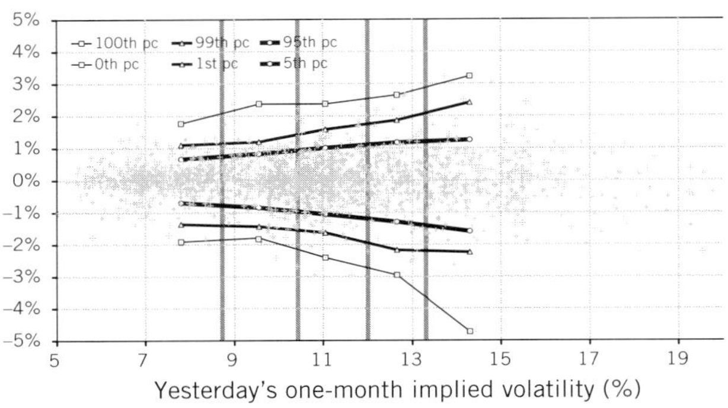

In this chart each + represents the actual move in the $/£ exchange rate versus yesterday's one-month implied volatility. We divide yesterday's implied volatility into five quintiles. The boundaries of these quintiles are indicated by the shaded vertical lines in the chart. For each quintile we compute the 0th, 1st, 5th, 95th, 99th and 100th percentile actual moves for all the points for which yesteday's implied volatility belonged to the quintile. We have plotted these percentiles at points in the middle of each quintile, ie, at the 10th, 30th, 50th, 70th and 90th percentiles of yesterday's implied volatility. We then join the corresponding percentile moves with straight lines.

As one would hope, the percentiles fan out as yesterday's implied volatility increases and we can use this chart to estimate the distribution of tomorrow's move *given the current level of the indicator* without assuming conditional normality and without explicitly specifying a transformation between implied volatility and one-day realised volatility.

10. $/£ January 1988–April 1996 daily log returns

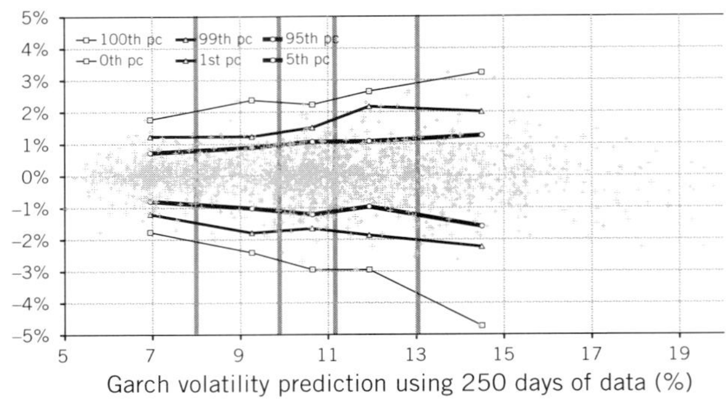

This chart is similar to Figure 8 except that now Garch is used as the volatility indicator instead of implied volatility. In general, if we have a choice as we do for $/£ we would use implied volatility as our indicator in preference to Garch since it has better predictive power (Jorion, 1995). However, implied volatility is unavailable for most financial time series so we will illustrate our approach to modelling a distribution conditional on the level of a volatility indicator with Garch as the indicator.

9. $/£ January 1988–April 1996 daily log returns: all data

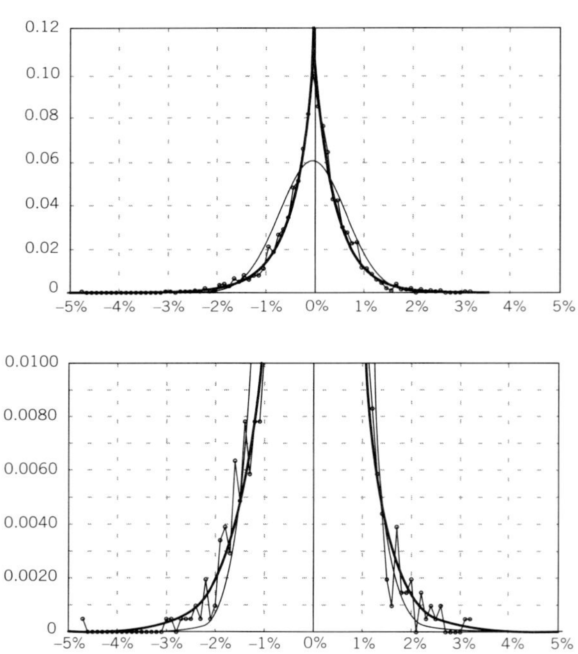

Two views of the same chart in which a double-Weibull distribution is fitted with maximum likelihood to the unconditional changes in the $/£ exchange rate. The normal distribution with the same mean and standard deviation as the data is also plotted for comparison. Note that the double-Weibull distribution with its taller, thinner peak and fatter tails fits the data much better.

11. $/£ January 1988–April 1996 daily log returns

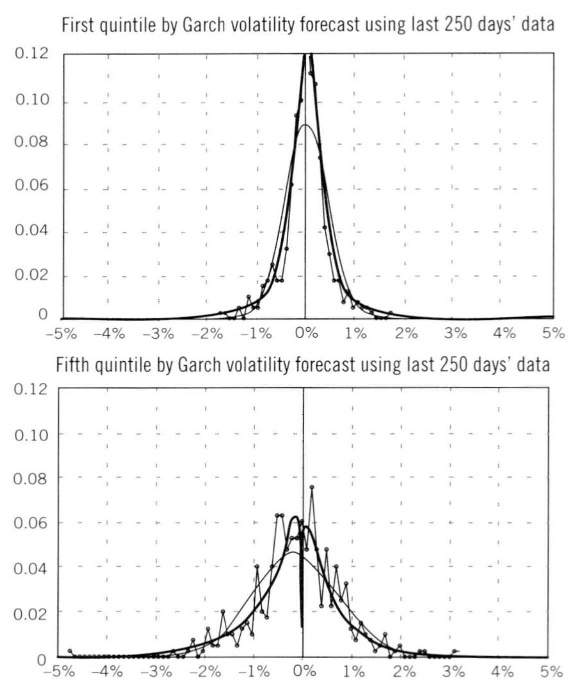

The top chart shows our fit to $/£ *conditional on* the Garch volatility estimate being in the *first* quintile of its range. The lower chart is our fit to the same series conditional on the Garch volatility esimate being in the *fifth* quintile of its range.

illustrated in Figure 12. Obviously, when K is large the sampling is very sparse even when a large number of samples is taken, simply because there is so much ground to cover.

12. A Latin hypercube sample
with N = 6 and K = 2 for X distributed uniformly on the unit square

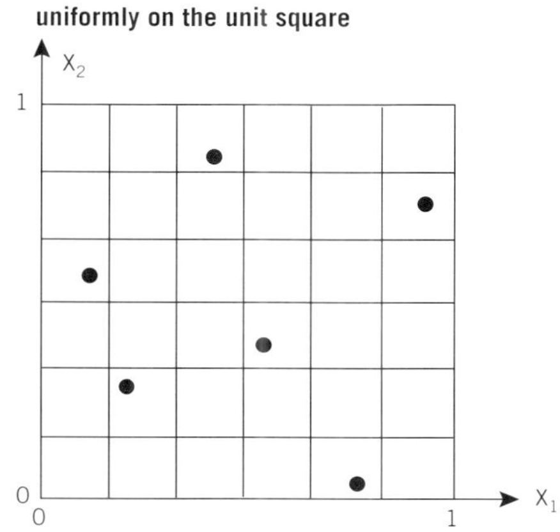

Fortunately, as Stein (1987) shows, this not a problem if the function to be evaluated is approximately *additive* (linear in each variable and orthogonal in the individual variables), in which case a Latin hypercube scheme will work well. In particular, if the function is exactly additive, then the variance of the average of that function decreases in proportion to the inverse of N itself, a huge improvement over the unstratified technique.

Typical option portfolios are not additive, so we cannot expect to achieve the full speedup, but we can concentrate our sampling effort on the worst-behaved ones.

Latin hypercubes can be generalised to orthogonal arrays, in which the samples are chosen so that we are stratified in each pair or triple of dimensions, not just each dimension individually (Owen, 1994, Tang, 1993). Orthogonal arrays in turn can be generalised to (t,m,s) nets (Owen, 1994).

Remember that we want our simulation not only to be efficient but also to preserve the "correlation structure" of our empirical multivariate distribution. So we need to do it without destroying our stratification.

To justify our choice of technique let us first consider some methods for multivariate simulation that will not solve our problem. Every simulation text shows how to sample from a correlated multivariate normal distribution; indeed, it is usually the only multivariate technique given. The trick is to apply the Cholesky decomposition of the correlation matrix to independent normal samples to get a correlated multivariate sample, a technique developed by Scheuer and Stoller (1962). This is no good to us for several reasons – we reject the assumption of normality, we do not want linear correlation and applying the Cholesky decomposition would destroy stratification of the individual variables.

Hull (1977) discards both the normality assumption and the assumption that the individual distributions are of the same form. This requires that rank correlation replace linear correlation, and Hull shows how to create a sample with approximately the same rank correlation matrix by using the standard Cholesky decomposition technique and mapping the actual distributions on to normal distributions percentile by percentile. Independently, Iman and Connover (1982) use a similar technique, except that their samples have a Latin hypercube stratification. This is better but, as discussed earlier, we cannot adequately represent the relations between factors by any correlation coefficient.

Stein's algorithm, by contrast, gives us almost everything we want (see Appendix). It creates a sample that is both hypercube stratified and has approximately the same distribution (not just the same rank correlation matrix) as the joint distribution when the sample is large.

The key idea is that the initial sampling is only to obtain a correct correlation structure. These samples are replaced by their ranks and then discarded. In turn, the ranks are used to force stratification in each dimension simultaneously, and the correlation between the stratification intervals is determined by the original sample.

In our application of Stein's algorithm, our initial sampling is from the empirical joint distribution of risk factor changes, so that is where our correlation structure comes from. However, our inverse cumulative distribution functions, F_{k-1}, will be the distributions we have estimated or fitted to the distributions of the individual risk factor changes.

Analysis of simulated univariate P&L

In this alternative approach, let us start by implicitly assuming that the set of historical changes in the risk factors is the best estimate of tomorrow's joint distribution of changes in risk factors. Each combination of risk factor changes observed historically is assumed to be equally likely tomorrow (and, implicitly, any combination of changes not observed in the history has probability zero). The corresponding simulated P&Ls can then be sorted to obtain a complete estimate of the P&L distribution. The effects of historical crises on the portfolio can also be noted.

This process is worth carrying out because it is simple, easily understood and reveals interesting information such as the maximum simulated loss, but it does discard temporal information in the simulated returns. We know that financial markets change over time and we can see changes in the volatility of the simulated P&L over time – but this information is lost if we just sort the P&Ls. The obvious solution is to use econometric methods on the simulated P&L time series to improve our estimate of tomorrow's P&L distribution.

We have already massaged the raw historical risk factor changes to obtain a better estimate of the joint distribution of those changes. Our alternative route is to use the raw historical data to generate a raw simulated P&L time series and then massage this P&L series to improve our estimate of tomorrow's P&L distribution.

An advantage of modelling individual risk factor changes is that they are usually, if not normal, at least unimodal, whereas there is no reason why the P&L of portfolios which contain derivatives should be either. The other advantage of modelling the individual risk factor changes is that extra information such as implied volatilities is often available, which can easily be incorporated into our estimates of the individual risk factor return. But there is no direct way of incorporating such information into the estimates of our P&L distribution. The big disadvantage of modelling the individual risk factor changes is that we have to deal explicitly with correlation structure, which we can avoid by simulating the historical P&L distribution directly.

So if we are going to model the P&L distribution directly, we must be prepared to model a complicated, albeit univariate, distribution. However, univariate distributions are much easier to deal with than multivariate distributions and we can use non- or semi-parametric methods on the simulated P&L to incorporate extra information.

We suggest four techniques for improving on simple historical simulation through analysis of the simulated P&L time series.

- *Parametric use of P&L volatility forecasts* Take the simple simulated P&Ls, divide each P&L by the corresponding Garch or other forecast of the volatility for that particular day, then multiply them all by today's Garch volatility forecast to get adjusted P&Ls. Then take these sorted adjusted P&Ls as our P&L distribution forecast. This normalises each day's P&L by its forecast volatility to get a time series of P&L expressed in units of the standard deviation of the forecast volatility.[8]
- *Non-parametric use of P&L volatility forecasts* Take the simple simulated daily P&Ls and use Garch or any other technique to estimate the time-varying parameters of the simulated P&L given its time series to date. We can then use non-parametric methods to estimate P&L percentiles given the level of the Garch forecast.

  For example, given that the Garch volatility forecast is, say, in the 4th quintile (of all such forecasts that would have been made given the simulated P&Ls to that date), we would look at the observed percentiles of simulated P&L on the days that the forecast was in the 4th quintile of forecasts. These P&L percentiles would be our estimate of the P&L distribution. This non-parametric use of volatility forecasts is appropriate when the P&Ls are skewed or otherwise deviate from normality.
- *Apply extreme-value theory to the P&L series to estimate the (albeit unconditional) probability of extreme losses* (Longin, 1997).
- *Generalisation to other factors* We can generalise this and analyse the effect on P&L of various factors, not just forecast volatility. Conceptually, the idea is to make a scatter diagram of the (simulated historical) P&Ls versus changes in the factor and find the ones that are associated with big losses – exploratory data analysis is a fancy name for this. Then, if we have a view on what these factors will be tomorrow, we can use that to forecast P&L percentiles.

This can be automated and hundreds or even thousands of factors considered using artificial intelligence techniques. The "factors" can be arbitrarily complex – the value of a basket of European currencies against the dollar, for example. It is also worth noting that the effects of these factors incorporate historical correlates, so if we plot the effect of, say, dollar interest changes versus our simulated P&L, we are seeing the effect of all the changes that have been historically associated with dollar interest rates as well.

This is the key to the method's effectiveness. It is easy to tabulate the effects of individual risk factors moving in isolation – but not very informative since the portfolio effect is not seen. It is also easy to tabulate the P&Ls of various arbitrarily chosen scenarios but, once again, the scenarios are confined to a few factors at a time and the portfolio effect is not seen.[9] Using our method, we can discover complex scenarios that have the biggest effects on our P&L taking the correlation structure into account. This transforms VAR from a static number into a tool for understanding the dynamics of our portfolios.

Appendix

STEIN'S ALGORITHM (1987)

Suppose there are K factors in the multivariate distribution we want to simulate, and that we want to create N Latin hypercube stratified samples from the distribution.

First, draw a sample of N unstratified K-vectors from the distribution. From these vectors construct an $N \times K$ matrix, Y, in which the ijth element, Y(i,j), is the jth element from the ith sample. Define another $N \times K$ matrix, R, in which the ijth element, R(i,j), is the rank within the jth column of Y of the element Y(i,j).

From R, define a third $N \times K$ matrix, Y*, in which the ijth element

$$Y^*(i,j) = F_k^{-1}\{[R(i,j) + e_{ij} - 1]/N\}$$

where e_{ij} is a uniform [0,1] variate and F_k^{-1} is the inverse cumulative distribution function for the kth factor.

The rows of Y* constitute the Latin hypercube stratified sample we require.

1 NORMSINV *and other capitalised functions in this paper are the standard Excel function, but note that* NORMSINV(x) *incorrectly returns –500,000 when* $x < 10^{-7}$. *If you really want to use the normal distribution for extremes, you should use Moro's algorithm (Moro, 1995).*

2 *Actually it says that the sum of a number of distributions tends toward a stable distribution of which the normal distribution is only one member. For example, a sum of Cauchy distributions (which are very fat tailed) is Cauchy, not normal.*

3 *Garch stands for generalised auto regressive conditional heteroskedasticity. For an introduction see Alexander (1996). For an in-depth survey see Bollerslev, Chou and Kroner (1992).*

4 *The assumption of conditional normality is common but not essential when using Garch techniques. For example, the sterling-Deutschmark exchange rate is modelled as a conditional Student's* t *distribution in Pesaran & Robinson (1993). Since Student's* t *is a fat-tailed distribution, this is promising for risk management applications, though for equities and other assets that are more likely to melt down than melt up we would prefer an* asymmetric *fat-tailed distribution.*

5 *I swear I heard this rationale given by a prominent speaker at a conference on risk management.*

6 *It is a function of risk management to check front-office software, but checking software is much different and easier than creating an industrial-strength replica of it.*

7 *This strategy was suggested independently by Barone-Adesi and Giannopoulos.*

8 *This idea, at least when applied to individual risk factors, is due to Jan Kwiatkowski of BZW.*

9 *In the case of a multivariate normal world with linear sensitivities our method would correspond to Garman's "VaR Delta" (Garman, 1996).*

BIBLIOGRAPHY

Alexander, C., 1996, "Volatility and Correlation Forecasting" in *The Handbook of Risk Management*, John Wiley.

Bagg, J., 1996, "Risk Management – Taking the Wider View," *International Derivative Review*, June, pp. 12–14

Barone-Adesi, G., and K. Giannopoulos, Undated, "A Simplified Approach to the,Conditional Estimation of Value at Risk", Manuscript, University of Alberta, City University and University of Westminster.

Bollerslev, T., R. Chou and K. Kroner, 1992, "ARCH Modeling in Finance", *Journal of Econometrics* 52, pp. 5–59.

Chew, L., 1994 "Shock Treatment", *Risk*, 7(9), pp. 63–70.

Chobanov G., P. Mateev, S. Mittnik and S. Rachev, 1996, "Modeling the Distribution of Highly Volatile Exchange Rate Time Series", Lecture Notes in Statistics 115: Athens Conference on Applied Probability and Time Series, Volume II: Time Series Analysis, Springer.

Cox, E., 1995, "Magic and Regulation", *Risk* 8(3), pp. 50–4.

Drake, K., and R. Yim, 1997, "Abductive Information Modeling Applied to Financial Time Series", in Caldwell, Randall, B., *Nonlinear Financial Forecasting – Proceedings of the First INFFC*, Finance and Technology Publishing, Haymarket, VA.

Garman, M., 1996, "Improving on VAR," *Risk* 9(5), pp. 61–3.

Hull, J.C., 1977, "Dealing with Dependence in Risk Simulations", *Operational Research Quarterly*, 28, No. 1, ii, pp. 201–13.

Iman, R.L., and W.J. Connover, 1982, "A Distribution-Free Approach to Inducing Rank Correlation Among Input Variables", *Communications in Statistics Part B – Simulation and Computation* 11, pp. 311–34.

Jackwerth, J., and M. Rubinstein, 1996, "Recovering Probability Distributions from Option Prices," *Journal of Finance*, LI, No 5, pp. 1611–31.

Jorion, P., 1995a, "Predicting Volatility in the Foreign Exchange Market", *Journal of Finance* 50, pp. 507–28.

Longin, F., 1997, "From Value at Risk to Stress Testing: The Extreme Value Approach", CERESSEC Working Paper 97-004.

Moro, B., 1995, "The Full Monte", *Risk* 8(2), pp. 57–8.

Owen, A., 1994, "Lattice Sampling Revisited: Monte Carlo Variance of Means over Randomized Orthogonal Arrays", *Annals of Statistics* 22, 2, pp. 930–45.

Owen, A., Undated, "Randomly Permuted (t,m,s)-Nets and (t,s)-Sequences", Manuscript, Stanford University.

Pesaran, B., and G. Robinson, 1993, "The European Exchange Rate Mechanism and the Volatility of the Sterling–Deutschmark Rate", *The Economic Journal* 103, No 421, pp. 1,418–31.

Ray, C., 1993, *The Bond Market*, Business One Irwin, Homewood Illinois.

Ramberg, J., P. Tadikamalla, E. Dudewicz and E. Mykytka, 1979, "A Probability Distribution and Its Uses in Fitting Data," *Technometrics* 21, No. 2, pp. 201–14.

Scheuer, E., and D. Stoller, 1962, "On the Generation of Random Normal Vectors", *Technometrics* 4, pp. 278–81.

Stein, M., 1987, "Large Scale Properties of Simulations Using Latin Hypercube Sampling", *Technometrics* 29, No. 2, pp. 143–51.

Taleb, N., 1997, "Against VAR," *Derivatives Strategy*, pp. 20–6.

Tang, B., 1993, "Orthogonal Array-Based Latin Hypercubes", *Journal of the American Statistical Association*, 88, No. 424, Theory and Methods pp. 1,392–7.

Using Non-Normal Monte Carlo Simulation to Compute Value-at-Risk

Gerald D. Quinlan
TrueRisk Inc

As value-at-risk (VAR) has come into widespread use for the measurement of market risks, there has been frequent debate about whether the methodologies used for calculating VAR make adequate allowance for large losses that can result from unexpected market changes – changes that fall in the tails of the distributions (eg Lawrence and Robinson, 1995; Longerstaey and Zangari, 1995). Calculations of VAR use statistical techniques to quantify the maximum loss a portfolio might suffer over a specified time horizon, usually one day, with a specified confidence level, usually in the range 95–99%. A probability distribution is assumed for the changes to the risk factors on which the portfolio depends, and the portfolio is re-valued under these assumed changes, with the VAR being the maximum loss that occurs with the specified probability. There are two problems that can cause VAR predictions made at high confidence levels to be unrealistically low.

First, some VAR methodologies compute the change to the portfolio value from an analytic, linear approximation, and then scale the change up to the desired confidence level assuming that the portfolio changes come from a normal distribution. This is not valid for a portfolio containing options or other non-linear instruments, as is well known. This problem is easily avoided by computing the portfolio changes using full valuation or a global approximating function instead of a linear approximation.

The second problem is more difficult: the changes to the underlying risk factors may come from a distribution that is not the same as assumed in the calculation of VAR. The assumption usually made is that the logarithmic changes in risk factors, when divided by their volatility forecasts, come from a multivariate normal distribution. The changes are said to be conditionally normal – "conditional" meaning that the changes are normal only after being rescaled by their volatility forecasts, which vary with time. But even when this rescaling is done, the assumption of normality is not satisfied in the tails of the distributions. The changes in real risk factors are not normal – not even conditionally normal.

Consider the data shown in Figure 1 for the New Zealand dollar (NZ$) two-year government zero rate (Z02). The figure shows data for the 1,100 trading days starting on March 1, 1993, and ending on May 16, 1997. The daily logarithmic changes in the rates have been converted into standardised form by dividing them by their volatility forecasts, computed from the recursive formula[1] used in the RiskMetrics data sets, with the decay constant λ set to 0.94 (Figure 1(b)). According to the assumption of conditional normality, these standardised changes should come from the standard normal distribution. The mean of the distribution is close to zero, as predicted, but the standard deviation is 1.088, larger than the predicted 1.0, and the skewness and kurtosis are $S = 1.29$ and $K = 17.27$, significantly different from the values $S = 0$ and $K = 3$ for a normal distribution.[2] There are seven days on which the rate changed by more than four times the volatility – a change that would be highly unlikely if the distribution were normal – and one day on which it changed by 12 times the volatility (November 8, 1993, day 181 in the figure, when the rate jumped from 5.40% to 6.08%).

Graph (c) in Figure 1 compares the standard normal distribution with the positive tail of the

1. New Zealand dollar two-year zero rate, March 1, 1993–May 16, 1997

(a) The rate versus time (counting only the business days)
(b) The daily logarithmic change in the rate divided by the volatility forecast
(c) The historical distribution of positive changes to the rate, together with the normal and non-normal (RTDM) fits

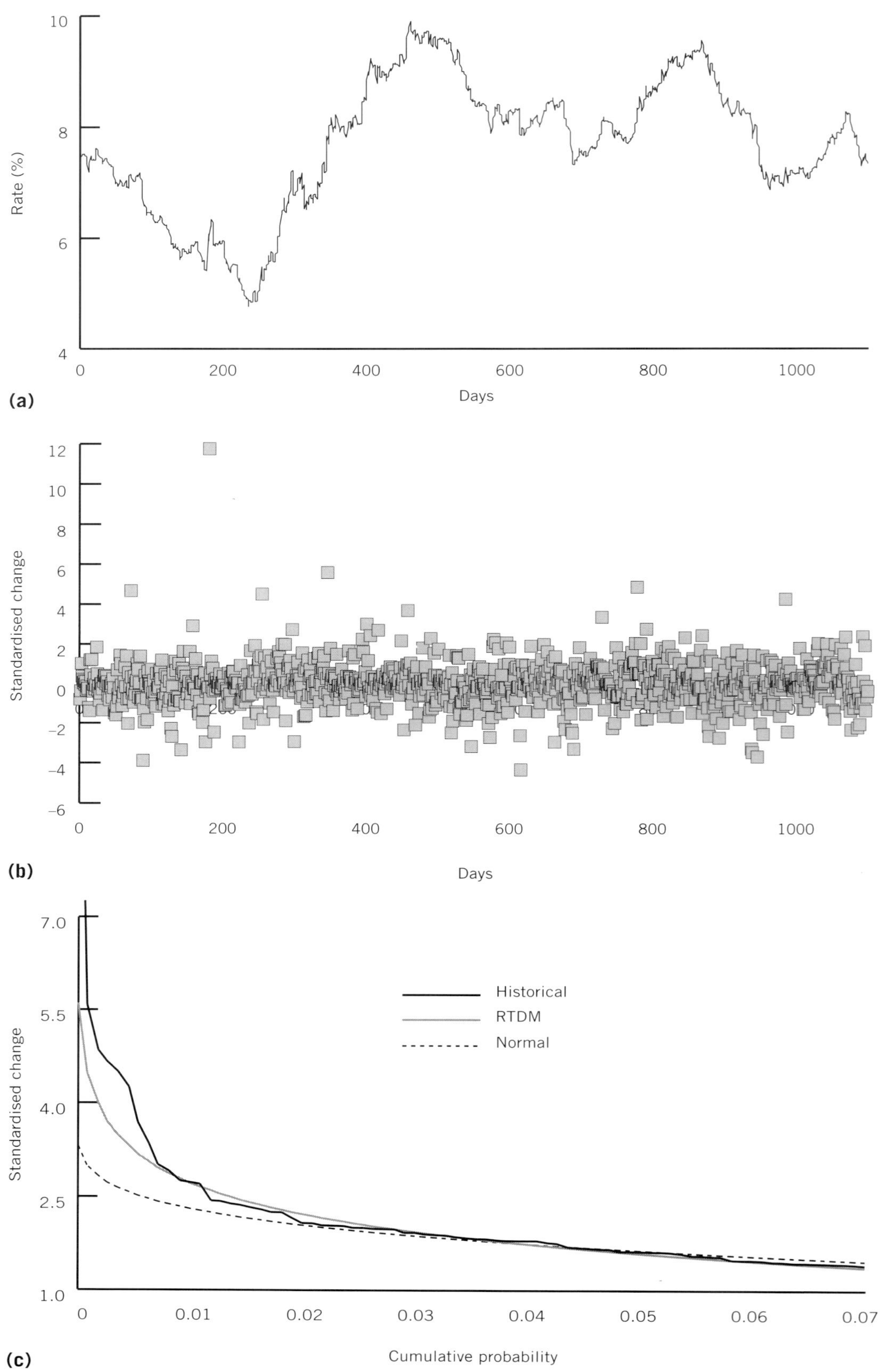

2. Distributions of skewness and kurtosis values for standardised changes in 83 zero rates and 39 money market rates from 10 currencies

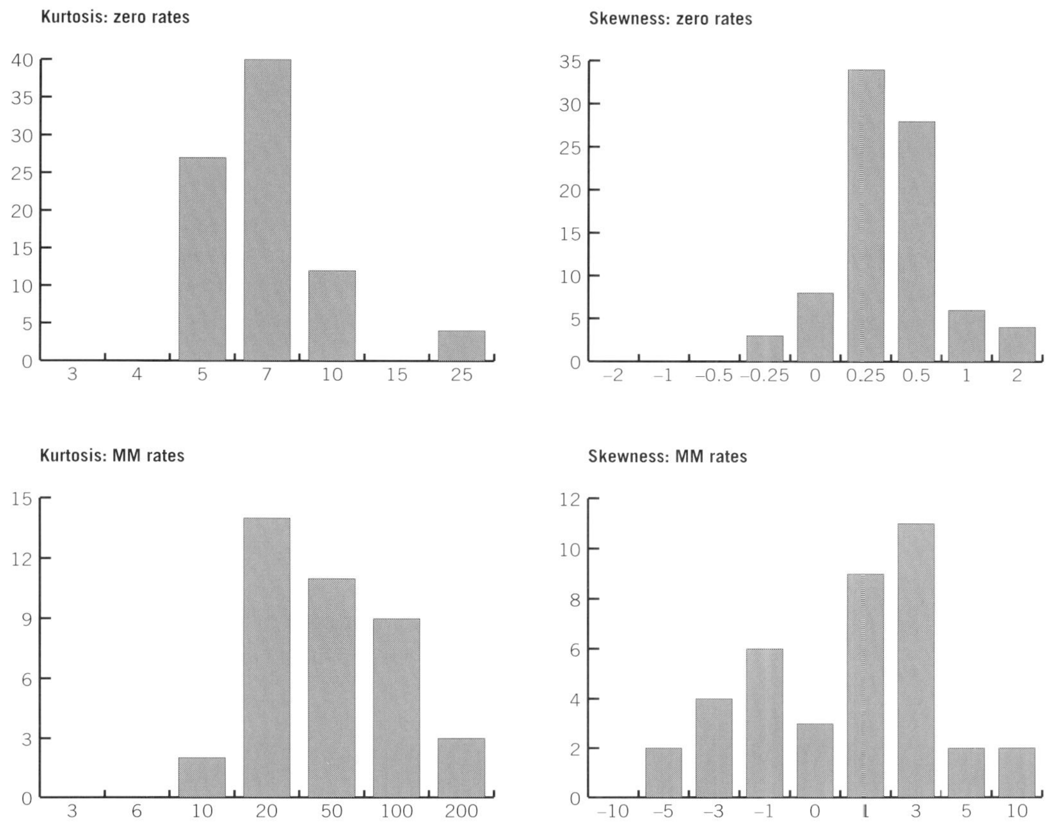

historical distribution of the standardised changes. The normal distribution fits reasonably well for confidence levels in the 95–97% range, but fails above that range. At the 99% level the normal distribution predicts the standardised change to be no larger than 2.33, while the historical distribution shows a change of 2.76. At higher confidence levels the discrepancy gets worse.

The distribution of standardised changes for the NZ\$.Z02 rate exhibits leptokurtosis, or, more colloquially, "fat tails". The kurtosis for the NZ\$.Z02 rate is higher than is typical for two-year interest rates, but it is not extreme. Figure 2 shows the distribution of skewness and kurtosis values for the standardised changes in 83 zero rates and 39 money market (MM) rates from 10 different currencies.[3] Kurtosis values of 4–10 are common for long-term interest rates; these are comparable with the values typically seen for stock indexes (4–6) or foreign exchange rates (4–10) from OECD countries. For money market rates the values are usually much higher. Interest rates tend to have a positive skewness – so that a large change is more likely to push the rates up than down – although this is not always the case.

For 53 of the 122 rates in Figure 2, the kurtosis of the standardised changes is larger than the kurtosis computed without rescaling the changes by their volatility forecasts. The volatility forecast can make the non-normality worse if large changes in the rates are preceded by quiet periods during which the volatility gets pushed towards zero (which often happens for the money market rates). Attempts to reduce the non-normality of risk factors by optimising the decay constant in the volatility forecast have not had much success (Finger, 1996). This is not to say that the volatility forecasts should be ignored: they are essential for making accurate VAR predictions at the confidence levels commonly used. But rescaling the risk-factor changes by their volatility forecasts does not necessarily improve the normal fit to the tails of the distributions.

Given that risk factors exhibit non-normal behaviour, how can this be included in calculations of VAR? One solution is to use a historical simulation, setting the changes in the risk factors to those that have occurred in the past. This has the advantage of reproducing the true historical distributions with the correct correlations but the disadvantage of ignoring the time-dependence of volatility. It is possible to rescale the historical changes to allow for time-dependent volatilities (Shaw, 1997; Hull and White, 1998b), but even when this is done other drawbacks remain. First, to perform historical simulation one needs access to the relevant historical data. Second, historical simulation can predict only what has occurred in the past. Third, it is not easy to test the sensitivity of a historical simulation to the various risk factors. For example, one may want to know how the VAR will change if the volatilities or correlations of selected risk factors increase or decrease, or if the skewness and kurtosis of those risk factors take on extreme values. Although questions like these can be studied by manipulating the historical data, it is easier to change the input parameters to a conventional Monte Carlo simulation.

There are thus advantages to having a simulation methodology that allows for non-normal behaviour without making direct use of historical data. Several authors have published papers describing how this can be done. Zangari (1996) has replaced the assumption of conditional normality with the assumption that the standardised changes in risk factors come from a mixture of two normal distributions chosen to fit the fat tails of historical distributions. Shaw (1997) - also Chapter 21 of this publication has proposed an ambitious methodology that uses Stein's algorithm (Stein, 1987) to fit both the fat tails and the rank correlations of historical distributions. Hull and White (1998a) have proposed a third methodology that has elements in common with those of Zangari and Shaw. The present contribution is intended not as a presentation of a fourth methodology but as a test of some of the ideas that have already been proposed, a review of the questions that arise when using these ideas in a multivariate simulation with correlated, highly non-normal variables, and a demonstration that even a simple non-normal simulation methodology can give accurate VAR predictions at confidence levels ranging up to 99%.

A methodology for non-normal Monte Carlo simulation

The methodology to be tested is a simplified version of that proposed by Shaw (1997). We first review the use of Stein's algorithm, and then consider the fitting of non-normal distributions to market data and two methods for matching the correlation structure.

STEIN'S ALGORITHM

Stein's algorithm provides a way of using a Latin hypercube to generate stratified, correlated, non-normal random variables. To generate N stratified samples of K correlated risk factors, one first generates N unstratified K-vectors from a distribution that has the desired correlation structure - eg the correct linear correlation or the correct rank correlation. These are combined into a N by K matrix whose element Y_{ij} stores the jth risk factor from the ith sample. This matrix is replaced by the rank matrix R, in which element R_{ij} is the rank of element Y_{ij} within the jth column of Y. The non-normal variables for the jth risk factor are then computed from

$$F_j^{-1}[(R_{ij} + e_{ij} - 1)/N]$$

where F_j^{-1} is the desired inverse cumulative distribution function for the jth factor and e_{ij} is a uniform random variable in the range 0–1. In some applications the extra randomisation provided by the e_{ij} is not needed, and the e_{ij} can be set equal to 0.5.

FITTING NON-NORMAL DISTRIBUTIONS TO MARKET DATA

Many choices are possible for the distribution function used to generate the non-normal random numbers in Stein's algorithm. The main requirement for our problem is that the function require only a small number of parameters to describe the deviation from a normal distribution and that these parameters be easy to fit to market data. It also helps if the inverse cumulative distribution can be computed easily. The distribution proposed by Ramberg *et al.* (1979) satisfies both of these requirements. This distribution, which we shall call the RTDM distribution, is specified by its inverse cumulative distribution function:

$$F^{-1}(p) = \lambda_1 + [p^{\lambda_3} - (1 - p)^{\lambda_4}]/\lambda_2$$

where p is the cumulative probability. There are four free parameters: λ_1 and λ_2 are fixed by the mean and standard deviation; λ_3 and λ_4 can be chosen to generate distributions with a wide range of values for the skewness and kurtosis, including both positive and negative skewness values. If the skewness and kurtosis are set to 0 and 3, the RTDM distribution fits the normal distribution to within 1% over the range 0.02–0.98 in p.

3. Distributions of capped skewness and kurtosis values for rates in Figure 2

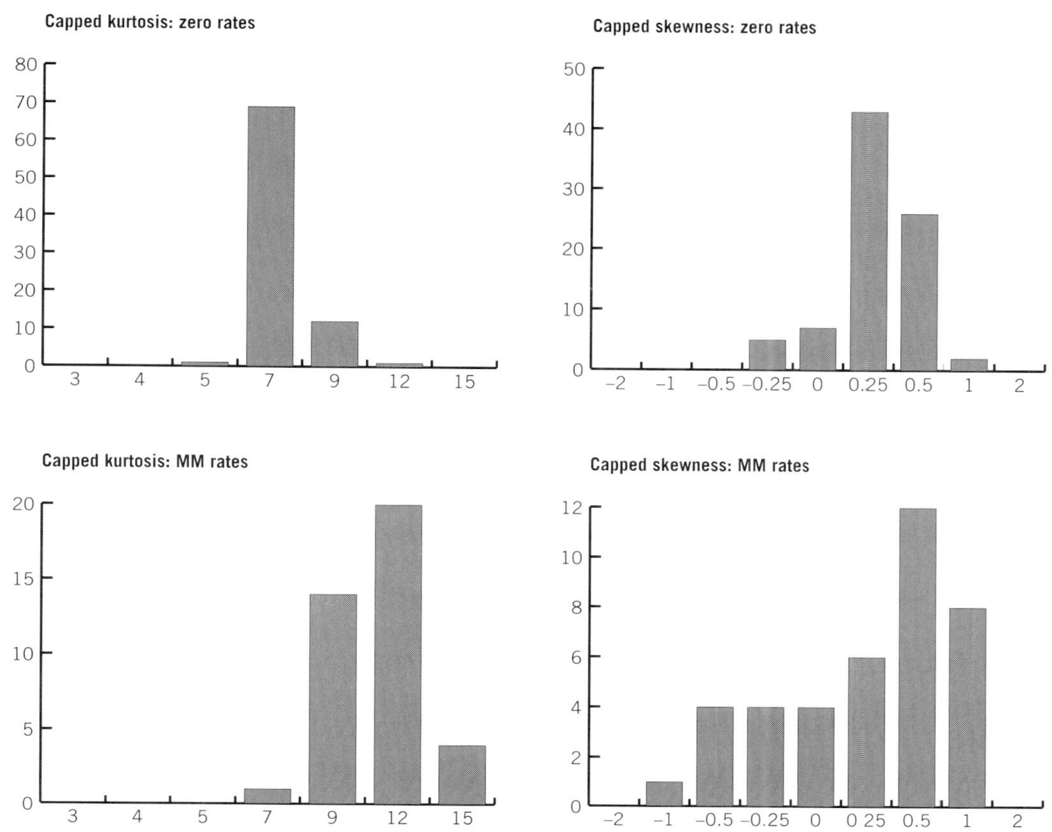

Fitting the RTDM distribution to the measured skewness and kurtosis values of risk factors, however, is not a good idea. The skewness and kurtosis are sensitive to extreme data points. If the single largest change is removed from the NZ$.Z02 data of Figure 1, for example, the kurtosis drops from 17.27 to 5.99 and the skewness from 1.29 to 0.19. These moments cannot be measured reliably for highly non-normal distributions. Another problem is that if the RTDM distribution is fitted to high values for the skewness and kurtosis, the distribution will probably fit badly at the 95–99% confidence level because it will be stretched out to fit the extreme points in the tail of the distribution. And even if distributions with high values for the skewness and kurtosis could be measured and fitted, they would be difficult to reproduce in a Monte Carlo simulation without using a large number of scenarios.

This fitting problem raises the question of what the goal of a non-normal simulation should be. It is unrealistic to expect a simple distribution function with a few adjustable parameters to provide an accurate fit to a non-normal empirical distribution at confidence levels ranging up to 99.9%. It is also unrealistic to expect a non-normal simulation to provide accurate VAR predictions at such high confidence levels. Even if this could be done, the simulation results could not be backtested. A more realistic goal for non-normal simulation is to improve the VAR predictions at confidence levels in the 95–99% range (a similar goal is adopted by Hull and White, 1998a). The risks associated with extreme events – market changes that are many times larger than the volatility forecasts – are best investigated with stress testing. A non-normal simulation can be of some help in predicting these risks, of course, and can be used as a tool for stress testing, but its limitations must be kept in mind.

Given this goal, one solution to the RTDM-fitting problem is to choose λ_3 and λ_4 to optimise the fit to the tails of the distribution at the 95–99% confidence level (considering both positive and negative changes). This is possible, but it involves a non-linear least-squares fit. One can obtain nearly the same result by choosing λ_3 and λ_4 to match capped values for the skewness and kurtosis, "capped" meaning that in the computation of the skewness and kurtosis the sizes of the standardised changes are capped at 5 in absolute value (if the change on a given day is above 5, it gets replaced by 5; and if it is below −5, it gets replaced by −5).[4]

For the NZ$.Z02 example in Figure 1, the capped skewness and kurtosis are S = 0.19 and K = 7.04 – much smaller than the true skewness and kurtosis but, for the kurtosis at least, still significantly non-normal. Figure 1(c) shows the positive tail of the RTDM distribution computed with these capped values. The distribution fits well at confidence levels up to 99%. It underestimates the probability of extreme changes, as expected, but it is still better than the normal distribution.

Figure 3 shows the distributions of the capped skewness and kurtosis values for the rates considered in Figure 2. The capped kurtosis does not take on the extreme values that are sometimes seen for the true kurtosis. It is nevertheless still significantly larger than 3, showing that the distributions remain non-normal even when the extreme data points are capped. The RTDM distributions with these capped skewness and kurtosis values typically reduce the root mean square discrepancy with the tails of the historical distributions by a factor of two over the 95–99% range in confidence levels. In most cases the capped skewness does not make a big difference: the distribution that uses only the capped kurtosis works nearly as well as the one that uses both the kurtosis and the skewness.

4. Correlations between standardised changes in three NZ$ zero rates

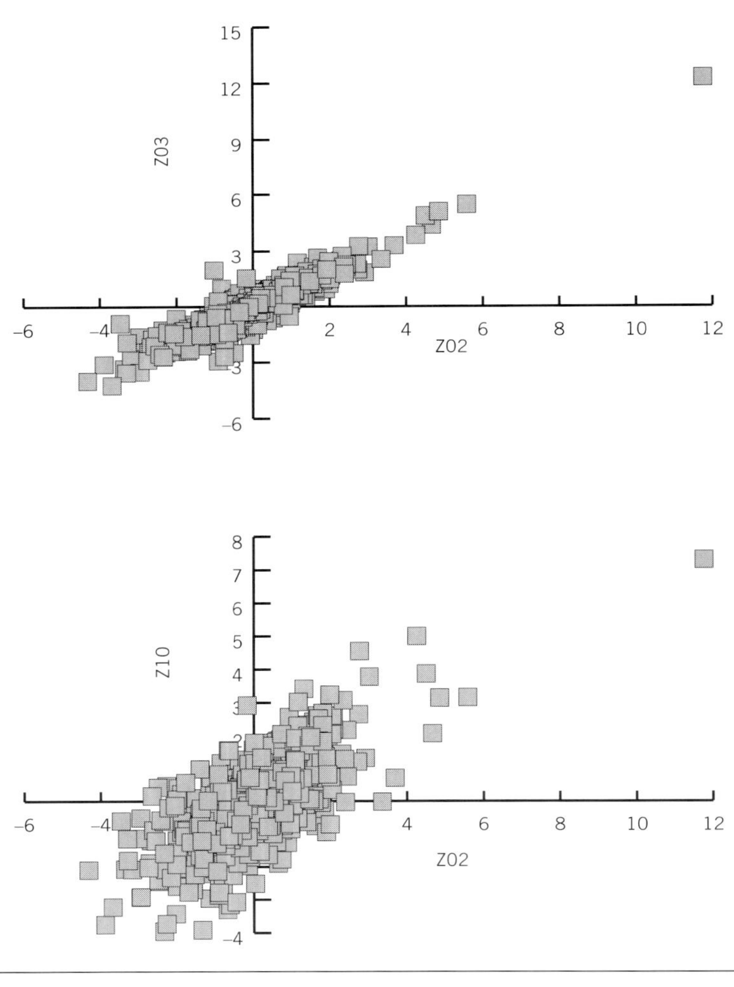

To check the stability of the capped skewness and kurtosis values, the values computed from the first 550 days of the data were compared with those computed from the second 550 days. The agreement was good for the kurtosis, but not for the skewness: the skewness values from the first and second halves sometimes had opposite signs (for the NZ$.Z02 rate, for example, the values for the first and second halves are 0.57 and −0.20). It may be possible to fit this time variation using a recursive forecast like that used for the volatilities, but with a larger decay constant λ (since more data are needed to compute a skewness than a volatility). For most risk factors the capped skewness is not large enough for this to be worth the effort. Perhaps the skewness should be reserved for stress testing – for example, to see how the VAR changes when the risk factors take on extreme skewness values. For the tests discussed in this chapter, the capped skewness is assumed to be zero.

TREATMENT OF CORRELATIONS

The correlations between risk factors can be more complicated than a simple linear correlation (Blythe, 1996). The stock indexes in different countries may be only moderately correlated during normal times, for example, but highly correlated when a crash occurs. Even ignoring crashes, it is not clear that linear correlation is adequate for all risk factors. Consider Figure 4, which shows the correlations between the standardised changes in three NZ$ zero rates. The two-year and three-year rates are more strongly correlated when there is a large rise in the rates than when there is a fall or a moderate rise. This is not true for all rate combinations, however; the correlation between the two-year and 10-year rates does not get stronger when the changes are large.

Stein's algorithm provides two methods for matching the correlation structure. The simpler of the two is to generate normal variables having the correct linear correlation matrix, and then to perform the non-normal transformation and accept whatever correlation results. Figure 5 shows how this works when 1,000 standard normal variables with a linear correlation of 0.9 (upper plot) are transformed to non-normal variables with a kurtosis of 10 (lower plot). In this example the linear correlation is nearly preserved (it can change if the variables are given different kurtosis values), but the scatter becomes larger in the tails of the distributions. This is all right for modelling a correlation like that between the NZ$ two-year and 10-year zero rates, but it is not as good for the two-year and three-year rates.

The second of the two methods is to choose the normal variables to have the same rank matrix as the historical distribution to be matched. This is the method advocated by Shaw (1997) and also by Hull and White (1988b). It has obvious advantages over the first method – one being that a rank correlation is more robust than a linear correlation – and would give a better fit to a correlation like that between the NZ$ two-year and three-year rates, but it requires direct use of historical data.

For the tests presented here we consider only the simpler of the two methods, realising that it may not work well in the extreme tails of the distributions. The test results show that it works satisfactorily at confidence levels ranging to 99%.

Tests of the methodology

DESCRIPTION OF THE TESTS

We compare three methodologies in the tests. The first is standard Monte Carlo simulation, which assumes the risk factors to be conditionally normal. The second is the non-normal simulation methodology that has just been described, using the capped kurtosis and the same linear correlation forecast employed in the standard simulation. The third is a modified historical simulation, similar to those suggested by Shaw (1997), Zangari (1997) and Hull and White (1998b).

The third methodology works as follows. The distribution of possible gains and losses for a portfolio on a given day is set equal to the distribution of actual gains and losses the portfolio would have suffered on days in the past, computed using historical data with the maturities of the instruments held fixed, and with each gain or loss rescaled by the ratio of the volatility the portfolio has today to the volatility it would have had in the past. For example, if the portfolio value would have changed by +1 on some day in the past, and if the portfolio volatility would have been 0.1 on that day and is 0.15 today, the change gets counted as +1.5. This third methodology takes account of time-dependent volatility, non-normality and possible non-linear correlation in the historical distributions, and should be the most accurate of the three methodologies. For practical risk management work, of course, it has the usual drawbacks of historical simulation.

For the risk factors we use the three NZ$ zero rates from Figure 4 (the two-, three- and 10-year rates) and, for simplicity, consider portfolios containing only zero-coupon bonds. There is nothing special about this choice of rates. The bonds underestimate the importance of non-normal behaviour for derivative portfolios: interest rate options would be more sensitive than bonds to large changes in the rates. The six portfolios to be tested are listed in

5. Non-normal transformation using linear correlation

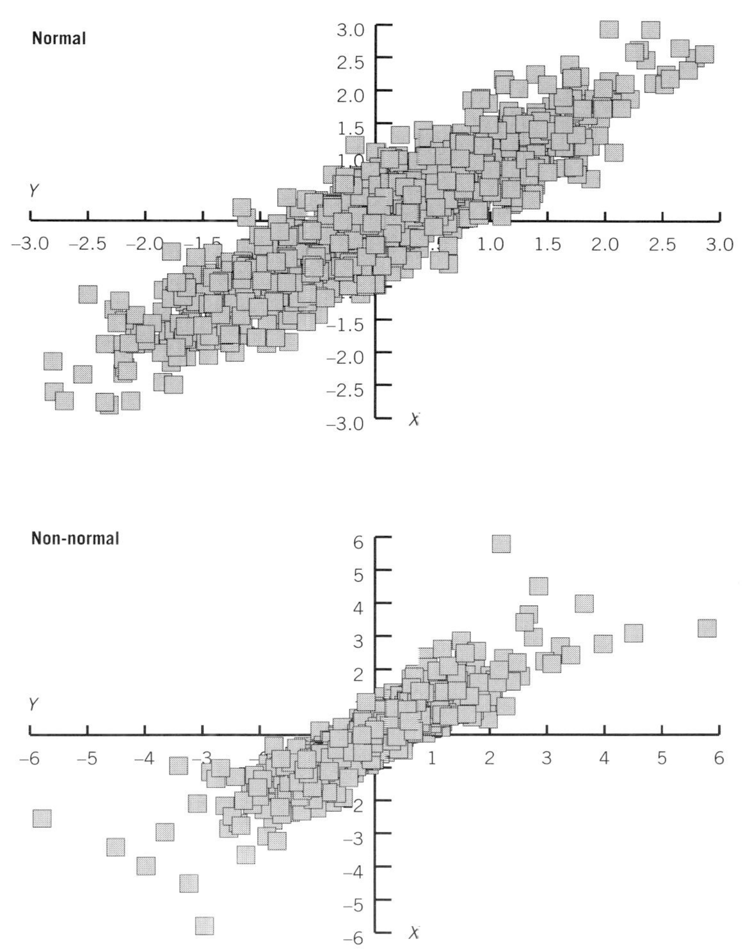

Table 1: Test portfolios

Portfolio	Weights			Value
	Z02	Z03	Z10	
A	1	0	0	100
B	0	1	0	100
C	0	0	1	100
D	1	1	0	100
E	1	1	1	100
F	1	–1	0.05	100

Table 1. Each portfolio has a value of 100, with the weights in the table indicating the proportion of the value that is invested in bonds with the three maturities. Portfolio F, for example, has NZ$2,000 invested in the two-year bond, -NZ$2,000 invested in the three-year bond (ie a short position in the bond) and NZ$100 invested in the 10-year bond.

SIMULATED PROFIT AND LOSS DISTRIBUTIONS

We first compare the simulated profit and loss (P&L) distributions on a single day, the last day of the 1,100-day data set. Figure 6 shows the tails of the distributions for three of the six portfolios. The historical simulations made use of all the data in the data set; the standard Monte Carlo and non-normal Monte Carlo simulations used 2,000 scenarios each, with stratified sampling. The axes in the graphs show the P&L (X) and the cumulative probability (Y). The VAR estimate at a specified confidence level, say 99%, for a long position in the portfolio is given by the loss corresponding to the cumulative probability, 0.01, and, for a short position, by the profit corresponding to the same cumulative probability.

The three portfolios in the figure are affected differently by correlation. Portfolio A is invested entirely in the two-year bond. Portfolio D has equal investments in the two- and three-year bonds, which are strongly correlated when there is a large rise in the rates. Portfolio F has offsetting positions in the two- and three-year bonds, which could cause difficulty for the non-normal simulation if large changes in rates are not given the correct correlation.

For all three portfolios the P&L distributions from the non-normal simulation and the standard Monte Carlo simulation agree at about the 97% confidence level. At higher confidence levels the non-normal simulation gives a better fit to the historical losses. For portfolios A and D the non-normal simulation underestimates the historical losses at confidence levels above 99%, because the RTDM distributions do not fit the extreme positive changes in the rates, but this was expected. For portfolio F, surprisingly, the P&L distribution from the non-normal simulation agrees with that from the historical simulation at all confidence levels. At the extreme confidence levels this is partly an accident, since for this portfolio the underestimate of extreme changes in the rates compensates for the overestimate of the change in the spread between the rates; but, together with the results for the other portfolios, it gives reassuring evidence that the non-normal methodology works up to the 99% confidence level.

BACKTEST RESULTS

Figure 6 compares the predictions of the three methodologies on just one day. Another way of comparing the predictions is to check how often the true loss to a portfolio exceeds the predicted VAR at a specified confidence level. This was done using the last 500 days of the historical data set. On each day the three methodologies were used to predict the VAR at confidence levels of 95%, 97.5% and 99% for the six portfolios listed in Table 1, making use only of the data that were available prior to the day under consideration. The number of days on which the portfolio loss

Table 2. Backtest results

	Confidence level (%)								
	95			97.5			99		
Portfolio	SMC	NNMC	HIST	SMC	NNMC	HIST	SMC	NNMC	HIST
A	0.055	0.059	0.056	0.036	0.030	0.029	0.022	0.011	0.009
B	0.051	0.056	0.052	0.032	0.031	0.027	0.018	0.011	0.011
C	0.059	0.062	0.052	0.031	0.027	0.022	0.017	0.013	0.010
D	0.056	0.059	0.059	0.035	0.028	0.025	0.019	0.009	0.010
E	0.050	0.057	0.047	0.033	0.031	0.020	0.017	0.012	0.010
F	0.052	0.059	0.051	0.033	0.028	0.025	0.019	0.009	0.009
Average	0.054	0.059	0.053	0.033	0.029	0.025	0.019	0.011	0.010

Simulations: SMC, standard Monte Carlo; NNMC, non-normal Monte Carlo; HIST, historical.

exceeded the VAR level was recorded for each of the methodologies, averaging the results for long and short positions in the portfolios to improve the statistics. Table 2 gives these numbers expressed as fractions for the six portfolios separately and, in the last row, the average results for the six.

If the VAR predictions are correct, then, ignoring statistical fluctuations, the fraction of losses that exceed the 95%, 97.5% and 99% VAR levels should be 0.05, 0.025 and 0.01. The fractions in the table for the historical simulation are consistent with this result; the slight disagreement at the 95% confidence level could be a fluctuation. For the standard Monte Carlo simulation the 99% VAR level is exceeded twice as often as it should be. The non-normal simulation does slightly worse than the normal simulation at the 95% confidence level but gives consistent results at the 97.5% and 99% levels.

The real advantage of the non-normal simulation over the normal simulation is obscured by this test, which simply counts the frequency at which VAR levels are exceeded without noting by how much they are exceeded; but the test nevertheless gives further evidence that the non-normal simulation works at confidence levels ranging up to 99%.

The capped kurtosis values used in this backtest were updated on each day, using the data available prior to the day under consideration. The test results turn out to be nearly the same if the capped kurtosis values are kept fixed at the values computed from the first 600 days of the data set – 7.3 for the two-year rate, 6.8 for the three-year rate, and 6.7 for the 10-year rate – and even if the same value, 7.0, is used for all three rates. The variation of the capped kurtosis with time and between risk factors does not matter much at the 99% confidence level; what does matter is that the kurtosis is significantly larger than the value for a normal distribution. This suggests that, at least for estimates at the 99% confidence level, it may be all right to hold the shape of the risk-factor distributions fixed, perhaps with a single capped kurtosis value for each risk-factor class (one for zero rates, another for money market rates, and so on), and to let the scale and structure of the distributions vary with the time-dependent volatilities and correlations. The non-normal simulation then becomes a simple and easy-to-implement extension of the standard Monte Carlo simulations already in use.

6. Simulated profit and loss distributions
(SMC, standard Monte Carlo; NNMC, non-normal Monte Carlo; hist, historical)

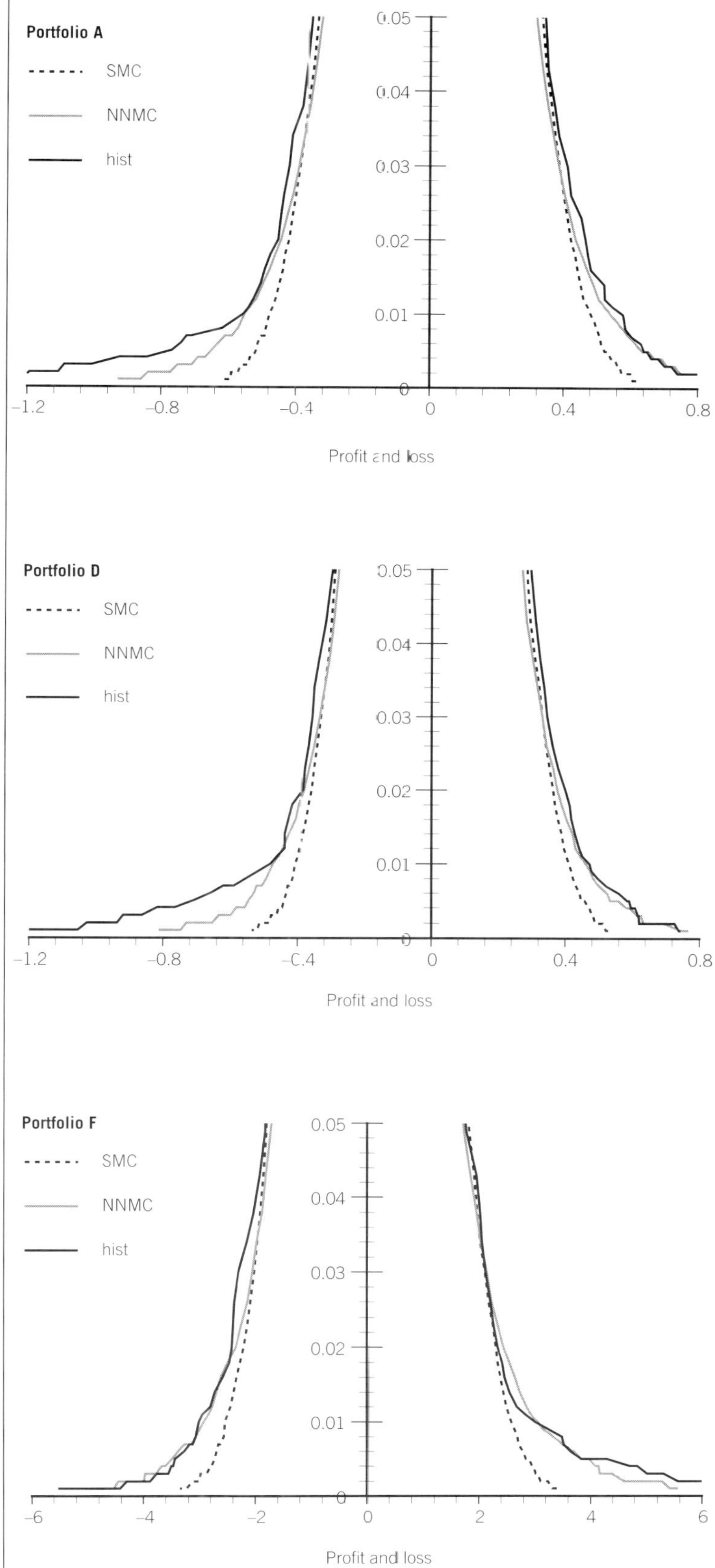

1 *See equation (5.4) of the RiskMetrics technical document (JP Morgan, 1996). The recursive forecast was started using data from before March 1, 1993, not shown in the figure.*

2 *If the data are analysed using the discount factors instead of the rates, the standard deviation, skewness and kurtosis of the standardised changes are 1.092, –1.42 and 17.76. A positive skewness for a rate corresponds to a negative skewness for the discount factor. Aside from this difference, the conclusions about the non-normality of the interest rates considered in this chapter do not depend on whether the analysis uses the rates or the discount factors.*

3 *The currencies are the Australian dollar, Belgian franc, Canadian dollar, Deutschmark, French franc, British pound, Italian lira, Japanese yen, New Zealand dollar and US dollar, and the rates are the one-month, three-month, six-month and one-year money-market rates, and the zero rates with maturities ranging from two to 30 years. The data are from the same 1,100-day time period used in Figure 1. The histograms are to be interpreted as follows: if* X_1, X_2, X_3, *etc, are the numbers on the axis (X is either the skewness or the kurtosis), the bar above* X_i *then shows the number of data points in the range* $X_{i-1} \le X < X_i$.

4 *When computing these capped values we assume that the true distribution of standardised changes has mean of 0 and a standard deviation of 1 – ie it differs from the standard normal distribution only through the capped skewness and kurtosis. (This explains why the capped kurtosis values in Figure 3 are sometimes larger than the un-capped values in Figure 2.) In other words, we assume that the volatility forecast is the best estimate we have of the standard deviation of the changes. We could adjust* λ_2 *to fit the true standard deviation, which is often larger than the volatility forecast when the distribution is non-normal, but for fitting the tail of the distribution the adjustments to* λ_3 *and* λ_4 *are far more important.*

BIBLIOGRAPHY

Blythe, S., 1996, "Out of Line", *Risk* 9(10), pp. 82–4.

Finger, C.C., 1996, "Testing RiskMetrics Volatility Forecasts on Emerging Markets Data", *RiskMetrics Monitor*, 4th Quarter, pp. 3–19.

Hull, J., and A. White, 1998a, "Value at Risk When Daily Changes in Market Variables Are Not Normally Distributed", *Journal of Derivatives* 5(3), pp. 9–19.

Hull, J., and A. White, 1998b, "Enhancements to the Historical Simulation Method for Calculating Value at Risk", *working paper*, University of Toronto.

JP Morgan, 1996, *RiskMetrics Technical Document*, 4th edition, New York

Lawrence, C., and G. Robinson, 1995, "How Safe is RiskMetrics?", *Risk* 8(1), pp. 26–9.

Longerstaey, J., and P. Zangari, 1995, "A Transparent Tool", *Risk* 8(1), pp. 30–2.

Ramberg, J.S., P.R. Tadikamalla, E.J. Dudewicz and E.F. Mykytka, 1979, "A Probability Distribution and Its Uses in Fitting Data", *Technometrics* 21(2), pp. 201–14.

Shaw, J., 1997, "Beyond VAR and Stress Testing", in *VAR: Understanding and Applying Value at Risk*, Risk Publications, pp. 211–4 (Chapter 21 of this publication).

Stein, M., 1987, "Large Sample Properties of Simulations Using Latin Hypercube Sampling", *Technometrics* 29(2), pp. 143–51.

Zangari, P., 1996, "An Improved Methodology for Measuring VaR", *RiskMetrics Monitor*, 2nd Quarter, pp. 7–25.

Zangari, P., 1997, "Streamlining the Market Risk Measurement Process", *RiskMetrics Monitor*, 1st Quarter, pp. 29–35.

23

Scrambled Nets for Value-at-Risk Calculations

Art Owen and Domingo Tavella
Stanford University and Align Risk Analysis Inc.

The value-at-risk of a financial contract may be represented by a suitable multiple of the standard deviation of the contract value. The value of a financial contract is a function of a number of underlying stochastic variables, such as forward rates or indexes. When the value of the contract is linear in the underlying variables, the standard deviation of the contract value can be computed from the covariance matrix of the underlying variables. In cases where the contract value is a non-linear function of the underlying variables, simulation techniques are widely used to estimate the standard deviation.

There are two types of simulation approach – Monte Carlo methods and quasi-Monte Carlo methods. The former are characterised by rates of convergence that are independent of the dimension. The latter have superior asymptotic convergence rates, though the superiority is dimensionally-dependent and error estimation is more difficult.

This paper aims to explore the effectiveness of a new type of quasi-random sequence, the Randomised Quasi-Monte Carlo (RQMC) sequence, in VAR problems. We have selected two simple financial contracts for study: a zero coupon bond and an interest rate cap. We have deliberately chosen low-dimensional problems as a first step in investigating RQMC for VAR problems. Some high-dimensional Quasi-Monte Carlo (QMC) simulations have been used in financial problems elsewhere. See, for instance, Paskov and Traub (1995) and Caflisch and Morokoff (1996).

Risk manager's summary

This paper describes improvements on Monte Carlo methods for estimating VAR. Improved estimates of VAR are possible, meaning that VAR can be estimated with greater accuracy. Of course this does not mean reducing the actual level of risk: no sampling technique can do that!

In the numerical examples, we find that scrambled nets are more accurate than plain Monte Carlo. In fact it would require a simulation about 50 times as long in plain Monte Carlo to get the same accuracy as we found with scrambled nets. The practical impact of this for some risk managers may simply be that it saves some computer time. But others may be able to calculate VAR for more portfolios or under more scenarios than they would be able to using other methods.

Another way to express this difference is that the VAR estimate will differ from the true VAR by less under randomised quasi-Monte Carlo than it will under Monte Carlo. The estimate can be higher or lower than the truth with either method. But, using the same number of runs, the error will tend to be greater under ordinary Monte Carlo than with randomised quasi-Monte Carlo. In our examples, a factor of 50 in variance translates into a factor of $\sqrt{50} \doteq 7$ in errors. The VAR error under Monte Carlo would then be about seven times as large.

The quasi-Monte Carlo and randomised quasi-Monte Carlo methods were of about equal accuracy. But randomised quasi-Monte Carlo has one advantage: the accuracy of the VAR can be estimated by comparing replicates. So risk managers can decide if the simulation should be larger or if it could be smaller. Under ordinary quasi-Monte Carlo, by contrast, there is no principled way of estimating the accuracy of a VAR result from the data used to produce it.

But it must be stressed that any improvement is dependent on the model being simulated and the portfolio whose value is being assessed. So the performance ratio could be higher or lower than 50 on another problem. There are some theoretical and empirical reasons to expect that the randomised quasi-Monte Carlo results should never be much, if any, worse than ordinary Monte Carlo.

Value-at-risk

The value of the financial contracts we consider is determined by the term structure of interest rates. We characterise the term structure of interest rates by correlated stochastic processes $B_i(t,t+t_i)$, $i=1,\ldots,I$. The quantity $B_i(t,t+t_i)$ denotes the value of an account continuously reinvested in zero coupon bonds of maturity $t+t_i$.

We can think of B_i as the value of a "constant maturity" zero coupon bond. The correlated stochastic processes are governed by

$$\frac{dB_i}{B_i} = \mu_i dt + \sigma_i dW_i \qquad (1)$$

where the μ_i are drift parameters determined from arbitrage constraints, the σ_i are volatilities, and the W_i are correlated Wiener processes.

The value of the kth financial contract at time t is a function

$$Y_k = Y_k(t) = g_k(B_1(t,t+t_1),\ldots,B_I(t,t+t_I)) \qquad (2)$$

of some or all of the B_i.

The financial contract's VAR is a suitable multiple of the standard deviation of Y_k.

We characterise the US term structure over 20 years by five stochastic processes. The bond we consider is a zero coupon bond with a 20-year maturity. The cap is a 20-year cap with semi-annual caplets and a cap rate close to the initial value of the forward rate.

Monte Carlo simulation

A simulation approach to this problem is as follows: the risk manager generates six independent uniformly distributed random variables, makes a polar transformation to normally distributed random variables, induces the means, variances and correlations appropriate to interest rate fluctuations over seven days, and then computes the resulting value of the derivative, subsuming the function g_k from equation (2). This is repeated independently many times, and the mean and standard deviation of the resulting histogram are reported.

1. Histogram of cap values after seven days

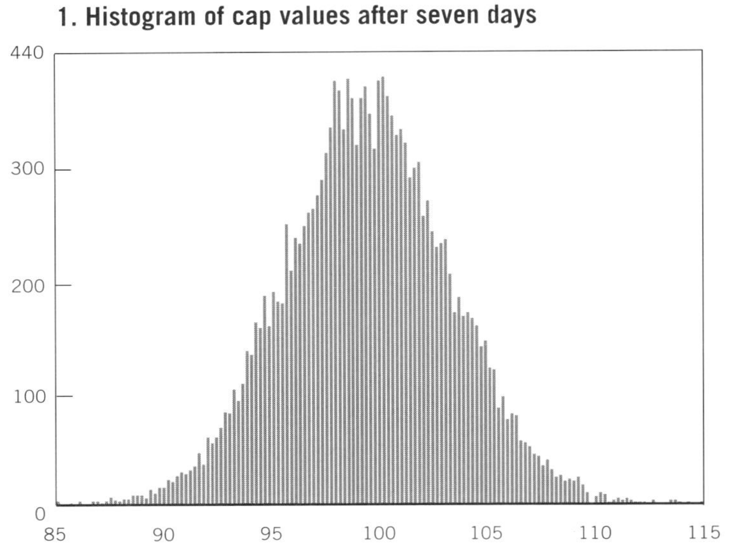

Figure 1 shows such a histogram for the cap. This histogram shows some "shoulders", which may represent caplets moving in or out of the money, though we should caution that this effect is quite subtle and may not be statistically significant. The histogram for the bond is similar but more nearly normally distributed.

Let n be the number of simulations used. Let U_i, $i=1,\ldots,n$ be n vectors of $d=6$ uniform random variables. That is, $U_i=(U_{i1},\ldots,U_{i6})$ and all of the U_{ij} are independent $U[0,1]$ random variables. Let Y_i be the value of the instrument over seven days, and let f be the function that turns uniform random variables into Gaussian random variables, induces means, variances and correlations, computes new interest rates and then finds the instrument's value.

Then $Y_i=f(U_i)$, the estimated mean is

$$\bar{Y} = \frac{1}{n}\sum_{i=1} Y_i$$

and the estimated standard deviation is

$$s_y = \sqrt{\frac{1}{n}\sum_{i=1}^{n}(Y_i - Y)^2} \qquad (3)$$

or perhaps

$$s_y^* = \sqrt{\frac{1}{n-1}\sum_{i=1}^{n}(Y_i - \bar{Y})^2} \qquad (4)$$

The difference between s_y and s_y^* is not important for large n in MC simulation. It can matter for RQMC simulation as described below, where s_y is to be preferred, at least for large n.

Standard statistical methods can be used to form approximate confidence intervals for these estimates. For example, $\bar{Y} \pm 1.96 s_y \sqrt{n}$ is an approximate 95% confidence interval for μ. Approximate confidence intervals for σ are less widely known. Miller (1986) mentions some. The bootstrap (see Efron and Tibshirani, 1993), the jackknife (see Miller, 1974) or non-parametric likelihood ratio methods (see Owen, 1990) can also be used.

Quasi-Monte Carlo

The value $\bar{Y}$ is estimating a sixfold integral

$$\mu = \int_{[0,1]^6} f(U)dU$$

and the value s_y is estimating

$$\sigma = \sqrt{\int_{[0,1]^6} (f(U) - \mu)^2 \, dU}$$

$$= \sqrt{\int_{[0,1]^6} (f(U))^2 dU - \left(\int_{[0,1]^6} (f(U))dU\right)^2}$$

Therefore if these six-dimensional integrals can be done more accurately than by Monte Carlo simulation, the VAR can be calculated more quickly or more accurately. Quasi-Monte Carlo integrals are a promising way to get better answers.

In QMC n vectors $U_1, \ldots, U_n$ are carefully constructed for use as above. Under Monte Carlo (MC), the points U_i tend to have some clumps and to leave some open spaces. Under QMC, the points U_i are constructed to have a much more even distribution.

Niederreiter (1992) gives an excellent account of Quasi-Monte Carlo methods. They can certainly be much more accurate than Monte Carlo methods, though it is still a research issue to determine what sort of integrands can be better integrated by QMC and for what value of n the improvement can be seen.

The QMC methods considered in this paper are known as (t,s)-sequences and (t,m,s)-nets. Their properties and some constructions are given by Niederreiter (1992). Some details are summarised below.

A drawback of deterministic QMC methods is that there is no practical way of estimating their accuracy from the values $Y_i = f(U_i)$.

Randomised QMC

By carefully randomising a QMC method, it is possible to produce random points $U_1, \ldots, U_n$ that are individually uniformly distributed on $[0,1]^d$ while avoiding the clumps and gaps of simpler random points.

Owen (1995) describes several such methods, including randomised (t,m,s)-nets and (t,s)-sequences. These are constructed by taking the (t,m,s)-net or (t,s)-sequence and scrambling their digits (in some base $b \geq 2$) at random. If the unscrambled points are a (t,m,s)-net or (t,s)-sequence, then the resulting points are still a (t,m,s)-net or (t,s)-sequence, respectively.

Moreover, the resulting points $U_1 \ldots, U_n$ individually have the uniform distribution over $[0,1]^d$.

This uniformity makes $\bar{Y}$ an unbiased estimator of μ. By reducing the gaps and clusters among the U_i, the variance of $\bar{Y}$ is usually smaller under randomised QMC than it would be under MC. Carefully stated and proved theorems on this point are given in Owen (1997a). In summary, the RQMC variance divided by the MC variance typically tends to zero as n tends to infinity. For finite n the QMC variance can never be much larger than the MC variance, and the largest ratios are obtained at quite implausible integrands. The improvements in RQMC are greatest for integrands that are well approximated by sums of smooth (or at least coarse) functions of small dimension.

Surprisingly, randomising nets can improve their accuracy for integration problems. Owen (1997b) shows that for smooth integrands the variance of randomised net estimates is eventually of order $n^{-3}(\log n)^{d-1}$, yielding an error rate that is better by a factor of $n^{-1/2}$ than the one for unrandomised nets.

For an estimate of the uncertainty, take $r > 1$ independent replications of the scrambled net. The results $\bar{Y}_j$, $j = 1,\ldots,r$, can be averaged to get $\bar{\bar{Y}}$. An unbiased estimate of the variance of $\bar{\bar{Y}}$ is

$$\frac{1}{r(r-1)} \sum_{j=1}^{r} (\bar{Y}_j - \bar{\bar{Y}})^2$$

Because r is not necessarily large the denominator $r(r-1)$ may be significantly better than r^2. That is, equation (4) is preferred to equation (3) for a small number r of independent replicates.

Owen (1997a) describes another method of constructing variance estimates. Instead of using r independently scrambled nets, a single large net is split into r equal length pieces. When this is done carefully, each of the pieces is also a net. Then $\bar{Y}$ estimated this way is simply the average of all of the Y values seen. Splitting one net into r subnets will usually make $\bar{Y}$ more accurate than using equally many function values taken from r independently randomised nets.

It will also usually be the case that the variance estimates obtained by using r pieces of a net are biased. On the average, over randomisations, these estimates overstate the true sampling variance of the statistic of interest, providing conservative estimates of uncertainty.

This section has described estimation of an average, not a standard deviation, even though standard deviations are of more importance here. The section on bias in standard deviations below considers the bias issues that arise when considering such statistics. For statistics such as the standard deviation, this bias is not ordinarily detrimental.

Bond example

The bond we consider is a zero coupon bond with a 20-year maturity. The value of this bond in seven days depends on the treasury curve as described above. We simulated via historically determined distributions using MC, QMC and RQMC.

The MC method used independent identically distributed observations. The QMC method used was a Sobol' sequence. The RQMC method used was a scrambled (t,m,s)-net – specifically, a $(0,5,6)$-net in base 7. This net has $7^5 = 16{,}807$ points in it, and is in fact the first 16,807 points of a $(0,6)$-sequence in base 7. For comparison, 16,807 MC and Sobol' points were also used, though using only the first $2^{14} = 16{,}384$ Sobol' points might be preferable, since they are generated in base 2.

Figure 2 shows the running standard deviation, s, based on the first n points of each series. The number of input points, n, is plotted on a logarithmic scale.

It is hard to discern the patterns in Figure 2 because the later points are so close together and the earlier points are erratic. Figure 3 shows the same running standard deviations for $n = \lambda 7^m$ for integers λ and m for which $49 \leq n \leq 16{,}807$. By dropping points with $n < 49$, the vertical scale of the plot can be refined. When using the first n points of a randomised (t,s)-sequence in base b, it is best to use $n = b^m$ for some power m or λb^m for integer λ. Thus the points plotted are the best ones for RQMC accuracy.

In plots like Figure 2, it can appear that two methods are settling down to different answers. Sometimes this is an optical illusion, at least for MC. In MC plots, consecutive points on the curve are typically in the order of $1/n$ apart, while they are in the order of $1/\sqrt{n}$ away from the true answer.

So two different MC plots can appear to have settled down to different answers, simply because the fluctuations within a curve are no guide to the error of the estimate at the end of the curve. Of course this could also arise if there were problems in the random number generator as reported by Paskov and Traub (1995). By spacing the points out nearly equally on the logarithmic scale, the fluctuations in the MC curve of Figure 3 are of a magnitude comparable to the error in the curve.

With RQMC the answers within a curve like Figure 2 also fluctuate by an amount of the order of $1/n$, but the errors in the estimate at points $n = \lambda b^m$ for large enough n (and smooth f) are of the order of $1/n\sqrt{n}$ (times logarithmic factors).

Thus with RQMC the fluctuations due to taking $n = \lambda b^m + 1$ can result in much greater error than taking $n = \lambda b^m$. For QMC the fluctuation in adding a point can be within logarithmic factors of the error magnitude.

It is possible to estimate the accuracy of the RQMC answer by independent repetitions. As we have already seen, there are reasons to prefer splitting the single series into subseries and treating them as repetitions. The original $7^5 = 16{,}807$ RQMC points were split into seven data sets of $7^4 = 2401$ consecutive observations. The same

2. Bond value standard deviation vs number of simulations

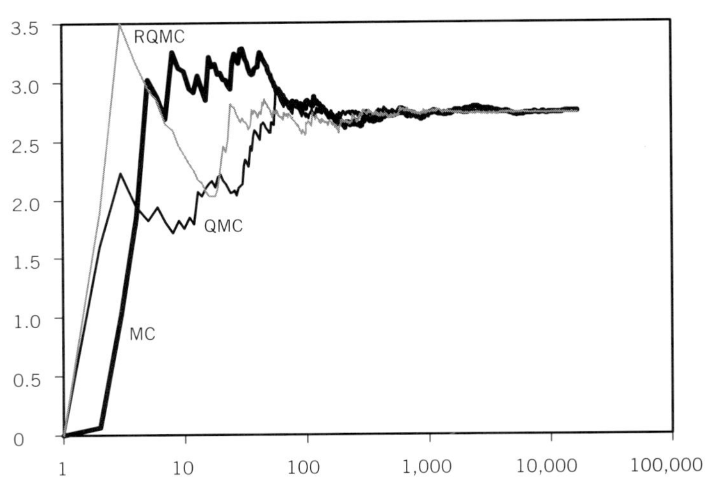

3. Bond value standard deviation versus number of simulations, at selected values of n

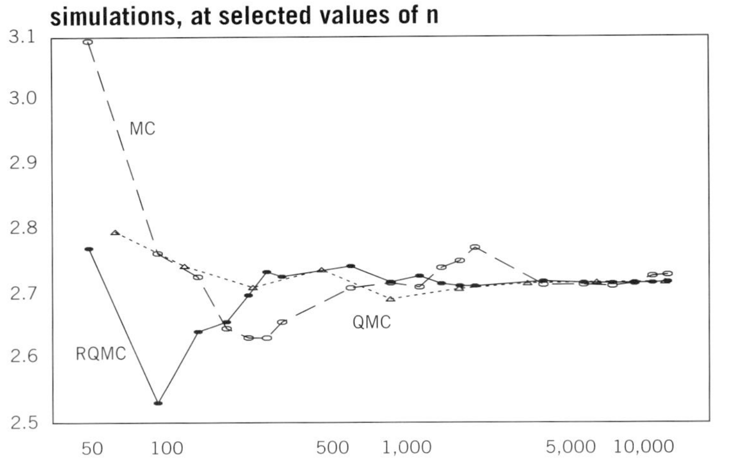

was done for the MC points. Figure 4 shows trajectories like the ones in Figure 3 for the seven data subsets generated by each method. Since the QMC points are Sobol′ points, there is no reason to consider them in groups of size equal to a power of seven. A better comparison is to split the first $2^{14} = 16{,}384$ Sobol′ points into eight groups of $2^{11} = 2048$ points.

It is clear from Figure 4 that the running standard deviations from the QMC and RQMC points settle down much faster than the ones from MC points. The RQMC points appear to be slightly better than the QMC ones.

The mean of the seven values of the RQMC points at $n = 2401$ is $\bar{s} = 2.7366$. All empirical values are given to five significant figures, except for a few of them, referred to as approximate. The standard deviation of all 16,807 RQMC values is 2.7367. The standard deviation of the seven RQMC values is 0.0058684. This allows us to estimate the sampling standard deviation of $\bar{s}$ by $0.0058684/\sqrt{7} = 0.0022180$. Adding plus or minus two of these standard errors to the estimate gives the interval (2.7322, 2.7411).

The MC points can be analysed in several ways, but for comparison we apply the same method as used for the RQMC points. The mean of the standard deviations from the seven data subsets is 2.7469. The standard deviation of them is 0.041770 and the corresponding interval is (2.7153, 2.7785).

4. Bond value standard deviation, on data subsets

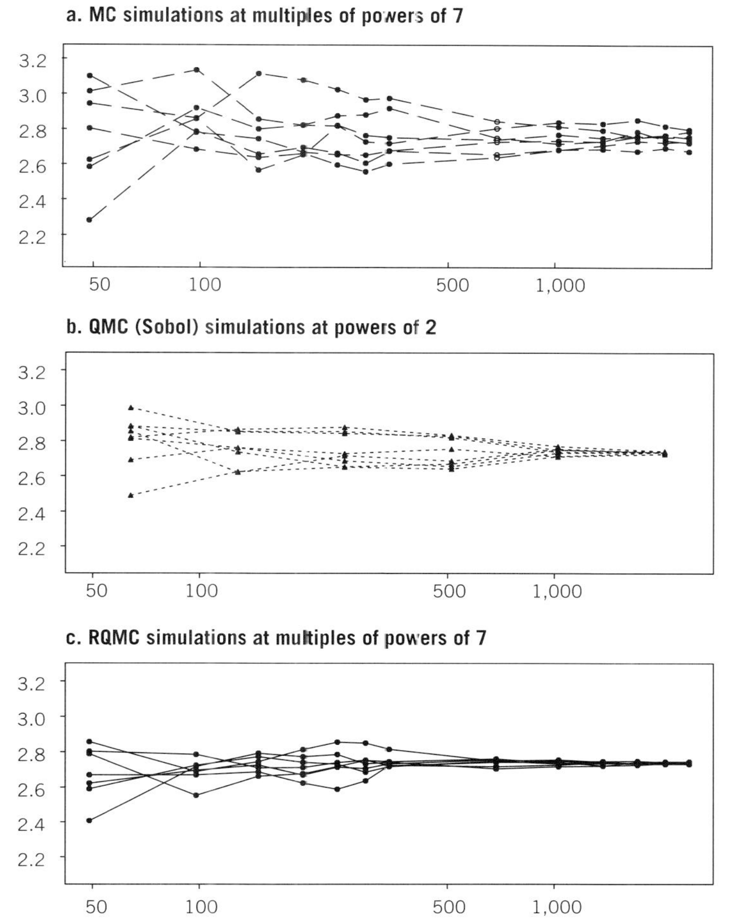

The variance of the MC values is just over 50 times as large as the RQMC points, indicating that for this problem at least RQMC has made a big improvement. Roughly speaking, comparable accuracy could be obtained with $50 \times 2401 \doteq 120{,}000$ MC simulations as with 2401 RQMC simulations. Theory suggests that the efficiency ratio is even more favourable to RQMC for the actual sample size of 16,807. As mentioned above there is reason to believe that the variation among the RQMC points overestimates their true variation, and hence the ratio might well be larger than 51.

The MC interval contains the RQMC interval and is approximately 7.1 times as wide. The centre of the MC interval does not lie inside the RQMC interval, and there is no reason to expect that it should. Similarly, the standard deviation of all 16,807 MC points is 2.7474, which is not in the RQMC interval, and neither is there any reason to expect that it should be in that interval. The standard deviation of the first 16384 Sobol′ points is 2.7366, and using all 16807 points only changes the final digit to 9 from 6. The QMC value is well within the RQMC interval, but the QMC method does not provide an interval of uncertainty.

Cap example

This section considers an interest rate cap related to the bond of the previous section.

The cap is a 20-year cap with semi-annual caplets and a cap rate close to the initial value of the forward rate. The value of the cap depends on the same interest rates as the bond and, as one would expect, it is negatively correlated with the bond.

The analysis of the cap data is the same as used for the bond data. Figures 5 and 6 are the same as Figures 3 and 4, with cap values substituted for bond values. Once again, the QMC and RQMC points show much more rapid convergence than the MC points, and the RQMC points appear slightly better than the QMC ones.

For RQMC the mean of the seven data subset values is 3.7194, the standard deviation among them is 0.0075330 and the plus or minus two standard error interval is (3.7137, 3.7251). For MC the mean of the seven data subset values is 3.7374, the standard deviation among them is

0.062510 and the interval is (3.6902, 3.7847). The ratio of variances is approximately 69, in favour of the RQMC method. The QMC value based on the first 16,384 observations is 3.7202.

Conclusions

For both VAR problems considered here, the Monte Carlo method was much slower to converge to the answer than either the QMC or RQMC method. The QMC method appeared in the plots to be somewhat slower than RQMC, while the RQMC method allowed data-based estimates of the uncertainty in the VAR.

The standard deviation of the bond value over seven days was estimated to be 2.7367%. To this number we can attribute a sampling standard deviation of about 0.0022%, a relative error smaller than one part in a thousand. The standard deviation of the cap value over seven days was estimated to be 3.7194% with a sampling standard deviation of about 0.0028%.

In this case the RQMC method appeared to be about 70 times better than MC judged by efficiency (≈ 50 for bond, ≈ 70 for cap). A ratio like this depends on the target function and the sample size. Theory says that as the sample size goes to infinity for a fixed integrand, this ratio does too.

For fixed n one can construct integrands for which the RQMC variance is higher than the MC variance, but that ratio cannot be large. On other integrands one has no reason to expect QMC methods to be much better than RQMC methods because RQMC methods are themselves QMC methods. More details of this theory are given below and in the bibliography.

5. Cap value standard deviation versus number of simulations, at selected values of n

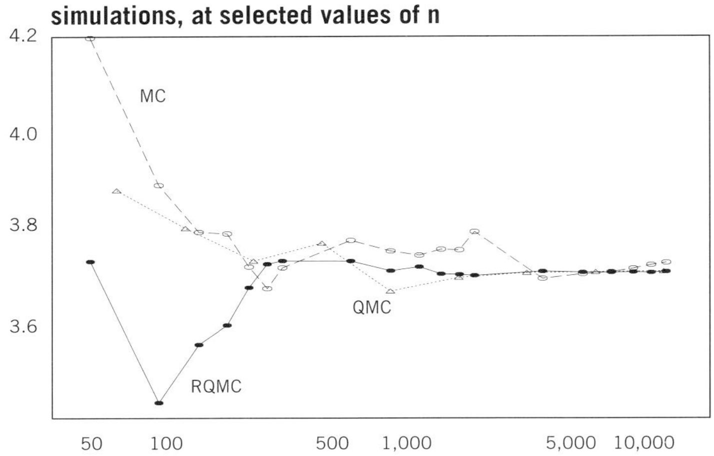

6. Cap value standard deviation, on data subsets

a. MC simulations at multiples of powers of 7

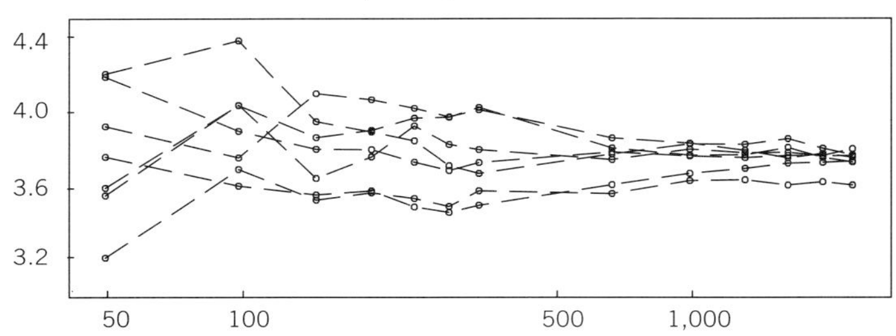

b. QMC (Sobol) simulations at powers of 2

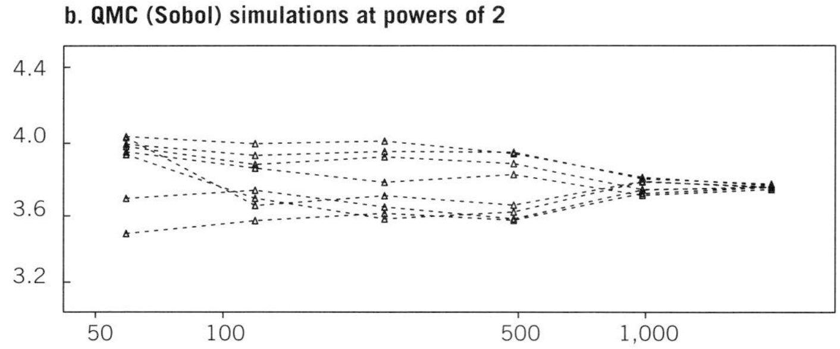

c. RQMC simulations at multiples of powers of 7

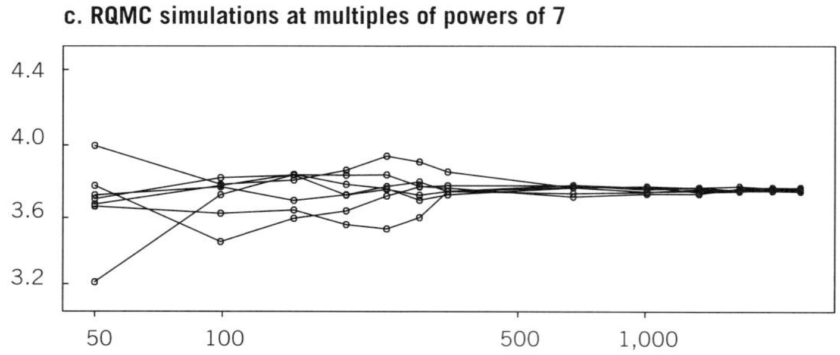

Statistical issues

We now address some statistical questions that arise in this paper. First we consider the differences that arise in simulation-based estimation of a standard deviation, as opposed to estimation of a mean. Then we consider why the usual practice of dividing by $n-1$ instead of n can in fact be harmful in RQMC simulations.

BIAS IN STANDARD DEVIATIONS

The description of RQMC and MC above assumed that the risk manager was interested in the average $\mu = E(Y)$, whereas in VAR calculations the risk manager is interested in the standard deviation $\sigma = \sqrt{V(Y)}$. It is well known that

$$\bar{Y} = \frac{1}{n}\sum_{i=1}^{n} Y_i$$

is unbiased for μ under MC, and this is also true under RQMC. Under MC s_y^{*2} is unbiased for σ^2, while both s_y and s_y^* have some bias as estimators of σ.

We now consider whether this bias is important, and what happens with RQMC. Let us consider the situation where the risk manager wants to estimate a smooth function $g(\mu)$ by $g(\bar{Y})$. For non-linear functions g, there is usually some bias in this estimate. Simple Taylor expansion arguments show that the bias is

$$E(g(\bar{Y})) - g(\mu) \doteq g''(\mu)V(\bar{Y})/2$$

and that the variance is

$$V(g(\bar{Y})) \doteq g'(\mu)^2V(\bar{Y})$$

where g' and g'' are the first two derivatives of g.

These both hold for MC and for RQMC. The squared error in $g(\bar{Y})$ is its variance plus the square of its bias. Because $V(\bar{Y})$ tends to zero as n increases, for large n the contribution of the bias squared ordinarily becomes negligible compared to the variance for both MC and RQMC. Any exceptions involve special cases with $g'(\mu)$ equal or nearly equal to zero or $g''(\mu)$ infinite or extremely large.

Because $V(\bar{Y})$ tends to zero faster under RQMC than under MC, the bias becomes negligible more quickly under RQMC than under MC.

The standard deviation is not of the form $g(E(Y))$ but it is of the form $g(E(Y),E(Y)^2)$ and we estimate it by $g(\bar{Y}, \overline{Y^2})$ where

$$\overline{Y^2} = (1/n)\sum_{i=1}^{n} Y_i^2$$

The same analysis applies except that the Taylor expansions have to become bivariate ones.

So, for either MC or RQMC, and large enough n, the bias ordinarily becomes negligible.

WHICH IS BETTER – s OR s*?

Should one estimate a standard deviation by s or by s^*? The ratio s/s^* is equal to

$$\sqrt{(n-1)/n} \doteq 1 - 1/(2n)$$

Therefore the choice of estimator makes a relative difference of roughly $1/2n$ in the value of the answer. For large enough n and MC sampling the issue is moot, because the estimate itself will differ from the truth by a larger amount of order $1/\sqrt{n}$.

For one of the data subsets, $1/2n$ is $1/2 \times 2401$ or approximately 0.0002 for each of the seven replicates. This results in a relative difference of less than 0.5% of the confidence interval width for each MC example.

Because RQMC is usually more accurate than MC, the difference between s and s^* can matter. In the examples the difference between using these two methods is approximately equal to 3% of the width of the intervals. For most purposes this is negligible, but in other settings the difference could be important.

As n goes to infinity, for smooth functions, the RQMC variance eventually decreases proportionally to n^{-3} times logarithmic factors. By the argument above the error in s is ordinarily of order $n^{-3/2}$, times logarithmic factors. Therefore the difference between s and s^* is eventually larger than the error in s, and so s is to be preferred for large enough n.

Why is it that the argument about bias in standard deviations above does not also apply to s^*? The reason is that s^* is not expressible as a function $g(\bar{Y}, \overline{Y^2})$, since it also depends on n. We can write $s^* = g(\bar{Y}, \overline{Y^2}, 1/n)$ with $\sigma = g(E(Y), E(Y^2), 0)$ for an appropriate function g but we cannot eliminate the effect of n.

Randomised (t,m,s)-nets and (t,s)-sequences

This section presents some background on (t,m,s)-nets and (t,s)-sequences based on Niederreiter (1992), and their randomisation based on Owen (1995, 1997a, 1997b.) For this section, we follow the conventions of using s to denote the dimension of the input space, and of integrating over the half-open space $[0,1)^s$.

DEFINITIONS FOR NETS

Let $d = s \geq 1$ and $b \geq 2$ be integers. An *elementary interval* in base b is a hyperrectangular "cell" of the form

$$E = \prod_{j=1}^{s}\left[\frac{c_j}{b_j^{k_j}}, \frac{c_j + 1}{b_j^{k_j}}\right)$$

for integers k_j, c_j with $k_j \geq 0$ and $0 \leq c_j < b^{k_j}$. The cell E has volume b^{-r}, where $r = \sum_{j=1}^{s} k_j$.

Let $m \geq 0$ be an integer. A finite sequence of points X_i, $i = 1, \ldots, b^m$ from $[0,1)^s$ is a (0,m,s)-net in base b if every elementary interval E in base b of volume b^{-m} has exactly one of the points. That is, every elementary interval that "should" have one point of the sequence does have one point of the sequence.

This is a very strong form of equidistribution and, by weakening it somewhat, constructions for more values of s and b become available. Let $t \leq m$ be a non-negative integer. A finite sequence of b^m points from $[0,1)^s$ is a (t,m,s)-net in base b if every elementary interval in base b of volume b^{t-m} contains exactly b^t points of the sequence. Elementary intervals that "should" have b^t points do have b^t points, though elementary intervals that "should" have one point might not.

For $t \geq 0$, an infinite sequence $(X_i)_{i \geq 1}$ of points from $[0,1)^s$ is a (t,s)-sequence in base b if for all

$k \geq 0$ and $m \geq t$ the finite sequence

$$(X_i)_{i=kb^m+1}^{(k+1)b^m}$$

is a (t,m,s)-net in base b.

The advantage of a (t,s)-sequence is that if the first b^m points are not found to be sufficient for an integration problem, another b^m points can be found that also form a (t,m,s)-net and tend to fill in places not occupied by the first net. If one continues to the point of having b such (t,m,s)-nets, then taken together they comprise a $(t,m+1,s)$-net.

For the case $s=6$ in this paper, we used a $(0,5,6)$-net in base 7 obtained from the first $7^5=16{,}807$ points of $(0,6)$-sequence in base 7, and a $(8,14,6)$-net in base 2 obtained from the first $2^{14}=16{,}384$ points of a $(8,6)$-sequence in base 2.

In the former, elementary intervals in base 7 of volume $7^{-5}=1/16{,}807$ each contain one point, and in the latter, elementary intervals in base 2 of volume $2^{8-14}=1/64$ each contain $2^8=128$ points. The $(0,6)$-sequence in base 7 is a randomised version of a Fauré (1982) sequence and the $(8,6)$-sequence in base 2 is a Sobol′ (1967) sequence.

PROPERTIES OF NETS

The theory of (t,m,s)-nets and (t,s)-sequences is given in Niederreiter (1992). A famous result of the theory is that integration over a (t,m,s)-net can attain an accuracy of order $O(\log(n)^{s-1}/n)$, while using (t,s)-sequences worsens this slightly to $O(\log(n)^s/n)$. These results hold for integrands of bounded variation in the sense of Hardy and Krause.

For large s, it takes unrealistically large n for these rates to be clearly better than $n^{-1/2}$, but in examples they seem to outperform simple Monte Carlo. The construction of (t,m,s)-nets and (t,s)-sequences is also described in Niederreiter (1992).

Here we remark that, for prime numbers s, a construction by Fauré (1982) gives $(0,s)$-nets in base s, and Niederreiter extended the method to prime powers s. See Niederreiter (1992) for details.

SCRAMBLING NETS

Owen (1995) describes a scheme to randomise (t,m,s)-nets and (t,s)-sequences. The points are written in a base b expansion and certain random permutations are applied to the coefficients in the expansion. The result is to make each permuted X_i uniformly distributed over $[0,1)^s$ while preserving the (t,m,s)-net or (t,s)-sequence structure of the ensemble of X_i.

Let X_i^j denote the jth component of the ith point in a (t,m,s)-net in base b. Now write $X_i^j=\sum_{k=1}^{\infty} x_{ijk} b^{-k}$, where x_{ijk} is an integer between 0 and $b-1$ inclusive. The scrambled version of X_i^j is to be $U_i^j=\sum_{k=1}^{\infty} u_{ijk} b^{-k}$, with u_{ijk} described below.

To scramble the first component, X_1^1 through X_n^1, replace each x_{i11} by $u_{i11}=\pi(x_{i11})$, where π is a randomly drawn permutation from the $b!$ possible permutations of digits 0 through $b-1$. This may be visualised as chopping the cube $[0,1)^s$ into b slices along the X^1 axis and scrambling them. Then take each of those b slices, slice it into b pieces and scramble the pieces as before. This has the effect of permuting the second digit x_{i12}. Continue in this way, using b^{k-1} independent random permutations for the kth base b digit of X_i^1, for $k=1,2,3, \ldots$. The other components X_i^j for $j=2,3, \ldots, s$ are scrambled the same way, independently of each other and of the first component. In practice one can usually stop scrambling after step k where b^{-k} is small compared to the floating point precision used.

One result of applying all these permutations is that each U_i individually has the uniform distribution on $[0,1)^s$, and so the sample estimate $n^{-1}\sum_{i=1}^{n} f(X_i)$ is unbiased for $\int f(X)dX$. Furthermore, if the original X_i are a (t,m,s)-net then so are the U_i and the result of scrambling a (t,s)-sequence is another (t,s)-sequence.

On some test integrands in Owen (1995), the randomised nets appeared to be somewhat more accurate than their unrandomised counterparts. A theoretical confirmation of this appears in Owen (1997b), including an explanation due to Hickernell (1996).

For any integrand f in $L^2[0,1)^s$ the variance of $\hat{I}$ under scrambled net integration is eventually smaller than any positive multiple of $1/n$, along a sequence of the form $n=\lambda b^m$, $1 \leq \lambda < b$, $0 \leq m$. See Owen (1997a). For smooth enough f, this variance is $O(n^{-3}(\log n)^{s-1})$, leading to typical errors of order $n^{-3/2}(\log n)^{(s-1)/2}$ in probability. This is superior to the rate $n^{-1}(\log n)^{s-1}$ obtained for unrandomised nets. See Owen (1997b). The former rate describes the average case over randomisation for a fixed function. The latter describes the worst case over integrands for a fixed sequence of points, but there is reason to believe that a typical fixed function will have the same rate, only a better constant. Thus, scrambling can improve asymptotic accuracy as well as allow data-based estimation of attained accuracy.

Hickernell (1996) has studied the rate achieved by the worst-case integrand (of bounded variation) averaged over randomisations of nets and finds a rate of $O(n^{-1}(\log n)^{(s-1)/2})$ for randomised (0,m,s)-nets. This rate is a factor of $n^{-1/2}$ worse than the asymptotic rate for scrambled nets on smooth integrands. Thus, if one picks the integrand pessimistically after observing how the net was randomised, one loses the asymptotic advantage of scrambling the nets.

BIBLIOGRAPHY

Caflisch, R.E., and W. Morokoff, 1996, "Quasi-Monte Carlo Computation of a Finance Problem", Technical Report CAM Report 96-16, Department of Mathematics, UCLA.

Efron, B.M., and R.J. Tibshirani, 1993, *An Introduction to the Bootstrap*, Chapman and Hall.

Fauré, H., 1982, "Discrépance de suites associées à un système de numération (en dimension s)", *Acta Arithmetica*, 41, pp. 337–51.

Hickernell, F.J., 1996, "The mean square discrepancy of randomised nets", Technical Report MATH-112, Department of Mathematics, Hong Kong Baptist University.

Miller, R.G., 1974, "The Jackknife – a Review", *Biometrika*, 61, pp. 1–15.

Miller, R.G., 1986, *Beyond ANOVA, Basics of Applied Statistics*, Wiley.

Niederreiter, H., 1992, *Random Number Generation and Quasi Monte Carlo Methods*, SIAM, Philadelphia, PA.

Owen, A.B., 1990, "Empirical Likelihood Ratio Confidence Regions", *Annals of Statistics* 18, pp. 19-120.

Owen, A.B., 1995, "Randomly Permuted (t,m,s)-nets and (t,s)-sequences", pp. 299–317 in *Monte Carlo and Quasi-Monte Carlo Methods in Scientific Computing* (H Niederreiter and P.J.S. Shiue, eds) Springer-Verlag: New York.

Owen, A.B., 1997, "Monte Carlo Variance of Scrambled Equidistribution Quadrature", *SIAM Journal of Numerical Analysis* 34.

Owen, A.B., 1997, "Scrambled Net Variance for Integrals of Smooth Functions", *Annals of Statistics* 25(4), pp. 1,541–62.

Paskov, S., and J. Traub, 1995, "Faster Evaluation of Financial Derivatives", *Journal of Portfolio Management* 22, 1, pp. 113–20.

Sobol', I.M., 1967, "The Distribution of Points in a Cube and the Accurate Evaluation of Integrals" (in Russian), *Zh. Vychisl. Mat. I Mat. Phys.* 7, pp. 784–802.

DETERMINISTIC METHODS

Introduction

Bruno Dupire

This section is devoted to deterministic methods, covering low-discrepancy sequences, lattice methods and numerical integration. It is natural to question the presence of such techniques in a book dedicated to Monte Carlo simulations, which are essentially concerned with randomness. The methods analysed in this section are not random – they always produce the same results. However, they address the same problems as Monte Carlo simulations and are favoured increasingly by practitioners. Low-discrepancy sequences are even commonly named "quasi-random" or "quasi-Monte Carlo".

All these methods attempt to compute expectations or integrals of high dimension as an average of the value function sampled at what is hoped are well-chosen points. Ideally, those points should be representative in some sense. The bulk of this section deals with low-discrepancy sequences, which attempt to fill the sample space evenly.

Chapter 24, by Joy, Boyle and Tan, is a good introduction to the subject. It describes the bit inversion construction of Faure sequences and compares them with Monte Carlo draws for Asian options in various dimensions, basket options and the Heath–Jarrow–Morton (HJM) model. The appendix provides a nice account of low-discrepancy convergence analysis.

A discussion of new methodologies for valuing derivatives, by Spassimir Paskov, follows (Chapter 25). It applies Sobol' and Halton low-discrepancy sequences to collateralised mortgage obligation (CMO) tranches valuation. These are complex path-dependent mortgage-backed securities, reputedly a high-dimension (360) problem. Paskov defines the concept of effective dimension, conducts numerical tests and concludes that Sobol' sequences outperform Halton sequences, which in turn are superior to standard or antithetic Monte Carlo simulations.

The following chapter, by Caflisch, Morokov and Owen, tackles the same problem and elaborates on the concept of effective dimension. They randomise low-discrepancy sequences (scrambled nets) and generate paths through the Brownian bridge construction, as opposed to the classical step-by-step procedure. They stress that the performance of low-discrepancy sequences depends critically on the integrand.

In Chapter 27, Moskowitz and Caflisch examine the performance of low-discrepancy sequences in high dimension. Though the theoretical order of the method is 1, the observed rate of convergence in high dimension is lower. It seems that the theoretical order is achieved only after a very long cycle has been completed. In high dimension, this takes it out of the reach of computer power. Another issue here is the smoothness of the integrand, which plays an important role in the error estimation through the variation that comes into play in the Koksma–Hlawka inequality. The Brownian bridge construction is used to reduce the dimension. This chapter takes a mathematical, rather than a financial, viewpoint but is still enlightening for financial applications.

Papageorgiou and Traub, in Chapter 28, apply modifications of Faure and Sobol' sequences to the CMO valuation. They define a speed-up measure and conclude that the generalised Faure sequence (suspected to be an IBM over-publicised method) is best.

In Chapter 29, Rupert Brotherton-Ratcliffe investigates the efficiency of Sobol' sequences applied to Asian option pricing in various dimensions. He finds that they always perform better than Monte Carlo methods, although the highest dimension tested is 48.

Recombining trees are sometimes termed lattices. The lattices examined in Chapter 30, by Stetson, Marshall and Loebell, and in Chapter 31, by Barrett, Moore and Wilmott, are of a different type. They are uniformly spaced sets of points wrapped around a torus of higher dimension. Ergodic theory says that, if the generators of the mesh are irrational, the points will be dense and uniformly distributed. The more difficult it is to approximate them by rational numbers, the lower the discrepancy. This method is easy to implement and is claimed to be specially efficient in high dimension.

Deterministic sequences are attractive to practitioners – they match the intuition of filling nicely the space of interest, and they always produce the same result. However, Monte Carlo simulations are extremely easy to implement, robust with regard to the dimension of the problem, and they have this subtle touch of mystery.

24

Quasi-Monte Carlo Methods in Numerical Finance

Corwin Joy, Phelim P. Boyle and Ken Seng Tan
Positron Energy Consulting, Houston, Texas; School of Accountancy, University of Waterloo; Department of Statistics and Actuarial Science, University of Waterloo

This chapter introduces and illustrates a new version of the Monte Carlo method that has attractive properties for the numerical valuation of derivatives. The traditional Monte Carlo method has proven to be a powerful and flexible tool for many types of derivatives calculations. Under the conventional approach pseudo-random numbers are used to evaluate the expression of interest. Unfortunately, their use yields an error bound that is probabilistic which can be a disadvantage. Another drawback of the standard approach is that many simulations may be required to obtain a high level of accuracy. There are several ways to improve the convergence of the standard method. This chapter suggests a new approach which promises to be very useful for applications in finance. Quasi-Monte Carlo methods use sequences that are deterministic instead of random. These sequences improve convergence and give rise to deterministic error bounds. The method is explained and illustrated with several examples. These examples include complex derivatives such as basket options, Asian options, and energy swaps.

Numerical methods have come to play an increasingly important role in modern quantitative finance. This is due to the fact that most finance models have analytical solutions in only a few special cases. In many cases the price of a given financial instrument can be assumed to evolve according to a second-order partial differential equation. Different techniques are used to solve this equation and compute the sensitivities of the price to different variables. One approach is to discretise the differential equation and solve the resulting set of difference equations. Another popular technique is to use a lattice or tree approximation. Monte Carlo methods are often used when these methods are difficult to implement due to the complexity of the problem.

In recent years there has been renewed interest in Monte Carlo methods. This is due to the flexibility of the method in handling increasingly complex[1] financial instruments and the advent of powerful workstations has significantly reduced the execution time required for simulation. For example, Monte Carlo algorithms are normally more efficient than competing procedures if there are several state variables or other complexities. In addition the relevant algorithms are often fairly straightforward to design. Boyle (1977) – also Chapter 1 of this publication, provides an early discussion of the use of Monte Carlo methods in pricing options. The disadvantages[2] of Monte Carlo methods are that the error term is probabilistic and that it can be computationally burdensome to achieve a high level of accuracy. To reduce the computational burden of standard (or crude) Monte Carlo methods a variety of variance reduction methods have been proposed. There are many such techniques and they include control variates, antithetic variates, stratified sampling and importance sampling.

The purpose of this chapter is to introduce another technique for improving the efficiency of the Monte Carlo method. This method is known as the quasi-Monte Carlo method and was pioneered in computational physics where large-scale Monte Carlo calculations are common. The basic idea is to use numbers that are deterministic rather than random. These deterministic sequences can be used in place of the usual random (or pseudo-random) numbers to obtain faster convergence with known error bounds. The use of quasi-Monte Carlo methods leads to more efficient numerical procedures with the added benefit of deterministic error bounds. The monograph by Niederreiter (1992) gives a good survey of quasi-Monte Carlo methods.

QUASI-MONTE CARLO METHODS IN NUMERICAL FINANCE

In this chapter we introduce quasi-Monte Carlo methods and discuss different applications in numerical finance. In the first section we introduce the problem and motivate the method by analysing a simple European call. The price of this call can be expressed in terms of an integral and we discuss the connection between numerical integration and standard Monte Carlo approaches. In the second section we discuss the procedure for valuing complex derivative securities such as Asian options using the Monte Carlo approach and we examine ways to improve the convergence. In the third section we introduce a particular type of quasi-random sequences such as Faure sequences. These sequences have attractive properties for many applications in finance. We show that they provide swifter convergence and better error bounds than standard Monte Carlo methods in several practical settings. In later sections we use Faure sequences to value a range of complex derivative securities. We demonstrate that quasi-Monte Carlo methods are more efficient for obtaining prices and price sensitivities than standard Monte Carlo methods.

Use of deterministic sequences to value standard options

We begin by illustrating the use of deterministic sequences to value a standard European option. The option price is written as a discounted expectation over the terminal price distribution. We discuss the numerical valuation of this integral and its relationship to the standard Monte Carlo procedure. Then we show how a judicious choice of the "random" numbers can improve the accuracy of the procedure. The main idea is that since the use of random numbers introduces error we can do better by preselecting the numbers in a deterministic fashion.

The current price of a European derivative security can be written as its discounted expectation under the equivalent martingale (or Q measure):

$$\text{Price} = e^{-rT} E_Q[g(A(T))] \tag{1}$$

where T is the time until maturity, A(T) is the asset price at time T,…g(·) is the payoff function, and r is the riskless rate.

Assume that the asset price dynamics follow geometric Brownian motion with constant drift, μ, and volatility, σ. The current price of the derivative security is obtained by integrating the terminal payoff under the risk-neutral measure. Hence we have

$$\begin{aligned}
&\text{Price} \\
&= e^{-rT}\int_{-\infty}^{\infty} g\left[A(0)e^{((r-(\sigma^2/2))T+\sigma\epsilon\sqrt{T})}\right]\frac{1}{\sqrt{2\pi}}e^{-\epsilon^2/2}d\epsilon \\
&= \int_{-\infty}^{\infty} h(\epsilon)\frac{1}{\sqrt{2\pi}}e^{-(\epsilon^2/2)}d\epsilon \\
&= \int_0^1 h(\Phi^{-1}(x))dx \\
&= \int_0^1 f(x)dx
\end{aligned} \tag{2}$$

In the first line of the above equations the argument of g (the term in square brackets) denotes the terminal distribution of the underlying asset. Under our assumptions this asset has a lognormal distribution which we can write in terms of the exponential of an approximate normal distribution. Thus the variable has a normal distribution and so the final term is the standard normal density function. In the second line we have used the function h to make the expression more compact. The critical step is in the third line where we map the normal variate ϵ into the interval [0,1] through the inverse transformation.

We transform the integral of h from minus infinity to plus infinity to the integral of f(x) over the uniform distribution with range [0,1]. This simplification permits us to obtain efficient approximations to the integral by carefully selecting sample points in [0,1]. This technique leads to more efficient approximations of the original integral.

We illustrate the standard Monte Carlo method using the same example. First we generate n unit normal variates, $\epsilon_1, \epsilon_2, \ldots, \epsilon_N$. The crude Monte Carlo estimate of the option price is

$$\frac{1}{N}\sum_{i=1}^{N} h(\epsilon_i) \quad \text{where } \epsilon_i \sim N(0,1) \tag{3}$$

This estimate has an error of $O(N^{-1/2})$. However, if we were to use a more uniformly spaced grid by choosing the N evenly spaced quantiles of N(0,1), we would obtain the N-panel method (or midpoint rule) for evaluating

$$\int_0^1 f(x)dx \tag{4}$$

This has an error of $O(N^{-1})$. For example, if N = 4 and we wish to use the four-panel method we split the interval [0,1] into four equal subintervals

$$\left[0,\tfrac{1}{4}\right], \left[\tfrac{1}{4},\tfrac{2}{4}\right], \left[\tfrac{2}{4},\tfrac{3}{4}\right], \left[\tfrac{3}{4},1\right]$$

and then use the midpoint of each subinterval to evaluate the integral. This would be equivalent to using the following values of ϵ in the Monte Carlo routine:

$$\bar{\epsilon} = \{\Phi^{-1}[\tfrac{1}{8}], \Phi^{-1}[\tfrac{1}{4} + \tfrac{1}{8}], \Phi^{-1}[\tfrac{2}{4} + \tfrac{1}{8}], \Phi^{-1}[\tfrac{3}{4} + \tfrac{1}{8}]\}$$
$$= [-1.15035, -0.31864, 0.31864, 1.15035]$$

One difficulty with the panel method is that we may not know in advance how many simulations will be needed to obtain the desired level of accuracy. In fact, in actual applications we may want to continue a simulation until we reach a given level of precision. In this situation we need a sequence of numbers that is evenly spaced and that somehow "knows" how to fill in the gaps that have been left by previous elements in the sequence. Sequences with this property are known as low-discrepancy[3] sequences. We can think of the discrepancy as a quantitative measure of the deviation of the set of numbers from a uniform distribution. As a crude example of such a sequence, suppose we bisect the initial interval $[0,1]$, and then go on to bisect recursively the resulting subintervals. We would then obtain the following sequence of nested N-panel methods for any $N = 2^k - 1$, $k \geq 1$, by taking the midpoints of these intervals:

$$\bar{\epsilon} = \{\underbrace{\underbrace{\underbrace{\Phi^{-1}[\tfrac{1}{2}]}_{N=1}, \Phi^{-1}[\tfrac{1}{4}], \Phi^{-1}[\tfrac{3}{4}]}_{N=3}, \Phi^{-1}[\tfrac{1}{8}], \Phi^{-1}[\tfrac{3}{8}], \Phi^{-1}[\tfrac{5}{8}], \Phi^{-1}[\tfrac{7}{8}]}_{N=7}\} \tag{5}$$

contracts. Standard or (crude) Monte Carlo methods can be used to value such contracts. The continuous distribution is sampled at discrete time steps. Duffie (1992, p. 200) points out that if we replace the continuous-time process with a discrete-time process using s evenly spaced time steps then the error for the usual Euler discretisation is $O(s^{-1})$. This error should be added to whatever error is incurred in evaluating the discrete-time version of the integral. So, unless the security has a set of natural discrete times on which its value depends, the number of time steps should be chosen with care. Ideally one ought to choose a time step that has an error which matches the error introduced by estimating the discrete integral.

Suppose that the time step has already been chosen to obtain the discrete-time version of the continuous-time model. Assume that the path-dependent security's terminal value at time T depends only on the prices $A_1, A_2, \ldots, A_s$ at s intermediate times

$$0 \leq t_1, t_2, \ldots, t_s \leq T$$

We assume the price increments are independent, take the expected discounted payoff under the risk-neutral measure, and transform the

Valuation of complex derivatives

Our discussion of European options on a single asset shows that by a judicious choice of the "random" points we can improve the accuracy of the estimate. We saw that the trick is not to have any randomness at all but to select deterministic sequences. In the case of path-dependent options or options involving several assets where there are no analytical solutions, we are more likely to require Monte Carlo approaches. For these cases we require s-dimensional sequences of "random" numbers. We discuss this below and later show how to generate such sequences.

Consider an Asian option where the payoff is based on the arithmetic average of the asset prices taken at certain dates known as reset points. We restrict attention to European

integral as before to obtain

$$\text{Price} = \int_{I^s} f(x)dx \tag{6}$$

where $I^s = [0,1)^s$.

We might be tempted to use the multidimensional version of the trapezoidal[4] rule to evaluate this integral. Unfortunately the trapezoidal rule does not work well for a large number of time steps because the error bound is $O(N^{-2/s})$. Niederreiter (1992) gives a clear discussion of the issue of error bounds. If we use standard Monte Carlo methods using pseudo-random numbers the error bound is $O(N^{-1/2})$. Quasi-Monte Carlo methods which use deterministic sequences provide deterministic error bounds $O(N^{-1}(\log N)^s)$ for suitably chosen points. Such sequences are known as quasi-random sequences. There are several types of quasi-random sequences, such as Halton sequences, Sobol' sequences and Faure sequences. Fox (1986, p. 374) explains the advantages of Faure sequences. In the next section we show how Faure sequences are generated.

Quasi-random numbers: Faure sequences

The basic idea of quasi-Monte Carlo is to use a set of points that are carefully selected in a deterministic fashion. We call these numbers quasi-random even though they are perfectly deterministic and have no random component. In this section we describe how to generate quasi-random sequences and describe a particular class of quasi-random sequences known as Faure sequences. Faure sequences have certain advantages for the valuation of high-dimensional integrals. This leads to efficient algorithms for the computation of prices and price sensitivities for complex derivative securities.

We first consider the case of a one-dimensional sequence of quasi-random numbers. Let r be any prime number (≥ 2). Any integer n has a unique expansion in terms of base r. We can generate a number in the interval $[0,1]$ by reflecting the expansion in base r about the "decimal point." An example will clarify the procedure. Suppose $r = 3$ and $n = 7$. We can write 7 in base 3 as

$$7 = 2(3^1) + 1(3^0) = 21$$

When we reflect 21 (in base 3) about the "decimal point" we obtain

$$\phi_3(7) = \frac{1}{3} + \frac{2}{9} = \frac{5}{9}$$

This is clearly a number in the interval $[0,1]$. The next number in the sequence is $\phi_3(8) = \frac{8}{9}$ and the first nine numbers in this sequence are

$$\frac{9}{27}, \frac{18}{27}, \frac{3}{27}, \frac{12}{27}, \frac{21}{27}, \frac{6}{27}, \frac{15}{27}, \frac{24}{27}, \frac{1}{27}$$

Notice how the new numbers that are added tend to fill in the gaps in the existing sequence. The general expression for n in terms of the base r is

$$n = \sum_{j=0}^{m} a_j(n) r^j \tag{7}$$

Only a finite number of these $a_j(n)$ will be non-zero. The corresponding quasi-random number according to this procedure is

$$\phi_r(n) = \sum_{j=0}^{m} a_j(n) r^{-j-1} \tag{8}$$

Note that if we have a sequence of integers running from r^k to $r^{k+1} - 1$ then the maximum distance between any two points in the corresponding quasi-random sequence must be given by r^{-k}. This happens because the most significant digit in the representation of n in base r becomes the least significant digit when we reflect about the decimal point. This means that as points are added they "know" how to fill the gaps evenly. One-dimensional quasi-random sequences of this type are known as van der Corput sequences and are described by Niederreiter (1992) and Faure (1982).

There are different ways to generalise the one-dimensional quasi-random numbers described above to s-dimensional quasi-random sequences. For our purposes Faure sequences have considerable advantages. The original discussion of these sequences by Faure (1982) is somewhat terse. Fox (1986) and Niederreiter (1992) give more detailed discussions.

We now describe how to generate an s-dimensional Faure sequence. Let r be the smallest prime number that is $\geq s$ and ≥ 2. We start as before by representing any integer n in terms of base r as

$$n = \sum_{j=0}^{m} a_j^1(n) r^j \tag{9}$$

The first Faure number is given by reflecting about the "decimal point" as before:

$$\phi_r^1(n) = \sum_{j=0}^{m} a_j^1(n) r^{-j-1} \tag{10}$$

The remaining elements of the sequence are found recursively. First assume we know all the $a_j^{k-1}(n)$. The $a_j^k(n)$ are generated as follows:

$$a_j^k(n) = \sum_{i \geq j}^{m} {}^iC_j a_i^{k-1}(n) \bmod r \tag{11}$$

where[5] ${}^iC_j = i!/j!(i-j)!$. Thus the next level of coefficients is obtained by multiplying by an upper triangular matrix with elements

$$\begin{matrix} {}^0C_0 & {}^1C_0 & {}^2C_0 & {}^3C_0 \cdots \\ 0 & {}^1C_1 & {}^2C_1 & {}^3C_1 \\ 0 & 0 & {}^2C_2 & {}^3C_2 \\ 0 & 0 & 0 & {}^3C_3 \end{matrix}$$

Table 1. Three-dimensional Faure sequences

n	$a_0(n)$	$a_1(n)$	$a_2(n)$	ϕ^1_n	ϕ^2_n	ϕ^3_n
1	1	0	0	1/3	1/3	1/3
2	2	0	0	2/3	2/3	2/3
3	0	1	0	1/9	4/9	7/9
4	1	1	0	4/9	7/9	1/9
5	2	1	0	7/9	1/9	4/9
6	0	2	0	2/9	8/9	5/9
7	1	2	0	5/9	2/9	8/9
8	2	2	0	8/9	5/9	2/9
9	0	0	1	1/27	16/27	13/27
10	1	0	1	10/27	25/27	22/27
11	2	0	1	19/27	7/27	4/27

The successive points in the Faure sequence are obtained via

$$\phi^k_r(n) = \sum_{j=0}^{m} a^k_j(n) r^{-j-1}, \quad 2 \le k \le s \qquad (12)$$

that is, we reflect $a^k(n)$ about the decimal point to obtain $\phi^k(n)$.

This recursive procedure permits us to generate the s points corresponding to the integer n in the Faure sequence based on r ($\ge s$).

As an example, suppose that $s = 3$. In this case we take $r = 3$. The first few elements of the three-dimensional Faure sequence in this case are given in Table 1. To get a visual feel for how Faure sequences fill up the unit square, we have prepared Figures 1 and 2. In this case we selected $r = 2 = s$. We denote the nth element of the series by

$$\phi_n = (\phi^1_n, \ldots, \phi^s_n)$$

To apply these results to a Monte Carlo simulation based on normal random variables, we map the point in the s-dimensional cube by applying the inverse normal function to the sequence[6]

$$\epsilon_n = \Phi^{-1}(\phi_n) = \{\Phi^{-1}(\phi^1_n), \ldots, \Phi^{-1}(\phi^s_n)\} \qquad (13)$$

These Faure values may be substituted for the normal vectors that would be returned by a pseudo-random number generator. This makes the implementation simple since very little effort is need to upgrade from pseudo-random numbers to quasi-Monte Carlo numbers. Code to implement the above algorithm is available in Fox (1986). We can also use this procedure to generate correlated random variables using appropriate combinations of uncorrelated random variables. This is useful when we are dealing with an instrument whose payoff depends on the values of several assets.

1. Points 1 to 128

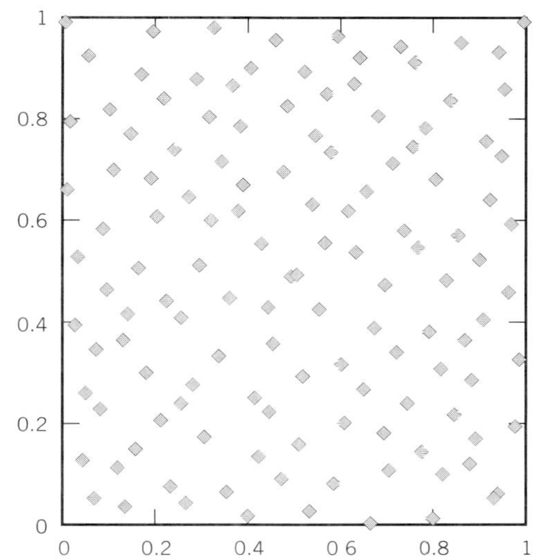

2. Points 1 to 512

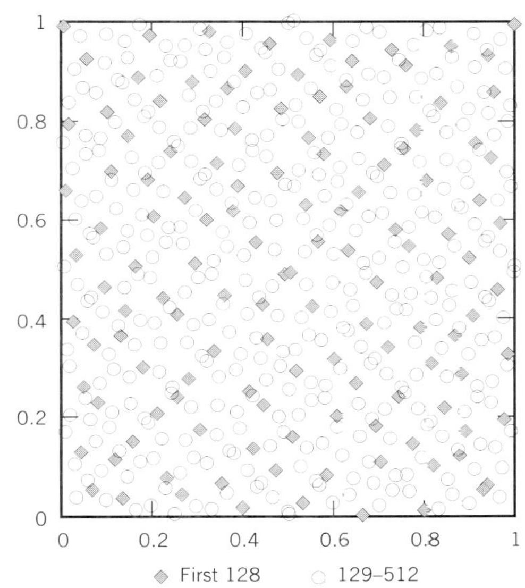

Applications: use of quasi-Monte Carlo for standard European options

In this section we illustrate the implementation of quasi-Monte Carlo methods in the case of plain vanilla European options. The exact values for the prices and sensitivities are available in these cases and we use these as benchmarks when examining convergence. We use graphs to compare the convergence properties of crude Monte Carlo with quasi-Monte Carlo. The graphs illustrate vividly the improvement in convergence when quasi-Monte Carlo methods are used.

Our base-case contracts are standard European calls and puts with the following parameters:

Current price, S	100
Strike price, S	100
Riskless rate, r	10%
Time to maturity, T	1 year
Volatility, σ	30%

The Black–Scholes option values for the standard call and put are

Call value 16.734 Put value 7.218

In Figure 3 we illustrate the convergence of the call option values under both the standard Monte Carlo approach and the quasi-Monte Carlo approach. The quasi-Monte Carlo numbers have been computed using Faure sequences (with r = 2). The numbers plotted represent the relative error in percentage between the estimated value and the true Black–Scholes value. Observe that the quasi-Monte Carlo estimates display better convergence and that the deviations from the true prices are much less erratic. In these applications we generally start the simulations at $n = r^4$ as recommended by Fox (1986).

3. Standard European call option

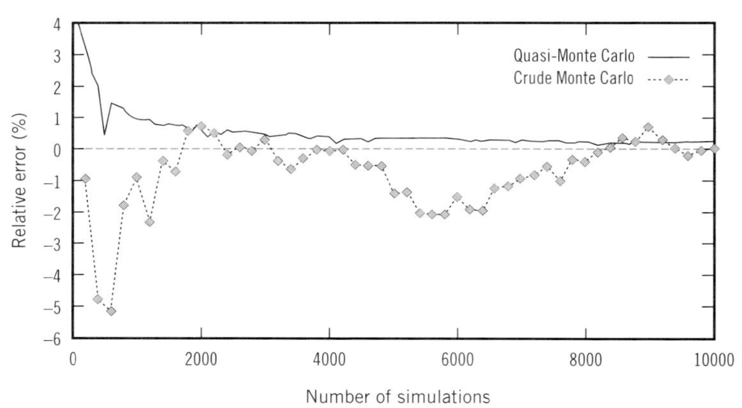

4. Standard European put option

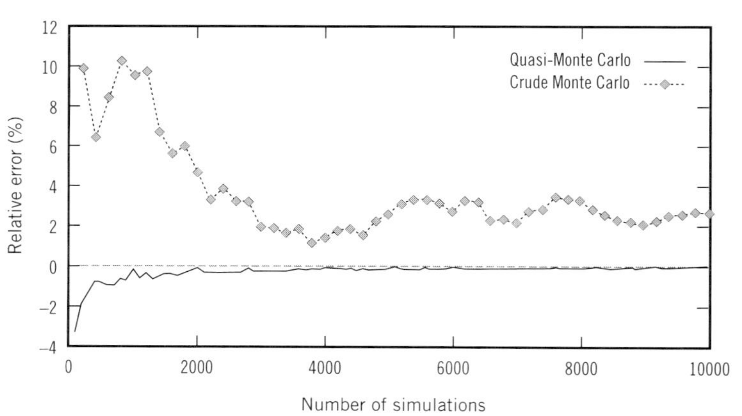

Figure 4 shows the same information for a standard put option. The quasi-Monte Carlo approach is clearly superior here as well. In many applications one is interested in the computation of the Greeks as well as the computation of the price. Broadie and Glasserman (1993) have recently discussed the different approaches to the computation of price sensitivities in the context of standard Monte Carlo methods. These different approaches can also be used with quasi-Monte Carlo methods, but in each case the quasi-Monte Carlo approach displays better convergence than the standard Monte Carlo approach. This is illustrated in Figures 5 and 6. Figure 5 shows the convergence of delta for the European call option and Figure 6 provides the same information for the gamma of the call option. These graphs demonstrate that the quasi-Monte Carlo approach outperforms the standard method.

Applications of quasi-Monte Carlo to value complex securities

In practice the most likely applications of the Monte Carlo approach concern situations where there is no known analytic solution or when the analytical solution involves prohibitive computation to produce the answer. Many contracts of current interest fall into this category. In this section we analyse several types of complex derivatives. These include basket options, Asian options, and options that require an HJM simulation framework. We start with a complex option for which there is a simple closed-form solution.

OPTIONS ON THE GEOMETRIC MEANS OF A PORTFOLIO

Our first example involves a European option on the geometric mean of several assets. Under the Black–Scholes assumptions there is a simple expression for the value of this contract. The detailed formula is given in Boyle (1993). The parameters used for this example are

Number of assets	3	
Initial asset prices, $S_i(0)$	100,	i = 1,2,3
Volatilities, σ_i	0.3,	i = 1,2,3
Correlations, ρ_{ij}	0.5,	i = 1,2, j > i
Strike price, K	100	
Riskless rate, r	10%	
Time to maturity, T	1 year	

With these parameters the value of a European

call on the geometric mean of the three assets is 13.771. We can also estimate the value of this option using Monte Carlo simulation methods. Figure 7 illustrates the convergence properties of standard Monte Carlo and quasi-Monte Carlo in this case. The quasi-Monte Carlo numbers were produced using a three-dimensional Faure sequence. The graph shows the superiority of the quasi-Monte Carlo approach in terms of convergence and error bounds. We now turn to an analysis of other types of complex derivatives.

BASKET OPTIONS

Table 2 shows the results of applying Faure sequences to basket options, which are options on a basket of different assets in pre-specified quantities (see Gentle, 1993, for a description). The spot price of a basket is the sum of these assets' prices multiplied by their respective quantities. The resulting distribution is thus a convolution of the individual assets' distribution. Because the resulting convolution is not easy to work with, these options have proved analytically intractable and hence numerical methods are used to value them. For the purposes of this study, a basket of 60% light sweet crude and 40% natural gas was chosen. In other words the spot price of the basket was given by

Basket spot
= 60% × Price of NYMEX light sweet crude/MMBTU[7]
+ 40% × Price of NYMEX natural gas/MMBTU

It was further assumed that the correlation between these two commodities was 20%. A call option on this basket was then priced by using both Monte Carlo simulation and Faure sequences. As expected, Faure sequences proved markedly superior in both speed and accuracy for pricing this option. This also provides a demonstration of the applicability of Faure sequences to correlated random variables.

ASIAN OPTIONS

The next type of option to be examined was a set of Asian options drawn from a comparative study

Table 2. Basket call option valuation

Initial price for light sweet crude: 3.09/MMBTU
Initial price for natural gas: 2.20/MMBTU
Volatility of light sweet crude: 20% NG: 30%
Correlation of both commodities: 20%
Strike price: 2.80/MMBTU Riskless Rate: 9%

Iterations	Crude Monte Carlo	Quasi-Monte Carlo
100	0.1386	0.1445
200	0.1562	0.1489
500	0.1621	0.1524
1,000	0.1455	0.1526
5,000	0.1545	0.1543
10,000	0.1493	0.1545
25,000	0.1553	0.1547
50,000	0.1569	0.1548
100,000	0.1552	0.1548
500,000	0.1544	0.1548

5. Standard European call option (delta)

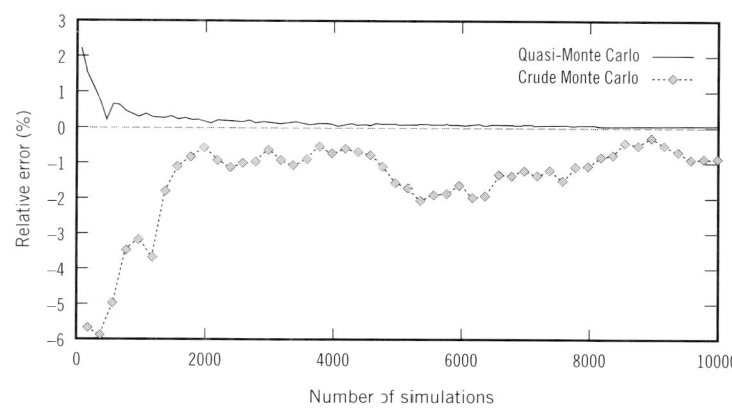

6. Standard European call option (gamma)

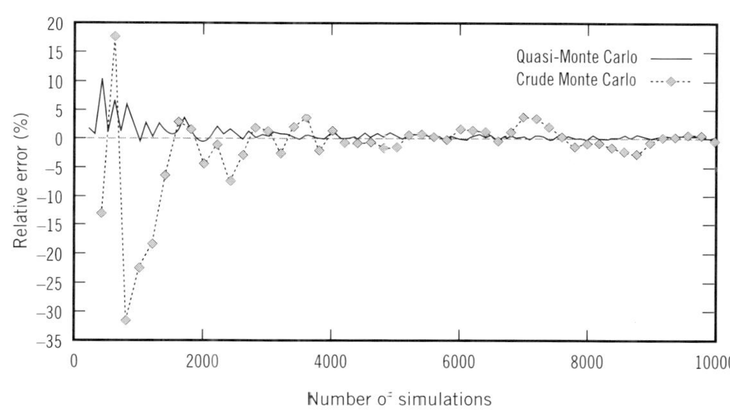

7. European call option on geometric mean of three stock prices

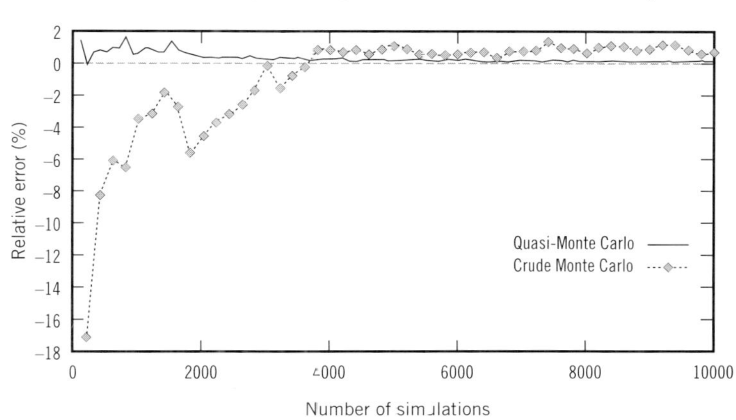

by Edmond Levy and Stuart Turnbull in *Risk* magazine (Levy and Turnbull 1992). The averaging period was quite long for these options (52 weekly prices) and as a result the dimension of the problem was much higher than in the previous example (53 time steps versus two random variables for the basket option). This caused slower convergence for both methods, but Faure sequences were still an improvement as shown in Table 3. To show that this slow convergence is due to high dimensionality, an example is given in Table 4 of a similar Asian option with an averaging period of five weeks. For the five-week option, convergence of the Faure sequence is very fast.

HJM FACTOR MODELS

The final type of option to be priced was the option to enter into a natural gas swap. Because swap prices are highly sensitive to the shape of the forward price curve, it is extremely important to use a model which can correctly capture price curve reshaping. One way to do this is through the use of a three-factor HJM model along the lines of Cortazar and Schwartz (1992). The factor loadings are estimated through historical data and then scaled to match historical volatilities (if this were being used for an actual trade, the factors would be scaled to match implied volatilities). In Table 5 we price a six-month option to enter into a one-year swap with a constant daily volume of 10,000 MMBTUs. Once again, Faure sequences are more accurate than plain Monte Carlo even though the dimensionality of the problem is fairly high ((5 months + 1 partial month) × 3 factors = 18). As a benchmark the last row compares the results to the price obtained by using Black's model for options on commodities. The volatility of 15% that is used here represents historical swap volatility for the last year and is well within the range of recent implied volatilities.

Table 3. Pricing of 52-week Asian options

52-week arithmetic average option with 53 reset points
Spot price: 100 Riskless rate: 9% Volatility: 50%

	Strike price = 110		Strike price = 100		Strike price = 90	
Iterations	Crude MC	Quasi-MC	Crude MC	Quasi-MC	Crude MC	Quasi-MC
100	9.207	13.131	10.206	17.043	19.493	21.942
200	11.778	10.224	15.405	14.219	18.902	19.298
400	10.156	9.180	11.818	13.327	16.854	18.642
800	8.761	10.965	12.593	14.991	17.140	20.118
1,600	9.706	8.953	13.206	12.928	17.873	18.171
3,200	9.280	8.820	12.934	12.801	18.135	18.074
4,800	9.191	8.747	12.965	12.673	18.326	17.919
9.600	9.009	8.894	12.684	12.804	18.121	17.990
19,200	9.195	9.101	12.754	12.993	18.018	18.164
38,400	9.152	9.057	13.033	12.949	18.147	18.116
76,800	8.987	9.052	13.039	12.948	18.214	18.114
153,600	8.999	9.062	12.948	12.962	18.110	18.128
307,200	9.026	9.063	12.914	12.963	18.106	18.128
614,400	9.002	9.065	12.922	12.965	18.085	18.131
1,228,800	9.030	9.063	12.885	12.963	18.064	18.128
2,457,600	9.040	9.062	12.906	12.962	18.100	18.128
Turnbull[a]	9.10 (0.03)		12.98 (0.03)		18.14 (0.03)	

[a] These results are given by Levy and Turnbull (1992), with the "standard error" bounds reported in brackets.

Table 4. Pricing of five-week Asian options

Five-week Asian option with six reset points
Spot price: 100 Riskless rate: 9%
Strike price: 100 Volatility: 50%

Iterations	Crude Monte Carlo	Quasi-Monte Carlo
100	2.869	3.876
200	3.899	3.878
400	3.706	3.590
800	3.435	3.612
1,600	3.512	3.618
3,200	3.559	3.605
4,800	3.484	3.607
9,600	3.584	3.605
19,200	3.625	3.612
38,400	3.645	3.611
76,800	3.590	3.611
153,600	3.592	3.612
307,200	3.618	3.612
614,400	3.617	3.612

Table 5. Pricing of swaptions under a three-factor HJM model

Cost to enter into an at-the-money natural gas swap option
Valuation date: Sept. 20, 1994
Inception of the swap: March 1, 1995
Maturity of the swap: Feb 28, 1996
Maturity of the swaption: Feb 28, 1995
Volume: 10,000 MMBTUs per day
Fixed price of the swap: current mid-market price for a swap of the above duration and volume

Iterations	Crude Monte Carlo ($)	Quasi-Monte Carlo ($)
100	231,593	268,027
200	236,071	288,062
400	271,236	300,659
800	267,513	270,850
1,600	271,635	268,295
3,200	258,667	268,025
4,800	265,364	266,580
9,600	267,193	267,198
19,200	265,433	267,894
38,400	266,137	266,910
76,800	263,069	267,538
153,600	265,279	267,302

Black–Scholes at 15% volatility: 260,930.

PORTFOLIO RISK AND SCIENTIFIC STRESS TESTS

One final application that Faure sequences are particularly well suited for is the simulation of risks in large portfolios. The first reason for this is that, since each simulation pass can take quite a long time, using Faure sequences to simulate prices is an excellent way to reduce the time needed to obtain good accuracy. The second reason for using Faure sequences is that they can be directly compared to a systematic set of stress tests based upon the simulation model. This analogy comes from the fact that Faure sequences are generated so as to form very evenly spaced points in the probability space. The result is that if one uses Faure sequences in conjunction with one's simulation module, one can be sure that every possibility has been "tried out" and that there is not some kind of disaster scenario that has not been simulated due to an uneven set of random numbers. In other words, what Faure sequences do in this context is to systematically try out different combinations of prices under the model being used and test each and every one of these combinations to determine the effect on the portfolio. This ensures that all reasonable combinations will be tried, and eliminates the possibility that a poor set of random numbers might fail to uncover a crucial weakness or a potential portfolio collapse. This realisation ends the dichotomy between ordinary stress tests and Monte Carlo simulation and helps one to understand how they relate to each other. Naturally, even Faure sequences cannot eliminate the risk that the model may fail, so ordinary stress tests are still very important.

Summary and conclusion

Quasi-Monte Carlo methods provide a way to improve the accuracy and reliability of Monte Carlo simulation. The key idea is to use deterministic sequences known as quasi-random sequences. This results in better convergence and deterministic error bounds. In this chapter we used the quasi-Monte Carlo approach to value a range of complex derivative securities that are of practical interest. This was done through the use of Faure sequences which have in general good convergence properties even when a large number of time steps are needed.

We illustrated the properties of the quasi-Monte Carlo approach in our numerical examples. It is evident from our graphs that the quasi-Monte Carlo approach leads to improved convergence and lower error bounds. The theoretical error bounds for the quasi-Monte Carlo approach are given in Appendix B. It turns out that these bounds are loose. In practice any option that has a reasonably regular payoff function will give far higher convergence than these loose bounds indicate.

Woźniakowski (1991) has derived average-case error bounds for real continuous functions on the unit cube. He shows that the average difference between the exact value of a multivariate integral and its Monte Carlo estimate is closely related to the discrepancy of the sample points that are used in the Monte Carlo approximation. This means that low-discrepancy sequences lead to low average-case error as well as low worst-case error. Paskov (1994) gives further discussion of these bounds.

We can also improve the performance of quasi-Monte Carlo methods by using traditional variance reduction techniques.[8] In addition we can apply Richardson extrapolation in conjunction with quasi-Monte Carlo methods to further enhance the accuracy. The above Faure sequences can be generated about as quickly as normally distributed random numbers. Furthermore, the theory shows that these sequences are robust in the sense that any sub-sequence of Faure numbers will give good convergence. Thus these sequences can be used much like pseudo-random number generators. Our conclusion is that this technique will prove to be a very powerful tool for many types of problems in computational finance.[9]

Although the present study has concentrated on Faure sequences, other low-discrepancy sequences may have advantages over Faure sequences. In particular Boyle, Broadie, and Glasserman (1997) show that Sobol' sequences outperform Faure sequences in many instances. Sobol' sequences also have advantages in that they tend to be more evenly dispersed[10] throughout the unit hypercube than Faure sequences. The application of low-discrepancy series to problems in finance is a topic of current interest. Some of the issues include the selection of appropriate termination criteria, how to exploit the regularity properties of the integrand and the performance of various low-dimensional sequences with each other and with standard Monte Carlo at very high dimensions for a range of situations.

Appendix A: Computing the normal inverse function

The authors are indebted to B. Moro for his generosity in providing the following algorithm.

This appendix explains one procedure for inverting the cumulative normal density function. The inverse, $\Phi^{-1}(x)$, of the cumulative normal density function

$$\Phi(x) = \frac{1}{\sqrt{2\pi}} \int_{-\infty}^{x} e^{-(t^2/2)} dt \tag{A1}$$

can generally be computed using the Newton-Raphson iterative method. However, this is too slow to be useful in practice. The procedure illustrated below has high accuracy for all values of x in the interval $[10^{-10}, 1 - 10^{-10}]$ while retaining the speed of a simple rational approximation.

If $|y| \leq 0.42$, where $y = x - 0.5$, $\Phi^{-1}(x)$ is computed using a rational approximation as given by Beasley and Springer (1977):

$$\Phi^{-1}(x) = y \frac{\sum_{n=0}^{3} a_n y^{2n}}{\sum_{n=0}^{4} b_n y^{2n}} \tag{A2}$$

where a_n and b_n are suitably chosen coefficients. For $|y| > 0.42$, we approximate $\Phi^{-1}(x)$ by a truncated Chebyshev series as

$$\Phi^{-1}(x) = \begin{cases} \sum_{n=0}^{8} c_n T_n(z) - \frac{c_0}{2} & \text{if } y > 0 \\ \frac{c_0}{2} - \sum_{n=0}^{8} c_n T_n(z) & \text{otherwise} \end{cases} \tag{A3}$$

Here $z = k_1 \times [2 \times \log(-\log(0.5 - |y|)) - k_2]$ and the constants k_1 and k_2 are chosen such that $z = -1$ when $\Phi(x) = 0.92$ and $z = 1$ when $\Phi(x) = 1 - 10^{-12}$. The upper limit was set rather arbitrarily. One can easily derive Chebyshev approximations which will be valid for $\Phi(x)$ up to $1 - 10^{-15}$, but that would come with increased computational cost. Once the coefficients c_n are computed (see Press and Teukolsky, 1992, equation (5.8.7)), equation (A3) can be efficiently evaluated using Clenshaw's formula (see Press and Teukolsky, 1992, equation (5.8.11)). The required constants a_n, b_n, c_n, k_1 and k_2 are reproduced in Table A1.

The rational approximation using equation (A2) has the largest absolute error of 3.0×10^{-9} in the interval $[0.5, 0.92)$. In the tails of the distribution, equation (A3) retains this accuracy for up to six standard deviations.

Table A1. Constants for inverting normal variates

n	a_n	b_n	n	c_n
0	2.50662823884	1.00	0	7.7108870705487895
1	–18.61500062529	–8.47351093090	1	2.7772013533685169
2	41.39119773534	23.08336743743	2	0.3614964129261002
3	–25.44106049637	–21.06224101826	3	0.0373418233434554
4		3.13082909833	4	0.0028297143036967
			5	0.0001625716917922
	k_1	k_2	6	0.0000080173304740
	0.4179886424926431	4.2454686881376569	7	0.0000003840919865
			8	0.0000000129707170

Appendix B: Error bounds for Faure sequences

This appendix discusses the error bounds for Faure sequences. We compare the error bounds for quasi-Monte Carlo based on Faure sequences with the error bounds from crude Monte Carlo. We begin with three definitions and an important theorem. First we definite the counting function of a subset of a set of points.

Definition 1. Let P be a point set consisting of points $x_1, x_2, \ldots, x_N$. For any subset B of $\bar{I}^s$ define

$$A(B; P) = \sum_{n=1}^{N} c_B(x_n)$$

where c_B is the characteristic function of B. Thus $A(B;P)$ is the counting function that indicates the number of n with $1 \leq n \leq N$ for which $x_n \in B$.

Next we define discrepancy.

Definition 2. Let $\mathcal{B}$ be a family of Lebesgue-measurable subsets of I^s. The discrepancy $D_N(\mathcal{B};P)$ of a point set P is defined as

$$D_N(\mathcal{B}; P) = \sup_{B \in \mathcal{B}} \left| \frac{A(B; P)}{N} - \lambda_s(B) \right|$$

where $\lambda_s(B)$ is the Lebesgue measure of $\mathcal{B}$.

Definition 3: Star discrepancy. The star discrepancy $D_N^*(P) = D_N^*(x_1, \ldots, x_N)$ of the point set P is defined by $D_N^*(P) = D_N(\mathcal{J}^*; \mathcal{P})$ where $\mathcal{J}^*$ is the family of all subintervals of I^s of the form $\prod_{i=1}^{s} [0, u_i)$.

Now comes the key theorem.

Theorem (Koksma–Hlawka). Let $I^s = [0,1)^s$ *and let* f *have bounded variation* $V(f)$ *on* I^s *in the sense of Hardy and Krause. Then, for any* $x_1, \ldots, x_N \in I^s$ *we have*

$$\left| \frac{1}{N} \sum_{n=1}^{N} f(x_n) - \int_{I^s} f(u)du \right| \leq V(f) D_N^*(x_1, \ldots, x_N)$$

Intuitively, the D-star discrepancy $D_N^*(x_1, \ldots, x_N)$ represents how "evenly" $x_1, \ldots, x_N$ are distributed within $\bar{I}^s$. It turns out that in the one-dimensional case, the best choice for $x_1, \ldots, x_N$ is the evenly spaced points $x_i = (2i - 1)/2N$ for $1 \leq n \leq N \in [0,1]$. This gives the minimum possible discrepancy of $D_N^*(x_1, \ldots, x_N) = 1/2N$ (Niederreiter, 1992, p. 23). However, this set has the drawback that in order to obtain this minimal discrepancy N must be specified in advance. For Faure sequences, the resulting sequence has low discrepancy *for all* N, which allows the simulation to be easily extended until the desired accuracy is obtained. Furthermore, the projection of Faure sequences on to lower-dimensional hypercubes I^d, $d \le s$, will still have discrepancy $O(N^{-1}(\log N)^d)$. This means that the first d terms from a higher-order Faure sequence of order s can be used for an integral of lower dimension d.

For the Faure sequences under consideration, we have the following result from Faure (1982):

Given the Faure sequence $x_1,\ldots,x_N$ *of dimension* s:

$$D_N^*(x_1,\ldots,x_N) \le A_s\frac{(\log N)^s}{N} + O\left(\frac{(\log N)^{s-1}}{N}\right)$$

$$\text{where } A_s = \frac{1}{s!}\left(\frac{q_s - 1}{2\log q_s}\right)^s$$

with q_s *as the smallest prime number that is greater than or equal to* s.

This series has nearly[11] the smallest known set of constants, A_s. When working with a fixed number of simulations, N, Niederreiter (1992, p. 97) suggests using the set $(n/N, x_n) \in I^s$, where $n = 0,1,\ldots,N-1$, which has $D^* \le O((\log N)^{s-1} \times N^{-1})$. Furthermore, in the s-dimensional case it is widely believed[12] that for any N-element point set $(x_1,\ldots,x_N)$ we must have

$$D_N^*(x_1,\ldots,x_N) \ge O(N^{-1}(\log N)^{s-1})$$

This suggests that $O(N^{-1}(\log N)^{s-1}$ is the best order of convergence that we can expect from any deterministic set.

If we compare the Faure error bound with theoretical Monte Carlo it may not seem that quasi-Monte Carlo methods are a significant improvement (see Table B1).

In practice, this table is deceptive because Faure sequences perform much better than this indicates. This is due to the fact that quasi-random Faure points are very evenly spaced while most crude Monte Carlo simulations are run using pseudo-random number generators whose D* discrepancies are higher than the D* discrepancies of Faure sequences (see Niederreiter, 1992, for a discussion of the discrepancies of pseudo-random number generators).

Table B1. Comparison of the theoretical error bound for both crude and quasi-Monte Carlo methods

s	N	$N^{-1/2}$	$A_sN^{-1}(\log N)^s$
5	10^4	0.01	0.163
10	10^8	0.0001	19.262

1 *Many of the new "exotic" securities involve several underlying assets, different currencies and path dependency.*

2 *Another disadvantage is that the method still has difficulties in handling the early exercise feature of American options despite notable progress by Tilley (1993) – also Chapter 14 of this publication and Grant, Vora and Weeks (1993).*

3 *We give a more technical definition of discrepancy in Appendix B.*

4 *See Press* et al. *(1992) for a discussion of the trapezoidal rule.*

5 *Note that for numerical accuracy,* ${}^iC_j \bmod r$ *is usually calculated via* ${}^iC_j \bmod r = ({}^{i-1}C_j \bmod r + {}^{i-1}C_{j-1} \bmod r) \bmod r$.

6 *Note that it is incorrect to use the standard Box-Muller transform (Press, 1992, p. 289) to map* ϵ_n *to the unit interval. This is because the Box-Muller transform does not preserve the matrix on* I^s *and thus fails to preserve the low discrepancy of the original Faure sequence. In other words, if we use the Box-Muller transform, then the even spacing of the Faure sequence will be scrambled, resulting in the loss of our low error bound.*

7 *MMBTU stands for million British thermal units and is the standard unit of measurement on the Nymex for natural gas. Light sweet crude and natural gas, however, are usually quoted in barrels and hence need to be converted to MMBTUs to make the two prices compatible in terms of units.*

8 *Quasi-Monte Carlo methods perform even better when combined with variance reduction techniques. For example, to apply the control variate method the original integral* $f(x)dx$ *is replaced with an integral of the form* $[f(x) - g(x)]dx$ *where* $g(x)$ *has the same type of behaviour as* $f(x)$. *Clearly if crude Monte Carlo can be used to value* $f(x) - g(x)$, *then quasi-Monte Carlo can value this difference even more accurately. Two common variance reduction techniques that do not improve upon standard quasi-Monte Carlo methods are antithetic variables and stratified sampling. Quasi-Monte Carlo methods are intrinsically designed to keep on subdividing the integration region into uniformly spaced parts with every additional simulation point. Quasi-Monte Carlo methods already incorporate the advantages associated with these two variance reduction techniques.*

9 *The authors are grateful to the referees and Robert Heinkel for constructive comments.*

10 *As we move to higher dimensions the Faure points are not evenly dispersed throughout the hypercube. If we project*

the 50th and 51st dimension we find strong patterns and large gaps. This point is discussed in Boyle, Broadie, and Glasserman (1996). We are also grateful to Alan Jung for discussions on this issue.

11 *A slight improvement on the Faure sequence has been made by Niederreiter (1987) but we prefer to stay with Faure sequences since their calculation is simpler. Sobol' (1967) sequences also have this same order of convergence but their constants are never better than Faure sequences and are in fact worse for all* $s \geq 8$ *(Niederreiter, 1992, p. 96). However, Sobol' sequences are at least better than Halton (1960) sequences, which have the same order of convergence but display superexponential growth in their constants as the dimension* s *increases.*

12 *This can be shown for dimensions* $s = 1$ *and* 2 *but the best general result that has been* proved *for* $s \geq 3$ *is* $D^*_N(x_1,\ldots,x_N) \geq O(N^{-1}(\log N)^{(s-1)/2})$ *(Niederreiter, 1992, p. 32).*

BIBLIOGRAPHY

Beasley, J.D., and S.G. Springer, 1977, "The Percentage Points of the Normal Distribution", *Applied Statistics* 23, pp. 118–21.

Boyle, P., 1977, "Options: A Monte Carlo Approach", *Journal of Financial Economics* 4, pp. 323–38 (Chapter 1 of this publication).

Boyle, P., 1992, "New Life Forms on the Option Landscape", *Journal of Financial Engineering* 2(3), pp. 217–52.

Boyle, P., M. Broadie and P. Glasserman, 1997, "Monte Carlo Methods for Security Pricing", *Journal of Economics, Dynamics and Control* 21(8–9), pp. 1,267–321 (Chapter 2 of this publication).

Bratley, P., and B.L. Fox, 1988, "ALGORITHM 659: Implementing Sobol's Quasi Random Sequence Generator", *ACM Trans. Math. Software* 14, pp. 88–100.

Broadie, M., and P. Glasserman, 1993, "Estimating Security Price Derivatives Using Simulation", Working paper, Columbia University, New York.

Cortazar, G., and E. Schwartz, 1992, "The Valuation of Commodity Contingent Claims", Working paper, Anderson Graduate School of Management, UCLA, Los Angeles, CA.

Duffie, D., 1992, *Dynamic Asset Pricing Theory*, Princeton University Press, Princeton, NJ.

Faure, H., 1982, "Discrépance de suites associées à un système de numération (en dimension *s*)", *Acta Arithmetica* 41, pp. 337–51.

Fox, P., 1986, "ALGORITHM 647: Implementation and Relative Efficiency of Quasirandom Sequence Generators", *ACM Trans. Math. Software* 12, pp 362–76.

Gentle, D., 1993, "Basket Weaving", *Risk* June, p. 51.

Grant, D., D. Vora and D. Weeks, 1993, "Path Dependent Options: Extending the Monte Carlo Approach", Working paper, University of New Mexico, Albuquerque, NM.

Halton, J. H., 1960, "On the Efficiency of Certain Quasi-random Sequences of Points in Evaluating Multi-dimensional Integrals", *Numerical Mathematics* 2, pp. 84–90.

Hammersley, J., and D. Handscomb, 1964, *Monte Carlo Methods*, John Wiley & Sons, New York.

Hlawka, E., 1961, "Funktionen von beschränkter Variation in der Theorie der Gleichverteilung", *Ann. Math. Pura Appl.* 54, pp. 325–33.

Hlawka, E., F. Firneis and P. Zinterhof, 1981, *Zahlentheoretische Methoden in der Numerischen Mathematik*, Oldenbourg, Vienna, Austria.

Hua, L., and Y. Wang, 1981, *Applications of Number Theory to Numerical Analysis*, Springer, Berlin.

Niederreiter, H., 1987, "Point Sets and Sequences with Small Discrepancy", *Monatsh. Math.* 104, pp. 273–337.

Niederreiter, H., 1988, "Quasi-Monte Carlo Methods for Multidimensional Numerical Integration", in H. Brass and G. Hammerlin (eds), *Numerical Integration III, International Series of Numerical Mathematics*, Birkhäuser, Basel, 85, pp. 157–71.

Niederreiter, H., 1992, *Random Number Generation and Quasi-Monte Carlo Methods*. SIAM, Philadelphia, PA.

Paskov, S., 1994, "Computing High Dimensional Integrals with Applications to Finance", Technical report CUCS-023-94, Dept. of Computer Science, Columbia University, New York.

Press, W., S. Teukolsky, W. Wetterling and B. Flannery, 1992, *Numerical Recipes in C, 2nd edn*, Cambridge University Press, New York, pp. 309–15.

Sobol', I.M., 1967, "The Distribution of Points in a Cube and the Approximate Evaluation of Integrals", *USSR Computational Math. and Math. Physics* 19, pp. 86–112.

Tilley, J.A., 1993, "Valuing American Options in a Path Simulation Model", *Transactions of the Society of Actuaries* 45, pp. 499–520 (Chapter 14 of this publication).

Turnbull, S., and E. Levy, 1992, "Average Intelligence", *Risk* 5(2), pp. 53–9.

Woźniakowski, H., January 1991, "Average Case Complexity of Multivariate Integration", *Bulletin (New Series) of the American Mathematical Society* 24, pp. 185–94.

New Methodologies for Valuing Derivatives

Spassimir H. Paskov
Barclays Capital

High-dimensional integrals are usually solved with Monte Carlo algorithms (on which vast sums are spent annually) although theory suggests that low-discrepancy algorithms are sometimes superior. We report on numerical testing which compares low-discrepancy and Monte Carlo algorithms on the evaluation of financial derivatives. The testing is performed on a collateralised mortgage obligation (CMO) which is formulated as the computation of 10 integrals of dimension up to 360.

We tested two low-discrepancy algorithms (Sobol' and Halton) and two randomised algorithms (classical Monte Carlo and Monte Carlo combined with antithetic variables). We conclude that for this CMO the Sobol' algorithm is always superior to the other algorithms. We believe that it will be advantageous to use the Sobol' algorithm for many other types of financial derivatives.

Our conclusion regarding the superiority of the Sobol' algorithm also holds when a rather small number of sample points are used – an important case in practice.

As mentioned at the outset, theory (see Neiderreiter, 1992, and Woźniakowski, 1991) suggests that low-discrepancy algorithms are sometimes superior to Monte Carlo algorithms. However, a number of researchers, such as Janse van Rensburg and Torrie (1993), Morokoff and Caflisch (1995) and Moskowitz and Caflisch (1996) – also Chapter 27 of this publication, report that their numerical tests show that this theoretical advantage decreases with increasing dimension. Furthermore, they report that the theoretical advantage of low-discrepancy algorithms disappears for rather modest values of the dimension, say, $d \leq 30$.

We decided to compare the efficacy of low-discrepancy and Monte Carlo algorithms on the valuation of financial derivatives. We use a CMO, provided by Goldman Sachs, with 10 bond classes (tranches) which is formulated as the computation of 10 integrals of dimension up to 360. The reasons for choosing this CMO are that it has fairly high dimension and that each integrand evaluation is very expensive, making it crucial to sample the integrand as few times as possible. We believe that our conclusions regarding this CMO will hold for many other financial derivatives.

The low-discrepancy sample points chosen for our tests are Sobol' and Halton points. We compared the algorithms based on these points with the classical Monte Carlo algorithm and also with the classical Monte Carlo algorithm combined with antithetic variables. We refer to the latter as the antithetic variables algorithm (see section on antithetic variables for a discussion of why this algorithm was tested).

An explanation of our terminology is in order here. *Low-discrepancy* points are sometimes referred to as *quasi-random* points. Although in widespread use, we believe the latter term to be misleading since there is nothing random about these deterministic points. We prefer to use the terminology *low discrepancy* or *deterministic*.

We assume that the finance problem has been formulated as an integral over the unit cube in d dimensions. We have built a software system called FINDER to compute high dimensional integrals. FINDER runs on a heterogeneous network of workstations under PVM (Parallel Virtual Machine) 3.2. Since workstations are ubiquitous, this is a cost-effective way to do large computations fast. The measured speed-up is at least $0.9N$ for N workstations, where $n \leq 25$. Of course, FINDER can also be used to compute high-dimensional integrals on a single workstation. The software system FINDER is available and interested readers should contact the author.

A routine for generating Sobol' points is given in Press *et al.*, (1992). However, major improvements have been incorporated in FINDER, and we emphasise that the results reported here were obtained using this software. One of the improvements was developing the table of primitive polynomials and initial direction numbers for dimensions up to 360.

This chapter is based on software construction and testing which began in the autumn of 1992. Preliminary results were presented to a number of New York City financial houses in the autumn of 1993 and the spring of 1994. A January, 1994, article in *Scientific American* discussed the theoretical issues and reported that "Preliminary results obtained by testing certain finance problems suggest the superiority of the deterministic methods in practice". Further results were reported at a number of conferences, including presentations by Paskov (1994), Traub (1994a), and Traub (1994b). An "extended abstract" of this chapter was published as Paskov and Traub (1995), and a slightly different version appeared as a Columbia University Technical Report in October, 1994.

Summary of findings

We summarise our main conclusions regarding the evaluation of the CMO. The conclusions may be divided into three groups.

DETERMINISTIC AND MONTE CARLO ALGORITHMS

The Sobol' algorithm consistently outperforms the Monte Carlo algorithm and the Halton algorithm. In particular:

❑ The Sobol' algorithm converges significantly faster than the Monte Carlo algorithm.

❑ The convergence of the Sobol' algorithm is smoother than that of the Monte Carlo algorithm. This makes automatic termination easier for the Sobol' algorithm.

❑ Using our standard termination criterion (see section on the comparison of the deterministic and Monte Carlo algorithms), the Sobol' algorithm terminates two to five times faster than the Monte Carlo algorithm, often with smaller error.

❑ The Monte Carlo algorithm is sensitive to the initial seed.

SOBOL', MONTE CARLO, AND ANTITHETIC VARIABLES ALGORITHMS

The Sobol' algorithm consistently outperforms the antithetic variables algorithm, which in turn, consistently outperforms the Monte Carlo algorithm. In particular:

❑ These conclusions also hold when a rather small number of sample points are used, an important case in practice. For example, for 4,000 sample points, the Sobol' algorithm running on a single Sun-4 workstation achieves accuracies in the range of one part in a thousand to one part in a million, depending on the tranche, within a couple of minutes.

❑ Statistical analysis on the small sample case further strengthens the case for the Sobol' algorithm over the antithetic variables algorithm. For example, to achieve similar performances with a confidence level of 95%, the antithetic variables algorithm needs from seven to 79 times more sample points than the Sobol' algorithm, depending on the tranche. In a similar comparison, the Monte Carlo algorithm needs 27 to 607 times more sample points than the Sobol' algorithm depending on the tranche. These speed-ups are measured conservatively (see section on statistical analysis).

❑ Statistical analysis for a large number (1,000,000) of sample points shows that the superiority of the Sobol' algorithm is greater than for a small number of points. For example, to achieve similar performances with a confidence level of 95%, the antithetic variables algorithm needs 16 to 230 times more sample points that the Sobol' algorithm, depending on the tranche. In a similar comparison, the Monte Carlo algorithm needs 110 to 1,769 times more points than the Sobol' algorithm depending on the tranche. These speed-ups are measured conservatively (see section on statistical analysis).

❑ The antithetic variables algorithm is sensitive to the initial seed. However, convergence of the antithetic variables algorithm is less jagged than that of the Monte Carlo algorithm.

NETWORK OF WORKSTATIONS

All the algorithms benefit by being run on a network of workstations. In particular:

❑ For N workstations, the measured speed-up is at least $0.9N$, where $N \leq 25$.

❑ A substantial computation which took seven hours on a Sun-4 workstation took 20 minutes on a network of 25 workstations.

We emphasise that we do not claim that the Sobol' algorithm is always superior to the Monte Carlo algorithm. We do not even claim that it is always superior for financial derivatives. After all, the test results reported here are only for one particular CMO. However, we do believe it will be advantageous to use the Sobol' algorithm for many other types of financial derivatives.

The Monte Carlo algorithm

In this section, we give a brief discussion of the theory underlying the Monte Carlo algorithm for the problem of multivariate integration. For more details, see, for example, Halton (1970), Hammersley and Handscomb (1964), Kalos and Whitlock (1986), and Sobol' (1973).

We now present the classical Monte Carlo algorithm. For brevity, we refer to it as the Monte Carlo algorithm. Let $t_1,\ldots,t_n$ be n randomly selected points which are independent and uniformly distributed over $D = [0,1]^d$, the d-dimensional unit cube. Consider a function, f, from the space $L_2(D)$ of L_2-integrable functions. The problem is to compute approximately

$$I(f) = \int_D f(x)dx$$

using function evaluations at randomly chosen points. The classical Monte Carlo algorithm is given by

$$I(f) \approx U_n(f) = U_n(f;t_1,\ldots,t_n) = \frac{1}{n}\sum_{i=1}^{n} f(t_i) \qquad (1)$$

The main idea underlying the Monte Carlo algorithm for multivariate integration is to replace a continuous average by a discrete average over randomly selected points.

The expected value of the estimate $U_n(f; t_1,\ldots,t_n)$ as a function of the random variables $t_1, t_2,\ldots,t_n$ is

$$E(U_n(f;t_1,\ldots,t_n)) = I(f)$$

The expected error of the Monte Carlo algorithm for a function f is defined by

$$E_n(f) = \left(\int_{D^n} (I(f) - U_n(f))^2 dt_1\ldots dt_n\right)^{1/2}$$

It is well known that

$$E_n(f) = \frac{\sigma(f)}{\sqrt{n}}$$

where the variance, $\sigma^2(f)$, of f is defined as

$$\sigma^2(f) = \int_D (f(t) - I(f))^2 dt = I(f^2) - I^2(f)$$

Clearly, reducing the variance of the integrand would also reduce the expected error. In fact, this is the main idea underlying the various variance reduction techniques which are often used in combination with the Monte Carlo algorithm. Examples of variance reduction techniques are importance sampling, control variates and antithetic variables (see, for example, Kalos and Whitlock, 1986). Antithetic variables will be discussed in a later section.

Let $B(L_2(D))$ denote the unit ball of $L_2(D)$. The expected error of the Monte Carlo algorithm with respect to the class $B(L_2(D))$ is measured for the worst function f in $B(L_2(D))$:

$$e_n^{wor-MC}(B(L_2(D)) = \sup_{f \in B(L_2(D))} E_n(f)$$

It can easily be concluded that

$$e_n^{wor-MC}(B(L_2(D))) = n^{-1/2} \qquad (2)$$

Since the rate of convergence is independent of d and the Monte Carlo algorithm is applicable to a very broad class of integrands, its advantages for high-dimensional integration are clear. Nevertheless, the Monte Carlo algorithm has several serious deficiencies (see, for example, Neiderreiter, 1992, and Spanier and Maize, 1994). We mention just three of them here. Even though the rate of convergence is independent of the dimension, it is quite slow. Furthermore, there are fundamental philosophical and practical problems with generating independent random points; instead, pseudo-random numbers are used (see Knuth, 1981, and Traub and Woźniakowski, 1992). Finally, the Monte Carlo algorithm provides only probabilistic error bounds, which is not a desirable guarantee for problems where highly reliable results are needed.

Low-discrepancy deterministic algorithms

In an attempt to avoid the deficiencies of the Monte Carlo algorithm, many deterministic algorithms have been proposed for computing high-dimensional integrals for functions belonging to various subsets of $L_2(D)$ (eg Neiderreiter, 1992; Novak, 1988; Traub, Wasilkowski and Woźniakowski, 1988; Woźniakowski, 1991; Paskov, 1993; and Tezuka, 1994). One class of such deterministic algorithms is based on low-discrepancy sequences. Discrepancy is a measure of deviation from uniformity of a sequence of points in D. Now, very briefly, we give the theoretical basis of the low-discrepancy sequences to be used as sample points for computing multivariate integrals. For a more detailed description and treatment of these results, see Neiderreiter (1992).

For $t = [t_1,\ldots,t_d] \in D$, define $[0,t_1) = [0,t_1) \times \ldots \times [0,t_d)$. Let $\chi_{[0,t)}$ be the characteristic (indicator) function of $[0,t)$. For $z_1,\ldots,z_n \in D$, define

$$R_n(t;z_1,\ldots,z_n) = \frac{1}{n}\sum_{k=1}^{n} \chi_{[0,t)}(z_k) - t_1,t_2\cdots t_d$$

The L_2 (or L_∞) discrepancy of $z_1,\ldots,z_n$ is defined as the L_2 (or L_∞) norm of the function $R_n(\cdot;z_1,\ldots,z_n)$, ie

$$\|R_n(\cdot\,;z_1,\ldots,z_n)\|_2 = \left(\int_D R_n^2(t;z_1,\ldots,z_n)dt\right)^{1/2}$$

$$\|R_n(\cdot\,;z_1,\ldots,z_n)\|_\infty = \sup_{t\in D} |R_n(t;z_1,\ldots,z_n)|$$

Roth (1954) and (1980) proves that

$$\inf_{z_1,\ldots,z_n} \|R_n(\cdot\,;z_1,\ldots,z_n)\|_2 = \Theta(n^{-1}(\log n)^{(d-1)/2}) \qquad (3)$$

Of special interest for numerical integration are infinite low-discrepancy sequences, $\{z_k\}$, in which the definition of z_k does not depend on the specific value of the number of points, n. Examples of such infinite sequences are the Halton (1960) and Sobol' (1967) sequences. For the reader's convenience, we give a short description of both sequences later in this section. The following bounds hold for the Halton and the Sobol' sequences:

$$\|R_n(\cdot\,;z_1,\ldots,z_n)\|_2 \le \|R_n(\cdot\,;z_1,\ldots,z_n)\|_\infty = O\left(\frac{(\log n)^d}{n}\right) \qquad (4)$$

This is the best upper bound known and it is widely believed that it is sharp for these sequences (Neiderreiter, 1992).

We stress that the constants in the bounds (3) and (4) depend on the dimension d and good estimates of these constants are not known. Bounds with known constants and n independent of d are studied in Woźniakowski (1994) and Wasilkowski and Woźniakowski (1995).

Sequences satisfying the upper bound in (4) are known as low-discrepancy sequences or quasi-random sequences. We will refer to them as low-discrepancy sequences or deterministic sequences.

Clearly, the idea behind the low-discrepancy sequences is that for any rectangular $[0,t)$ the fraction of the points within $[0,t)$ should be as "close" as possible to its volume. That way, the low-discrepancy sequences cover the unit cube as "uniformly" as possible by reducing gaps and clustering of points. This idea is illustrated in Figures 1 and 2.

1. 256 random points in the unit square

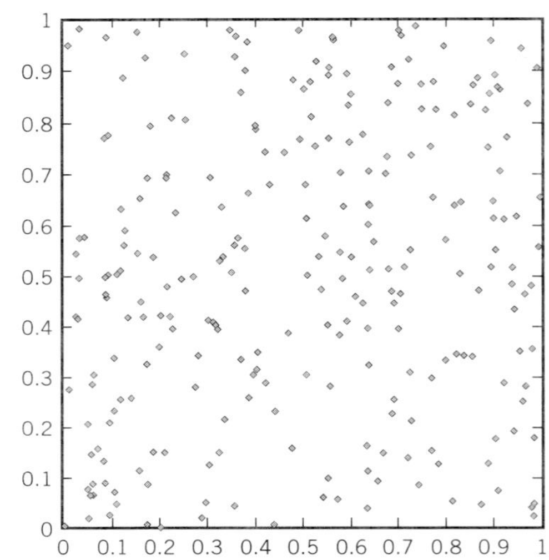

We now state the theoretical bases for the use of low-discrepancy sequences as sample points for multivariate integration. Let $V(f) < \infty$ be the variation of f on D in the sense of Hardy and Krause (see Neiderreiter, 1992), and let $\{z_k\}$ be a low-discrepancy sequence. The Koksma–Hlawka inequality guarantees that

$$\left| I(f) - \frac{1}{n}\sum_{k=1}^{n} f(z_k)\right| \le V(f)\,\|R_n(\cdot\,;z_1,\ldots,z_n)\|_\infty \qquad (5)$$

Upper bounds in terms of L_2 discrepancy have also been proven (Neiderreiter, 1992). Therefore, (4) and (5) provide a worst-case assurance for the use of low-discrepancy sequences as sample

2. 256 Sobol' points in the unit square

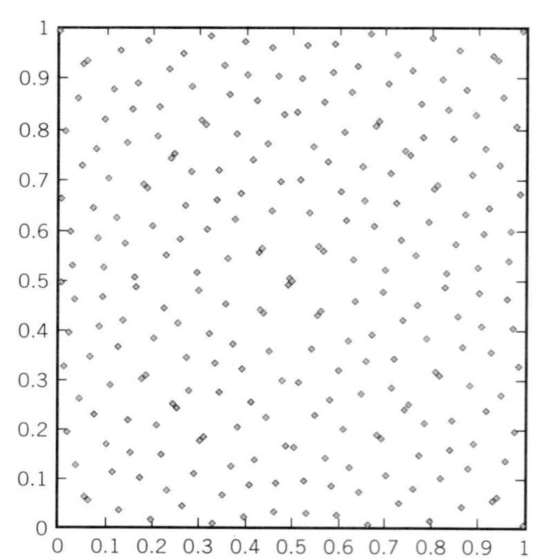

points for numerical integration of functions with bounded variation in the sense of Hardy and Krause. Furthermore, the deterministic algorithms that are based on low-discrepancy sequences have better rates of convergence than the Monte Carlo algorithm.

In addition, low-discrepancy sequences have good properties in the average case setting. Indeed, let $F = C_d$ be the class of real continuous functions defined on D and equipped with the classical Wiener sheet measure, w. That is, w is Gaussian with mean zero and covariance kernel R_w defined as

$$R_w(t,x) \stackrel{\text{def}}{=} \int_{C_d} f(t)f(x)w(df) = \min(t,x) \stackrel{\text{def}}{=} \prod_{j=1}^{d} \min(t_j,x_j)$$

where $t = (t_1,\ldots,t_d)$, $x = (x_1,\ldots,x_d)$ for $t, x \in D$.

Define

$$x_k = 1 - z_k \qquad k = 1,2,\ldots,$$

where $\{z_k\}$ is a low-discrepancy sequence. We approximate the integral of f from C_d by the arithmetic mean of its values at x_k:

$$I(f) = \int_D f(t)dt \approx U_n(f) = \frac{1}{n}\sum_{k=1}^{n} f(x_k), \qquad \forall f \in C_d$$

Woźniakowski (1991) relates the average case error of $U_n(f)$ to the L_2 discrepancy:

$$\int_{C_d} (I(f) - U_n(f))^2 w(df) = \int_D R_n^2(t;z_1,\ldots,z_n)dt \qquad (6)$$

Thus, (3), (4), and (6) provide an average case assurance for the use of the sequence $\{x_k\}$ as a set of sample points for numerical integration of the functions in C_d equipped with the classical Wiener sheet measure. Using the identity $\chi_{[0,t)}(z_k) = \chi_{(1-t,1]}(x_k)$ and Proposition 2.4 in Niederreiter (1992), we conclude that if $\{z_k\}$ is a low-discrepancy sequence, then $\{x_k\}$ is also a low-discrepancy sequence.

HALTON POINTS

We give a short description of the Halton low-discrepancy sequence.

Let $p_1,p_2,\ldots,p_d$ be the first d prime numbers. Any integer $k \ge 0$ can be uniquely represented as $k = \sum_{i=0}^{[\log k]} a_i p_j^i$ with integers $a_i \in [0,p_j - 1]$. The radical inverse function ϕ_{p_j} is defined as

$$\phi_{p_j}(k) = \sum_{i=0}^{[\log k]} a_i p_j^{-i-1} \qquad (7)$$

The Halton d-dimensional points are defined as

$$z_k = (\phi_{p_1}(k),\phi_{p_2}(k),\ \ldots,\ \phi_{p_d}(k)), \qquad k \ge 0$$

SOBOL' POINTS

We give a short description of the Sobol' low-discrepancy sequence.

Assume first that $d = 1$. A one-dimensional Sobol' sequence is generated as follows. For $i = 1,2,\ldots,w$, let $v_i = m_i/2^i$ be a sequence of binary fractions with w bits after the binary point, where $0 < m_i < 2^i$ are odd integers. The numbers v_i are called *direction numbers*. Assume for a moment that they are already generated; we will discuss their generation later.

In Sobol's original algorithm, a one-dimensional Sobol' sequence is generated by

$$x_k = b_1v_1 \oplus b_2v_2 \oplus \ldots \oplus b_wv_w, \qquad k \ge 0$$

where $k = \sum_{i=0}^{[\log k]} b_i 2^i$ is the binary representation of k and $\oplus$ denotes a bit-by-bit exclusive-or operation. For example, $k = 3$ is 11 in base 2. If $v_1 = 0.1$ and $v_2 = 0.11$, then

$$b_1v_1 \oplus b_2v_2 = 0.1 \oplus 0.11 = 0.01$$

Antonov and Saleev (1979) suggest a shuffling of the original Sobol' sequence which preserves good convergence properties and can be generated much faster. This version of the Sobol' sequence is used here.

The sequence of direction numbers v_i is generated by a primitive polynomial (see Knuth, 1981; and Sobol', 1967) with coefficients in the field Z_2 with elements $\{0,1\}$. Consider, for example, the primitive polynomial

$$P(x) = x^n + a_1x^{n-1} + \ldots + a_{n-1}x + 1$$

of degree n in $Z_2[x]$. Then the direction numbers are obtained from the recurrence formula

$$v_i = a_1v_{i-1} \oplus a_2v_{i-2} \oplus \ldots \oplus a_{n-1}v_{i-n+1} \oplus v_{i-n} \oplus (v_{i-n}/2^n), \qquad i > n$$

where the last term v_{i-n} is shifted right n places. The initial numbers $v_1 = m_1/2,\ldots,v_n = m_n/2^n$ are such that m_i is odd and $0 < m_i < 2^i$ for $i = 1,2,\ldots,n$.

Consider now an arbitrary $d \ge 1$. Let $P_1, P_2,\ldots,P_d$ be d primitive polynomials in $Z_2[x]$. Denote by $\{x_k^i\}_{k=1}^{\infty}$ the sequence of one-dimensional Sobol' points generated by the polynomial

P_i. Then the sequence of d-dimensional Sobol' points is defined as

$$x_k = (x_k^1, x_k^2, \ldots, x_k^d)$$

Software system for computing high-dimensional integrals

ALGORITHMS

The theory presented in the previous sections suggests that the low-discrepancy deterministic algorithms provide an interesting alternative to the Monte Carlo algorithm for computing high-dimensional integrals. We have developed and tested a distributed software system for computing multivariate integrals on a network of workstations. The deterministic algorithms and the Monte Carlo algorithm are implemented. The software utilises the following sequences of sample points:

- Halton points;
- Sobol' points; and
- uniformly distributed random points.

The user can choose the sequence of sample points from a menu. The software is written in a modular way so that other kinds of deterministic and random number generators can be easily added. One or several multivariate functions defined over the unit cube of up to 360 variables can be integrated simultaneously. The number of variables could be extended as well.

A routine for generating Sobol' points is given in Press *et al.* (1992). However, we have made major improvements and the results reported in this chapter were obtained using FINDER and not the routine given by Press *et al.* One of the improvements was to develop the table of primitive polynomials and initial direction numbers for dimensions up to 360.

The software uses various kinds of random number generators. More specifically, the random number generators ran1 and ran2 from the first edition of *Numerical Recipes in C* (1988), and RAN1 and RAN2 from the second edition are used because of their wide availability and popularity. All of the above random number generators are based on linear congruential generators with some additional features. For more details on these random number generators we refer readers to Press *et al.* (1988) and (1992).

SYSTEMS

Since workstation clusters and networks provide cost-effective means of performing large-scale computation, we have built and debugged a software system under PVM (Parallel Virtual Machine) for computing multivariate integrals. The system runs on a heterogeneous network of machines. PVM 3.2 is a software package that allows a network of heterogeneous UNIX computers to be used as a single, large-parallel computer. Thus, large computational problems can be solved by using the aggregate power of many computers.

A master/slave model is used as the programming paradigm for computing multivariate integrals. In this model, the master program spawns and directs a number of slave processes. Each of the slave processes generates a sequence of sample points specified by the master and evaluates the integrand at those points. A partial sum of integrand values is returned to the master by each of the slave processes. The master combines partial sums as specified by an algorithm, computes an approximate value of the integral, and checks a termination criterion. If the termination criterion is not satisfied, the master spawns a new round of computations. This process continues until the termination criterion is satisfied or some prespecified upper bound of the number of sample points is reached. At the end, the master returns the final result of computation and timing information. In addition, the master process keeps information about each spawned process. If a host dies, the master reallocates its job to some other host.

In a multi-user network environment the load-balancing method can be one of the most important factors for improving performance. A dynamic load-balancing method is used. Namely, the pool of tasks paradigm is especially suited for a master/slave program. In the pool of tasks method the master program creates and holds the "pool" and sends out tasks to slave programs as they become idle. The fact that no communication is required between slave programs and the only communication is to the master makes our integration problem suitable for the pool of tasks paradigm.

Finance (CMO) problem

We now consider a finance problem which is a typical CMO. A CMO consists of several bond classes, commonly referred to as tranches, which derive their cashflows from an underlying pool of mortgages. The cashflows received from the pool of mortgages are divided and distributed to each of the tranches according to a set of prespecified rules. The cashflows consist of interest and repayment of the principal.

The technique of distributing the cashflows transfers the prepayment risk among different tranches. This results in financial instruments with varying characteristics which might be more suitable to the needs and expectations of investors. For more details on CMOs, we refer to Fabozzi (1992). We emphasise that the amount of the obtained cashflows will depend on the future level of interest rates. Our problem is to estimate the expected value of the sum of present values of future cashflows for each of the tranches.

We now give some details of the studied CMO and the way it is reduced to the problem of multivariate integration. The CMO[1] consists of 10 tranches. Denote them by A, B, C, D, E, G, H, J, R, Z. Throughout this section, we describe results for tranche A in more detail. Results for other tranches are similar unless stated explicitly. The underlying pool of mortgages has a 30-year maturity and cashflows are divided and distributed according to prespecified rules. The actual rules for distribution are rather complicated and are given in detail in the prospectus describing the financial product.

For $1 \le k \le 360$, denote by C the monthly payment on the underlying pool of mortgages; by i_k the appropriate interest rate in month k; by w_k the percentage prepaying in month k; and by $a_{360-k+1}$ the remaining annuity after month k. Recall that the remaining annuity, a_k, is given by

$$a_k = 1 + v_0 + \cdots + v_0^{k-1}, \quad k = 1,2,\ldots,360$$

with $v_0 = 1/(1 + i_0)$ and i_0 the current monthly interest rate. In the notation above, C and $a_{360-k+1}$ are constants, and i_k and w_k are stochastic variables to be determined as set out below.

We now describe the interest rate model. Assume that the interest rate, i_k, is of the form

$$i_k = K_0 e^{\xi_k} i_{k-1} = K_0^k i_0 e^{\xi_1 + \cdots + \xi_k}$$

where $\{\xi_k\}_{k=1}^{360}$ are independent normally distributed random variables with mean 0 and variance σ^2, and K_0 is a given constant. In our case σ^2 is chosen to be 0.0004.

Suppose that the prepayment model w_k, as a function of i_k, is computed as

$$\begin{aligned} w_k &= w_k(\xi_1,\ldots,\xi_k) = K_1 + K_2 \arctan(K_3 i_k + K_4) \\ &= K_1 + K_2 \arctan(K_3 K_0^k i_0 e^{\xi_1 + \cdots + \xi_k} + K_4) \end{aligned}$$

where K_1, K_2, K_3, K_4 are given constants.

The cashflow in month k, $k = 1, 2,\ldots,360$, is

$$\begin{aligned} M_k &= M_k(\xi_1,\ldots,\xi_k) \\ &= C(1 - w_1(\xi_1))\cdots(1 - w_{k-1}(\xi_1,\ldots,\xi_{k-1})) \\ &\quad \times (1 - w_k(\xi_1,\ldots,\xi_k) + w_k(\xi_1,\ldots,\xi_k) a_{360-k+1}) \end{aligned}$$

This cashflow is distributed to the tranches according to the rules of the CMO under consideration. Let $G_{k;T}(\xi_1,\ldots,\xi_k)$ be the portion of the cashflow M_k for month k directed to the tranche T. The form of the function $G_{k;T}$ is very complex. Here, it suffices to say that it is a continuous function which is a composition of min functions and smooth functions. By min functions we mean a function which is a minimum of two functions.

To find the present value of the tranche T for month k, $G_{k;T}(\xi_1,\ldots,\xi_k)$ has to be multiplied by the discount factor

$$u_k(\xi_1,\ldots,\xi_{k-1}) = v_0 v_1(\xi_1)\cdots v_{k-1}(\xi_1 \ldots \xi_{k-1})$$

with

$$\begin{aligned} v_j(\xi_1,\ldots,\xi_j) &= \frac{1}{1 + i_j(\xi_1,\ldots,\xi_j)} \\ &= \frac{1}{1 + K_0^j i_0 e^{\xi_1 + \cdots + \xi_j}}, \quad j = 1,2,\ldots,359 \end{aligned}$$

Summing up the present values for every month k, $k = 1,2,\ldots,360$, for tranche T will give us the present value, PV_T:

$$PV_T(\xi_1,\ldots,\xi_{360}) = \sum_{k=1}^{360} G_{k;T}(\xi_1,\ldots,\xi_k) u_k(\xi_1,\ldots,\xi_{k-1})$$

We want to compute the expected value $E(PV_T) = E(PV_T(\xi_1,\ldots,\xi_{360}))$. By change of variables, it is easy to see that

$$E(PV_T) = \int_{[0,1]^{360}} PV_T(y_1(x_1),\ldots,y_{360}(x_{360}))dx_1\ldots dx_{360}$$

where $y_i = y_i(x_i)$ is implicitly given by

$$x_i = \frac{1}{\sqrt{2\pi\sigma}} \int_{-\infty}^{y_i} e^{-t^2/(2\sigma)} dt \tag{8}$$

Therefore, our problem is reduced to a problem of computing 10 multivariate integrals over the 360-dimensional unit cube. It should be noted that after generating a point $(x_1,\ldots,x_{360})$ in the 360-dimensional unit cube, the point $(y_1,\ldots,y_{360})$ has to be computed by finding the value of the inverse normal cumulative distribution function at each x_k, $k = 1,2,\ldots,360$. We use software available through NETLIB to compute the points $(y_1,\ldots,y_{360})$. Then the function values $PV_T(y_1,\ldots,y_{360})$ are computed for all tranches, T.

We now discuss some of the smoothness properties of the integrand PV_T, where T is one of the tranches. Since the change of variables in (8) is based on smooth functions, we may restrict our discussion to PV_T as a function of $y_1,\ldots,y_{360}$. Recall that

$$PV_T(y_1,\ldots,y_{360}) = \sum_{k=1}^{360} G_{k;T}(y_1,\ldots,y_k)u_k(y_1,\ldots,y_{k-1})$$

Clearly, u_k is a smooth function. As mentioned before, $G_{k;T}$ is a composition of min functions and smooth functions. That is why we believe that the function PV_T has finite variation in the sense of Hardy and Krause, and that is why we decided to use low-discrepancy sequences for these types of integrands. Of course, it would be desirable to have a good bound on the variation of PV_T. Unfortunately, due to the complex form of PV_T, this is a very difficult task.

As reported at the beginning of this chapter, it is widely believed that the theoretical advantage of low-discrepancy sequences degrades with increase of dimension and that such sequences are effective, in general, only for $d \le 30$. Although the CMO problem is apparently of dimension 360, some of the tranches are of lower dimension. To understand the performance of the different algorithms for the CMO problem, it is important to know the dimensions of the various tranches.

The integrand PV_T of tranche T depends on the cashflows $G_{k;T}$. The cashflow $G_{k;T}$ has an important property. If the tranche T is retired at month k_T, ie the whole principal is paid off, then $G_{k;T}$ is zero for month $k > k_T$. More precisely:

$$G_{k;T}(\xi_1,\ldots,\xi_k) = 0 \quad \forall k > k_T = k_T(\xi_1,\ldots,\xi_{k_T})$$

with $\xi_1,\ldots,\xi_k$ corresponding to a particular interest rate scenario.

We checked computationally that k_T is typically smaller than 360. We generated $n = 3{,}000{,}000$ random sample points and determined the maximum k_T for each tranche T. These are reported in Table 1. As we see, one tranche is of dimension 360. However, for the other tranches the dimension is still high.

The concept of the dimension of a tranche can be relaxed further. Intuitively, if the dependence of PV_T on the variables $x_{k+1},x_{k+2},\ldots,x_{360}$ is negligible, then only the first k variables

Table 1. The maximum of the numbers k_T obtained by generating 3,000,000 random sample points for each tranche of the CMO problem

Tranche	Maximum of k_T
A	189
B	250
C	278
D	298
E	309
G	77
H	91
J	118
R	360
Z	167

$x_1,x_2,\ldots,x_k$ contribute substantially to the integral of PV_T. This concept of *effective dimension* can be defined rigorously as follows.

Let f be a function with domain $[0,1]^d$ and let $\varepsilon > 0$. The effective dimension $K(\varepsilon)$ of f is the smallest integer $k \in [1,d]$ such that

$$\left| \int_{[0,1]^k} f(x_1,\ldots,x_k,0,\ldots,0)dx_1\cdots dx_k - \int_{[0,1]^d} f(x)dx \right| \le \varepsilon \left| \int_{[0,1]^d} f(x)dx \right|$$

We apply this definition of effective dimension to the 10 tranches of the CMO problem. Clearly, the effective dimension is a function of ε. What value of ε should we choose? In practice, the CMO problem does not have to be solved with high accuracy (see also the discussion in the later section on small number of sample points), and so we choose $\varepsilon = 0.001$. We checked computationally that the values of the integrals of all tranches are about $a \times 10^6$ with $a \in [2,42]$. Hence, by taking $\varepsilon = 0.001$ we introduce an absolute error which is on the order of several thousand dollars.

We estimate $K_T(\varepsilon)$ for $\varepsilon = 0.001$, where $K_T(\varepsilon)$ denotes the effective dimension of tranche T. To compute $K_T(\varepsilon)$ we need a fairly accurate approxi-

Table 2. The approximate values of $K_T(\varepsilon)$ for $\varepsilon = 0.001$

Tranche	$K_T(\varepsilon)$
A	131
B	175
C	212
D	239
E	261
G	42
H	63
J	84
R	338
Z	114

mation of the integral $I(PV_T) = \int_{[0,1]^{360}} PV_T(x)dx$. We approximate $I(PV_T)$ by using 20,000,000 sample points, as explained in the section on small number of points. Using these results we compute approximately the effective dimension for each of the tranches. The results are summarised in Table 2. As we see, the effective dimension, although smaller, is still high for most of the tranches.

Comparison of deterministic and Monte Carlo algorithms

We now present the results of extensive testing of the deterministic and Monte Carlo algorithms for the CMO problem. For the reader's convenience, the results are summarised in a number of figures.

Figure 3 shows the results for tranche A of Sobol', Halton, and Monte Carlo runs with two randomly chosen initial seeds. The pseudo-random generator RAN2 from Press *et al.* (1992) is used to generate random sample points for the Monte Carlo runs. Figure 3 illustrates behaviour that is common in all our comparisons of the Sobol' algorithm with the Monte Carlo algorithm:

❑ The Monte Carlo algorithm is sensitive to the initial seed.

❑ The deterministic algorithms, especially the Sobol' algorithm, converge significantly faster than the Monte Carlo algorithm.

❑ The convergence of the deterministic algorithms, especially of the Sobol' algorithm, is smoother than the convergence of the Monte Carlo algorithm. This makes automatic termination easier for the Sobol' algorithm (see the discussion below).

❑ The Sobol' algorithm outperforms the Halton algorithm.

Figure 4 shows the same Sobol' and Halton runs but with a different random number generator: here ran1 from Press *et al.* (1988) was used to generate four Monte Carlo runs using four randomly chosen initial seeds. The plot also indicates that there are some problems with this random number generator since the Monte Carlo results seem to lie on horizontal lines when the number of sample points exceeds 300,000 and it seems unlikely that they will converge to the same value. This claim is also supported by the fact that the same effect has been observed for 16 additional Monte Carlo runs. We assume that this is why ran1 has been replaced by a different random number generator in Press *et al.* (1992). Figure 4 is included to show that the results can be affected by the poor performance of the random number generator.

Figure 5 plots the same Sobol' and Halton runs versus the arithmetic mean of 20 Monte Carlo runs using RAN2 from Press *et al.* (1992). Note that the number of sample points on the X-axis is correct only for the deterministic algorithms. The actual number of sample points for the averaged Monte Carlo graph is 20 times the number of sample points on the X-axis. The results of the deterministic algorithms and the averaged Monte Carlo result are approximately the same. We thus conclude that to achieve similar performances we have to take about 20 times more random than deterministic sample points.

3. Sobol' and Halton runs for tranche A and two Monte Carlo runs using RAN2

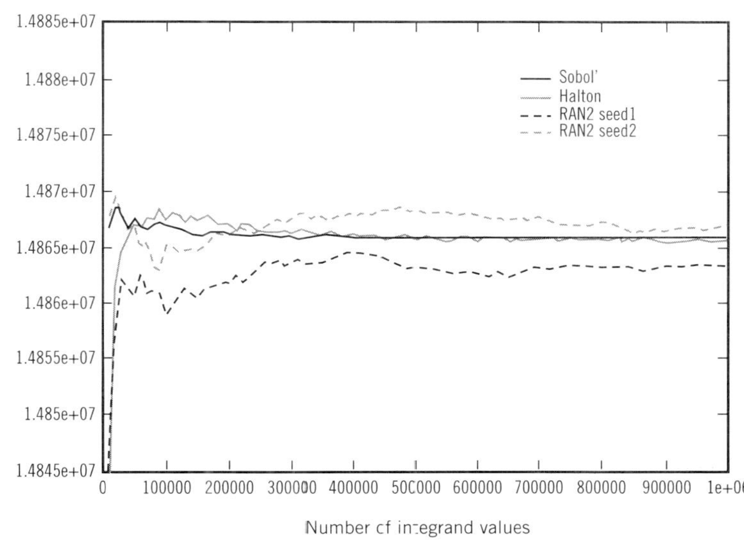

4. Sobol' and Halton runs for tranche A and four Monte Carlo runs using ran1

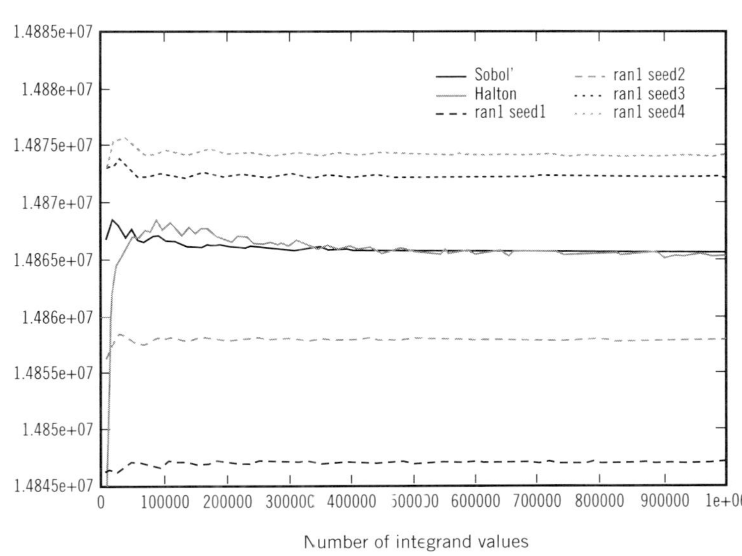

In Figure 6, an automatic termination criterion is applied to Sobol', Halton, and three Monte Carlo runs using RAN2 from Press *et al.* (1992) as the pseudo-random generator. We choose a standard automatic termination criterion. Namely, when two consecutive differences between consecutive approximations using $10{,}000i$, $i = 1,2,\ldots,100$, sample points become less than some threshold value for all of the tranches of the CMO, the computational process is terminated. With the threshold value set at 250, the Sobol' run terminates at 160,000 sample points, the Halton run terminates at 700,000 sample points, and the three Monte Carlo runs terminate at 410,000, 430,000, and 780,000 sample points, respectively. Hence, the Sobol' run terminates two to five times faster than the Monte Carlo runs.

5. Sobol' and Halton runs for tranche A and an average of 20 Monte Carlo runs using RAN2

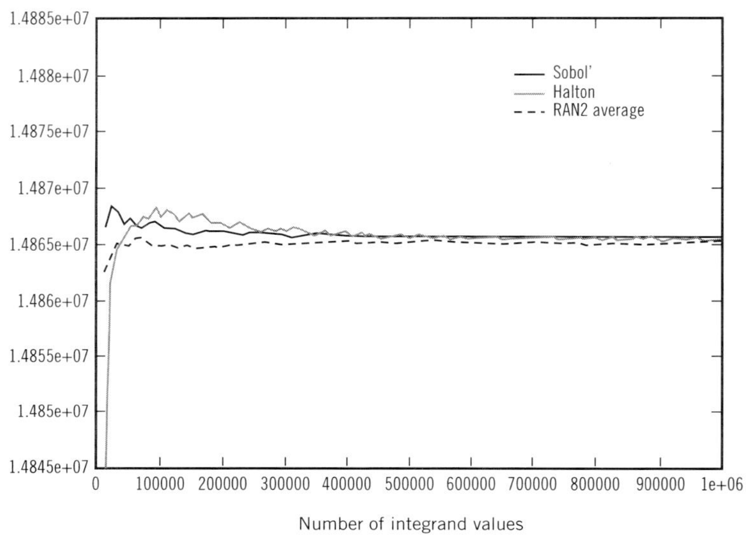

6. Automatic termination criterion applied to Sobol', Halton, and three Monte Carlo runs using RAN2 for tranche A

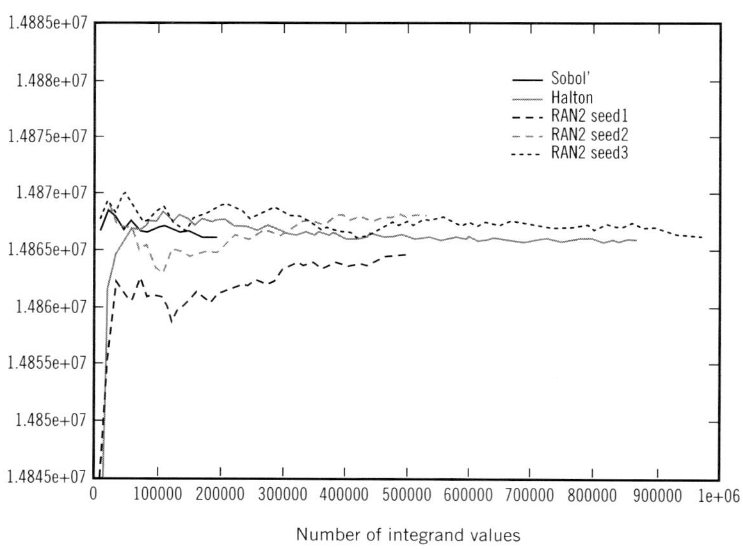

Moreover, even though the Sobol' algorithm terminates faster, the result is more accurate than two of the results achieved by the Monte Carlo algorithm. In the section on small number of sample points, we show that the value of the integral for tranche A using 20,000,000 sample points is about $m = 14{,}865{,}801$. Using this value of m, we compute the absolute error for each of the results at the point of termination:

- the error for 160,000 Sobol' sample points is 379;
- the error for 700,000 Halton sample points is 69;
- the error for 410,000 random sample points with initial seed 1 is 1,298;
- the error for 430,000 random sample points with initial seed 1147902781 is 2,177; and
- The error for 780,000 random sample points with initial seed 1508029952 is 180.

Automatic termination criteria are often used in computational practice. It is of interest to study the relation between the threshold value and the actual error of approximation. See Paskov (1995) for the approximation of linear operators in the average case setting assuming that arbitrary linear continuous functionals can be computed. It is proved that standard termination criteria work well. The corresponding problem for the approximation of linear operators in the average case setting with information consisting only of function values is still open.

The conclusions presented in this section for the Monte Carlo and Sobol' algorithms also hold for the rest of the tranches. Although the Halton algorithm performs better than the Monte Carlo algorithm for tranche A and a few other tranches, it does not perform well for most of them. Therefore our emphasis for the remainder of the chapter will be on Sobol' rather than Halton points.

Antithetic variables

As already mentioned in our discussion of the Monte Carlo algorithm, the expected error of the classical algorithm depends on the variance of the integrand. Therefore, by reducing the variance of the integrand, the expected error can also be reduced. Various variance reduction techniques – such as importance sampling, control variates, antithetic variables and others (see, for example, Kalos and Whitlock, 1986) – are often used with the classical Monte Carlo algorithm.

For the low-discrepancy algorithms, the error bound depends on the variation of the integrand (see section on low-discrepancy deterministic algorithms). Therefore, the error bound will be reduced by reducing the variation. Reducing the variation of the integrand has not been studied as extensively as reducing the variance (see, however, Chelson, 1976; Maize, 1981; Moskowitz and Caflisch, 1996; and Spanier and Maize, 1994). This is a major area for further research.

An important advantage of the classical Monte Carlo algorithm and of the deterministic algorithms studied here is that they can be utilised very generally. This is important in a number of situations – for example: if a financial house has a book with a wide variety of derivatives, it is advantageous to use algorithms which do not need to be tuned to a particular derivative; and if a new derivative has to be priced, there is no immediate opportunity to tailor a variance reduction technique to a particular integrand.

Although variance reduction techniques can be very powerful, they may require considerable analysis before being applied. We will therefore limit ourselves here to just one variance reduction technique: antithetic variables. The advantage of the technique is that it can be easily utilised. Tests reveal that it is superior to the classical Monte Carlo algorithm for our CMO problem, although it should be emphasised that antithetic variables is not a palliative as it can be inferior to the classical Monte Carlo algorithm.

For brevity, we refer to the antithetic variables variance reduction technique combined with the Monte Carlo algorithm as the antithetic variables algorithm. This algorithm is based on the identity

$$\int_D f(x)dx = \int_D g(x)dx$$

where

$$g(x) = \tfrac{1}{2}(f(x) + f(1 - x)) \quad \text{for } f \in L_2(D)$$

Clearly, if the variance, $\sigma^2(g)$, is much smaller than the variance, $\sigma^2(f)$, of f the Monte Carlo algorithm for g will have a much smaller expected error.

We must, of course, remember that the cost of one evaluation of g is equal to the cost of two function evaluations of f. Hence, the antithetic variables algorithm is preferable to the Monte Carlo algorithm only if the reduction of the variance is by at least a factor of two. In general, let

$$r = \frac{\sigma(f)}{\sqrt{2}\sigma(g)}$$

Then the expected error of the antithetic variables algorithm is r times smaller than the expected error of the Monte Carlo algorithm. Furthermore, the cost of both algorithms is about the same. Since r can be less than 1 for some functions, the antithetic variables variance reduction technique, although simple to use, does not work in general and should be used with care (see also Kalos and Whitlock, 1986).

We tested the antithetic variables algorithm for the CMO problem with the 10 tranches. Let f_T be the integrand for tranche T. We computed $\sigma(f_T)$, $\sigma(g_T)$, and $r_T = \sigma(f_T)/(\sqrt{2}\sigma(g_T))$ approximately for all tranches. The results for the ratios r_T are given in Table 3, from which it can be seen that the antithetic variables algorithm has at least 1.55 and at most 4.70 smaller expected error than the Monte Carlo algorithm for the 10 tranches of the CMO. Therefore the antithetic variables algorithm works better than the Monte Carlo algorithm for the CMO problem.

We also tested the antithetic variables technique combined with Sobol' points for the CMO problem. Since it did not perform as well as the Sobol' algorithm we omit the results.

We now present graphs of the results obtained for the deterministic and antithetic variables algorithms for the CMO problem.

Figure 7 is analogous to Figure 3 and it shows the results of Sobol', Halton and antithetic variables runs with two randomly chosen initial seeds. Again, Figure 7 illustrates the sensitivity of the antithetic variables algorithm to the initial seed. However, as should be expected, the antithetic variables algorithm works better than the Monte Carlo algorithm. That is why the spread

Table 3. The ratios r_T for the 10 tranches of the CMO

Tranche	r_T
A	2.82
B	2.78
C	3.09
D	3.61
E	4.70
G	2.80
H	2.81
J	3.14
R	1.55
Z	2.50

and the jaggedness of the antithetic variables runs in Figure 7 are smaller than the corresponding ones for Monte Carlo runs in Figure 3.

Figure 8 plots the same Sobol' and Halton runs versus the arithmetic mean of 20 antithetic variables runs using RAN2 from Press *et al.* (1992). Note that the number of sample points on the x-axis is correct only for the deterministic algorithms: the actual number of sample points for the averaged antithetic variables graph is 20 times the number of sample points on the x-axis. Clearly, the Sobol' and the averaged antithetic variables plots are very close after 450,000 points.

7. Sobol' and Halton runs for tranche A and two antithetic variables runs using RAN2

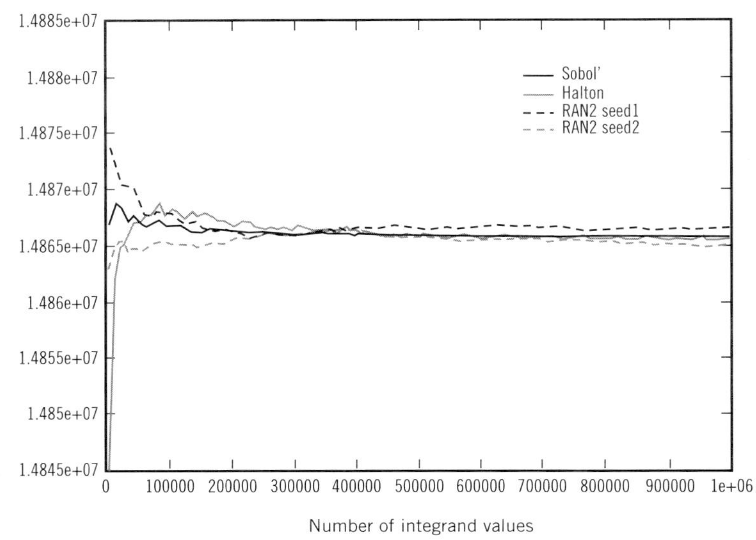

8. Sobol' and Halton runs for tranche A and an average of 20 antithetic variables runs using RAN2

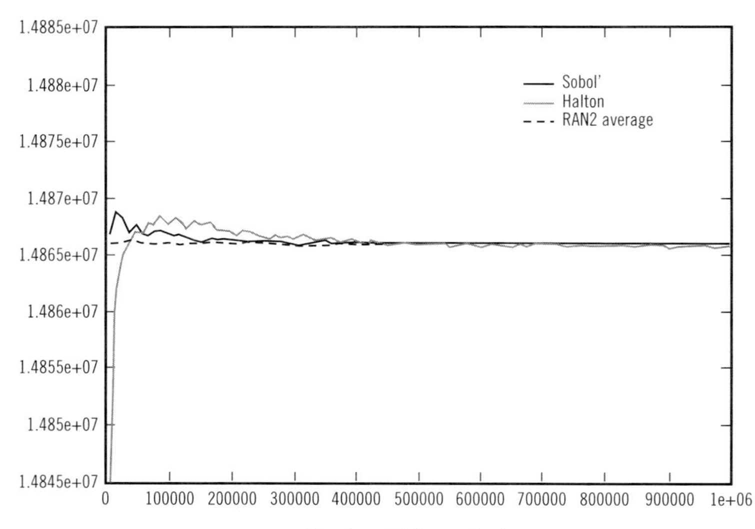

Small number of sample points

In this section we compare the performance of the deterministic algorithms with the Monte Carlo and antithetic variables algorithms for a small number of sample points. Results for a small number of points are sometimes of special importance for people who evaluate CMOs and other derivative products. They need algorithms which can evaluate a derivative in a matter of minutes. Rather low accuracy, on the order of 10^{-2} to 10^{-4}, is often sufficient. The integrands are complicated and computationally expensive. Furthermore, many may have to be evaluated on a daily basis with limited computational resources, such as workstations. We must, therefore, limit ourselves to only a small number of sample points.

By a small number of points we mean 4,000 sample points. As we shall see, this still leads to reasonable results and takes less than a couple of minutes of workstation CPU time. We believe that this sample size may yield comparable results for other mortgage-backed securities and interest rate derivatives.

We drop the Halton algorithm from consideration in this section since it is outperformed by both the Monte Carlo and the antithetic variables algorithms for 4,000 sample points.

From now on, let f be the integrand for tranche A and let

$$g(x) = \tfrac{1}{2}(f(x) + f(1 - x)) \quad \text{for } x \in D$$

Consider the 20 results

$$U_{MC}^{(1)}(f), U_{MC}^{(2)}(f), \ldots, U_{MC}^{(20)}(f) \qquad (9)$$

given by the Monte Carlo algorithm and the 20 results

$$U_{AV}^{(1)}(g), U_{AV}^{(2)}(g), \ldots, U_{AV}^{(20)}(g) \qquad (10)$$

given by the antithetic variables algorithm with different initial seeds, each using 4,000 f function evaluations. For 4,000 Sobol' points, we obtained

$$U_S(f) = 14{,}868{,}261 \qquad (11)$$

To compute the relative errors of all presented results, we need to know the true or almost true value, I(f), of the integral. In the previous section, we showed that the antithetic variables algorithm works better than the Monte Carlo algorithm for the CMO problem. That is why it is reasonable to compute an approximation, m, to

Table 4. Number of "wins" of the antithetic variables algorithm and the Sobol' algorithm

Tranche	Antithetic variables	Sobol'
A	9	11
B	1	19
C	6	14
D	10	10
E	11	9
G	2	18
H	3	17
J	2	18
R	8	12
Z	9	11

Table 5. The smallest and the largest relative error of 20 antithetic variables results and the relative error of the Sobol' results, each using 4,000 sample points

Tranche	Smallest error	Largest error	Sobol' error
A	9.850831e-06	1.078310e-03	1.654482e-04
B	1.684136e-06	3.965597e-04	1.766718e-06
C	2.351182e-06	1.017674e-03	2.154940e-04
D	3.762572e-06	8.231360e-04	2.626482e-04
E	2.111560e-05	5.663271e-04	2.243416e-04
G	1.179616e-06	1.677086e-04	7.305784e-06
H	2.368361e-06	6.108946e-04	3.811441e-05
J	6.517283e-06	7.988188e-04	3.172316e-05
R	3.910856e-05	4.748614e-03	1.179118e-03
Z	1.456121e-05	1.979723e-03	3.784033e-04

$I(f)$ using 20,000,000 function evaluations with the antithetic variables algorithm. For tranche A, we obtain

$$I(f) \approx m = 14{,}865{,}801$$

We now compute the relative error for each of the results in (9), (10), and (11). In Table 4, we indicate how many times out of 20 the Sobol' algorithm has a smaller relative error than the antithetic variables algorithm. As we see, the Sobol' algorithm wins for eight of the tranches, it ties for one, and it loses for one. However, for five of the tranches the Sobol' algorithm wins decisively. In total, the Sobol' algorithm wins 139 times and loses 61 times. Therefore, the antithetic variables algorithm sometimes performs better than the Sobol' algorithm but in most cases it does not.

In Table 5 we report the smallest and the largest relative errors of the 20 antithetic variables runs in (10). The last column of Table 5 reports the relative errors for the Sobol' algorithm.

Note that each of these runs performs very well since the relative error is always at most 0.005. The Sobol' algorithm performs even better; it achieves accuracies within a range of from one part in a thousand to one part in a million, depending on the tranche. It takes approximately 103 seconds to compute the Sobol' results and about 113 seconds to compute the antithetic variables results for the 10 tranches; each algorithm uses 4,000 function evaluations and it is run on a Sun-4/630-M140 workstation.

The results of a comparison of the Monte Carlo and the Sobol' algorithms are summarised in Tables 6 and 7. As would be expected, the performance of the Sobol' algorithm relative to the Monte Carlo algorithm is even better than its performance relative to the antithetic variables algorithm. In this case, the Sobol' algorithm wins decisively for all tranches. In total, the Sobol' algorithm wins 177 times and loses 23 times.

Statistical analysis

In this section, we perform statistical analysis of the Monte Carlo and antithetic variables algorithms for a small number of points. We also compare these two randomised algorithms with the Sobol' algorithm.

Again, we study tranche A. Consider the 1,000 results

$$U^{(1)}_{MC}(f), U^{(2)}_{MC}(f), \ldots, U^{(1000)}_{MC}(f) \tag{12}$$

Table 6. Number of "wins" of the Monte Carlo algorithm and the Sobol' algorithm

Tranche	Monte Carlo	Sobol'
A	3	17
B	0	20
C	3	17
D	3	17
E	2	18
G	0	20
H	0	20
J	0	20
R	8	12
Z	4	16

Table 7. The smallest and the largest relative error of 20 Monte Carlo results and the relative error of the Sobol' results, each using 4,000 sample points

Tranche	Smallest error	Largest error	Sobol' error
A	3.339877e-05	2.633738e-03	1.654482e-04
B	1.037307e-04	3.040330e-03	1.766718e-06
C	3.329284e-05	3.341837e-03	2.154940e-04
D	1.099033e-04	3.536105e-03	2.626482e-04
E	7.673315e-05	3.665118e-03	2.243416e-04
G	3.094756e-05	5.107036e-04	7.305784e-06
H	9.123779e-05	1.454661e-03	3.811441e-05
J	5.152820e-05	1.768061e-03	3.172316e-05
R	3.564157e-05	9.429284e-03	1.179118e-03
Z	5.896356e-05	3.749949e-03	3.784033e-04

given by the Monte Carlo algorithm and the 1,000 results

$$U_{AV}^{(1)}(g), U_{AV}^{(2)}(g), \ldots, U_{AV}^{(1000)}(g) \quad (13)$$

given by the antithetic variables algorithm with different initial seeds, each using 4,000 evaluations of f. Let

$$m_{MC} = \frac{1}{1000} \sum_{i=1}^{1000} U_{MC}^{(i)}(f) = 14,864,881$$

$$m_{AV} = \frac{1}{1000} \sum_{i=1}^{1000} U_{AV}^{(i)}(g) = 14,865,749$$

9. Density function based on 1,000 Monte Carlo results each using 4,000 sample points generated by RAN2 and the normal density function with mean m_{MC} and variance s_{MC}

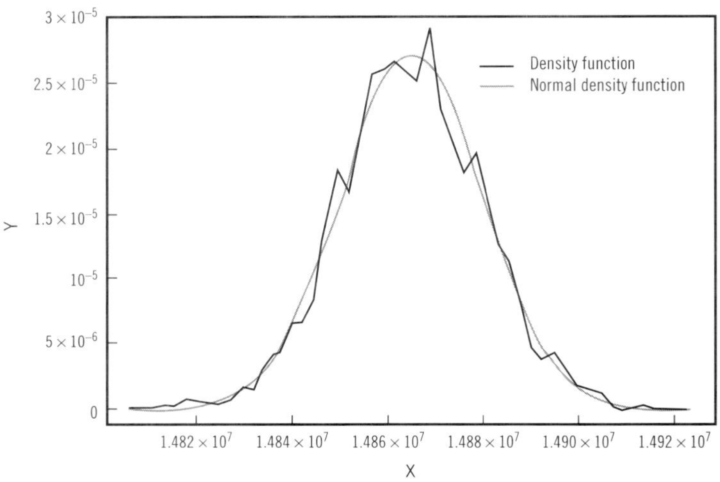

10. Density function based on 1,000 antithetic variable results each using 4,000 sample points generated by RAN2 and the normal density function with mean m_{AV} and variance s_{AV}

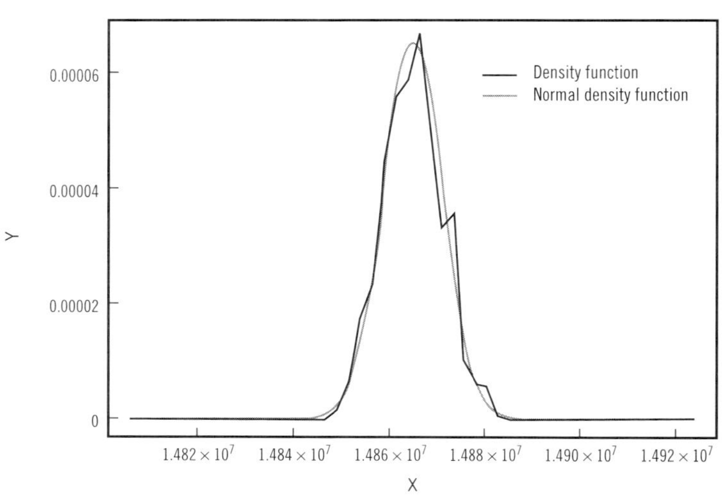

be the sample means, and let

$$s_{MC} = \frac{1}{999} \sum_{i=1}^{1000} (U^{(i)}(f) - m_{MC})^2 = 216,929,148$$

$$s_{AV} = \frac{1}{999} \sum_{i=1}^{1000} (U^{(i)}(g) - m_{AV})^2 = 37,331,468$$

be the sample variances of these 1,000 Monte Carlo and antithetic variables results, respectively.

According to the central limit theorem (see, for example, Kalos and Whitlock, 1986), the distributions of $U_n(f)$ and $U_n(g)$ are normal for $n \to \infty$ with the same mean $\int_D f(x)\,dx = \int_D g(x)dx$ and variances $\sigma^2(f)/n$ and $\sigma^2(g)/n$, respectively. To check how far we differ from the normal distribution we plotted the density functions of these 1,000 results (Figures 9 and 10) using the statistical package S-Plus. We conclude that these are very good approximations of the normal distribution.

Further statistical analysis requires knowledge of the true value, m, of the integral. In the previous section, we showed that it is reasonable to assume that $m \approx 14,865,801$. We seek the number of antithetic variables runs, each using 4,000 sample points, which after averaging would give an error δ with probability η. This procedure is widely used in statistical analysis (Hoel, 1976).

That is, let $\xi_1, \xi_2, \ldots, \xi_k$ be the results for k antithetic variables runs each using 4,000 f-function evaluations. As already discussed, we may assume that they are approximately normally distributed with mean m and variance $s = s_{AV}$. Hence, we want to find k such that

$$P\left\{\left|\frac{1}{k}\sum_{i=1}^{k} \xi_i - m\right| \leq \delta\right\} \geq 1 - \eta$$

Note that the sum $\frac{1}{k}\sum_{i=1}^{k} \xi_i$ is normally distributed with mean m and variance s/k, ie with standard deviation $\sqrt{s/k}$. To guarantee an error δ with confidence level 68%, ie, with $\eta = \sqrt{2/\pi}\int_0^1 e^{-t^2/2}dt \approx 0.68269$, which corresponds to one standard deviation from the mean, we need to determine k_1 such that $\sqrt{s/k_1} = \delta$. In this case

$$k_1 = s/\delta^2 = 37,331,468/\delta^2 \quad (14)$$

To guarantee an error δ with confidence level 95%, ie with

$$\eta = \sqrt{2/\pi}\int_0^2 e^{-t^2/2}\,dt \approx 0.9545$$

which corresponds to two standard deviations

from the mean, we need to determine k_2 such that $2\sqrt{s/k_2} = \delta$. Therefore, in this case,

$$k_2 = 4s/\delta^2 = 149{,}325{,}872/\delta^2 \qquad (15)$$

We now compare the performance of the Sobol' algorithm with the antithetic variables algorithm. Since $U_S(f) = 14868261$ is the result for 4,000 Sobol' points, the error in this case is

$$|U_S(f) - m| = 2460$$

Set $\delta = 2460$. Due to (14), to achieve an error at most δ of 4,000 Sobol' points with confidence level 68%, we need to obtain approximately seven antithetic variables runs each using 4,000 sample points and then to average them. Due to (15), to achieve error at most δ with confidence level 95%, we need to obtain approximately 25 antithetic variables runs.

Clearly, the above analysis depends on the assumption about the true answer of the integral. Furthermore, one may argue that by chance the result from Sobol' points might happen to be very close to the true answer. That is why we performed a similar analysis for the 10 Sobol' results that use $n = 4100 - 20i$, $i = 0,1,\ldots,9$, sample points and we set δ as the worst error of the 10 Sobol' results for tranche A. Then the seven antithetic variables runs obtained before are replaced by three runs.

Results for the other tranches are summarised in Table 8. Note that the number of antithetic variables runs are for the worst error of n Sobol' points for $n = 4100 - 20i$, $i = 0,1,\ldots,9$. Hence, for a fixed i, the number of antithetic variables runs could be even higher than the corresponding number in Table 8. From Table 8 we conclude that we need from seven to 79 antithetic variables runs each using 4,000 sample points to achieve an error with confidence level 95% comparable with the worst error of n Sobol' points, $n = 4100 - 20i$, $i = 0,1,\ldots,9$.

For the Monte Carlo algorithm, we proceed analogously. The results are summarised in Table 9, from which we conclude that we need from 27 to 607 Monte Carlo runs each using 4,000 sample points to achieve an error with confidence level 95% comparable with the worst error of n Sobol' points, $n = 4100 - 20i$, $i = 0,1,\ldots,9$.

Similar statistical analysis may also be performed for a large number of sample points. Due to the large computational cost of these simulations, we compare Sobol' results with 20 antithetic variables results (instead of 1,000 results as for a small number of sample points), each with different initial seed and using 1,000,000 function evaluations. Results for all tranches are summarised in Table 10, where the number of antithetic variables runs are for the worst error of n Sobol' points for $n = 1{,}050{,}000 - 10{,}000i$, $i = 0,1,\ldots,9$.

Table 8. Number of antithetic variables runs each using 4,000 random points needed to achieve the worst error of the n Sobol' points with n = 4100 – 20i, i = 0,1,...,9

Tranche	With confidence 68%	With confidence 95%
A	3	10
B	20	79
C	4	15
D	2	8
E	2	7
G	9	36
H	7	27
J	12	47
R	4	15
Z	2	8

Table 9. Number of Monte Carlo runs each using 4,000 random points needed to achieve the worst error of the n Sobol' points with n = 4100 – 20i, i = 0,1,...,9

Tranche	With confidence 68%	With confidence 95%
A	14	56
B	153	607
C	44	173
D	35	137
E	53	212
G	64	255
H	38	153
J	67	266
R	13	52
Z	7	27

Table 10. Number of antithetic variables runs each using 1,000,000 random points needed to achieve the worst error of the n Sobol' points with n = 1,050,000 – 10,000i, i = 0,1,...,9

Tranche	With confidence 68%	With confidence 95%
A	55	219
B	58	230
C	26	103
D	27	108
E	15	60
G	8	32
H	4	16
J	5	18
R	13	50
Z	5	18

Table 11. Number of Monte Carlo runs each using 1,000,000 random points needed to achieve the worst error of the n Sobol' points with n = 1,050,000 – 10,000i, i = 0,1,...,9

Tranche	With confidence 68%	With confidence 95%
A	434	1735
B	443	1769
C	246	982
D	350	1399
E	331	1324
G	62	245
H	31	122
J	44	176
R	30	120
Z	28	110

By comparing Table 8 and Table 10, we conclude that the superiority of the Sobol' algorithm over the antithetic variables algorithm for a large number of sample points is greater than for a small number of sample points.

A similar conclusion also holds for the Monte Carlo algorithm. We summarise the results of this comparison in Table 11.

Timing results and parallel speed-up for the CMO problem

In this section we report the timing results of the different generators on a single workstation. Then the speed-up of the distributed software system on a network of workstations is also measured for the CMO problem

The CPU time in seconds for simultaneous evaluation of the 10 tranches is given in Table 12 for the Sobol', Halton, RAN2, RAN1, ran1 and ran2 generators. The real time in seconds, as would be expected, is slightly higher than the CPU time. It is given in the second column because it is later compared with the real time for the network of workstations to derive the parallel speed-up. These results were obtained on a single Sun-4/630-M140 workstaton at the Department of Computer Science, Columbia University. All programs were compiled with the

Table 12. Timing results in seconds for evaluation of the CMO using 1,000,000 sample points on a single workstation

Generator	CPU time	Real time
Sobol'	25,634	25,654
Halton	29,814	29,881
RAN2	31,501	31,607
RAN1	28,092	28,126
ran1	30,911	30,946
ran2	28,320	28,332

Table 13. Timing results in seconds for evaluation of the CMO using 1,000,000 sample points generated by RAN2

Number of machines	Real time for RAN2	Real time for Sobol'
5	6382	5360
10	3170	2625
15	2113	1750
20	1705	1408
25	1398	1125

gcc v2.4 compiler with the highest level (O2) of optimisation. Clearly, the generator of Sobol' points is the fastest. However, we do not regard this as the most significant reason for the use of Sobol' points.

Since the speed of the Sun-4/630-M140 is 4.2 Mflops, we note from the above table that it takes approximately 107,500 floating point operations to generate one 360-dimensional Sobol' point and to evaluate integrands for this problem. Since the cost of generating one 360-dimensional Sobol' point is significantly smaller than the evaluation of the integrands, it takes about 100,000 floating point operations to evaluate only integrands. Clearly, this is a rather rough estimate. Some interest rate and prepayment models used in practice are significantly more complicated and more time-consuming to evaluate than those considered in this chapter.

We now discuss the parallel speed-up for the finance problem achieved when the software system is distributed over a network of workstations. The software system for computing multivariate integrals was tested on up to 25 Sun workstations and the timing results were measured. The random number generator RAN2 and the Sobol' generator were used. The results are given in Table 13. We now compare the real-time entry of RAN2 in Table 12 with entries in Table 13. Let N be the number of workstations. In our case N is 5, 10, 15, 20 and 25. The speed-up for the first three entries is about $0.99N$. Then it degrades slightly to $0.92N$ and $0.9N$ for the fourth and the fifth entries, respectively. This degradation is partially due to the fact that the last several machines added for the test, were slightly slower than the Sun-4/630-M140 that was used to measure the time on a single workstation.

Similar speed-ups were measured for the Sobol' generator. Those for the first three entries are $0.96N$, $0.98N$ and $0.98N$, respectively. The speed-up for the last two entries is $0.91N$.

It is evident that networks of workstations are a cost-effective way of performing these computations. The measured speed-up for the parallel/distributed software system for both the Monte Carlo and deterministic algorithms is at least 0.9N for $N \leq 25$.

Discussion and future directions

In this section we present some thoughts about why the Sobol' points are so successful. We end with some directions for future research.

We give two reasons which we believe explain, at least in part, the success of Sobol' points in solving the CMO problem.

First, let us assume that there are no prepayments. This implies that the cashflow from the underlying pool of mortgages for every month is constant. Due to the time value of the money, this means that the present value of the cashflow is a decreasing function of the number of the month. In other words, the present value of the cashflow for earlier months is greater than the present value of the cashflow for later months. The presence of prepayments and rules of distribution to different tranches will distort this monotonicity property. However, this property will be reflected to some extent in the cashflows of the 10 tranches. It is a well-known property of Sobol' points[2] (see Sobol', 1967) that the low-dimensional components are more uniformly distributed than the high-dimensional components. As pointed out in Sobol' (1989), the Sobol' points will be more efficient for numerical integration of functions for which the influence of the x_jth variable decreases as j increases. Since this property also holds for many other financial derivatives, we expect that Sobol' points will provide a powerful alternative to the Monte Carlo and antithetic variables algorithms.

Second, appropriate choices for the initial direction numbers increase the uniformity of the Sobol' points in the 360-dimensional unit cube. We believe that our choice of the initial direction numbers contribute to the successful performance of the Sobol' points.

The Halton algorithm did not perform as well as the Sobol' algorithm. We believe that this is due to the fact that the Halton points are less uniformly distributed than the Sobol' points, especially in high dimensions and for sample sizes of 1,000,000 or less (see Morokoff and Caflisch, 1994).

We end this section by suggesting some directions for future work.

- ❑ Compare the performance of low-discrepancy and Monte Carlo algorithms on other financial derivatives.
- ❑ Test the performance of other known low-discrepancy sequences on various derivatives.
- ❑ Because the results for a small number of samples are often of special interest in finance, it would be attractive to design new deterministic sequences which are very uniformly distributed for a small number of points.
- ❑ Characterise the analytic properties of classes of financial derivatives and design new algorithms tuned to these classes.
- ❑ Study error reduction techniques for deterministic algorithms.
- ❑ There are numerous unsolved theoretical problems concerning high-dimensional integration and low-discrepancy sequences; see Wasilkowski and Woźniakowski (1995) for some of them. We believe that their solution will aid in the design of better algorithms for finance problems.

Reprinted by permission, Paskov, Spassimir, "New Methodologies for Valuing Derivatives", in Pliska and Dempster (eds), *Mathematics of Derivative Securities*, Cambridge University Press

This research was supported in part by the National Science Foundation and the Air Force Office of Scientific Research. I am grateful to Goldman Sachs for providing the finance problem.

An earlier report on this work was presented by Professor J. Traub at the Bank of England Conference at the Isaac Newton Institute, Cambridge, UK, in May 1995.

I express my gratitude to I. Vanderhoof for his valuable help, numerous discussions and suggestions, and introducing me to Goldman Sachs.

I also express my gratitude to J. Traub and H. Woźniakowski for numerous discussions, suggestions and guidance on research and on the preparation of this paper.

I would like to thank A. Belur for his help, advice and cooperation in providing the finance problem.

I am also grateful to T. Boult and S. Baker for their numerous discussions and suggestions during the preparation of this paper.

I appreciate discussions with N. Marinovich, J. Tilley, J. Langsam, P. Karasinski, D. Schutzer, B. Bojanov, S. Tezuka, A. Werschulz, I.M. Sobol', B. Shukhman, V. Temlyakov, S. Heinrich, E. Novak and T. Warnock.

I would like sincerely to thank the Mortgage Research Group at Goldman Sachs and people associated with it – particularly P. Niculescu, R. Wertz and J. Davis, who gave me a chance to greatly improve my knowledge of mortgage-backed securities and interest rate derivatives.

I also appreciate the help of P. Cheah in building the distributed platform used in the software system for computing integrals.

1 *It is labelled as "Fannie Mae REMIC Trust 1989-23".*

2 *Other low-discrepancy sequences also satisfy this property (Neiderreiter, 1992).*

BIBLIOGRAPHY

Adler, S.J., 1981, *The Geometry of Random Fields*, Wiley.

Antonov, I.A., and V.M. Saleev, 1979, "An Economic Method of Computing LP_τ-Sequences", *USSR Computational Mathematics and Mathematical Physics* 19, pp. 252–6.

Bratley, P., and B.L. Fox, 1988, "Algorithm 659, Implementing Sobol's Quasirandom Sequence Generator", *ACM Transactions Mathematical Software* 14, pp. 88–100.

Chelson, P., 1976, "Quasi-random Techniques for Monte Carlo Methods", PhD dissertation, The Claremont Graduate School.

Fabozzi, F.J., 1988, *Fixed Income Mathematics*, Probus Publishing Co.

Fabozzi, F.J., 1992, *Handbook of Mortgage Backed Securities*, Probus Publishing Co.

Fox, B.L., 1986, "Algorithm 647, Implementation and Relative Efficiency of Quasirandom Sequence Generators", *ACM Transactions Mathematical Software* 12, pp. 362–76.

Geweke, J., 1996, "Monte Carlo Simulations and Numerical Integration", in H. Amman, D. Kendrick, and J. Rust (eds), *Handbook of Computational Economics*, Elsevier, pp. 731–800.

Halton, J.H., 1960, "On the Efficiency of Certain Quasi-Random Sequences of Points in Evaluating Multi-Dimensional Integrals", *Num. Math.* 2, pp. 84–90.

Halton, J.H., 1970, "A Retrospective and Prospective Survey of the Monte Carlo Method", *SIAM Review* 12, pp. 1–63.

Halton, J.H., and G.B. Smith, 1964, "Algorithm 247, Radical-Inverse Quasi-Random Point Sequence", *Communications of the ACM* 7, pp. 701–2.

Hammersley, J., and D. Handscomb, 1964, *Monte Carlo Methods*, Methuen.

Hoel, P.G., 1976, *Elementary Statistics*, Wiley.

Janse van Rensberg, E.J., and G.M. Torrie, 1993, "Estimation of Multidimensional Integrals: Is Monte Carlo the Best Method?" *Journal of Physics A: Mathematical and General* 26, pp. 943–53.

Kalos, M.H., and P.A. Whitlock, 1986, *Monte Carlo Methods*, Vol. 1, Wiley.

Knuth, D.E., 1981, *Seminumerical Algorithms*, Vol. 2 of *The Art of Computer Programming*, Addison Wesley.

Maize, E., 1981, "Contributions to the Theory of Error Reduction in Quasi-Monte Carlo Methods", PhD dissertation, The Claremont Graduate School.

Morokoff, W.J., and R.E. Caflisch, 1994, "Quasi-Random Sequences and their Discrepancies", *SIAM Journal of Scientific Computing* 15, pp. 1251–79.

Morokoff, W.J., and R.E. Caflisch, 1995, "Quasi-Monte Carlo Integration, *Journal of Computational Physics* 122, pp. 218–30.

Moskowitz, B., and R.E. Caflisch, 1996, "Smoothness and Dimension Reduction in Quasi-Monte Carlo Methods", *Math. Comp. Modeling* 23, pp. 37–54 (Chapter 27 of this publication).

Niederreiter, H., 1992, *Random Number Generation and Quasi-Monte Carlo Methods*, CBMS-NSF 63, SIAM.

Novak, E., 1988, *Deterministic and Stochastic Error Bounds in Numerical Analysis*, Lecture Notes in Mathematics, Springer Verlag.

Paskov, S.H., 1993, "Average Case Complexity of Multivariate Integration for Smooth Functions", *Journal of Complexity* 9, pp. 291–312.

Paskov, S.H., 1994, "Computing High Dimensional Integrals with Applications to Finance", Paper presented to Joint Summer Research Conference on Continuous Algorithms and Complexity, June, Mount Holyoke College.

Paskov, S.H., 1995, "Termination Criteria for Linear Problems", *Journal of Complexity* 11, pp. 105–37.

Paskov, S.H., and J.F. Traub, 1995, "Faster Valuation of Financial Derivatives", *Journal of Portfolio Management* 22, pp. 113–20.

Press, W., S. Teukolsky, W. Vetterling and B. Flannery, 1988, *Numerical Recipes in C: The Art of Scientific Computing*, First Edition, Cambridge University Press.

Press, W., S. Teukolsky, W. Vetterling and B. Flannery, 1992, *Numerical Recipes in C: The Art of Scientific Computing*, Second Edition, Cambridge University Press.

Roth, K.F., 1954, "On Irregularities of Distribution", *Mathematika* 1, pp. 73–9.

Roth, K.F., 1980, "On Irregularities of Distribution, IV", *Acta Arithmetica* 37, pp. 67–75.

Rust, J., 1994, "Using Randomization to Break the Curse of Dimensionality", Social Systems Research Institute Working Paper Series, No. 9429.

Sarkar, P.K., and M.A. Prasad, 1978, "A Comparative Study of Pseudo and Quasi Random Sequences for the

Solution of Integral Equations", *Journal of Computational Physics* 68, pp. 66–88.

Sobol', I.M., 1967, "On the Distribution of Points in a Cube and the Approximate Evaluation of Integrals", *USSR Computational Mathematics and Mathematical Physics* 7, pp. 86–112.

Sobol', I.M., 1973, *Numerical Monte Carlo Methods* (in Russian), Izdat "Nauka", Moscow.

Sobol', I.M., 1989, "Quadrature Formulas for Functions of Several Variables Satisfying General Lipschitz Condition", *USSR Computational Mathematics and Mathematical Physics* 29, pp. 935–41.

Spanier, J., and E. Maize, 1994, "Quasi-Random Methods for Estimating Integrals Using Relatively Small Samples", *SIAM Review* 36, pp. 19–44.

Tezuka, S., 1994, "A Generalization of Faure Sequences and its Efficient Implementation", Technical report, IBM Research, Tokyo.

Traub, J.F., 1994a, "Average Case Computational Complexity of High-Dimensional Integration with Applications to Finance", Paper presented to NSF Symposium on Simulation and Estimation, August, Department of Economics, University of California, Berkeley.

Traub, J.F., 1994b, "Solving Hard Problems with Applications to Finance", *Thirteenth World Computer Congress, IFIP 94*, Hamburg.

Traub, J.F., G.W. Wasilkowski and H. Woźniakowski, 1988, *Information-Based Complexity*, Academic Press.

Traub, J.F., and H. Woźniakowski, 1992, "The Monte Carlo Algorithm with a Pseudorandom Generator", *Mathematics of Computation* 58, pp. 323–39.

Wasilkowski, G.W., and H. Woźniakowski, 1995, "Explicit Cost Bounds of Algorithms for Multivariate Tensor Product Problems", *Journal of Complexity* 11, pp. 1–56.

Woźniakowski, H., 1991, "Average Case Complexity of Multivariate Integration", *Bulletin of the AMS (New Series)* 24, pp. 185–94.

Woźniakowski, H., 1994, "Tractability and Strong Tractability of Linear Multivariate Problems", *Journal of Complexity* 10, pp. 96–128.

26

Valuation of Mortgage-Backed Securities using Brownian Bridges to Reduce Effective Dimension

Russel E. Caflisch, William Morokoff and Art Owen
UCLA; Goldman Sachs; Stanford University

The quasi-Monte Carlo method for financial valuation and other integration problems has error bounds of size $O((\log N)^k N^{-1})$, or even $O((\log N)^k N^{-3/2})$, which suggests significantly better performance than the error size $O(N^{-1/2})$ for standard Monte Carlo. But in high-dimensional problems this benefit might not appear at feasible sample sizes. Substantial improvements from quasi-Monte Carlo integration have, however, been reported for problems such as the valuation of mortgage-backed securities in dimensions as high as 360. The authors believe that this is due to a lower effective dimension of the integrand in those cases. This paper defines the effective dimension and shows in examples how the effective dimension may be reduced by using a Brownian bridge representation.

Introduction

Simulation is often the only effective numerical method for the accurate valuation of securities whose value depends on the whole trajectory of interest rates or other variables. Standard Monte Carlo simulation using pseudo-random sequences can be quite slow, however, because its convergence rate is only $O(N^{-1/2})$ for N sample paths. Quasi-Monte Carlo simulation, using deterministic sequences that are more uniform than random ones, holds out the promise of much greater accuracy, close to $O(N^{-1})$ in optimal cases. Randomised versions of quasi-Monte Carlo simulation can in some cases bring the typical error close to $O(N^{-3/2})$.

This dramatic improvement in convergence rate has the potential for significant gains both in computational time and in range of application of simulation methods for finance problems. An optimistic reading of the results suggests an effective squaring or even cubing of the sample size, N. Large improvements have in fact been found in a number of earlier studies (Caflisch and Morokoff, 1996, Ninomiya and Tezuka, 1996, and Paskov and Traub, 1995), which were all motivated by the results of Paskov (1996) on mortgage-backed securities.

Quasi-Monte Carlo simulation is not a magic bullet, however. The asymptotic error magnitudes are the ones it is 'close to' above, multiplied by $(\log N)^k$, where k depends on the dimension, s, of the simulation. In high dimensions these powers of $\log N$ do not become negligible at any computationally possible sample size. This loss of effectiveness has been documented for a series of test problems in Morokoff and Caflisch (1993, 1994, 1995). When simulations are cast as integration problems, the resulting integral is often of very high dimension (eg dimension 360 for a mortgage of length 30 years), so any loss of effectiveness at high dimensionality can affect them.

Our first goal in this paper is to reconcile two apparently conflicting truths. The first is that quasi-Monte Carlo is not much better than Monte Carlo in high dimensions with practical sample sizes. The second is that quasi-Monte Carlo has been seen to far surpass Monte Carlo in some high-dimensional examples. It is our view that success in high-dimensional problems tells us more about the integrand than about the method of integration. Some high-dimensional integrands are indeed amenable to quasi-Monte Carlo simulation. Integrands of low 'effective dimension', which we define in two ways below, are of this type. Our second goal is to give an example of a financial simulation in which one can reduce the effective dimension of an integrand, thereby making quasi-Monte Carlo much more effective.

The outline of this paper as follows. The second (following) section gives a brief introduction to quasi-random sequences and their properties, including the Koksma–Hlawka inequality, which is the basic estimate on integration error for quasi-Monte Carlo. The dependence on dimension and the character of two-dimensional projections of quasi-random sequences are also discussed. The third section introduces some useful decompositions of integrand, and uses them to define two notions of the effective dimension of an integrand. The mortgage-backed security problem is formulated in the fourth section. Our main technical tool for formulating the problem with reduced effective dimension is the Brownian bridge representation of a random walk, which is described in the fifth section. Computational results for the mortgage-backed security problem are presented in the sixth section, and conclusions are discussed in the final section.

Quasi-random sequences: discrepancy and integration error

BASIC PROPERTIES

The origin of the improved accuracy of quasi-Monte Carlo methods is the improved uniformity of quasi-random sequences. Figure 1 shows two plots, each of 4096 points in two dimensions. The top is a pseudo-random sequence and the bottom is a quasi-random (Sobol') sequence. In the pseudo-random sequence there is clumping of points, which limits their uniformity. The cause of this clumping is that, since points in a pseudo-random sequence are (nearly) independent, they have a certain chance of landing very near to each other. The constructions used for points in a quasi-random sequence, on the other hand, prevent them from clumping together.

The uniformity of a sequence of points in the s-dimensional unit cube $I^s = [0,1]^s$ can be measured in terms of its discrepancy. This is defined by considering the number of points in rectangular subsets of the cube. For a set $J \subseteq I^s$ and a sequence of N points $(x_n)_{n=1}^N$ in I^s, define

$$R_N(J) = \frac{1}{N}\sum_{n=1}^{N} \mathcal{X}_J(x_n) - m(J)$$

Here $\mathcal{X}_J$ is the characteristic function of the set J, and m(J) is its volume. If E^* is the set of subrectangles with one corner at $(0,\ldots,0)$, then the star discrepancy is defined as

$$D_N^* = \sup_{J \in E^*} |R_N(J)| \tag{1}$$

Some other discrepancies do not treat the origin differently from the other corners of I^s, and there are L^2 counterparts to the L^∞ definition given above. See Niederreiter (1992) for some of these other definitions and Hickernell (1998) for some more recent generalisations.

The importance of discrepancy can be seen from the Koksma–Hlawka inequality for integration error. For the integral of a function f on the s-dimensional unit cube, the simulation-based estimate of the integral is

$$I_N(f) = \frac{1}{N}\sum_{n=1}^{N} f(x_n) \tag{2}$$

and the integration error is

$$e_N(f) = \frac{1}{N}\sum_{n=1}^{N} f(x_n) - \int_{I^s} f(x)\,dx \tag{3}$$

The Koksma–Hlawka inequality says that

$$|e_N(f)| \le V(f)D_N^* \tag{4}$$

where V(f) is the variation of f.

In one dimension, the variation is $V(f) = \int_0^1 |df|$. The definition in higher dimension is more complicated. Define for all $k \le s$ and all sets of k integers $1 \le i_1 < \cdots < i_k \le s$ the quantity

$$V^{(k)}(f; i_1,\ldots,i_k) = \int_{I^k} \left| \frac{\partial^k f}{\partial t_{i_1} \ldots \partial t_{i_k}} \right|_{t_j = 1, j \ne i_1,\ldots,i_k} dt_{i_1} \ldots dt_{i_k}$$

The variation of f (in the sense of Hardy and Krause) is defined as

$$V(f) = \sum_{k=1}^{s} \sum_{1 \le i_1 < \cdots < i_k \le s} V^{(k)}(f; i_1,\ldots,i_k)$$

The Koksma–Hlawka inequality (4) should be compared with the formula for root-mean-square error of Monte Carlo integration using a random sequence. If the sequence (x_n) is uniformly distributed on I^s, then

$$\mathbf{E}\left[e_N(f)^2\right]^{1/2} = \sigma(f)N^{-1/2} \tag{5}$$

in which **E** is the usual expectation and σ(f) is the square root of the variance of f, given by

$$\sigma(f) = \left(\int_{I^s} \left[f(x) - \bar{f}\,\right]^2 dx \right)^{1/2} \tag{6}$$

with $\bar{f} = \int_{I^s} f\,dx$.

1. Two-dimensional projection of a pseudo-random sequence (top) and a Sobol' sequence (bottom)

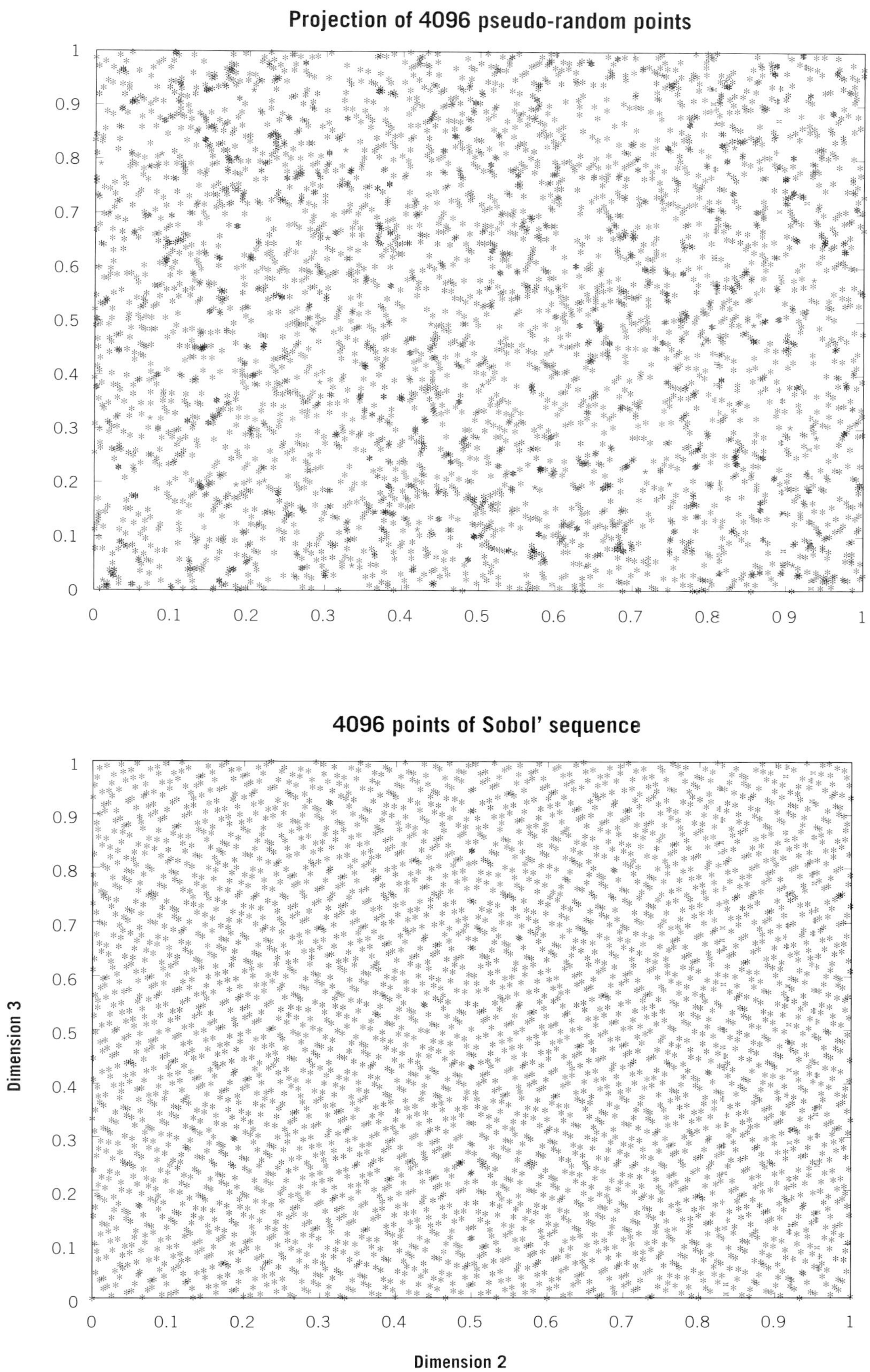

The error magnitudes (4) and (5) are similar in that the bound is a product of one term depending on properties of the integrand function and a second term depending on properties of the sequence. The Koksma–Hlawka inequality is an absolute bound, which is more satisfying theoretically than (5), an equality in expectation that holds only probabilistically. For practical purposes the preference is reversed. Each factor in (4) is incredibly hard to compute, whereas the Monte Carlo variance can be estimated from the same data used to compute $\bar{f}$. Furthermore, the Koksma–Hlawka bound is an inequality that is only tight for a worst-case function f, whose fluctuations are exquisitely matched to the discrepancies in the sequence (x_n), while the Monte Carlo variance estimates the error for the actual f being sampled.

The infinite sequence $(x_n)_{n=1}^{\infty}$ is said to be *quasi-random* if

$$D_N^* \leq c(\log N)^k N^{-1}$$

in which the constant c and the logarithmic exponent k may depend on the dimension s. For integration by quasi-random sequences, the Koksma–Hlawka inequality says that the integration error is size $O((\log N)^k N^{-1})$, which for large N is much more accurate than standard Monte Carlo simulation.

Examples of quasi-random sequences have been constructed by Halton, Faure, Sobol', Niederreiter, and others. For a comprehensive discussion, see the monograph of Niederreiter (1992).

Although the variation of f requires s derivatives of f, we have found in practice that only a minimal amount of smoothness of f is needed for effectiveness of quasi-Monte Carlo integration. For problems in which f is discontinuous, however, the improvements of quasi-Monte Carlo integration are diminished.

The effectiveness of quasi-Monte Carlo in high-dimensional problems was studied by Morokoff and Caflisch (1994, 1995). There it was shown that the faster convergence rate for quasi-Monte Carlo is generally lost for problems of high dimension. The simplest evidence for this conclusion is seen in the discrepancy as a function of N for various values of the dimension s. For small dimensions, the discrepancy appears to be $O(N^{-1})$, ignoring logarithmic factors, for all N. For large dimensions, the discrepancy behaves initially like $O(N^{-1/2})$ as for a random sequence, taking on an apparent $O(N^{-1})$ rate only for very large values of N.

The transition value of N appears to grow exponentially with the dimension. This has not been rigorously proved, but for quasi-random sequences based on nets as described below, it is in keeping with the definition of nets. In high dimensions, unless one uses a very large number of points, quasi-random sequences are no more uniform than random sequences. Thus the Koksma–Hlawka bound does not imply any advantage for quasi-Monte Carlo with moderate values of N and large dimension s. On the other hand, we have found almost no problems for which quasi-Monte Carlo is worse than standard Monte Carlo.

It is difficult to evaluate the uniformity of a sequence in a high-dimensional space. A necessary but not sufficient condition for uniformity is uniformity of low-dimensional coordinate projections of the sequence. The Sobol' sequence used in the computations below (based on the sequence generated by FINDER) has excellent one-dimensional projections, and in the 360-dimensional case we test, many, but not all, of its two-dimensional projections are very uniform. In the third section we show how such partial uniformity is sufficient for the task of integrating functions of low effective dimension.

The one- and two-dimensional projections of (x_n) are easily graphed. Three-dimensional projections can be investigated, with some difficulty, with dynamic graphics tools, such as XGobi (Swayne, Cook and Buja, 1991). The graph at the bottom of Figure 1 shows a 'good' pairing of dimensions using Sobol's second and third dimensions with his recommended starting values. A 'bad' pairing of dimensions is presented in Figure 2, which shows two higher dimensions (following Sobol's convention for associating dimension with generating polynomial) based on the polynomials $x^7 + x^5 + x^4 + x^3 + 1$ and $x^7 + x^5 + x^4 + x^3 + x^2 + x + 1$ and the starting values (1,3,5,11,3,3,35) and (1,1,7,5,11,59,113), respectively. (The polynomials correspond to dimensions 27 and 32, respectively, in FINDER, but the starting values are different.) Although this non-uniformity could go away if these starting values were changed, we have found that this type of non-uniformity is fairly typical of some of the two-dimensional projections of high-dimensional Sobol' and Halton sequences.

The bad behaviour seen in the second plot of Figure 2 can be explained in terms of filling in holes. If 8192 (2^{13}) points are used, the plot looks almost identical to what is shown for 4096. However, the next 8192 points fall only where

2. Two-dimensional projection of a Sobol' sequence

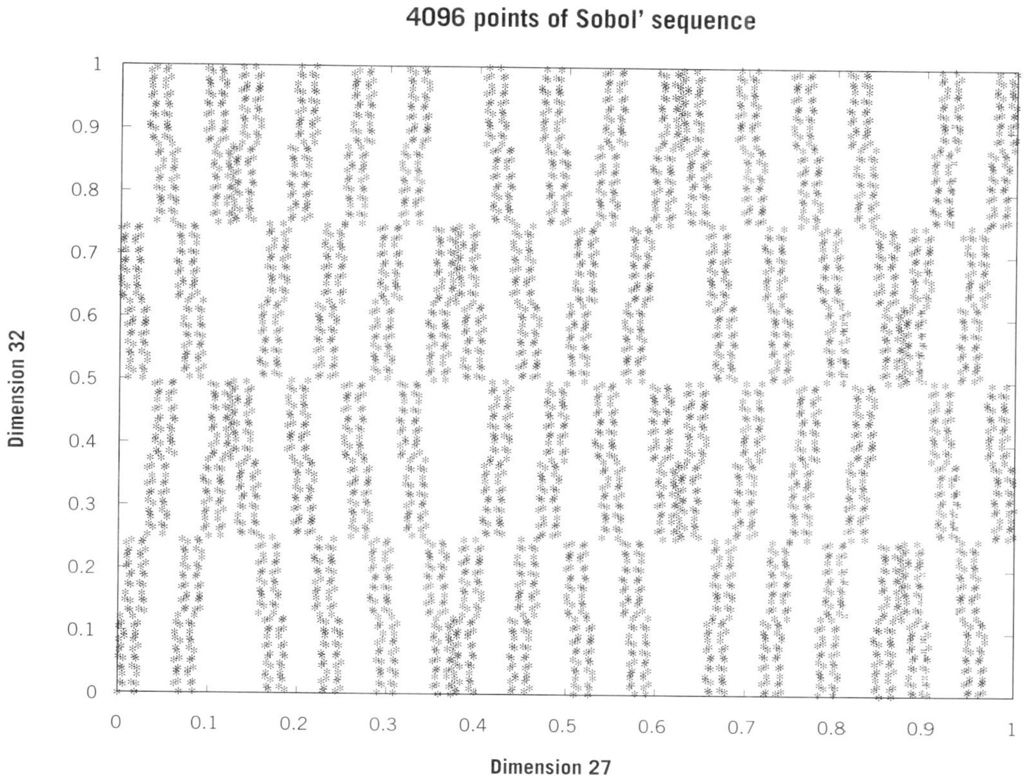

the gaps appear. Thus, by $N = 16{,}384$ the projection plot is almost perfectly uniform. The problem is that the cycle for filling in such holes can be too long.

(t,m,s)-NETS AND (t,s)-SEQUENCES

The quasi-random points we consider are, or are based on, (t,m,s)-nets and (t,s)-sequences. The definitive reference on this topic is chapter 4 of the monograph by Niederreiter (1992).

An elementary interval in base b is a set of the form

$$J = \prod_{j=1}^{s} \left[\frac{a_j}{b^{k_j}}, \frac{a_j + 1}{b^{k_j}}\right) \subseteq I^s$$

for non-negative integers a_j and k_j, with $a_j < b^{k_j}$. For $t \geq 0$, a sequence of $N = b^m$ points x_n is a (t,m,s)-net in base b if every elementary interval J in base b of volume b^{t-m} has $R_N(J) = 0$. Given m, s, and b, the smaller t is, the better.

An infinite sequence $(x_n)_{n=1}^{\infty}$ is a (t,s)-sequence in base b if each finite sequence $(x_n)_{n=Ab^m+1}^{(A+1)b^m}$, with $A \geq 0$, is a (t,m,s)-net in base b for all $m \geq 0$.

The net property becomes relevant for $m > t$, that is, $N \geq b^{t+1}$. Below this value of N, any sequence of points in I^s, even identical points, is consistent with the net property. The smallest N at which the net property constrains some fully s-dimensional elementary sub-interval (one with all $k_j > 0$) is b^{t+s}. The asymptotic rate for the discrepancy of nets should therefore start at around $N = b^{t+s}$.

The most widely used constructions of (t,s)-sequences are due to Sobol', Faure and Niederreiter. The Sobol' sequences are (t,s)-sequences in base $b = 2$. For $s = 360$ the value of t is quite large for Sobol' sequences - in the thousands according to Niederreiter (personal communication, 1996). The Faure sequences are (0,s)-sequences in base b, where $b \geq s$ is a prime number. For $s = 360$, the smallest value of b is 361.

For Faure sequences the net property starts to be relevant at $N = 361$, but the asymptotics are not relevant until $N = 361^{360} > 10^{920}$. For Sobol' sequences the net property starts to be relevant at $N = b^t$ and the asymptotics are relevant at $N = b^{t+s}$. Even if t is as small as 1000, these two values are larger than 10^{300} and 10^{409}, respectively.

Recent constructions of Niederreiter and Xing (1995) hold the promise of (t,s)-sequences in base 2 with $t \doteq s$, a significantly smaller value than previously known possible. Even here, though, this suggests that the net property becomes relevant at $N = 2^{361} > 10^{108}$ and the asymptotics become relevant at $N = 2^{720} > 10^{216}$.

Despite the impractically large values of N mentioned above, quasi-Monte Carlo works well on some high-dimensional integrands, using realistic sample sizes. The reason is that low-dimensional views of the quasi-Monte Carlo points can have small discrepancy and some integrands are dominated by low-dimensional structure, as described in the third section. For example, the points in Figure 1 appear to be a $(t,m,2)$-net with a much smaller value of t than holds for all 360 dimensions of the points.

SCRAMBLED (t,m,s)-NETS AND (t,s)-SEQUENCES

A scrambled net is a hybrid of quasi-Monte Carlo and Monte Carlo methods in which a sequence (x_n) is randomised. These are defined and analysed in a sequence of papers by Owen (1995, 1997a and b); see also Hickernell (1996). The randomisation is carefully constructed in the base b so that (t,m,s)-nets map on to (t,m,s)-nets, (t,s)-sequences on to (t,s)-sequences, and each x_n individually has the uniform distribution on I^s. As a result $I_N(f)$ becomes a random variable with mean f and at least quasi-random accuracy.

By replicating the randomisation, one can estimate the sampling variance statistically from the same data used to estimate the integral. The variance of $I_N(f)$ is $o(N^{-1})$ for any f with $\int f(x)^2 dx < \infty$.

The precise improvement over Monte Carlo depends on properties of f. If f is well approximated by sums of characteristic functions of elementary intervals of large volume, then $V(I_N(f))$ decreases rapidly. In particular, if f is sufficiently smooth that $h(x) = \partial^s f(x)/\partial x^s$ is Lipschitz-continuous, $|h(x) - h(x')| \leq B \| x - x' \|^\beta$ for some $B \geq 0$, and $\beta \in (0,1]$, then $V(I_N(f)) = O(N^{-3}(\log N)^{s-1})$ for scrambled nets on N points. This provides for errors of $O(N^{-3/2}(\log N)^{(s-1)/2})$ in probability, a better rate than that achieved by unscrambled nets. The extra $N^{-1/2}$ factor of accuracy may be attributed to cancellation of some integration errors, which does not happen with deterministic sequences.

The asymptotic rate can be expected to commence at $N = b^{t+s}$, as for unscrambled nets, though with realistic sample sizes significant benefits can be seen on integrands of lower effective dimension.

Integral decompositions and effective dimension

Suppose that we can write

$$f(x) = \sum_{m=0}^{M} f_m(x) \tag{7}$$

Then $\bar{f} = \sum_{m=0}^{M} \bar{f}_m$ and $I_N(f) = \sum_{m=0}^{M} I_N(f_m)$, and $e_N(f) = \sum_{m=0}^{M} e_N(f_m)$ in obvious notation.

In decompositions like equation (7) we will customarily arrange that $f_0(x) = \bar{f}$ is constant in x and that $\bar{f}_m = 0$ for $m \geq 1$. Then

$$e_N(f) = \sum_{m=1}^{M} e_N(f_m) \tag{8}$$

because $I_N(f_0) = \bar{f}_0$. If $\int f_m(x) f_k(x) dx = 0$ for $m \neq k$, the decomposition (7) is said to be orthogonal.

In simple Monte Carlo simulation, the x_n are taken independently from the uniform distribution on $[0,1]^s$. For an orthogonal decomposition, $\sigma^2(f) = \sum_{m=1}^{M} \sigma^2(f_m)$. Many Monte Carlo methods have the effect of changing the sampling variance of $\bar{f}$ to

$$\frac{1}{N} \sum_{m=1}^{M} \Gamma_m \sigma^2(f_m) \tag{9}$$

for some "gain" constants $\Gamma_m \geq 0$. Getting every $\Gamma_m < 1$ implies an improvement over simple Monte Carlo. Even when the Γ_m are not all smaller than 1, a method in which Γ_m is small whenever σ_m^2 is large can be very effective.

A case in point is antithetic sampling. Write $f = f_0 + f_1 + f_2$, where $f_0(x) = \bar{f}$, $f_1(x) = \frac{1}{2}[f(x) + f(1 - x)] - \bar{f}$, and $f_2(x) = \frac{1}{2}[f(x) - f(1 - x)]$. Here $1 - x$ is interpreted componentwise. This is an orthogonal decomposition into constant, even (symmetric), and odd (antisymmetric) parts. Antithetic sampling involves taking the points x_n and $x_n' = 1 - x_n$ in pairs, for $1 \leq n \leq \frac{1}{2}N$. One can show that in antithetic sampling $\Gamma_2 = 0$ and $\Gamma_1 = 2$. Antithetic sampling can be anywhere from half as good as simple Monte Carlo (when $f_2 = 0$) to infinitely better than simple Monte Carlo (when $f_1 = 0$). When it is thought that $\sigma_2^2 \gg \sigma_1^2$, as for nearly linear functions, antithetic sampling becomes attractive.

In an analysis of variance (ANOVA) decomposition, $M = 2^s - 1$, and there is one term in (7) for each subset of the s components of x. The empty set corresponds to f_0. The term for each subset is a function only of the components within the subset. It is convenient to replace the labels $0,\ldots,M$ by subsets $u \subseteq \{1,\ldots,s\}$. Thus $f(x) = \sum_u f_u(x)$. The

ANOVA decomposition is an orthogonal one, so $\sigma^2(f) = \sum_u \sigma_u^2$, where $\sigma_u^2 = \sigma^2(f_u)$. See Owen (1992) for definitions, Takemura (1983) for some history of this decomposition, and Hickernell (1998) for some recent generalisations.

Let $g_t(x) = \sum_{|u|=t} f_u(x)$ for $0 \le t \le s$. Then g_t describes the part of f that is exactly t-dimensional and $\sum_{d=1}^{t} g_d$ describes the part that is at most t-dimensional. The variance of g_t is $\sigma^2(g_t) = \sum_{|u|=t}\sigma_u^2$. If, for example, $\sigma^2(g_1) = 0.99\sigma^2(f)$, then we can conclude that 99% of the variance of f is due to the components of X taken one at a time. In such a case we might say that f is effectively one-dimensional. Likewise, if $\sigma^2(g_1) + \sigma^2(g_2) + \sigma^2(g_3)$ is close enough to σ^2, we might consider f to be effectively three-dimensional. For such an f, a set of points with good uniformity in every triple of variables will produce an accurate integral estimate, even if the points are not particularly uniform in some quadruples of variables.

Definition 3.1 The effective dimension of f, in the superposition sense, is the smallest integer d_S such that $\sum_{0<|u|\le d_S}\sigma^2(f_u) \ge 0.99\sigma^2(f)$.

This notion of effective dimension differs from the one implicitly used in Paskov (1996), which we state formally as:

Definition 3.2 The effective dimension of f, in the truncation sense, is the smallest integer d_T such that $\sum_{u \subseteq \{1,\ldots,d_T\}}\sigma^2(f_u) \ge 0.99\sigma^2(f)$.

The threshold 0.99 is an arbitrary choice, and one might well prefer other values in some settings. For each $d = 1,\ldots,s$, we can define how d-dimensional f is, in the senses above, by the ratios $\sum_{0<|u|\le d}\sigma_u^2/\sigma^2$ and $\sum_{u\subseteq\{1,\ldots,d\}}\sigma_u^2/\sigma^2$. Clearly the fraction of variance that is d-dimensional is higher in the superposition sense than in the truncation sense.

To distinguish the two definitions, consider the function $f = \sum_{i=1}^{s} 2^i x_i$. For this function $d_S(f) = 1$ and $d_T(f) = s$. In general, arranging for all d_S-dimensional projections of $(x_n)_{n=1}^N$ to have low discrepancy ensures an accurate simulation. Alternatively, arranging for the first d_T components of $(x_n)_{n=1}^N$ to have low discrepancy will suffice. The dimension d_S does not depend on the order in which the input variables are indexed, while d_T does.

If f is nearly one-dimensional in the superposition sense, then Latin hypercube sampling will be an extremely effective simulation technique. In Latin hypercube sampling (McKay, *et al.* 1979), the jth component of x_n is $x_{nj} = [\pi_j(n) - u_{nj}]/N$, where π_j is a (uniform) random permutation of the integers 1 through N and the u_{nj} are independent $U[0,1]$ random variables. Latin hypercube sampling stratifies each individual dimension, but imposes no higher-dimensional stratification. In fact, a Latin hypercube sample is a scrambled (0,1,s)-net in base N. Stein (1987) shows that Latin hypercube sampling is essentially equivalent to having $\Gamma_u = 0$ for subsets u of cardinality $|u| = 1$ and $\Gamma_u = 1$ for $|u|>1$. The variance contribution of each f_u with $|u| = 1$ is $o(N^{-1})$ and, if that f_u is smooth, the variance contribution is in fact $O(N^{-3})$. We use Latin hypercube sampling below to investigate the fraction of the variance that comes from one-dimensional parts of an integrand.

Some orthogonal array sampling schemes (Owen, 1992; Owen, 1994; Tang, 1993) balance all margins up to order t, randomise the higher ones, and have $\Gamma_u \doteq 1_{|u|>t}$. These should be effective on integrands that are, or are nearly, of effective dimension t or less, in the superposition sense.

For non-randomised quasi-Monte Carlo methods, the decomposition (7) does not lead to a variance interpretation, but we may still write

$$|e_N(f)| \le \sum_{|u|\ge 1}^{M} |e_N(f_u)| \le \sum_{|u|\ge 1}^{M} D_{N,u}^{*} V_u(f_u) \tag{10}$$

where $D_{N,u}^{*}$ is the discrepancy of the $|u|$-dimensional points obtained by keeping only those components $(x_n)_{n=1}^N$ in u, and $V_u(f_u)$ is the variation of f_u taken as a $|u|$-dimensional function. We believe that many successes of quasi-Monte Carlo methods on high-dimensional problems can be attributed to a low effective dimension of the integrand, in one or both of the senses above. In such cases, arranging for small values of $D_{N,u}^{*}$ to coincide with the few large values of σ_u^2 and large values of $D_{N,u}^{*}$, if any, to coincide with small values of σ_u^2 (which ordinarily implies small $V(f_u)$) should produce accurate results.

As N increases, more and more of the terms in equation (10) should switch from the Monte Carlo rate, $N^{-1/2}$, to a quasi-Monte Carlo rate $N^{-1}(\log N)^k$. While this is taking place, there may be an appearance of an asymptotic rate better than $N^{-1/2}$. If, however, the very highest dimensional term is non-zero, then one can expect that the errors are bounded below by a small multiple of $N^{-1/2}$ until N reaches the sometimes astronomical sample sizes b^{t+s} as described earlier.

Scrambled net sampling can be analysed in terms of a square-wavelet decomposition applied to each term in the ANOVA decomposition (Owen, 1997a). The result is that the contribution of f_u decreases very rapidly after $N = b^{t+|u|}$. This explains how the scrambled net variance is $o(N^{-1})$. Large gains can be expected if f has sufficiently low dimension in the superposition sense.

Classical Monte Carlo techniques often replace the integrand f by another one with the same integral and a smaller variance. The ANOVA decomposition above suggests two more ways to improve integration: reducing the effective dimension in either of the two senses above. In high-dimensional problems reducing both sorts of effective dimension pays off. It is, for example, easier to arrange low discrepancy for all three-dimensional projections of 16 variables than for all three-dimensional projections of 360 variables. We believe that the Brownian bridge discretisation in the next section but one succeeds for these reasons on the mortgage-backed securities problem introduced in the following section.

Mortgage-backed securities

Consider a security backed by mortgages of length M months with fixed interest rate i_0, which is the current interest rate at the beginning of the mortgage. The present value of the security is then

$$PV = \mathbf{E}[\upsilon] = \mathbf{E}\left[\sum_{k=1}^{M} u_k m_k\right] \tag{11}$$

in which **E** is the expectation over the random variables involved in the interest rate fluctuations. The variables in the problem are as follows:

u_k = discount factor for month k,
m_k = cashflow for month k,
i_k = interest rate for month k,
w_k = fraction of remaining mortgages prepaying in month k,
r_k = fraction of remaining mortgages at month k,
c_k = (remaining annuity at month k)/c,
c = monthly payment,
ξ_k = an $N(0,\sigma)$ random variable.

This notation follows that of Paskov (1996), except that our c_k corresponds to his a_{M-k+1}. Several of these variables are easily defined:

$$u_k = \prod_{j=0}^{k-1}(1 + i_j)^{-1}$$

$$m_k = cr_k[(1 - w_k) + w_k c_k]$$

$$r_k = \prod_{j=1}^{k-1}(1 - w_j)$$

$$c_k = \sum_{j=0}^{M-k}(1 + i_0)^{-j}$$

Following Paskov, we use models for the interest rate fluctuations and the prepayment rate given by

$$\begin{aligned} i_k &= K_0 e^{\xi_k} i_{k-1} \\ &= K_0^k e^{\xi_1 + \dots + \xi_k} i_0 \\ w_k &= K_1 + K_2 \arctan(K_3 i_k + K_4) \end{aligned} \tag{12}$$

in which K_1, K_2, K_3, K_4 are constants of the model. The constant $K_0 = e^{-\sigma^2/2}$ is chosen to normalise the lognormal distribution, i.e. so that $\mathbf{E}[i_k] = i_0$. The initial interest rate, i_0, is an additional constant that must be specified.

In this study we do not divide the cashflow of the security among a group of tranches, as in Paskov (1996), but only consider the total cashflow. Nevertheless, the results should be indicative of a more general computation involving a number of tranches.

The expectation PV can be written as an integral over $\mathbb{R}^M$ with Gaussian weights

$$g(\xi) = (2\pi\sigma^2)^{-1/2}\, e^{-\xi^2/2\sigma^2} \tag{13}$$

This is transformed into an unweighted integral by a mapping $\xi = G(x)$ with $G'(x) = g(\xi)$, which takes a uniformly distributed variable, x, to an $N(0,\sigma)$ variable, ξ. The formula for PV is

$$\begin{aligned} PV &= \int_{\mathbb{R}^M} \upsilon(\xi_1,\dots,\xi_M) g(\xi_1)\dots g(\xi_M)\, d\xi_1 \dots d\xi_M \\ &= \int_{[0,1]^M} \upsilon(G(x_1),\dots,G(x_M))\, dx_1 \dots dx_M \end{aligned} \tag{14}$$

Note that, in quasi-Monte Carlo evaluation of an expectation involving a stochastic process with M time steps, the resulting integral is M-dimensional.

The parameter values i_0, K_1, K_2, K_3, K_4, σ used in Paskov (1996) are proprietary and not known to us. In the numerical study below, we have used two sets of values for them. The first set, chosen to be plausible to us, turned out by inspection to be very nearly a linear function of the Gaussian increment random variables ξ_k. We refer to this set of parameters as the "nearly linear example". For this case, the parameters are

$$(i_0, K_1, K_2, K_3, K_4, \sigma^2) = (0.007, 0.01, -0.005, 10, 0.5, 0.0004) \quad (15)$$

In the second example, the "nonlinear example", the parameters are

$$(i_0, K_1, K_2, K_3, K_4, \sigma^2) = (0.007, 0.04, 0.0222, -1500.0, 7.0, 0.0004) \quad (16)$$

This example also has a large linear component, but the effect is less extreme. While perhaps less plausible, it represents a more challenging integrand. The monthly interest rate i_0 corresponds to a yearly rate of 8.4%. The variance in interest rate increments, σ^2, leads to yearly fluctuations of size 0.5%. In the nearly linear example, the prepayment rate is nearly linear in the interest rate in the range of interest, whereas for the nonlinear example the prepayment rate has a step increase when the interest rate falls much below i_0. In both examples the length of the loans is taken to be 30 years ($M = 360$).

Brownian bridge

Since Brownian motion is a Markov process, it is most natural to generate its value, $b(t + \Delta t_1)$, as a random jump from a past value $b(t)$ as

$$b(t + \Delta t_1) = b(t) + \sqrt{\Delta t_1}\,\nu \quad (17)$$

in which ν is an $N(0,1)$ random variable. On the other hand, the value $b(t + \Delta t_1)$ can also be generated from knowledge of both a past value $b(t)$ and a future value $b(T = t + \Delta t_1 + \Delta t_2)$, with $0 \leq \Delta t_i$, according to the *Brownian bridge* formula

$$b(t + \Delta t_1) = ab(t) + (1 - a)b(T) + c\nu \quad (18)$$

in which

$$a = \Delta t_2/(\Delta t_1 + \Delta t_2), \quad c = \sqrt{a\Delta t_1} \quad (19)$$

Note that $c \leq \sqrt{\Delta t_1}$, so that the variance of the random part of the Brownian bridge formula (18) is less than that in (17).

The standard method of generating a random walk $y_k = \sigma b(k\Delta t)$ is based on the updating formula (17). The initial value is $y_0 = 0$. Each subsequent value y_{k+1} is generated from the previous value y_k using formula (17), with independent normal variables ν_k.

Another method, which we refer to as the *Brownian bridge discretisation*, can be based on (18). Suppose we wish to determine the path $y_0,\ldots,y_M$, and for convenience assume that M is a power of 2. The initial value is $y_0 = 0$. The next value generated is $y_M = \sigma\sqrt{M\Delta t}\,\nu_0$. Then the value at the midpoint $y_{M/2}$ is determined from the Brownian bridge formula (18). Subsequent values are found at the successive midpoints, i.e. $y_{M/4}$, $y_{3M/4}$, $y_{M/8}$,.... The procedure is easily generalised to general values of M.

Although the total variance in this representation is the same as in the standard discretisation, much more of the variance is contained in the first few steps of the Brownian bridge discretisation due to the reduction in variance in the Brownian bridge formula. This reduces the effective dimension of the random walk simulation, which increases the accuracy of quasi-Monte Carlo. Moskowitz and Caflisch (1996) – also Chapter 27 of this publication applied this method to the evaluation of Feynman–Kac integrals and showed the error to be substantially reduced when the number of time steps, which is equal to the dimension of the corresponding integral, is large. Since the mortgage-backed securities problem described above depends on a random walk, and can be written as a discretisation of a Feynman–Kac integral, we were naturally led to apply the Brownian bridge discretisation to this problem.

Numerical results

NEARLY LINEAR EXAMPLE

The value PV for this example was calculated to be 131.78706. The mean length of a mortgage in this case is 100.9 months and the median length is 93 months. The Monte Carlo variance of PV is 41.84 and the variance in antithetic sampling is 0.014. This suggests that the function is very nearly an odd function of the Gaussian increments. In fact, solving $41.84 = \sigma_1^2 + \sigma_2^2$, with $0.014 = 2\sigma_1^2$, provides the rough estimates $\sigma_1^2 \doteq 0.007$ and $\sigma_2^2 = 41.833$, so that the odd or antisymmetric part of this integrand provides about 99.98% of the variation.

Similarly, the variance in Latin hypercube sampling is about 0.0155, from which we find that roughly $(41.84 - 0.0155)/41.84 \doteq 99.96\%$ of the variation comes from one-dimensional structure. This function is effectively one-dimensional in the superposition sense, and it is nearly antisymmetric. The percentages quoted above are based on ratios of sampling variances and may not be exact, but both of these findings agree with what we found by numerical inspection: this function is very nearly linear in the Gaussian increment variables. Application of Latin hypercube sampling to the antithetic integrand leads to only a slight decrease in variance compared to the non-antithetic case, which we interpret to mean that the multidimensional part of the integrand is not predominantly odd.

We now describe the accuracy of various integration methods for this problem as a function of N, the number of paths. For each of these results, we present the root mean square of the error among 25 computations. For Monte Carlo methods the 25 computations are statistically independent, at least to the level possible using pseudo-random numbers. The Sobol' calculations for each of the 25 runs and for each value of N were computed using different non-overlapping sub-sequences of the Sobol' sequence. Because they are not a sample from any population, the root-mean-square error presented is the difference between the values obtained and a 'gold standard' obtained from using the quadratic terms of a Taylor-series expansion of the integrand about the origin in the Gaussian coordinates as a control variate and $N \approx 3.2$ million. The results are plotted in terms of error versus N, both in log base 10.

In these plots, for the antithetic computations N refers to the number of times the antithetic integrand $\frac{1}{2}[f(x) + f(1 - x)]$ is evaluated. This corresponds to 2N function evaluations. Plotting the antithetic runs versus 2N would be more appropriate when function evaluation is the dominant cost and plotting versus N when generating the x_n is the dominant cost. We do not attempt to plot CPU time on the axis, as this can depend on how efficiently a method is implemented.

3. Error versus N (log base 10) for the nearly linear problem in the original representation

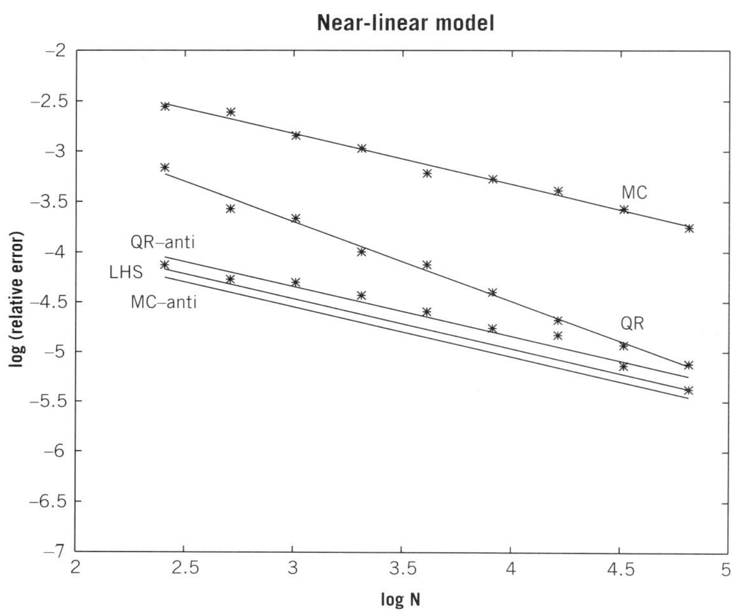

First, we perform straightforward Monte Carlo evaluation, with results plotted in Figure 3. The top curve shows results from Monte Carlo using standard pseudo-random points, with the error decreasing at the expected rate of $N^{-1/2}$. The second curve shows a dramatic improvement using the 360-dimensional Sobol' sequence (generated with part of the code FINDER). In a separate calculation (not plotted), it was found that if only the first 50 dimensions (using the standard discretisation) were taken to be quasi-random and the rest pseudo-random, the size of the error decreased slightly compared with the purely random case, and the apparent convergence rate remained $N^{-1/2}$. So, in the truncation sense, the dimensionality is not below 50.

The third curve in Figure 3 shows the results of combining the quasi-random sequence with antithetic sampling. This leads to a sizeable reduction in the error size as the dominant anti-symmetric part has been removed. However, the improved convergence rate, characteristic of low-dimensional quasi-Monte Carlo methods, also disappears. Finally, reference lines for Latin hypercube sampling and antithetic random Monte Carlo are shown. These both effectively remove the one-dimensional linear elements of the integrand, with antithetic variates killing it off exactly, while this error decreases as $O(N^{-3/2})$ for the Latin hypercube case, leaving the remaining errors to converge at the standard $N^{-1/2}$ rate. The quasi-random error, while outperforming simple random Monte Carlo on the one-dimensional elements, still can only achieve the $O(N^{-1})$ rate in the optimal case. Both Latin hypercube and antithetic random outperform the antithetic quasi-random sequence slightly. This may be because the quasi-random sequence is performing at worse than the Monte Carlo level on the higher-dimensional part of the integrand.

A scrambled (0, 360)-sequence in base 361, with N a small multiple of 361, behaves like Latin hypercube sampling. When N nears $361^2 = 130{,}321$, some bivariate sources of variance disappear. For intermediate multiples of 361, some of the bivariate effects are reduced by a multiple $N/361^2$. On this function, the results for the scrambled (0, 360)-sequence were essentially the same as for Latin hypercube sampling, and thus they are not plotted in Figure 3. When applied to the antithetic integrand, the scrambled sequence showed a slight improvement over the antithetic Latin hypercube sampling; however, no significant gains are achieved in either case with antithetic sampling because of the increased computation time required.

Our results are consistent with the results of Paskov (1996), Paskov and Traub (1995) and Ninomiya and Tezuka (1996). They contradict the observation that the effectiveness of quasi-Monte Carlo is lost in high dimensions. However, we argue below that the improvement is almost entirely due to improved integration of the one-dimensional parts of the integrand.

Next we consider the Brownian bridge version of the integrand. The reformulated integrand has the same mean and the same variance as the original, but more of the structure is packed into the first few dimensions. This should help the Sobol' sequence because it would only need to have small discrepancy among the first few dimensions. This encoding reduces the variance in Latin hypercube sampling from 0.0155 to 0.00963, suggesting that the Brownian bridge encoding has made the integrand even more inherently one-dimensional.

Figure 4 shows the results from Sobol' sequence integration in the Brownian bridge representation, with and without antithetic sampling. Also shown are reference lines for simple Monte Carlo, which is not affected by the change of representation, and for Latin hypercube sampling with the Brownian bridge. Again, antithetic sampling does not substantially improve Latin hypercube sampling here. For the Brownian bridge Sobol' sequence without antithetic sampling, the results are essentially the same as for the standard representation. This is because, for the dominant linear one-dimensional elements, the Brownian bridge representation simply rearranges the weights on the elements but the sum remains constant. Because the errors associated with each one-dimensional projection of the Sobol' sequence are nearly identical, no improvement is seen.

In the Brownian bridge formulation, the Sobol' sequence with antithetic sampling is much better than either antithetic variables or Latin hypercube sampling. This suggests that it must be capturing some higher-dimensional anti-symmetric structure, most probably among the first few variables.

The theory of scrambled (0, 360) sequences in base 361 predicts that they will not be much better than Latin hypercube sampling until $N = 361^2 = 130{,}321$, which is beyond the range we explore here. But by packing most of the structure into the first few dimensions of the integrand, we can consider methods in which scrambled nets are used on the most important dimensions and something else is used on the rest. Such methods have been proposed before: Spanier (1995) describes a scheme in which quasi-Monte Carlo methods are used on the first few dimensions of an integrand and simple Monte Carlo is used on the rest, and Owen (1994) considered augmenting a randomised orthogonal array with further dimensions taken from Latin hypercube samples.

We considered using a scrambled (0, 32)-sequence in base 32 for the first 32 dimensions and Latin hypercube sampling on the last 328 dimensions. For $N = 32^m$ one gets a scrambled $(0, m, 32)$-net in base 32 for the first 32 dimensions and Latin hypercube sampling for the rest. This will integrate the one-dimensional part of the function and much of the m-dimensional structure (superposition sense) of the first 32 dimensions, with the rest of the structure being integrated at the Monte Carlo rate. But (0, 32)-sequences can be stopped early or extended as necessary, whereas Latin hypercube sampling requires a prespecified number, N, of runs. As a compromise, we ran a scrambled (0, 32)-sequence for the first 32 dimensions and took repeated independent Latin hypercube samples of size 1024 for the last 328 dimensions. Such a

4. Error versus N (log base 10) for the nearly linear problem in the Brownian bridge representation

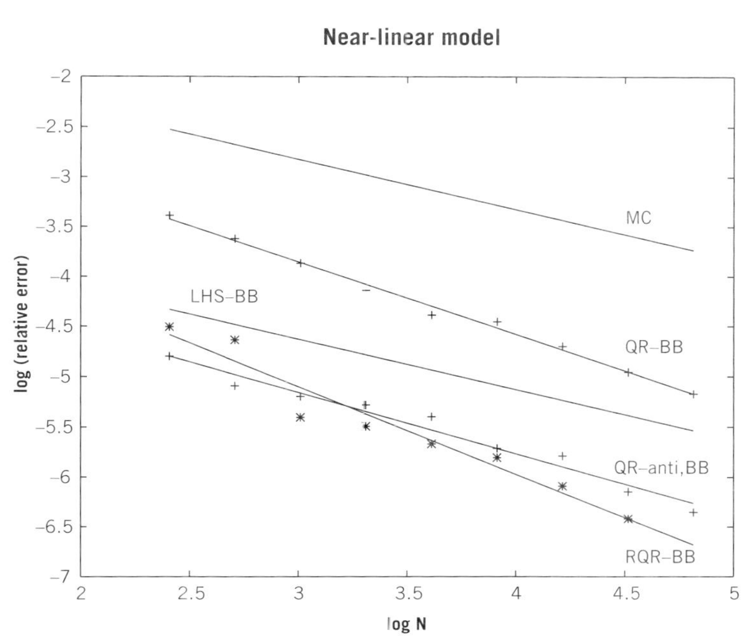

simulation can be conveniently stopped at any multiple of 1024 runs. The results are shown in Figure 4, labelled as RQR-BB for randomised quasi-random with Brownian bridge. All pairs of two variables among the first 32 variables start to become balanced at sample size $N = 32^2 = 1024$, and, similarly, all triples of variables among the first 32 variables start to become balanced at sample size $N = 32^3 = 32{,}768$. This leads to results which are similar to the antithetic Sobol' with Brownian bridge results. However, the convergence rate for the scrambled net in the Brownian bridge representation appears to be larger, leading to greater accuracy at large N.

We were able to achieve still better results than those shown here by approximating the integrand by a quadratic function about the origin in the Gaussian coordinates and using this function as a control variate with antithetic random Monte Carlo. The quadratic terms were calculated by evaluating all $360 \times 361/2$ second derivatives of the integrand.

5. Error versus N (log base 10) for the non-linear problem, in the original representation

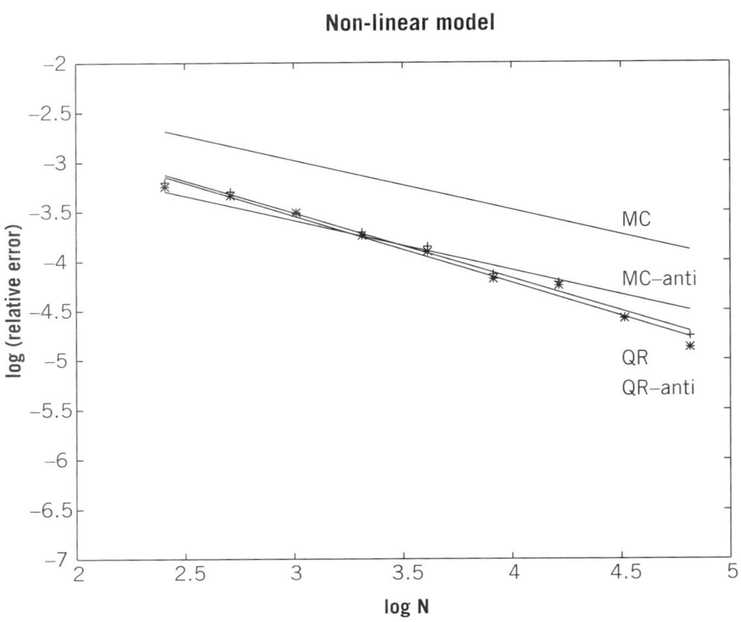

6. Error versus N (log base 10) for the non-linear problem in the Brownian bridge representation

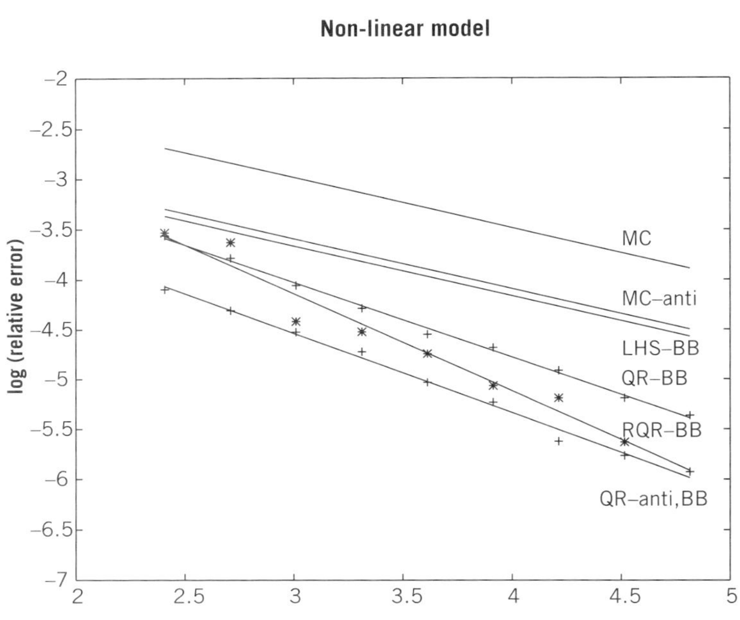

NON-LINEAR EXAMPLE

We next consider the non-linear example and display the same error curves. The mean value of PV for this problem is 130.712365. This value was obtained by averaging 200 independent Cranley–Patterson randomisations of 2^{19} fixed paths, generated by using the Sobol' sequence with the Brownian bridge ordering. These randomisations are due to Cranley and Patterson (1976), and Tuffin (1996) appears to be the first to have realised their utility on nets. The certainty of this answer can be estimated by considering a six standard deviation range, which was determined to be (130.712348, 130.712382). The error curves in Figures 5 and 6 are based on using this value as the exact solution.

The mean length of a mortgage in this case is 76.5 months and the median length is 58 months. The variance of PV in this value is 18.54 and the variance in the antithetic computation of this value is 1.127. Thus the function is about 97.7% antisymmetric. The variance under Latin hypercube sampling is 1.087, so that the function is only about 94.1% one-dimensional. This may seem like a lot of one-dimensional structure, but, compared with the previous example, the proportion of higher-dimensional structure is greatly increased.

As in the nearly linear example, antithetic sampling and Latin hypercube sampling in combination do not work better than separately. For this integrand as for that one, they each appear to remove the same source of error. In fact, for the non-linear example, both give almost exactly the same errors.

Figure 5 shows the error reference lines from Monte Carlo and antithetic Monte Carlo sampling. Both Latin hypercube sampling and the randomised (0, 360)-sequence in base 360, as well as their antithetic counterparts, give roughly the same accuracy as the antithetic Monte Carlo sampling. For simplicity, these error curves have therefore not been included in this graph. Superimposed are errors and lines for Sobol' and those for Sobol' with antithetic sampling. In this case the Sobol' sequence is seen to catch up with the antithetic random sampling, but little is gained by combining antithetic sampling with the quasi-random sequence. The Sobol' sequence outperforms Latin hypercube sampling on this, probably because the one-dimensional parts of it are no longer so dominant.

Figure 6 shows the errors from the Brownian bridge representation of the integrand. Reference lines are shown for Monte Carlo, antithetic sampling, and Latin hypercube sampling. In this case also, Latin hypercube sampling does a bit better after the Brownian bridge transformation, suggesting that the function has become somewhat more one-dimensional. As before, combining Latin hypercube sampling with antithetic sampling does not do much good.

The Sobol' sequences perform especially well on the non-linear function in the Brownian bridge representation with antithetic sampling. In terms of equation (10), this may be due to good equidistribution among the leading Sobol' dimensions, matched with their greater importance to the integrand.

Conclusions

Our main conclusions are as follows.

❑ Quasi-Monte Carlo methods provide significant improvements in accuracy and computational speed for problems of small to moderate dimension.

❑ While the effectiveness of quasi-Monte Carlo can be lost on problems of high dimension, this does not happen if the integrand is of low effective dimension in the superposition sense.

❑ Some problems that have a large nominal dimension can be reformulated to have a moderate-sized effective dimension, so that the effectiveness of quasi-Monte Carlo is increased.

❑ The Brownian bridge representation reduces the effective dimension for problems like the mortgage-backed security problem described here.

Instead of straightforward use of high-dimensional quasi-random sequences, our recommendations are:

❑ First analyse the problem, mathematically or numerically, to determine the most important input dimensions.

❑ Where possible, reformulate the problem to concentrate the variation in fewer dimensions.

❑ When a small number of dominant dimensions can be identified or induced, apply quasi-random or randomised quasi-random (when sample-based error estimates are desired) sequences to those dimensions.

❑ For the remaining dimensions use pseudo-random or Latin hypercube sampling.

❑ Consider applying classical variance reduction techniques, such as antithetic sampling, control variates, stratification, and importance sampling in conjunction with the above.

We believe that high-dimensional integration problems can range in difficulty from completely intractable to quite simple. In some cases it is possible to turn the former into the latter by carefully engineering the integrand. It is too early to say whether such manageable integrands are rare or dominant in financial applications. However, the results here indicate that, for the valuation of securities which depend on a single stochastic factor modelled as a Gaussian process with nonstochastic drift and volatility, the Brownian bridge representation may be extremely effective in reducing the dimension of the simulation.

The authors are grateful to Spassimir Paskov and Joseph Traub for a number of discussions and for providing us with their quasi-random number generator FINDER. Partial support for this work came from the following sources: NSF (Owen), an NSF Postdoctoral Fellowship (Morokoff), ARO grant no. DAAH04-95-1-0155 (Caflisch and Morokoff).

BIBLIOGRAPHY

Caflisch, R.E., and W. Morokoff, 1996, "Quasi-Monte Carlo Computation of a Finance Problem" in: K.T. Fang and F.J. Hickernell, eds., *Proc. Workshop on Quasi-Monte Carlo Methods and their Applications*, pp. 15–30, and UCLA CAM Report 96-16.

Cranley, R., and T.N.L. Patterson, 1976, "Randomization of Number Theoretic Methods for Multiple Integration", *SIAM Journal of Numerical Analysis* 13, pp. 904–14.

Hickernell, F.J., 1996, "The Mean Square Discrepancy of Randomized Nets", *ACM Transactions in Modeling and Computer Simulation* 6, pp. 274–96.

Hickernell, F.J., 1998, "A Generalized Discrepancy and Quadrature Error Bound", *Mathematics of Computation* 67, American Mathematical Society, pp. 299–322.

McKay, M.D., R.J. Beckman and W.J. Conover, 1979, "A Comparison of Three Methods for Selecting Values of Input Variables in the Analysis of Output from a Computer Code", *Technometrics* 21(2), pp. 239–45.

Morokoff, W., and R.E. Caflisch, 1993, "A Quasi-Monte Carlo Approach to Particle Simulation of the Heat Equation", *SIAM Journal of Numerical Analysis* 30, pp. 1558–73.

Morokoff, W., and R.E. Caflisch, 1994, "Quasi-random Sequences and their Discrepancies". *SIAM Journal of Scientific and Statistical Computing* 15, pp. 1251–79.

Morokoff, W., and R.E. Caflisch, 1995, "Quasi-Monte Carlo Integration", *Journal of Computational Physics* 122, pp. 218–30.

Moskowitz, B., and R.E. Caflisch, 1996, "Smoothness and Dimension Reduction in Quasi-Monte Carlo Methods". *Journal of Mathematical Computer Modelling* 23, pp. 37–54 (Chapter 27 of this publication).

Niederreiter, H., 1992, "Random Number Generation and Quasi-Monte Carlo Methods". SIAM, Philadelphia.

Niederreiter, H., and Xing, C., 1995, "Low Discrepancy Sequences Obtained from Global Function Fields" in K.T. Fang and Hickernell, F.J., eds., *Proceedings, Workshop on Quasi-Monte Carlo Methods and their Applications*, pp. 81–112.

Ninomiya, S., and S. Tezuka, 1996, "Toward Real-time Pricing of Complex Financial Derivatives", *Applied Mathematical Finance* 3, pp. 1–20.

Owen, A.B., 1992, "Orthogonal Arrays for Computer Experiments, Integration and Visualization", *Statistica Sinica* 2, pp. 439–52.

Owen, A.B., 1994, "Lattice Sampling Revisited: Monte Carlo Variance of Means over Randomized Orthogonal Arrays". *Annals of Statistics* 22, pp. 930–45.

Owen, A.B., 1995, "Randomly permuted (t,m,s)-nets and (t,s)-sequences", in: H. Niederreiter and P. Jau-Shyong Shiue, eds., Monte Carlo and Quasi-Monte Carlo Methods in Scientific Computing, pp. 299–317, Springer, New York.

Owen, A.B., 1997a, "Monte Carlo Variance of Scrambled Equidistribution Quadrature", *SIAM Journal of Numerical Analysis* 34(5), pp. 1884–910.

Owen, A.B., 1997b, "Scrambled Net Variance for Integrals of Smooth Functions", *Annals of Statistics* 25(4), pp. 1,541–6.

Paskov, S.H., 1996, "New Methodolgies for Valuing Derivatives", in: S. Pliska and M. Dempster, eds., *Mathematics of Derivative Securities*, Isaac Newton Institute/ Cambridge University Press (Chapter 25 of this publication).

Paskov, S.H., and J.F. Traub, 1995, "Faster Valuation of Financial Derivatives", *Journal of Portfolio Management* 22(1), pp. 113–20.

Spanier, J., 1995, "Quasi-Monte Carlo Methods for Particle Transport Problems", in H. Niederreiter and P. Jau-Shyong Shiue, eds., *Monte Carlo and Quasi-Monte Carlo Methods in Scientific Computing*, Springer, New York, pp. 121–48.

M. Stein, M., 1987, "Large Sample Properties of Simulations using Latin Hypercube Sampling", *Technometrics* 29(2), pp. 143–51.

Swayne, D.F., D. Cook and A. Buja, 1991, "XGobi: Interactive dynamic graphics in the X window system with a link to S", in: H. Niederreiter and P. Jau-Shyong Shiue, eds., *ASA Proceedings of Statistical Graphics Section*, pp. 1–8.

Takemura, A., 1983, "Tensor Analysis of ANOVA Decomposition", *Journal of the American Statistical Association* 78, pp. 894–900.

Tang, B., 1993, "Orthogonal Array-based Latin Hypercubes", *Journal of the American Statistical Association* 88, pp. 1392–97.

Tuffin, B., 1996, "On the Use of Low Discrepancy Sequences in Monte Carlo Methods", *Technical Report Publication Interne* 1060, IRISA.

27

Smoothness and Dimension Reduction in Quasi-Monte Carlo Methods

Bradley Moskowitz and Russel E. Caflisch[1]
Bettis Laboratory and UCLA

Monte Carlo integration using quasi-random sequences has theoretical error bounds of size $O(N^{-1} \log^d N)$ in dimension d, as opposed to the error of size $O(N^{-1/2})$ for random or pseudo-random sequences. In practice, however, this improved performance for quasi-random sequences is often not observed. The degradation of performance is due to discontinuity or lack of smoothness in the integrand and to large dimension of the domain of integration, both of which often occur in Monte Carlo methods. In this chapter modified Monte Carlo methods are developed, using smoothing and dimension reduction, so that the convergence rate of nearly $O(N^{-1})$ is regained. The standard rejection method, as used in importance sampling, involves discontinuities, corresponding to the decision to accept or reject. A smoothed rejection method, as well as a method of weighted uniform sampling, is formulated below and found to have error size of almost $O(N^{-1})$ in quasi-Monte Carlo. Quasi-Monte Carlo evaluation of Feynman–Kac path integrals involves high dimension, one dimension for each discrete time interval. Through an alternative discretisation, the effective dimension of the integration domain is drastically reduced, so that the error size close to $O(N^{-1})$ is again regained.

The loss of the theoretically expected accuracy for Monte Carlo integration using quasi-random sequences was found by Morokoff and Caflisch (1993, 1994 and 1995) to be due to two causes.

The first is discontinuity or lack of smoothness in the integrand. The $O(N^{-1} \log^d N)$ error bound quoted above is established using the Koksma–Hlawka inequality, which states that for Monte Carlo integration of a function f on the unit cube in R^d using N points, the integration error, $E_N(f)$ is bounded by

$$E_N(f) \leq D_N V(f) \qquad (1)$$

in which D_N is the *discrepancy* of the sequence and V(f) is the variation of f in the Hardy–Krause sense. Moreover, for quasi-random sequences the discrepancy, D_N, is of size $O(N^{-1} \log^d N)$. For V(f) to be finite in dimension $d > 1$, in general, f must be smooth (see Neiderreiter, 1992, for a precise statement). On the other hand, discontinuous integrands arise often in practical Monte Carlo methods. For example, many methods involve a decision process, for which some component of the integrand is 1 or 0 if the decision is "yes" or "no". For such a discontinuous integrand, the inequality (1) is inapplicable, and the computation in Morokoff and Caflisch (1995) indicates that the error bounds revert to size $O(N^{-1/2})$.

A second limitation on the accuracy of quasi-Monte Carlo integration occurs for integration problems in which the dimension d of the integration domain is large. For large d, Morokoff and Caflisch (1994) found that the discrepancy of the quasi-random sequence is of size $O(N^{-1/2})$ for intermediate values of N, even though D_N is $O(N^{-1} \log^d N)$ for N sufficiently large. The transition from $O(N^{-1/2})$ to $O(N^{-1} \log^d N)$ behaviour for D_N appears to occur at approximately $N = e^d$, although this has not been proven.

There are a range of applications, such as scattering from rough surfaces, collisionless particle transport (Lecot, 1991, and Schuhman, 1990), and simulation of small networks (Adlakha,

1992), for which the dimension is not too large and quasi-Monte Carlo is directly applicable. On the other hand, integration in high dimensions is one of the main reasons for using Monte Carlo in the first place, and high-dimensional integration occurs in many problems coming from atomic physics, neutron transport, computational chemistry and molecular biology. These results indicate that it will be difficult to adapt quasi-Monte Carlo methods to such problems.

The purpose of this chapter is to show how these two fundamental limitations – smoothness and dimension – can be overcome, at least in several examples. This will be accomplished through modification of standard Monte Carlo methods, by smoothing discontinuities and by reducing the effective dimension of the problem.

The standard rejection method, as used to sample from a density function in importance sampling, can be phrased as an integration involving a discontinuous integrand, with the discontinuity coming from the decision to accept or reject. As shown below, this leads to large errors in a quasi-random method. A *smoothed rejection method* will be formulated, for which the optimal error size of nearly $O(N^{-1})$ is attained (here and in the subsequent discussion the logarithmic factors $\log^d N$ are omitted, since they cannot be reliably detected numerically). This is also true of the method of *weighted uniform sampling*.

Quasi-random simulation of a stochastic process, such as Brownian motion, requires a high-dimensional quasi-random sequence, with one dimension for each time step that is simulated. This easily leads to loss of the optimal $O(N^{-1})$ convergence rate. This will be illustrated below for Feynman–Kac path integrals. Nevertheless, by a modified representation of Brownian motion, or equivalently a modified quadrature rule for the path integral, the effective dimension of the problem can be drastically reduced, so that the optimal error size of approximately $O(N^{-1})$ is attained in this problem as well.

These applications of quasi-Monte Carlo to test problems may serve as a starting point for the application of such methods to a range of problems in which the integrand is smooth and the effective dimension is small, after some modification. For example, Moskowitz (1993) has applied these methods to the determination of the ground-state energy for the square-well potential and the hydrogen atom.

Integrand smoothing

In this section, quasi-Monte Carlo integration is applied to problems in which the integrand is first smoothed to remove discontinuities that would naturally appear. The first application is an illustrative example of integration of a characteristic function. Then the rejection method and smoothed versions of the rejection method are presented.

EXAMPLE: SMOOTHING A CHARACTERISTIC FUNCTION

The characteristic function χ_E of the set E has infinite variation, unless the set E is rectangular with sides parallel to the co-ordinate axes. Consider the characteristic function, f_0, of a rotated cube, defined as follows.

Example 1: Given $x_1, x_2, x_3 \sim U\left[-\frac{1}{2}, \frac{1}{2}\right]$, let

$$u_1 = \frac{\sqrt{2}\,x_1 + x_2 + x_3}{2}$$

$$u_2 = \frac{-\sqrt{2}\,x_1 + x_2 + x_3}{2}$$

$$u_3 = \frac{-\sqrt{2}\,x_2 + \sqrt{2}\,x_3}{2}$$

Define discontinuous characteristic function

$$f_0(x_1,x_2,x_3) = \prod_{k=1}^{3} \chi\{|u_k| \le 0.2\}$$

where

$$\chi\{|u_k| \le 0.2\} = \begin{cases} 1 & \text{if } |u_k| \le 0.2 \\ 0 & \text{otherwise} \end{cases}$$

and smoothed functions f_δ (δ is the level of smoothing)

$$f_\delta(x_1,x_2,x_3) = \prod_{k=1}^{3} \chi_\delta\{|u_k| \le 0.2\}$$

where

$$\chi_\delta\{|u_k| \le 0.2\} = \begin{cases} 1 & \text{if } |u_k| \le 0.2 - \delta \\ 0 & \text{if } |u_k| \ge 0.2 + \delta \\ \frac{\delta - (|u_k| - 0.2)}{2\delta} & \text{otherwise} \end{cases}$$

The integrals of the functions f_0 and f_δ, $0 < \delta \ll 1$, over the unit cube in three dimensions all equal 0.064.

The results of crude Monte Carlo estimation in three different cases are compared: first, the discontinuous case; second, the continuous case with weak smoothing ($\delta = 0.025$); and third, the continuous case with strong smoothing ($\delta = 0.15$). Results are obtained by computing the root mean square error over 75 repetitions. (See below, on computational examples, for more details). Log–log plots are used so that the slopes (which are presented parenthetically in the figure keys) correspond to convergence rate. Numerical results are plotted in Figure 1. As expected, the quasi-random sequences in this example from Bratley, Fox and Niederreiter (1992) perform better as the level of smoothing is increased, while pseudo-random sequences are not affected by the smoothing.

The results for this simple example show how the effectiveness of quasi-Monte Carlo integration is lost if the integrand is discontinuous but can be regained if the integrand is smoothed without changing the value of the integral. Similar tests have been performed by Morokoff and Caflisch (1994), Press and Teukolsky (1988), and Berblinger and Schlier (1991) with consistent results.

IMPORTANCE SAMPLING USING THE REJECTION METHOD

Variance reduction through importance sampling is one of the most commonly used Monte Carlo methods. For the integral $A = \int_D f(\mathbf{x})d\mathbf{x}$, introduce an importance function, h, which mimics the behaviour of f over D but is either integrable analytically or easily numerically integrated. The sampling procedure is then altered to generate points distributed according to h instead of points which are uniformly distributed. Then, instead of evaluating $f(\mathbf{x})$ at each sample point, one evaluates $f(\mathbf{x})/h(\mathbf{x})$, which provides an unbiased estimate of the integral since

$$E_h\left(\frac{f}{h}\right) = \int_D \frac{f(\mathbf{x})}{h(\mathbf{x})} \cdot h(\mathbf{x})d\mathbf{x} = \int_D f(\mathbf{x})d\mathbf{x} = A$$

The importance-sampled Monte Carlo estimate can be written as follows:

$$\tilde{A}_N = \frac{1}{N}\sum_{i=1}^{N} \frac{f(\mathbf{x}_i)}{h(\mathbf{x}_i)}, \quad \mathbf{x}_i \sim h(\mathbf{x}) \tag{2}$$

where $\mathbf{x}_i \sim h(\mathbf{x})$ indicates that the sample points $\mathbf{x}_1,\ldots,\mathbf{x}_N$ have a probability density given by h (note that h should be normalised so that $\int_D h\, d\mathbf{x} = 1$). The variance of this estimate is substantially reduced if f/h has a smaller variance than f.

For a limited class of density functions, h, the sample points may be generated by a transformation of the uniform distribution. For more general h, some other generation procedure must be used, such as the rejection method.

Standard rejection method Here is the basic algorithm for the rejection method when the original sample is uniformly distributed in the d-dimensional unit cube. Note that one additional dimension is added for the extra variable y, which acts as a "decision-maker":

1. Select $\gamma \geq \sup_{\mathbf{x}\in D} h(\mathbf{x})$.
2. Repeat until N points have been accepted:
   (a) Sample $(\mathbf{x}_t, y_t)$ from $U([0,1]^{d+1})$.
   (b) If $y_t < \gamma^{-1} h(\mathbf{x}_t)$, accept trial point $\mathbf{x}_t$. Otherwise, reject the trial point.

This algorithm produces a sequence of accepted points in d dimensions which are distributed according to h, as required for importance sampling.

The sum (2), in which the points $\mathbf{x}_i$ are chosen using the rejection method, can be reinterpreted as a Monte Carlo evaluation of the following integral:

$$A = \gamma \int_D \int_0^1 \frac{f(\mathbf{x})}{h(\mathbf{x})} \chi(y < \gamma^{-1} h(\mathbf{x}))dyd\mathbf{x} \tag{3}$$

The direct Monte Carlo estimation of this integral is

$$\tilde{A}_N = \frac{\gamma}{N}\sum_{i=1}^{N} \chi\left\{y_i < \frac{h(\mathbf{x}_i)}{\gamma}\right\} \cdot \frac{f(\mathbf{x}_i)}{h(\mathbf{x}_i)} \tag{4}$$

The variance of this integral is reduced by replacing the total number of points, N, by the number

1. Results, 75 runs, example 1

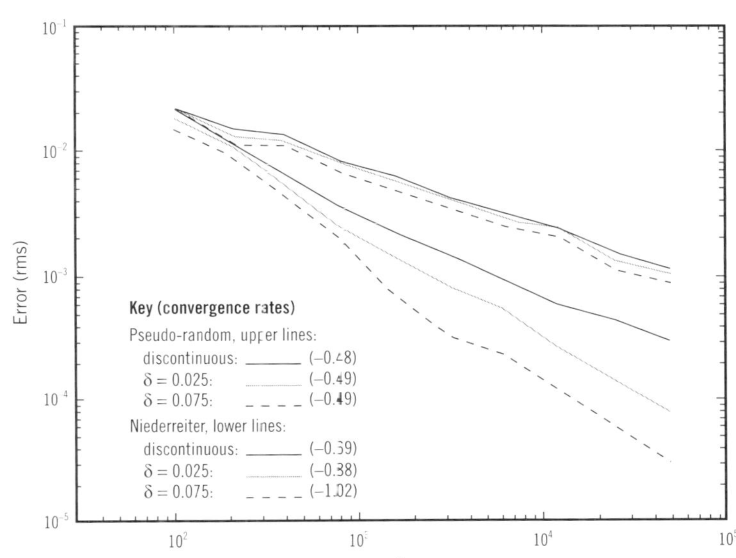

of accepted points to get the following Monte Carlo estimator:

$$\tilde{A}_N = \frac{1}{\sum_{i=1}^{N} \chi\{y_i < \frac{h(\mathbf{x}_i)}{\gamma}\}} \times \sum_{i=1}^{N} \chi\left\{y_i < \frac{h(\mathbf{x}_i)}{\gamma}\right\} \cdot \frac{f(\mathbf{x}_i)}{h(\mathbf{x}_i)} \quad (5)$$

One difference between the sum (5) and a straightforward Monte Carlo evaluation of (3) is that in (5) the number N is taken to be the number of accepted points rather than the total number of trial points. This accounts for the factor γ in (4).

Quasi-Monte Carlo cannot be effectively applied to the integral (3) because the integrand contains a characteristic function, corresponding to the decision to accept or reject. This will be demonstrated in computational examples below.

Smoothed rejection The rejection method is discontinuous as a result of the inherently discontinuous nature of the binary decision: accept or reject. This can be eliminated by allowing sample points to have associated "acceptance" weights as described below. A smoothed rejection method that retains the flexibility and advantages of the rejection method but does not involve a discontinuous integrand will now be formulated. First, replace the integral in (3) by an equivalent smooth integral, as in

$$A = \gamma \int_D \int_0^1 \frac{f(\mathbf{x})}{h(\mathbf{x})} \chi_\delta(y, \gamma^{-1} h(\mathbf{x})) dy d\mathbf{x} \quad (6)$$

in which the smooth function χ_δ satisfies

$$\int_0^1 \chi_\delta(y, \gamma^{-1} h(\mathbf{x})) dy = \gamma^{-1} h(\mathbf{x}) \quad (7)$$

The function χ_δ, which will also be referred to as the weight w, will be chosen to be piecewise linear below.

The corresponding Monte Carlo approximation corresponds to the following smoothed rejection procedure:

1. Select $\gamma \geq \sup_{\mathbf{x} \in D} h(\mathbf{x})$ and $0 < \delta \ll 1$.
2. Repeat until weight of accepted points is within one unit of N:
   (a) Sample $(\mathbf{x}_t, y_t)$ from $U([0,1]^{d+1})$.
   (b) If $y_t < \frac{h(\mathbf{x}_t)}{\gamma} - \frac{1}{2}\delta$, then acceptance weight is $w = 1$.
   Or if $y_t > \frac{h(\mathbf{x}_t)}{\gamma} + \frac{1}{2}\delta$, then $w = 0$.
   Otherwise $w = \frac{1}{\delta}(\frac{h(\mathbf{x}_t)}{\gamma} + \frac{1}{2}\delta - y_t)$.

The density of accepted points x is $f_{accept}(\mathbf{x})$, given by (with $w = \chi_\delta$)

$$f_{accept}(\mathbf{x}) = \frac{1 \cdot \int_0^1 w(\mathbf{x},y)dy}{\int_{I^d} \left[1 \cdot \int_0^1 w(\xi,y)dy\right] d\xi} = \frac{h(\mathbf{x})/\gamma}{1/\gamma}$$

$$f_{accept}(\mathbf{x}) = h(\mathbf{x})$$

which shows that the density function h is correctly sampled.

There is some extra work required by smoothed rejection as compared to ordinary rejection. First, there is the work associated with assigning each sample point a weight, and storing and using these weights. This is typically small enough to be considered insignificant. Second, there is additional work which comes from the acceptance of points with weights of less than 1; ie in order to reach a total acceptance weight of size N, more than N evaluations of f are required. This extra work can be minimised by setting the constant δ sufficiently small. On the other hand, if δ is made too small, the advantages of continuity will be effectively lost.

Weighted uniform sampling Another alternative to the rejection method can be formulated by eliminating the accept/reject decision entirely, and instead assigning each sample point a weight equal to its acceptance probability, given by $\gamma^{-1} h(\mathbf{x}_i)$.

A new Monte Carlo estimate is obtained as follows:

$$\breve{A}_N = \frac{\sum_{i=1}^{N} \frac{f(\mathbf{x}_i)}{h(\mathbf{x}_i)} \cdot \frac{h(\mathbf{x}_i)}{\gamma}}{\sum_{i=1}^{N} \frac{h(\mathbf{x}_i)}{\gamma}} = \frac{\sum_{i=1}^{N} f(\mathbf{x}_i)}{\sum_{i=1}^{N} h(\mathbf{x}_i)}$$

This sum is essentially the ratio of two crude Monte Carlo integration estimates – an estimate of the original function, f, in the numerator, and an estimate of the importance function, h, in the denominator. The resulting estimate is biased. Nevertheless, Powell and Swann (1966) and Spanier and Maize (1994) have shown that this bias is negligible in comparison with the root mean square error as $N \to \infty$. Moreover, the positive correlation between f and h, when h has been well chosen to closely mimic the behaviour of f, provides substantial variance reduction.

The bias and rms error are given as follows:

$$\text{bias}(\breve{A}_N) = \frac{A\,\text{var}(h)}{N} - \frac{\text{cov}(f,h)}{N} + O(N^{-3/2}) \quad (8)$$

$$\text{rmse}(\breve{A}_N) = \frac{\sqrt{\text{var}(f) + A^2\,\text{var}(h) - 2A\,\text{cov}(f,h)}}{\sqrt{N}} + O(N^{-3/4}) \quad (9)$$

One advantage of weighted uniform sampling, as emphasised in Powell and Swann (1966), is that sample points need not be generated for the density h. A second advantage is that the weighted uniform sampling estimate is continuous (assuming that f and h are continuous), so that quasi-random sequences may be effectively used in the sum. This will be demonstrated in computational examples below.

On the other hand, a disadvantage of weighted uniform sampling is that for problems with large regions of low importance, many more function evaluations will be performed within such regions than if importance sampling were used instead. This is reflected in a greater amount of variance reduction in the latter case, and for such problems smoothed rejection may be preferable to weighted uniform sampling.

Computational examples The rejection, smoothed rejection and weighted uniform sampling methods will now be compared on several examples. Consider the following integral:

$$A = \int_{I^d} f(\mathbf{x})d\mathbf{x}$$

where I^d is the unit cube in d dimensions, and evaluate the following four estimates of this integral:

Crude Monte Carlo:

$$\hat{A}_N^{(1)} = \frac{1}{N}\sum_{i=1}^{N} f(\mathbf{x}_i),\ \mathbf{x}_i \sim U(I^d)$$

Weighted uniform sampling method:

$$\hat{A}_N^{(2)} = \frac{\sum_{i=1}^N f(\mathbf{x}_i)}{\sum_{i=1}^N h(\mathbf{x}_i)},\ \mathbf{x}_i \sim U(I^d)$$

Rejection method:

$$\hat{A}_N^{(3)} = \frac{1}{N}\sum_{i=1}^{N} \frac{(\mathbf{x}_i)}{h(\mathbf{x}_i)},\ \mathbf{x}_i \sim h(\mathbf{x}),\ \text{accepted point}$$

Smoothed rejection method:

$$\hat{A}_N^{(4)} = \frac{1}{N}\sum_{i=1}^{N^*} w_i \frac{f(\mathbf{x}_i)}{h(\mathbf{x}_i)},\ (\mathbf{x},w_i) \sim h(\mathbf{x})$$

For the last estimate, $(\mathbf{x}_i,w_i)$ represents a weighted sample point with acceptance weight w_i, and N^* is chosen such that the sum of the acceptance weights is within one unit of N.

For a given value of N, take M samples of each of these estimates, denoted by $\hat{A}_{N,k}^{(j)}$ for $1 \le k \le M$ (for quasi-random sampling, using successive points from a single quasi-random sequence.) Two measurements of the average integration error are the empirical root mean square error and the empirical standard deviation, defined as

$$\text{rmse}(\hat{A}_N^{(j)}) \approx \sqrt{\frac{1}{M}\sum_{k=1}^{M} (\hat{A}_{N,k}^{(j)} - A)^2},\quad j = 1,\ldots,4 \quad (10)$$

$$\text{sd}(\hat{A}_N^{(j)}) \approx \sqrt{\frac{1}{M-1}\sum_{k=1}^{M} (\hat{A}_{N,k}^{(j)} - \overline{\hat{A}_N^{(j)}})^2},\quad j = 1,\ldots,4 \quad (11)$$

in which $\overline{\hat{A}_N^{(j)}} = \frac{1}{M}\sum_{k=1}^{M} \hat{A}_{N,k}^{(j)}$. Note that the formula for rms error uses the exact integral A, which is known for the examples below, whereas the sd error uses an empirical average.

Since successive estimates $\hat{A}_{N,k}^{(j)}$, $k = 1,\ldots,M$, are not independent for quasi-Monte Carlo, there is no theoretical basis for these error expressions. This is in contrast to standard Monte Carlo, for which the central limit theorem implies that the $\hat{A}_{N,k}^{(j)}$ come approximately from a Gaussian distribution with variance given by the rms error. Nevertheless, in the examples below these two error estimates give consistent results never differing by more than 2%.

Finally, the bias of weighted uniform sampling proved to be insignificant, as expected, when using either pseudo-random or quasi-random sequences of points. (Note: In examples 2 and 3, we use the base-2 Niederreiter quasi-random sequence (Bratley, Fox and Neiderreiter, 1992).

Example 2: The first example is Monte Carlo integration over $I^5 = [0,1]^5$ of the function

$$f_2(\mathbf{x}) = \exp\left(\sum_{i=1}^{5} a_i x_i^2 \frac{2 + \sin(\sum_{j=1,\, j\neq i}^{5} x_j)}{2}\right)$$

using the positive definite importance function

$$h_2(\mathbf{x}) = \frac{1}{\eta}\exp\left(\sum_{i=1}^{5} a_i x_i^2\right)$$

where $\mathbf{a} = (1,\frac{1}{2},\frac{1}{5},\frac{1}{5},\frac{1}{5})$ and $\eta = \int_{I^5} \exp(\sum_{i=1}^5 a_i x_i^2)d\mathbf{x}$ so that h_2 is normalised (ie its integral is one).

The resulting rms error for example 2 using pseudo-random and quasi-random points are presented in Figures 2 and 3, respectively. Note that η is easily computed with high accuracy as the product of five one-dimensional integrals, using, for example, Gaussian quadrature.

Example 3: The second example is Monte Carlo integration over $I^7 = [0,1]^7$ of the function

$$f_3(x_1,\ldots,x_7) = e^{1-(\sin^2(\frac{\pi}{2}x_1)+\sin^2(\frac{\pi}{2}x_2)+\sin^2(\frac{\pi}{2}x_3))} \times \arcsin\left(\sin(1) + \frac{x_1 + \cdots + x_7}{200}\right)$$

using the positive definite importance function

$$h_3(x_1,\ldots,x_7) = \frac{1}{\eta}\, e^{1-(\sin^2(\frac{\pi}{2}x_1)+\sin^2(\frac{\pi}{2}x_2)+\sin^2(\frac{\pi}{2}x_3))}$$

where η is

$$\eta = \int_{I^7} h_3(x_1,\ldots,x_7)dx_1\ldots dx_7 = e\cdot\left(\int_0^1 e^{-\sin^2(\frac{\pi}{2}x)}dx\right)^3$$

which is easily approximated to high accuracy as a one-dimensional integral.

2. Pseudo-random results, 70 runs, example 2 – pseudo-random sequence

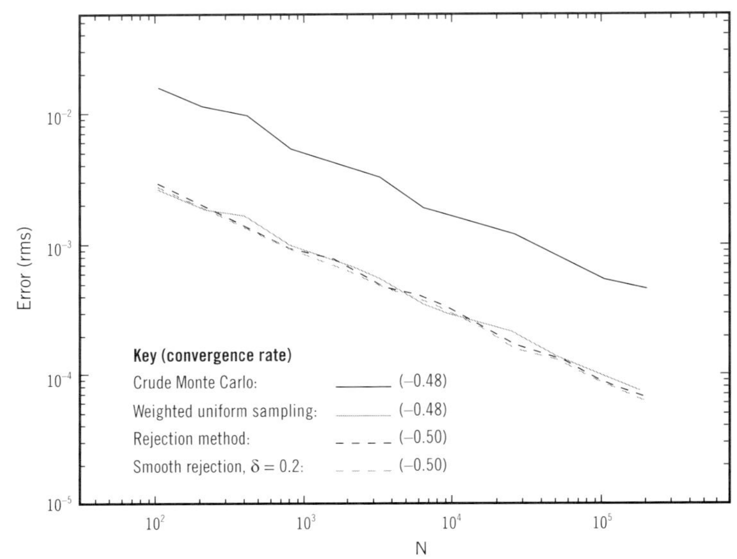

3. Quasi-random results, 70 runs, example 2 – Niederreiter sequence

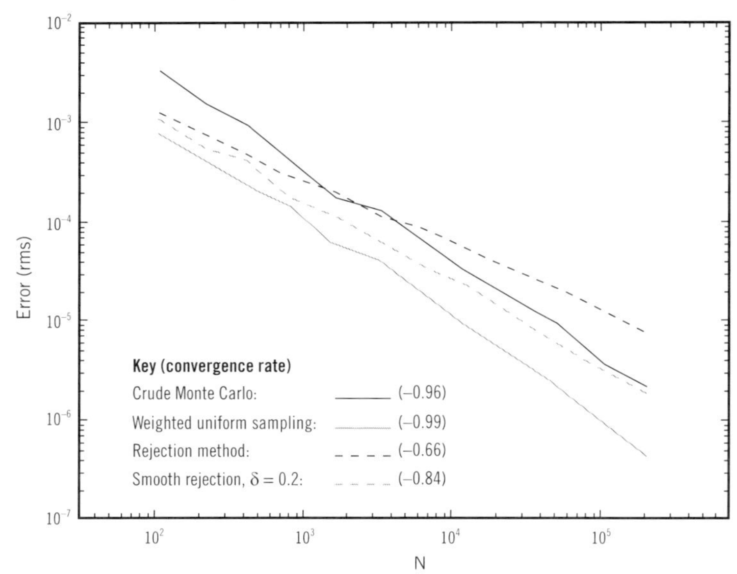

The resulting rms errors for example 3 using pseudo-random and quasi-random points are presented in Figures 4 and 5, respectively.

These computational examples show that quasi-Monte Carlo works well for the smoothed rejection method, but even better for weighted uniform sampling.

The results confirm that smoothed rejection is an improvement over ordinary rejection when using quasi-random sequences. Furthermore, weighted uniform sampling produces results that are far superior to those obtained using importance sampling with the rejection method, even with smoothing. We believe the reason for this to be that, although the smoothed rejection method involves a continuous integrand, there is still a relatively sharp slope in acceptance weights in the transition from acceptance to rejection regions. This leads to higher variation and poorer quasi-random performance, in general, than for weighted uniform sampling, where the sharp transition is completely eliminated. Another, admittedly minor, advantage to weighted uniform sampling is that the extra "decision" variable y is no longer needed, reducing the effective dimension by one.

Nevertheless, smoothed rejection may be superior to weighted uniform sampling for problems in which there are large regions of low importance, as discussed at the end of the section on weighted uniform sampling.

Reduction of dimensions

Unlike standard Monte Carlo, which is independent of the number of dimensions, quasi-random methods tend to become less effective as the number of dimensions in the integrand increases (see Morokoff and Caflisch, 1994; Morokoff and Caflisch, 1995; Sobol', 1990; Levitan, Markovich, Rozin and Sobol', 1988), although much less so than numerical quadrature methods such as the trapezoidal rule. Therefore, while numerical quadrature methods are best for low dimensional integrals, and pseudo-random Monte Carlo appears to be best for very high dimensional integrals, quasi-random methods can often be optimal for moderate-dimensional integrals. Our experience indicates that quasi-random methods, with N in the range of 10^4 to 10^6, are effective for integrals of dimension four to 30.

In this section, we discuss methods to reduce the effective dimension of a quasi-Monte Carlo integration problem.

A SIMPLE EXAMPLE OF DIMENSION REDUCTION

The following example shows that quasi-random integration error may be reduced by rearrangement of the variables so that the principal variations of the integrand occur over the lower dimensions.

Example 4: Compare the results of approximating these two equal integrals using pseudo-random and Halton points (Halton, 1960):

$$A = \int \cdots \int_{I^d} 5\cos\left(4(x_4 + x_5 + x_6) + \frac{(x_1 + x_2 + x_3)}{5}\right) dx_1 \ldots dx_6$$

$$B = \int \cdots \int_{I^d} 5\cos\left(4(x_1 + x_2 + x_3) + \frac{(x_4 + x_5 + x_6)}{5}\right) dx_1 \ldots dx_6$$

These are equivalent, but the first integrand varies more as a function of the higher three dimensions while the second varies more as a function of the lower three dimensions. Therefore we should expect that quasi-random estimates will be more effective in estimating B than A. On the other hand, using pseudo-random points there should be little difference between the two cases. The results, shown in Figure 6, confirm this.

DIMENSION REDUCTION FOR THE FEYNMAN-KAC FORMULA

The Feynman-Kac Formula provides a connection between linear parabolic differential equations and stochastic path integrals, which is similar to the method of characteristics for solving hyperbolic differential equations. In the simple case when the stochastic paths are Brownian motion, a modification of the discretisation of the Brownian path allows one to concentrate most of the variance into the first few dimensions, which significantly improves the accuracy of quasi-Monte Carlo integration.

Consider the following initial value problem:

$$\frac{\partial u}{\partial t} = \frac{1}{2}\frac{\partial^2 u}{\partial x^2}(x,t) + v(x,t)\,u(x,t),\ t \geq 0 \qquad (12)$$

with initial condition $u(x,0) = f(x)$. Its solution is given by the Feynman-Kac formula as

$$u(x,t) = E_{x,0}\left[f(\xi(t))e^{\int_0^t v(\xi(r),t-r)\,dr}\right] \qquad (13)$$

where ξ represents a Wiener path starting at $\xi(0) = x$.

4. Pseudo-random results, 75 runs, example 3 – pseudo-random sequence

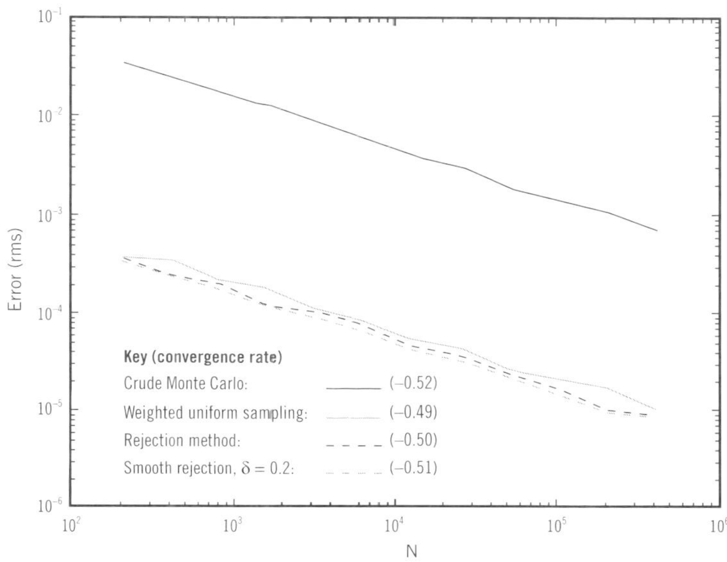

5. Quasi-random results, 75 runs, example 3 – Niederreiter sequence

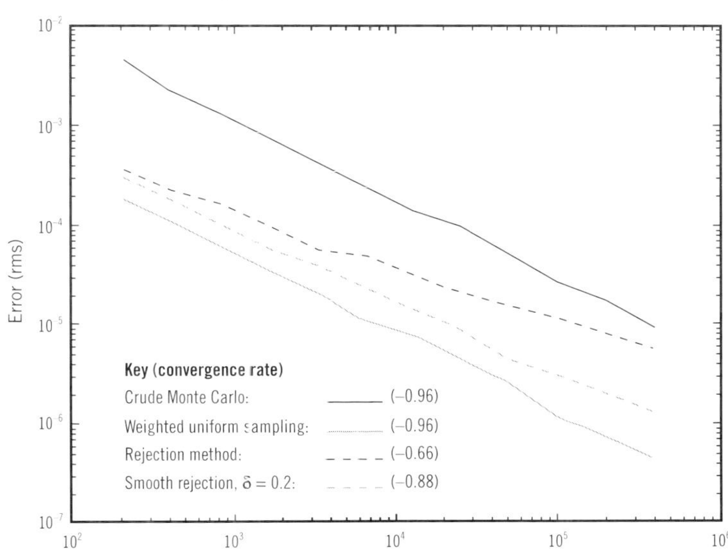

This solution can be approximated by discretising time, generating Wiener-process sample paths and averaging over the set of sample paths to obtain Monte Carlo estimates of the expectation above. The integral in the exponent is itself approximated for each sample path using the trapezoidal rule.

Standard discretisation The standard discretisation of the Wiener process is defined as follows:

1. Choose (x,t) at which the solution is to be approximated.
2. Choose the number of equal time steps, m, and set $\Delta t = \frac{t}{m}$ and $t_i = i\Delta t$ for $i = 0,\ldots,m$.
3. Generate Gaussian independent variables, $\Delta W^{(i)}$, from the distribution $\mathcal{N}(0,\Delta t)$ (the normal distribution with mean 0 and variance Δt) for $i = 1,\ldots,m$.
4. The Wiener path is sampled exactly at each t_i as the following sum:

$$\xi^{(i)} = x + \sum_{j=1}^{i} \Delta W^{(j)} \tag{14}$$

The sample paths are exact at the discrete times t_i, but an approximation is involved when the path integral of v is estimated. For example, using the trapezoidal rule the Wiener path integral is approximated as

$$\int_0^t v(\xi(r),t-r)\,dr$$

$$\approx \frac{\Delta t}{2}\left[v(x,t) + 2\sum_{i=1}^{m-1} v(\xi^{(i)},t-t_i) + v(\xi^{(m)},0)\right] = \tilde{I} \tag{15}$$

The non-differentiability of the paths leads to an error (bias) of order $O(\sqrt{\Delta t})$ for this quadrature rule. In the computations presented below, however, Δt is sufficiently small so that discretisation errors are negligible in comparison to statistical errors.

Using the sample paths above, the Monte Carlo estimate of the Feynman–Kac path integral is

$$\hat{u}_N(x,t) = \frac{1}{N}\sum_{k=1}^{N} f(\xi_k^{(m)})\, e^{\tilde{I}_k} \tag{16}$$

in which $\tilde{I}_k$ and $\xi_k^{(m)}$ for $1 \le k \le N$ are N realisations of $\tilde{I}$ and $\xi^{(m)}$. If the $\xi_k^{(m)}$ are normally distributed, as described above, then

$$E(\hat{u}_N(x,t)) = u(x,t) + \text{bias error}$$

in which the bias error comes from the quadrature rules and is insignificant.

For application of quasi-random Monte Carlo, this average is expressed as a multidimensional integral. First, each normal variable $\Delta W^{(j)}$ can be represented as a transformation, G, of a uniformly distributed variable x_j by $\Delta W^{(j)} = G(x_j)$. Then define

$$\tilde{f}(x_1,x_2,\ldots,x_m) = f(\xi^{(m)})$$
$$\tilde{I}(x_1,x_2,\ldots,x_m) = \tilde{I}$$

in which $\xi^{(i)} = x + \Sigma_{j=1}^{i} G(x_j)$ and x_j are uniformly distributed, eg pseudo-random or quasi-random points. The expectation of the Monte Carlo estimate in (16) can then be written as the following integral:

$$E(\hat{u}_N(x,t)) = \int_0^1 \ldots \int_0^1 \tilde{f}(x_1,\ldots,x_m)\, e^{\tilde{I}(x_1,\ldots,x_m)}\, dx_1 \ldots dx_m$$

The Monte Carlo estimate of this integral is

$$\tilde{u}_N(x,t) = \frac{1}{N}\sum_{k=1}^{N} \tilde{f}(x_1^k,\ldots,x_m^k)\, e^{\tilde{I}(x_1^k,\ldots,x_m^k)} \tag{17}$$

The transformation G used here is that of Marsaglia (1991), which is continuous and monotonic. By contrast, Box–Muller, which is discontinuous, gives poorer results for quasi-random sequences (Morokoff and Caflisch, 1993).

6. Convergence results and log–log plot for example 4, 100 trials

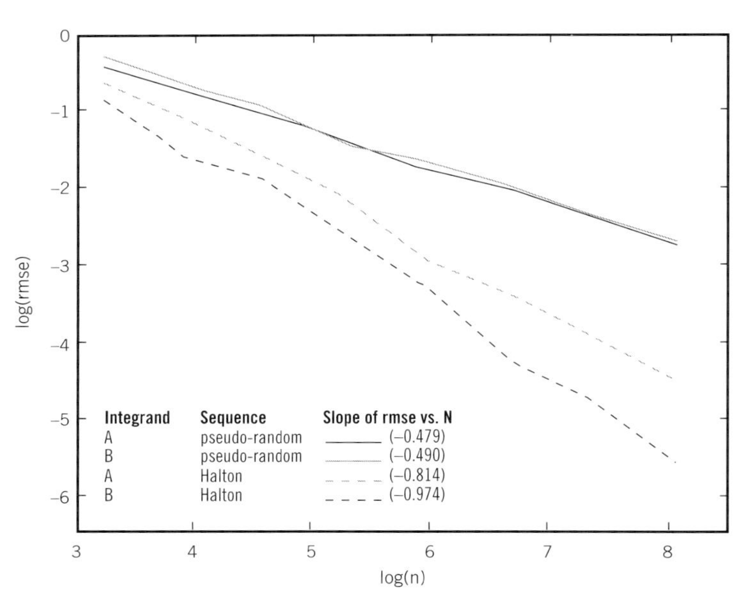

Alternative discretisation method When the number of dimensions, m, is moderately large, an alternative discretisation method leads to significant improvements in the quasi-random integration estimates by concentrating most of the variance into the lowest dimensions.

Assume that n is a power of two, and define the alternative discretisation as follows:

1. Choose (x,t) at which the solution is to be approximated.
2. Choose the number of equal time steps $m = 2^p$. Then let $\Delta t = \frac{t}{m}$ and $t_i = i\Delta t$ for $i = 0,\ldots,m$.
3. Generate Gaussian variables, $\Delta W^{(0)}$, distributed according to $\mathcal{N}(0,t)$ and $\Delta W^{(i)}$ distributed according to $\mathcal{N}(0,t/\alpha)$, for $i = 1,\ldots,m$, in which $\alpha = 4 \cdot 2^{\text{int}(\log_2 i)}$.
4. The Wiener path $\xi^{(i)}$ is sampled at each t_i as follows:

$$
\begin{aligned}
\xi^{(0)} &= x \\
\xi^{(m)} &= \xi^{(0)} + \Delta W^{(0)} \\
\xi^{(m/2)} &= \frac{\xi^{(0)} + \xi^{(m)}}{2} + \Delta W^{(1)} \\
\xi^{(m/4)} &= \frac{\xi^{(0)} + \xi^{(m/2)}}{2} + \Delta W^{(2)} \\
\xi^{(3m/4)} &= \frac{\xi^{(m/2)} + \xi^{(m)}}{2} + \Delta W^{(3)} \\
&\vdots
\end{aligned}
$$

For this discretisation method, the first step is directly from 0 to t. Then the intermediate steps are filled in by taking successive subdivisions of the time intervals into halves. Each new intermediate path position is determined using the following rule:

$$
\xi^{(j)} = \frac{\xi^{(j_1)} + \xi^{(j_2)}}{2} + \Delta W, \quad \Delta W \in \mathcal{N}\left(0, \frac{t_{j_2} - t_{j_1}}{4}\right) \tag{19}
$$

where j_1, j_2 are the indices of the nearest prior and later time steps, respectively, for which the positions have already been determined. This representation of the Wiener path in terms of past and future positions is referred to as the Brownian bridge (also called a tied-down Wiener process).

The representation (18) can be conveniently rewritten as follows:

$$
\begin{aligned}
\xi^{(0)} &= x \\
\xi^{(m)} &= x + \Delta W^{(0)} \\
\xi^{(m/2)} &= x + \frac{\Delta W^{(0)}}{2} + \Delta W^{(1)} \\
\xi^{(m/4)} &= x + \frac{\Delta W^{(0)}}{4} + \frac{\Delta W^{(1)}}{2} + \Delta W^{(2)} \\
\xi^{(3m/4)} &= x + \frac{3\Delta W^{(0)}}{4} + \frac{\Delta W^{(1)}}{2} + \Delta W^{(3)} \\
&\vdots
\end{aligned} \tag{20}
$$

The variance contributed by each dimension to the Wiener sample paths is compared for the standard and the alternative discretisations in Table 1 for the case $m = 8$. Note that variance due to x_k slowly decreases for the standard

7. Graphical comparison of discretisation schemes

Upper plot: standard scheme, 4 steps; lower plot: alternative scheme, 4 steps

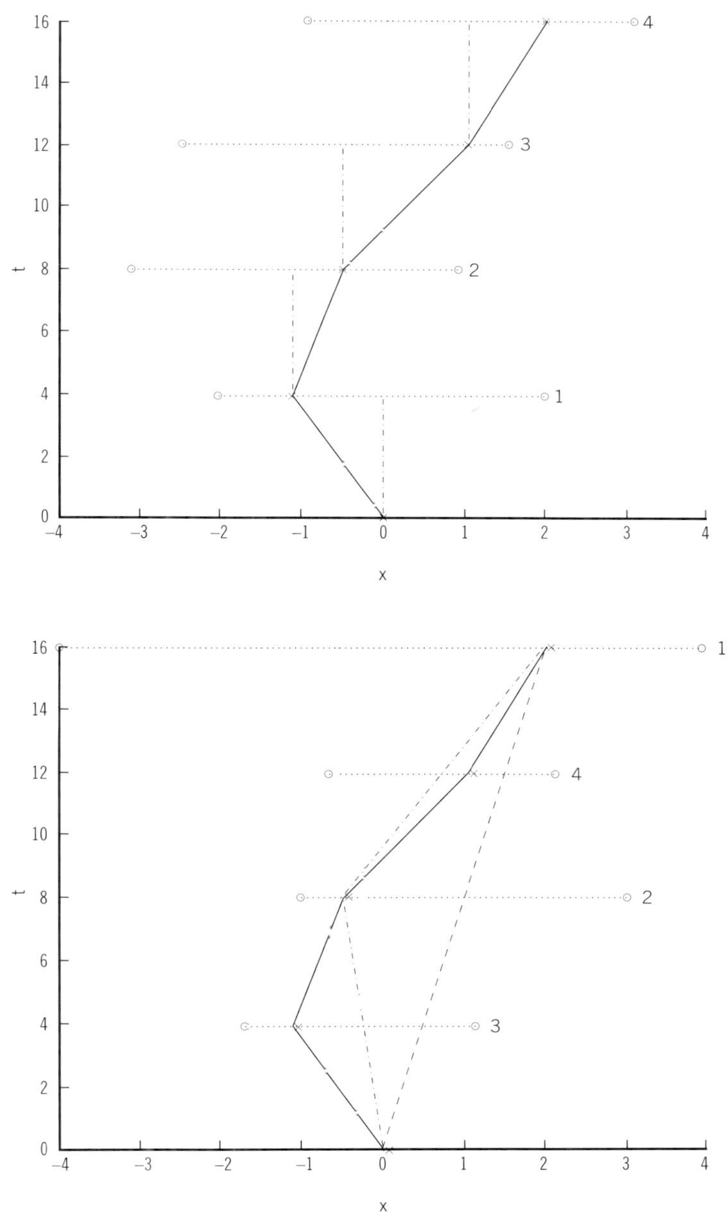

Table 1. Variances for standard and alternative discretisation schemes, m = 8

	Variance	
	Standard	Alternative
x_1	1.0000t	3.1875t
x_2	0.8750t	1.6875t
x_3	0.7500t	0.1875t
x_4	0.6250t	0.1875t
x_5	0.5000t	0.0625t
x_6	0.3750t	0.0625t
x_7	0.2500t	0.0625t
x_8	0.1250t	0.0625t

8. Log–log plot for example 5, 75 runs, T = 0.02, m = 8

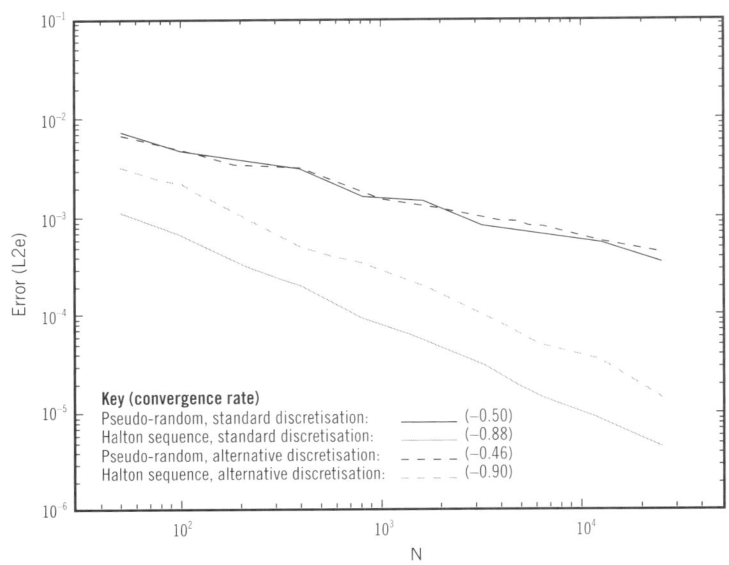

9. Log–log plot for example 5, 75 runs, T = 0.04, m = 16

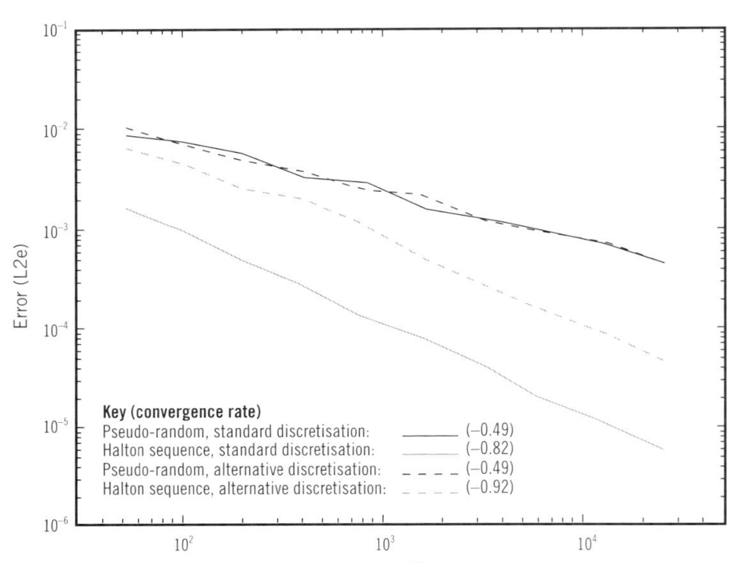

discretisation, since early values of the Wiener process affect more parts of the path than later values, but that this decay is much more rapid for the alternative discretisation.

Figure 7 is a graphical comparison of the standard and alternative discretisation points for $m = 4$, showing how more of the variance is concentrated in the lower dimensions for the alternative method. The dotted horizontal lines represent one standard deviation. The solid lines show the actual discretised stochastic paths. In the upper figure, the dash/dot line indicates the additive nature of the successive steps. In the lower figure, the dashed line shows the effects of the large first step on all of the intermediate positions, while the dash/dot lines show the effects of the second step on intermediate positions. Note that the two plots show the same resultant paths for convenience only. There is no connection between the particular path generated by the same pseudo-random or quasi-random point using each of the two different methods. There is only an equivalence in terms of the expectations over all such paths.

Computational example for the Feynman-Kac formula This example demonstrates the accuracy of Monte Carlo estimates of the solution of a simple linear parabolic differential equation using (16) at selected values of x and a fixed (small) time t. A comparison will be made of pseudo-random and quasi-random sequences with either the standard discretisation (14), or the alternative discretisation (19).

Example 5: Consider the linear parabolic differential equation

$$\frac{\partial f}{\partial t}(x,t) = \frac{1}{2}\frac{\partial^2 f}{\partial x^2}(x,t) + \left(\frac{1}{t+1} + \frac{1}{x^2+1} - \frac{4x^2}{(x^2+1)^2}\right)\frac{\partial f}{\partial x}(x,t)$$

with initial condition $f(x,0) = \frac{1}{x^2+1}$. The exact solution is: $f(x,t) = \frac{t+1}{x^2+1}$.

Estimates are computed at a fixed (small) time, T, and at eight equally spaced positions between $x = -3$ and $x = 3$ using a discretisation of time into m equal steps and a sample size of N. The estimates are then compared with the exact solution using the following L^2 measure of error:

$$L2e = \sqrt{\frac{1}{8}\sum_{k=0}^{7}\left[f(x_k,T) - \hat{f}_N(x_k,T)\right]^2}$$

where $x_k = -3 + 6k/7$.

T is selected to be very small so that discretisation errors are insignificant relative to the Monte Carlo errors.

A measure of the rate of decrease of the values of L2e is obtained by a least-squares fit of the function $cN^{-\alpha}$ to the calculated values of L2e. The resulting values of α for pseudo-random and quasi-random (Halton was used here) points used in both the standard and alternative discretisations are presented along with log-log plots of the data in Figures 8–10 for three values of T with equal time increments in each case; ie T = 0.02 (m = 8), T = 0.04 (m = 16) and T = 0.08 (m = 32). The results show that the convergence rate for pseudo-random is independent of T and of the discretisation scheme. On the other hand, the performance of quasi-random Monte Carlo rapidly degrades in the standard discretisation scheme, but is maintained at α nearly equal to 1 in the alternative discretisation. It is important to notice that the quasi-random improvement from the alternative scheme grows as the dimension increases. This is expected, as is the fact that at 32 dimensions standard quasi-random is just barely better than pseudo-random.

For standard Monte Carlo pseudo-random sequences, there is no difference between the two discretisations since the total variance of the two methods is the same. One the other hand, the alternative discretisation method provides significantly more accurate results than the standard discretisation when using quasi-random sequences in the Feynman–Kac formula. This improvement is due to reduction of the effective dimension of the problem.

10. Log–log plot for example 5, 75 runs, T = 0.08, m = 32

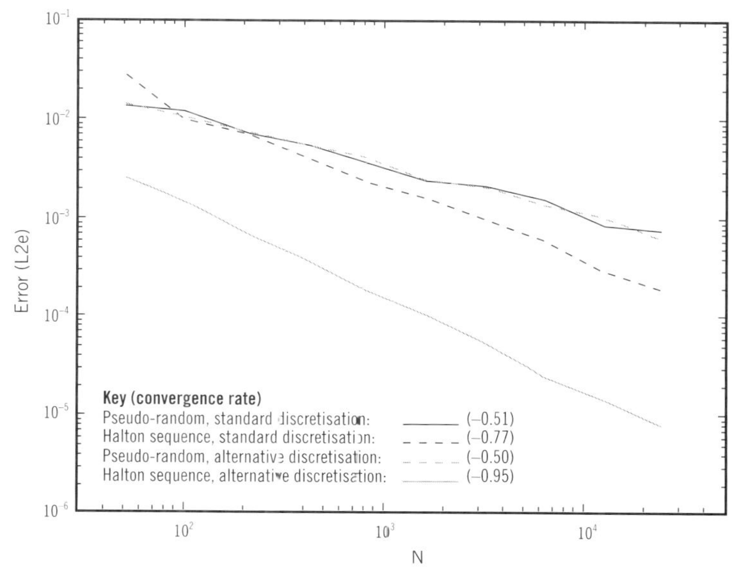

Conclusions

The analysis and computations presented above demonstrate the need to modify standard Monte Carlo methods for effective implementation of quasi-random sequences. The advantages of quasi-random sequences over pseudo-random sequences are lost for problems with discontinuous integrands or large dimension. These advantages can be regained, at least for some problems, by smoothing or dimension reduction.

This has been successfully implemented for the rejection method through a smoothed rejection method, and for the evaluation of Feynman–Kac integrals through an alternative discretisation. In a separate paper (Moskowitz, 1994), one of the authors (BM) showed how quasi-random sequences may also be used in diffusion Monte Carlo for computation of the ground-state energy of various systems. On the other hand, we have not yet been successful in applying quasi-random methods to the Metropolis algorithm.

We consider the results presented in these examples to be an important step forward in the continuing and promising application of quasi-random sequences to Monte Carlo methods. While quasi-random sequences will probably not be useful in all Monte Carlo methods, we expect that a wide range of effective applications may be developed.

Reprinted from *Journal of Mathematical Computer Modelling* 23, pp. 37–54, 1996, with permission from Elsevier Science.

1 *Research supported in part by the Air Force Office of Scientific Research under grant number F49620-94-1-0091, and in part by the Army Research Office under grant number DAAL03-91-G-0162.*

BIBLIOGRAPHY

Adlakha, V., 1992, "An Empirical Evaluation of Antithetic Variates and Quasi-Random Points for Simulating Stochastic Networks", *Simulation* 58(1), pp. 23–31.

Berblinger, M., and C. Schlier, 1991, "Monte Carlo Integration with Quasi-Random Numbers: Some Experience", *Computer Physics Communications* 66, pp. 157–66.

Bratley, P., B.L. Fox and H. Niederreiter, 1992, "Implementation and Tests of Low-Discrepancy Sequences", *ACM Transactions on Modeling and Computer Simulation* 2(3), pp. 195–213.

Halton, J.H., 1960, "On the Efficiency of Certain Quasi-Random Sequences of Points in Evaluating Multi-Dimensional Integrals", *Numerische Mathematik* 2, pp. 84–90.

Lecot, C., 1991, "A Quasi-Monte Carlo Method for the Boltzmann Equation", *Mathematics of Computation* 56(194), pp. 621–44.

Levitan, Y.L., N.I. Markovich, S.G. Rozin and I.M. Sobol', 1988, "Short Communications on Quasirandom Sequences for Numerical Computations", *Zh. vŷchisl. Mat. mat. Fiz.*, 28(5), pp. 755–59.

Marsaglia, G., 1991, "Normal (Gaussian) Random Variables for Supercomputers", *The Journal of Supercomputing* 5, pp. 49–55.

Morokoff, W., and R.E. Caflisch, 1993, "A Quasi-Monte Carlo Approach to Particle Simulation of the Heat Equation", *SIAM Journal on Numerical Analysis* 30, pp. 1558–73.

Morokoff, W., and R.E. Caflisch, 1994, "Quasi-random Sequences and Their Discrepancies", *SIAM Journal of Scientific Statistical Computing* 15(6) pp. 1251–79.

Morokoff, W., and R.E. Caflisch, 1995, "Quasi-Monte Carlo Integration", *Journal of Computational Physics* 122(2), pp. 218–30.

Moskowitz, B., 1993, "Application of Quasi-Random Sequences to Monte Carlo Methods", PhD thesis, University of California, Los Angeles.

Moskowitz, B., 1994, "Quasirandom Diffusion Monte Carlo", CAM report 94-39, UCLA Mathematics Department.

Niederreiter, H., 1992, *Random Number Generation and Quasi-Monte Carlo Methods*, Philadelphia, SIAM.

Powell, M.J.D., and J. Swann, 1966, "Weighted Uniform Sampling – A Monte Carlo Technique for Reducing Variance", *Journal of the Institute of Mathematical Applications*, 2, pp. 228–36.

Press, W.H., and S.A. Teukolsky, 1988, "Quasi- (that is, Sub-) Random Numbers", *Computers in Physics* 2(6), pp. 76–9.

Shuhman, B.V., 1990, "Application of Quasi-Random Points for Simulation of Gamma Radiation Transfer", *Progress in Nuclear Energy* 24, pp. 89–95.

Sobol', I.M., 1990, "Quasi-Monte Carlo Methods", *Progress in Nuclear Energy* 24, pp. 55–61.

Spanier, J., and E.H. Maize, 1994, "Quasi-Random Methods for Estimating Integrals Using Relatively Small Samples", *SIAM Review* 36, pp. 18–44.

28

Beating Monte Carlo

Anargyros Papageorgiou and Joseph Traub
Department of Computer Science, Columbia University

Monte Carlo simulation is widely used to price complex financial instruments, and much time and money have been invested in the hope of improving its performance. However, recent theoretical results and extensive computer testing indicate that deterministic methods, such as simulations using Sobol' or generalised Faure points, may be superior in both speed and accuracy.

In this chapter, we refer to a deterministic method by the name of the sequence of points it uses, eg the Sobol' method. We tested the generalised Faure sequence due to Tezuka (1995) and a modified Sobol' method which includes additional improvements to those documented in Paskov and Traub (1995). We compared these two low-discrepancy deterministic methods with basic Monte Carlo in the valuation of a collateralised mortgage obligation (CMO).

We found that deterministic methods beat Monte Carlo:

- by a wide margin. In particular:
  - — Both the generalised Faure and modified Sobol' methods converge significantly faster than Monte Carlo.
  - — The generalised Faure method always converges at least as fast as the modified Sobol' method and often faster.
  - — The Monte Carlo method is sensitive to the initial seed.
- for a small number of sample points:
  - — Deterministic methods achieve a low error level with a small number of points.
  - — For the most difficult CMO tranche, generalised Faure achieves accuracy of 10^{-2} with 170 points, while modified Sobol' uses 600 points. The Monte Carlo method, on the other hand, requires 2,700 points for the same accuracy.
  - — Monte Carlo tends to waste points due to clustering, which severely compromises its performance when the sample size is small.
- as the sample size and the accuracy demands grow. In particular:
  - — Deterministic methods are 20 to 50 times faster than Monte Carlo (the speed-up factor) even with moderate sample sizes (2,000 deterministic points or more).
  - — When high accuracy is desired, deterministic methods can be as much as 1,000 times faster than Monte Carlo.

There are two ways of valuing financial derivatives: via paths or as a high-dimensional integral. For simplicity, we will restrict ourselves to a discussion of the integral formulation and, without loss of generality (Paskov, 1996), the integral over the unit cube in d dimensions. For most finance problems, this integral cannot be analytically computed; we have to settle for a numerical approximation.

The basic Monte Carlo method obtains this approximation by computing the arithmetic mean of the integrand evaluated at randomly chosen points. More precisely, only pseudo-random points can be generated on a digital computer and these are used in lieu of random points. There are sophisticated variations of this method, but we refer only to the basic version in this chapter.

If pseudo-random points from a flat distribution are plotted on the unit square in two dimensions (see Figure 1), there are some regions where no sample points occur and others where the points are more concentrated. This is clearly undesirable. Random point samples are wasted due to clustering. Indeed, Monte Carlo simulations with very small sample sizes cannot be trusted. It would be better to place our sample points as uniformly as possible, which is the idea behind low-discrepancy sequences. Discrepancy is a measure of deviation from uniformity; hence low-discrepancy points are desirable. Figure 2 is a plot of certain low-discrepancy points on the unit square in two dimensions.

A low-discrepancy method approximates the integral by computing the arithmetic mean of the integrand evaluated at low-discrepancy points.

Low-discrepancy sequences have been extensively studied.[1] In contrast to the Monte Carlo method, these use deterministic points. They are sometimes said to be quasi-random.

In 1992, the conventional wisdom was that although theory suggested that low-discrepancy methods were sometimes superior to Monte Carlo, this theoretical advantage was not seen for high-dimensional problems. Joseph Traub and a PhD student, Spassimir Paskov, decided to compare the efficacy of low-discrepancy and Monte Carlo methods in valuing financial derivatives.

They used a CMO (Federal National Mortgage Association REMIC Trust 1989–2023), provided by Goldman Sachs, with 10 tranches requiring the evaluation of 10 integrals, each over 360 dimensions, and a particular low-discrepancy sequence due to Sobol'. The values of the tranches depended on the interest rate and prepayment models used. Paskov and Traub made major improvements in the Sobol' points, which led to a more uniform distribution. The improved Sobol' method consistently outperformed Monte Carlo (Paskov and Traub, 1995, and Paskov, 1996).

Software construction and testing of low-discrepancy deterministic methods for pricing financial derivatives began at Columbia University in autumn 1992. Preliminary results were shared with a number of New York City financial houses in autumn 1993 and spring 1994. The first published announcement was Traub and Woźniakowski's January 1994 article in *Scientific American*. A more detailed history is given in Paskov and Traub (1995).

In September 1995, IBM announced a product called the Deterministic Simulation Blaster, which uses a low-discrepancy deterministic method.[2] The company claimed a very large improvement over Monte Carlo. However, IBM has not revealed the method for choosing the sample points and the methodology for calculating the speed-up. IBM acknowledges that Columbia University pioneered the use of low-discrepancy methods to price financial derivatives.

1. Distribution of 512 pseudo-random points

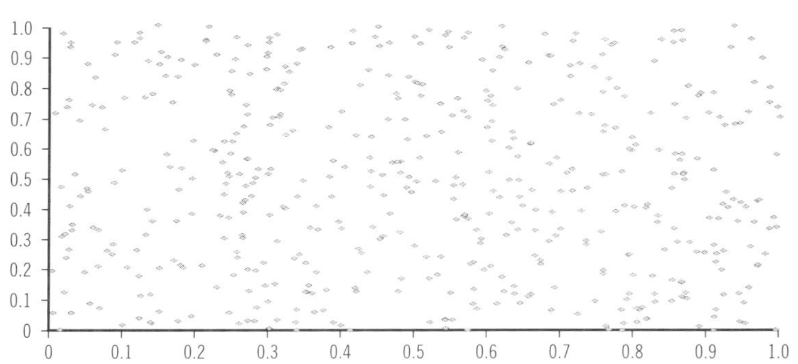

2. Distribution of 512 low-discrepancy points

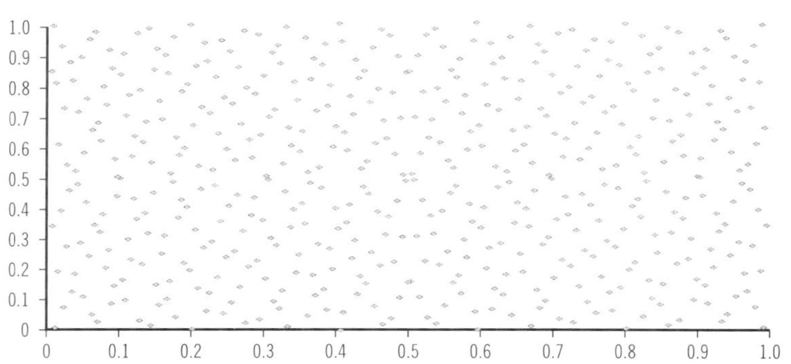

Results

We have built a software system called FINDER[3] for computing high-dimensional integrals which has modules for generating generalised Faure points and modified Sobol' points. It includes major improvements. Indeed, a number of financial institutions have informed us that they could not replicate our results using, for example, the Sobol' point generator found in Press *et al* (1992).

We used FINDER to price the CMO and to compare low-discrepancy methods with Monte Carlo simulation. As both types of method compute the arithmetic mean of the integrand evaluated at a number of points, the difference in performance depends on the number of points that each method uses to achieve the same accuracy. We observe the least number of points a method needs to achieve and maintain a relative error below a specified level, say, 10^{-2}. This is gauged by the speed-up of one method relative to another, which we define as the ratio of the least number of points required by one method divided by the least number of points required by the other method. This ensures that both methods maintain the same level of accuracy.

This definition of speed-up is new: we study the convergence and the error of a method throughout a simulation. We believe that this has advantages over speed-up calculations which are based only on the error values at the end of a simulation. Note that our definition of speed-up is a more rigorous requirement than only computing the confidence level of Monte Carlo.

Our extensive testing has shown that fixed accuracy requires different tranches to be treated with different numbers of points. Again, we emphasise that deterministic methods beat Monte Carlo for every tranche. Our results are reported using the residual tranche of the CMO, which depends on all 360 monthly interest rates, as the reference point because it is the most difficult to price. If this tranche can be priced with a given accuracy using a certain number of samples, the same number of samples will yield at least the same accuracy for the rest of the tranches.

Since pricing models for complicated derivatives are subject to uncertainty, financial houses are often content with relative errors of one part in a hundred. Furthermore, if they wish to price a book of instruments, it is critical to use a small number of samples. Deterministic methods achieve a relative error of one part in a hundred using a small number of points. In Figures 3 and 4 we see that some 170 generalised Faure points or 600 Sobol' points are sufficient for a relative error equal to 10^{-2}, whereas Monte Carlo requires 2,700 points for the same level of accuracy. A very small number of generalised Faure points thus yields an accurate price 16 times faster than Monte Carlo.

A further reduction of the error by a factor of 20 (equal to $10^{-3}/2$) requires about 16,000 generalised Faure points, whereas Monte Carlo may require up to 800,000 random points, which yields a speed-up factor of up to 50. In general, samples using as few as 2,000 generalised Faure points can price the CMO 20 to 50 times faster than Monte Carlo. As far as convergence rates are concerned, for $n \leq 10^4$ generalised Faure points the error is proportional to $n^{-0.82}$. This error estimate is conservative since a much higher convergence rate is frequently attained but, in any case, it is very much superior to the $n^{-0.5}$ expected Monte Carlo error.

Monte Carlo also exhibits a great sensitivity to the seed of the pseudo-random number generator. So, unless we are dealing with the result of a fairly long simulation, we cannot have much confidence. The long simulations needed yield a deterministic method speed-up of about 1,000.

Conclusion

The best of the deterministic methods we have tested is the one based on generalised Faure points. These usually achieve the same accuracy as the Sobol' points but 2.5 to 6.5 times faster.

3. Generalised Faure and Monte Carlo errors

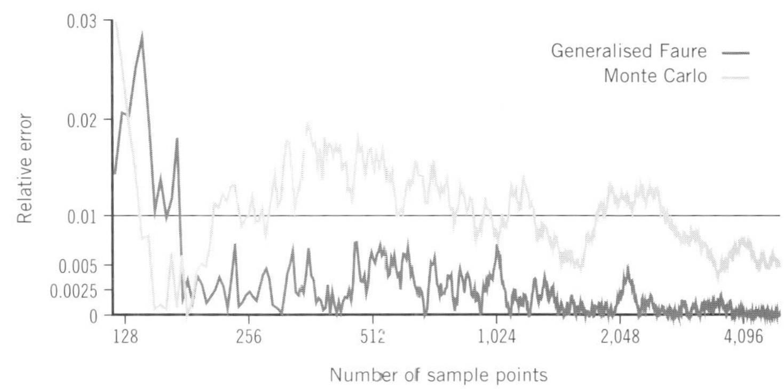

4. Generalised Faure and Sobol' errors

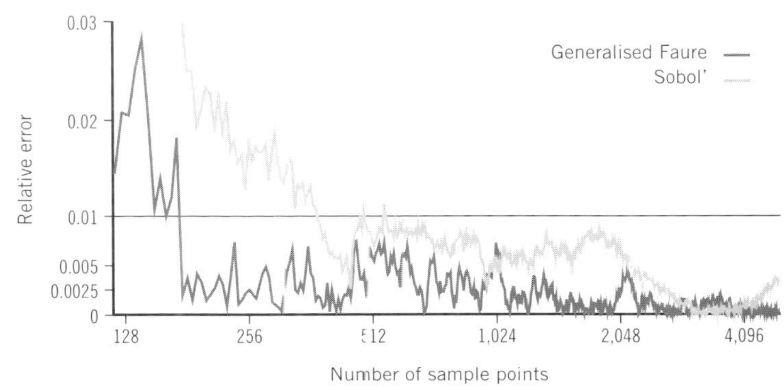

They can also be produced efficiently, at a cost similar to that for random points, and a only a small number of points is needed to price the CMO. In contrast to some other deterministic sequences, generalised Faure points can easily be produced in very high dimensions. It is much more complicated to obtain the improved Sobol' points that we have been using in very high dimensions.

FINDER contains features that further improve the quality of the approximation given by the generalised Faure method without any additional computational costs. Finally, preliminary but very encouraging results indicate that generalised Faure points can efficiently price financial derivatives modelled in more than 1,500 dimensions. Future work will include: further improvements in FINDER; a comparison of the performance of low-discrepancy and Monte Carlo methods on other financial derivatives; studying the possibilities that low-discrepancy methods offer for risk management; and designing new low-discrepancy methods tailored for use in financial computations.

A version of this paper was presented on April 15, 1996, at Mathematical Problems in Finance, *organised by the Institute for Advanced Study, Princeton, New Jersey. This work was supported in part by the National Science Foundation.*

1 *See Paskov (1996) for the formal definition of discrepancy and an extensive bibliography.*

2 *Risk Technology Supplement, August 1995, pp. 23–24;* Risk *November 1995, p. 47.*

3 *FINDER may be obtained by contacting Dr Fred Kant, Director, Columbia Innovation Enterprise, Columbia University (e-mail: fk3@columbia.edu).*

BIBLIOGRAPHY

Paskov, S.H., and J.F. Traub, 1995, "Faster Valuation of Financial Derivatives", *Journal of Portfolio Management* Fall 1995, pp 113–20.

Paskov, S.H., 1997, "New Methodologies for Valuing Derivatives", in S. Pliska and M. Dempster (eds), *Mathematics of Derivative Securities*, Isaac Newton Institute, Cambridge University Press, Cambridge, pp. 545–82 (Chapter 25 of this publication).

Press, W., S. Teukolsky, W. Wetterling and B. Flannery, 1992, *Numerical Recipes in C: The Art of Scientific Computing*, second edition, Cambridge University Press, Cambridge.

Tezuka, S., 1995, *Uniform Random Numbers: Theory and Practice*, Kluwer Academic Publishers, Boston.

Traub, J.F., and H. Woźniakowski, 1994, "Breaking Intractability", *Scientific American* January 1994, pp 102–7.

29

Monte Carlo Motoring

Rupert Brotherton-Ratcliffe
General Re Financial Products

Monte Carlo simulation provides a simple numerical procedure for valuing complex financial instruments. It is widely used to price so-called path-dependent instruments, such as collateralised mortgage obligations, and securities with variable cash flows which depend on financial indices. Despite its usefulness and wide applicability, a significant drawback of the method is that convergence is sometimes slow. Typically, if N random numbers are used to value a structure, the numerical error is of order $N^{-1/2}$, so that if one wishes to reduce a numerical error by a factor of 10, the computation time increases by a factor of 100.

There are a number of standard approaches, called variance reduction techniques, which can improve rate convergence. In an early application of the Monte Carlo method to option valuation, Boyle (1977) – also Chapter 1 of this publication discusses two variance reduction techniques, the antithetic variate method and the control variate method. Here, we investigate the potential of using quasi-random sequences, and in particular the Sobol′ sequences, to reduce errors in option valuation. We find that quasi-random sampling not only increases the efficiency of Monte Carlo simulation but is also more efficient than other simple variance reduction techniques, including the antithetic method and schemes based on stratified sampling. Quasi-random sampling may also be used with ad hoc, or problem-specific, techniques, such as the control variate method, to increase accuracy further.

A quasi-random sequence of numbers is one which fills up a region of the real line (typically between zero and one) more uniformly than a truly random sequence of numbers. More generally, a quasi-random sequence of n-tuples (or vectors with n elements) fills up n-dimensional space more uniformly than a random sequence of n-tuples. To illustrate this, in Figure 1a we have plotted points generated by a quasi-random sequence of 2-vectors (the first element being the x co-ordinate of each point, and the second being the y co-ordinate). For comparison, Figure 1b shows points generated by a simple random number generator, or what is more correctly termed a pseudo-random number generator, which uses a standard linear congruential recurrence relation. Each even term of the pseudo-random sequence is an x co-ordinate of a point, and the subsequent odd term is its corresponding y co-ordinate. While both sequences cover the whole unit square, the pseudo-random sequence tends to produce dense clusters of points in certain areas and is more sparsely distributed in other areas.

1a. First 512 points of a pair of Sobol′ sequences

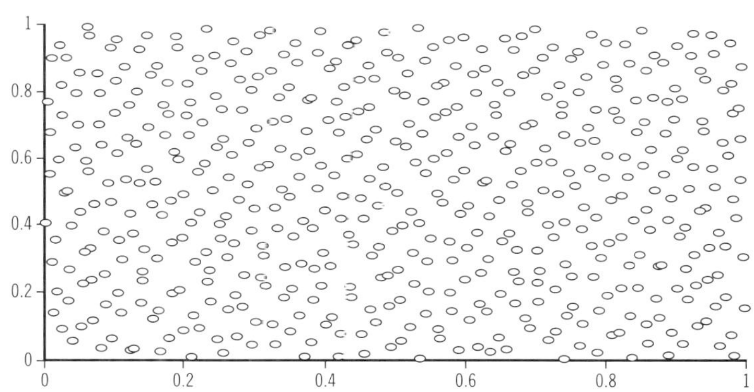

1b. First 512 points of a pair of pseudo-random sequences

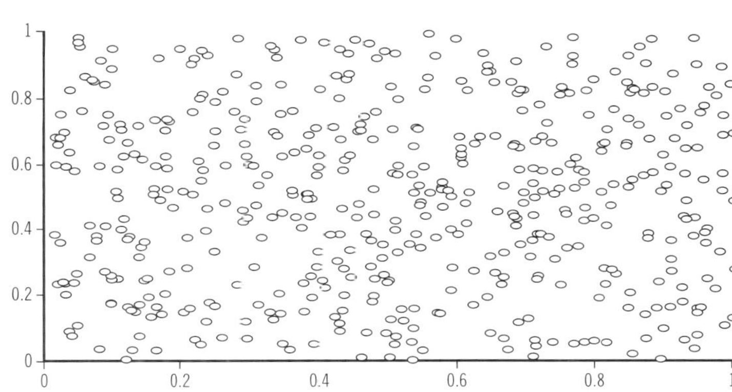

PANEL 1

GEOMETRIC MEAN STOCK OPTION VALUATION

We make the usual Black–Scholes assumptions, that the stock price is lognormally distributed with a constant volatility, σ, and that there is a constant risk-free interest rate, r. We also assume that the stock pays no dividend over the life of the option. Suppose that the option expires at time T_M, and that its payoff at time T_M is given by

$$C(T_M) = \max(G - K, 0)$$

where K is the strike price and G is the geometric mean of the stock prices S_i at times T_i (for i = 1, 2, ..., M, and $T_1 < T_2 < ... < T_M$):

$$G = \left\{\prod_{i=1}^{M} S_i\right\}^{1/M}$$

The value of this option at time t prior to the first observation (ie, when $t < T_1$) is

$$C(t) = e^{-r(T_M - t)}\{e^{\mu_G + \frac{1}{2}\sigma_G^2} SN(d + \sigma_G) - KN(d)\}$$

where μ_G and σ_G are the cumulative drift and volatility of the geometric mean over the life of the option:

$$\sigma_G = \frac{\sigma}{M}\sqrt{\sum_{i=1}^{M}(2M - 2i + 1)(T_i - t)}$$

$$\mu_G = \left(r - \frac{1}{2}\sigma^2\right)\frac{1}{M}\sum_{i=1}^{M}(T_i - t)$$

and

$$d = \frac{\ln(S/K) + \mu_G}{\sigma_G}$$

Here S is the stock price at time t. The price at later times, (ie, for $t \geq T_1$) is trivially related to the above formula

The quasi-random sequence, on the other hand, has points which are far more evenly distributed over the unit square. Moreover, subsequent terms of the quasi-random sequence will tend to fall in the gaps between previous points, whereas subsequent terms of the pseudo-random sequence will fall with equal probability in any sub-region of a given area, irrespective of the proportion of points that have already fallen in that sub-region.

The quasi-random sequence plotted in Figure 1a is a two-dimensional Sobol′ sequence. A description of how to generate Sobol′ sequences is given in Press *et al.* (1992), and we have used the routine sobseq( ) listed in that source.

When Monte Carlo simulation is used to price an option, sample paths of the underlying variables are generated according to their risk-neutral processes and option payoffs are computed along each sample path. For example, for an option on a stock, we must simulate paths of the future stock price. If the future stock price is assumed to be lognormally distributed, we can generate a path of future stock prices by using the recurrence relation

$$S_i = S_{i-1} e^{(r - \frac{1}{2}\sigma^2)\Delta + \sigma\sqrt{\Delta} Z_i}$$

for i = 1, 2, ..., with S_0 being the initial stock price. Here, Δ is the time between samples of the stock price, r is the risk-free interest rate and σ the stock price volatility. Finally, Z_i (i = 1, 2, ...) is a path of independent normally distributed random deviates, with zero mean and unit variance, which is computed by first using a random number generator to generate a path of uniformly distributed random deviates between zero and one, and then converting the path of uniform deviates into normal deviates using the Box–Muller transformation, for example. The normal course of action is to use a pseudo-random sequence of numbers as the uniform deviates. Instead, however, we shall use the quasi-random Sobol′ sequence.

For problems of dimension M, the fractional errors of Monte Carlo valuation with quasi-random numbers decrease with N, the number of samples, as $(\ln N)^M/N$. This is in contrast to normal Monte Carlo simulation, where errors decrease as $N^{-1/2}$. So while it appears that using quasi-random numbers will increase the rate of convergence for large N, the factor $(\ln N)^M$ in the numerator of the convergence rate is a cause for concern for problems of large dimension.

Table 1. Values of four-year ITM, ATM and OTM call options on the geometric mean stock price ($)

	Sampling frequency			
Option strike	**Annual**	**Semi-annual**	**Quarterly**	**Monthly**
In-the-money	25.870	23.660	22.570	21.840
At-the-money	11.570	10.430	9.875	9.510
Out-of-the-money	3.016	2.673	2.509	2.402

Note: initial stock price is $100, volatility is 20%, and risk-free interest rate is 5%. All figures are in dollars.

In order to test the usefulness of Sobol′ sequences for problems of both small and large dimension, we have applied them to the valuation of options on the geometric mean of a stock price. These have been chosen because they are path-dependent options which have an exact solution, given the usual Black–Scholes framework; a pricing formula is given in Panel 1. We consider four-year in-the-money, at-the-money and out-of-the-money options on the geometric mean, with annual, semi-annual, quarterly and monthly sampling frequencies, thus yielding problems of dimension 4, 8, 16 and 48, respectively. The at-the-money strikes correspond to the forward implied geometric mean, and the in- and out-of-the-money strikes correspond to means one standard deviation below and above the forward implied mean, respectively. These strikes depend on the sampling frequency, the initial stock price, the risk-free interest rate and the stock price volatility. In these examples, we take $S_0 = \$100$, $r = 5\%$ and $\sigma = 20\%$. The exact option prices are given in Table 1.

Table 2 shows the resulting root mean square errors, in dollars, based on 100 trials, for the valuation of the 12 options, using quasi-random Sobol′ sequences for a range of N from 64 to 8,192. For comparison, Table 3 shows the corresponding errors using a pseudo-random number generator.

The results are very encouraging. For an annual sampling frequency, the dimension of the problem is M = 4, and errors are approximately one-tenth of those for the regular Monte Carlo simulation using only 1,024 paths. This would imply that regular simulation would require about 100,000 simulations to match the accuracy of the Sobol′ sequence method with only 1,000 simulations. Also, one basis-point accuracy is achieved, ie one-cent accuracy for an initial stock price of $100, with 8,192 quasi-random simulations. Regular simulation has a root mean square error of 11 bp for the out-of-the-money option and 23 bp for in-the-money options.

For monthly sampling, the dimension of the problem is large (M = 48), and the Sobol′ sequence method is less accurate than regular simulation when N = 1,024. But even in this extreme case, the convergence of the Sobol′ sequence method improves rapidly as N increases. When N = 8,192, the error is less than half that of the regular error for out-of-the-money options and less than one-seventh for in-the-money options.

Compared with regular Monte Carlo simulation with a pseudo-random number generator, the quasi-random Sobol′ sequences yield a dramatic improvement in accuracy for the valuation of the path-dependent option on the geometric mean over a wide range of option strikes for problems of low dimension, ie, low sampling frequency of the geometric mean, and somewhat less dramatic,

Table 2. Root mean square of absolute errors, based on 100 trials, using quasi-random sampling

Option strike	Number of paths	Annual	Semi-annual	Quarterly	Monthly
		Sampling frequency			
In-the-money	64	0.83	0.95	1.07	1.55
	128	0.41	0.47	0.64	1.25
	256	0.27	0.34	0.38	1.07
	512	0.10	0.10	0.17	1.05
	1,024	0.06	0.07	0.08	1.02
	2,048	0.03	0.04	0.06	0.08
	4,096	0.02	0.02	0.04	0.05
	8,192	0.01	0.01	0.02	0.03
At-the-money	64	1.32	1.16	1.20	1.42
	128	0.34	0.44	0.75	1.10
	256	0.21	0.29	0.35	0.82
	512	0.09	0.10	0.22	0.78
	1,024	0.06	0.08	0.12	0.69
	2,048	0.03	0.04	0.07	0.11
	4,096	0.02	0.02	0.06	0.07
	8,192	0.01	0.01	0.02	0.05
Out-of-the-money	64	0.51	0.64	0.77	0.94
	128	0.24	0.34	0.55	0.67
	256	0.14	0.22	0.27	0.49
	512	0.08	0.11	0.17	0.39
	1,024	0.04	0.07	0.11	0.28
	2,048	0.03	0.04	0.07	0.10
	4,096	0.02	0.02	0.05	0.06
	8,192	0.01	0.01	0.02	0.04

Note: Figures in last four columns are in dollars.

Table 3. Root mean square of absolute errors, based on 100 trials, using pseudo-random sampling

Option strike	Number of paths	Annual	Semi-annual	Quarterly	Monthly
		Sampling frequency			
In-the-money	64	3.51	2.87	2.73	2.72
	128	2.36	2.23	1.98	1.65
	256	1.70	1.61	1.31	1.16
	512	1.19	1.10	0.94	0.89
	1,024	0.81	0.68	0.62	0.59
	2,048	0.52	0.42	0.40	0.38
	4,096	0.33	0.30	0.27	0.29
	8,192	0.23	0.20	0.21	0.24
At-the-money	64	2.70	2.27	2.08	2.02
	128	1.83	1.74	1.50	1.21
	256	1.31	1.27	0.96	0.87
	512	0.92	0.88	0.68	0.67
	1,024	0.67	0.53	0.47	0.44
	2,048	0.42	0.32	0.31	0.30
	4,096	0.26	0.23	0.22	0.22
	8,192	0.19	0.16	0.16	0.18
Out-of-the-money	64	1.34	1.23	1.14	1.05
	128	0.96	0.91	0.82	0.66
	256	0.69	0.65	0.52	0.48
	512	0.47	0.44	0.36	0.35
	1,024	0.34	0.28	0.26	0.24
	2,048	0.21	0.17	0.17	0.15
	4,096	0.15	0.13	0.12	0.11
	8,192	0.11	0.09	0.09	0.09

Note: Figures in last four columns are in dollars.

but still significant, improvement for problems of high dimension or high sampling frequency.

We have found that using quasi-random sequences also compares favourably with other variance reduction techniques. The antithetic variate method (see Boyle, 1977) is a very simple method which gives a lower error rate than regular simulation. Errors are reduced by a factor which, unlike quasi-random sampling, does not depend on the sampling frequency but does depend on the strike price. In general, the antithetic method works better for in-the-money options than for out-of-the-money options. For the examples given here, the antithetic method is roughly 5.5 times as efficient as regular simulation for in-the-money options, but this efficiency factor drops to only about 1.6 and 1.2 for the at-the-money and out-of-the-money options, respectively. Corresponding efficiency factors for worst-case Sobol′ sequence results ($M = 48$) for $N = 8{,}192$ are roughly 50, 14 and five.

Another variance reduction method is that of Latin hypercubes. A description of this method is given in Press *et al.* (1992), and Curran (1994) – also Chapter 7 of this publication uses this method to price options on the arithmetic mean of stock prices. For the options considered here, this method is about twice as efficient as the antithetic method, and so is also less efficient than the current approach. We have also looked at the efficiency of recursive stratified sampling (see Press *et al.*, 1992), but this turns out to be less efficient than regular Monte Carlo simulation in most cases.

It is worth noting that the current method focuses on the way in which sample paths of random variables are drawn. Using quasi-random sequences does not preclude the use of certain other variance reduction techniques, such as the control variate or stratified sampling. In some problems, the combination of quasi-random sequences with another variance reduction technique could provide a further increase in accuracy.

BIBLIOGRAPHY

Boyle, P.P., 1977, "Options: A Monte Carlo Approach", *Journal of Financial Economics* 4, pp. 323–38 (Chapter 1 of this publication).

Curran, M., 1994, "Strata Gems", *Risk* 7(3), 1994, pp. 70–1 (Chapter 7 of this publication).

Press, W.H., S.A. Teukolsky, W.T. Vetterling and B.P. Flannery, 1992, *Numerical Recipes in C: The Art of Scientific Computing*, 2nd edition, Cambridge University Press.

30

Laudable Lattices

Craig Stetson, Steve Marshall and David Loebell
Arizona Public Service; Chase Manhattan; Chase Manhattan

Analytic techniques for pricing path-dependent options require the assumption of a continuous-time price process with constant parameters. This assumption can prove problematic for longer-maturity options, where the path-dependency involves discrete, unevenly spaced reset points.

An alternative approach is to simulate the path of asset prices numerically. This paper illustrates how the use of lattice methods, a relatively recent innovation in numerical integration theory, can be applied to this problem. Lattice methods offer better rates of convergence than classical methods and compare favourably with other quasi-random Monte Carlo techniques. The technique suggested here is also fast and easy to implement.

Discrepancy and Monte Carlo

The classical Monte Carlo method or "method of statistical trials" is an integration model based on random sampling. It estimates the integral, I(f), of a function, f, over some subset, E, of the N-dimensional space of real numbers, R^N, by evaluating f at a point randomly sampled from E. The set of N points, x_i, in the sample is denoted $\{x_n\}$; the estimate of the integral, Q(f), is then taken to be the sample mean:

$$Q(f) = \frac{1}{N}\sum_{n=1}^{N} f(x_n) \approx \int_E f(t)\,dt = I(f) \qquad (1)$$

The error of this method is statistical and proportional to the square root of N. A random number generator is used to create the components of the sample vector points $\{x_n\}$, but such generators are not efficient at filling the subspace E.

However, as researchers soon realised, $\{x_n\}$ does not have to be chosen randomly. It is possible to use a more deterministic scheme, but such a scheme must account for sample size, N, the dimensionality of the problem, s, and possibly also the nature of the function f itself. Several methods for finding optimal points $\{x_n\}$ have been suggested.

By making suitable transformations, we can restrict the problem to functions over the unit cube in s dimensions. We then seek "low discrepancy" as our criterion for the cube to be filled efficiently, since discrepancy is a measure of the uniformity of the distribution of points.

Discrepancy also contributes to the error in quasi-Monte Carlo integration. This error is bounded by the product of a function's bounded variation, V(f), a measure of the function's "niceness" over the area of integration, and the discrepancy of the point set. Practitioners need to consider both of these.

It has been shown that there exist sets of points $\{x_n\}$ whose discrepancy converges to zero. Integration error converges to zero at a faster rate than classical methods. Low-discrepancy points can be described as either "open" or "closed" sequences, the difference being that closed sequences are optimised for a given sample size, while open sequences are not. Lattice rules are closed sequences, as are the well-known rectangular rule method and the Faure sequence (1982).

Some empirical evidence suggests that closed methods perform better than even the best of open sequences (Bratley and Fox, 1988). This is especially true for higher dimensionalities.

Lattice methods

A lattice may be described as a set of points in the unit cube generated by some deterministic scheme. A well-known lattice rule in two dimensions is the product rectangle rule, written with N^2 points as:

$$Q(f) = \left(1/N^2\right)\sum_{j=1}^{N}\sum_{k=1}^{N} f\left(\frac{j}{N}\frac{k}{N}\right) \qquad (2)$$

This rule requires two summations and is therefore described as a rank-two lattice. It is not practical for higher dimensions since it requires N^s computations. So we propose using a more gen-

PANEL 1

PRICING PATH-DEPENDENT DERIVATIVE SECURITIES

To use lattice methods to price path-dependent securities, we suggest the following method:

1) Construct an L(a) lattice using equations (3) and (4) for an arbitrary sample size 4r where r is prime.
2) Calculate the implied forward drift and forward volatility from the observed forward and volatility curve.
3) Using a normal deviate transformation, map the lattice points generated by step 1 into Gaussian increments.
4) Using the forward drift and volatility for each time step, as calculated in step 2, construct a set of price paths. Each price path has dimension s points. This defines a two-dimensional array (4r by s) of asset prices.
5) Using the array defined in step 4, find the discounted payout value at each reset point and add up the coupons as you iterate to s dimensions. Do this 4r times, until you find the sum of all the path payouts, and divide by 4r to get the discounted expectation value. First try a path-dependent security with a known closed-form solution (such as a knock-out option with constant drift and volatility over an s-day period). Adjust the value of a until comparison of the numeric and analytic solutions shows that the error in integration has been minimised.
6) Now perform step 5 for the security you wish to price, using the appropriate forward and volatility curves. Variance reduction techniques, including antithetic deviates, can be used to reduce any bias accumulating from discretisation or convexity of the integrand.

eral rank-two lattice rule that has been investigated by Sloan and Walsh (1990). This creates a one-parameter family of lattices denoted L(a), where a can be described as a seed for a generated lattice.

$$Q(f) = \left(1/n^2 r\right)\sum_{j=1}^{r}\sum_{k1=1}^{n}\sum_{k2=1}^{n} f\left\{\left(j\frac{z}{r} + k1\frac{y_1}{n} + k2\frac{y_2}{n}\right)\right\} \qquad (3)$$

We use n = 2 and a sample size $N = n^2r$, where n and r are relatively prime. The vectors z (a Korobov vector), y_1 and y_2 (independent shifting vectors) are defined as:

$$z = \left(1, a, a^2, \ldots, a^{s-1}\right)(\text{mod } r) \quad 1 \le a \le r/2$$
$$y_1 = (1,0,0,\ldots,0)$$
$$y_2 = (0,1,0,\ldots,0) \qquad (4)$$

The {} brackets used in equation (3) indicate

1. A poorly spaced lattice

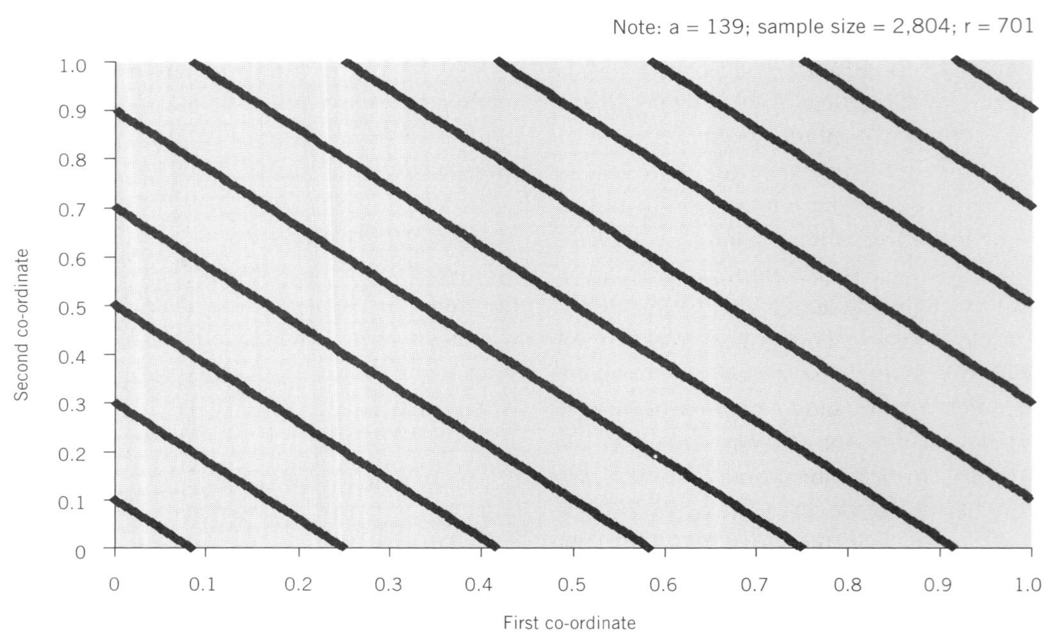

2. An equidistributed lattice

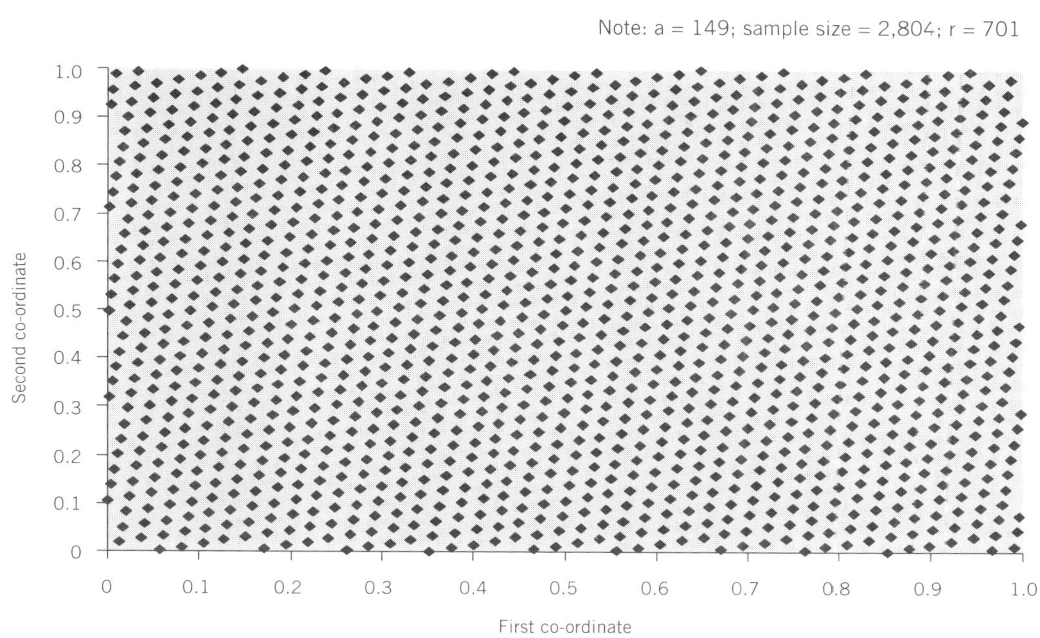

the fractional part of the number, ie, modulo 1. Since the one-parameter family of lattices L(a) can be written as double summation after suitable transformations, it is described as a rank-two rule. Sloan and Walsh (1990) showed that lattice rules of the type L(a) are often optimal for multi-dimensional integration on the unit cube.

We must now optimise a for a given integration problem. This is achieved by a search for a which minimises the integration error $|Q(f) - I(f)|$ but does not neglect it altogether.

Figures 1 and 2 show two of the dimensions of two different lattices with different choices of a. They show that the choice of a has a significant effect on discrepancy. The regularity of the points in Figure 2 can be compared with uneven dispersion of points generated by classical methods. This regularity (or low discrepancy) ensures that the integrand is sampled throughout the area of integration.

At this point we depart from the existing lattice theory by considering the transformation from the unit cube to R^S. It is usual in integration theory to use the inverse cumulative normal to map from U^2 to R^2 and perhaps a polynomial transformation to make the integrand periodic. This requires the probability function to be used when performing the integration, and consequently the payout functions must be programmed in terms of the transformed variables.

However, we dispensed with the probability density requirement and used instead normal deviates that are a function of the uniform point. One can use a Box–Muller or inverse cumulative normal transformation to map points from U^S to R^S. Care should be taken with the cumulative inverse transformation in the neighbourhood of points 0 and 1.

Despite this theoretical departure, we find that our method performs well. Transformed normal variates are illustrated in Figure 3.

3. Uniform points after transformation: N(0,1) space

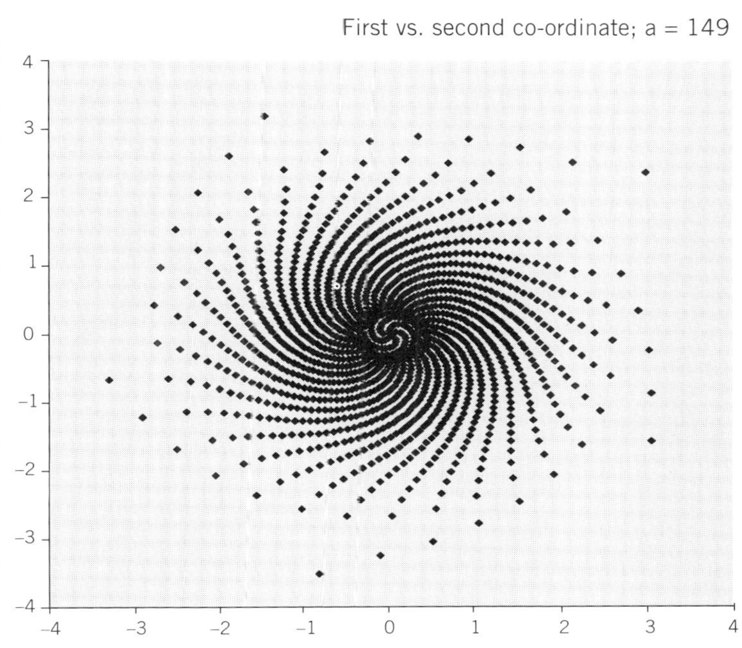

Comparison of methods

Both classical Monte Carlo and lattice methods can be used to generate uniform points that can be transformed into normal deviates in R^s. Each point is a vector whose components are identically and independently distributed normal variables with mean 0 and variance 1. Therefore we are not restricted to a particular stochastic process as long as the process uses Gaussian increments.

To test the application of lattice methods, we used a geometric Brownian motion process which is consistent with the assumptions of the Black-Scholes model.

Our test used increasing forwards and decreasing volatility curves for time steps equally spaced at 31 days apart for 10 dimensions. We evaluated European call prices at each dimension for four different strikes for four different samples.

In each case, the set of points generated by our lattice had a significantly lower level of error for each strike than the classical Monte Carlo method with antithetics (see Table 1). Points generated from both classical and lattice methods show higher levels of error when the options are far out of the money.

This suggests that this method should be used with caution when pricing derivatives that are conditional on small-probability events. The use of control variate or stratified sampling techniques can help with this problem.

Table 1. Percentage error of two types of Monte Carlo

Sample size (N)	r	Lattice	Classical Monte Carlo
10,012	2,503	0.000910%	0.058280%
5,036	1,259	0.003266%	0.191630%
916	229	0.016420%	0.358628%
508	127	0.092029%	0.382056%

Lattice methods appear to provide a useful tool for pricing derivative securities with discrete reset points. Lattice points share many of the advantages of quasi-random sequences but have the advantage that they can be optimised for the function in hand. Moreover, since the lattice $L(a)$ is generated from the single parameter a, the optimisation procedure is a simple search over points in the set $(1, r/2)$.

The above methods can be extended quite easily to multi-asset, path-dependent securities. Other areas of possible application include the use of lattice points in the solution of partial differential equations (Morokoff and Caflisch, 1993) and in the optimisation of multivariate functions (Niederreiter, 1992).

BIBLIOGRAPHY

Bratley, P., and B.L. Fox, 1988, "Implementing Sobol's Quasi-random Sequence Generator", *ACM Transactions on Mathematical Software* 14(1), pp. 91-8.

Faure, H., 1982, "Discrépance de suites associées à un système de numération (en dimensions s)", *ACTA Arithmetica* XLI, pp. 337-51.

Morokoff, W., and R. Caflisch, 1993, "A Quasi-Monte Carlo Approach to Particle Simulation of the Heat Equation", *SIAM Journal of Numerical Analysis* 30(6), pp. 1,558-73.

Niederreiter, H., 1992, *Random Number Generation and Quasi-Monte Carlo Methods,* Philadelphia, Pennsylvania SIAM.

Sloan, I.H., and P.J. Kachoyan, 1987, "Lattice Methods for Multiple Integration: Theory, Error Analysis and Examples", SIAM *Journal of Numerical Analysis* 24(1), pp. 116-28.

Sloan, I.H., and L. Walsh, 1990, "A Computer Search of Rank Two Lattice Rules for Multidimensional Quadrature", *Mathematics of Computation* 54(189), pp. 281-302.

31

Inelegant Efficiency

John W. Barrett, Gerald Moore and Paul Wilmott[1]
Imperial College; Imperial College; Imperial College and University of Oxford

A large literature is devoted to formulae for solving exotic option valuation problems. These formulae are usually the easiest and quickest route to a numerical answer to a specific problem. However, for several types of option for which explicit solutions exist it is still not clear what is the best practical numerical method of valuation. In this context, "practical" means easy to program, applicable to many problems and quick to generate results.

Indeed, the explicit solution can be something of a red herring; the most efficient way to compute the solution is definitely not to derive an elegant formula.

The options in question are of the European multi-asset or multi-factor type - for example, an option to exchange one asset for another or to purchase the highest-valued of a basket of assets. Such options are usually valued by taking the present value of the expected future payoff in a risk-neutral environment.[2] The initial assumption is often that each factor follows a normal/lognormal random walk. In this case simple representations exist for the fair value of some of these options in terms of sums of the cumulative distribution function for standardised multivariate normal distributions; for the single-asset option this reduces to the Black–Scholes formula.[3]

So far, so good. But how does one then calculate some real numbers? Tables, polynomial approximations and computer software exist for evaluating these multivariate functions but usually only for the bivariate normal distribution. And if one has n underlying assets, then one generally needs the n-variate normal distribution. So the first problem is numerical. The reason why such special functions are hard to evaluate will be appreciated when one considers the rapid growth in computational time as the number of underlying assets increases; this will be the main topic of this article.

A second problem is that each option with a different payoff at expiry has a different formula when written in terms of special functions. There are also the accompanying formulae for the delta, gamma, theta, vega … . It would be convenient if a single approach would unify many multi-factor option problems, requiring only minor changes to a program sub-routine to apply it to different payoffs. Such an approach will be outlined here, based on efficient numerical multiple integration, which is robust enough to cope with many valuation problems.

For straightforward option valuation problems the binomial method is popular. It is easy to program - it mimics the random walk - but not particularly efficient. The method we are advocating is as simple to program as the binomial method, but with the further advantage of being numerically sophisticated.

Let us consider a European[4] option whose payoff depends on n assets with prices S_i, $i = 1, \ldots, n$. The starting point is the random walk

$$dS_i = \mu_i S_i dt + \sigma_i S_i dZ_i$$

where σ_i is the annualised standard deviation of the ith asset and the dZ_i are Wiener processes. Suppose at expiry, time $t = T$, that the payoff is given by $G(S'_1, \ldots, S'_n)$ (where S' denotes a value at expiry), then the fair value, V, of the option (assuming risk-neutrality) will be

$$V = e^{-r(T-t)}(2\pi(T-t))^{-r/2}(\mathrm{Det}\,\Sigma)^{-1/2}(\sigma_1 \ldots \sigma_n)^{-1}$$

$$\times \int_0^{\infty} \ldots \int_0^{\infty} \frac{G(S'_1, \ldots, S'_n)}{S'_1 \ldots S'_n} \exp\left(-\tfrac{1}{2}\, \underline{\alpha}^T \Sigma^{-1} \underline{\alpha}\right) dS'_1 \ldots dS'_n \quad (1)$$

where

$$\alpha_i = \frac{1}{\sigma_i (T-t)^{1/2}} \times \left(\log\left(\frac{S'_i}{S_i}\right) - \left(r - \frac{\sigma_i^2}{2}\right)(T-t) \right),$$

r is the risk-free interest rate and Σ the correlation matrix for the n assets. Deltas, etc., can then be calculated by differentiating this expression.

We will recommend numerically sophisticated ways in which (1) is calculated directly. This is a better approach than rewriting (1) in terms of supposedly known functions. Our final recommended integration method is shown in formula (2) below. We conclude with a computer program (written in Fortran) which illustrates how simple the technique is to program.

The difficulties in evaluating the required integral numerically may be grouped under three headings:

❑ *The infinite region of integration* Since the integrand does not decay exponentially, standard integration methods for infinite regions are not appropriate, nor is the simple replacement of the infinite region with a large finite region. We recommend mapping the infinite region on to 2^n unit hypercubes.

❑ *The lack of smoothness of the integrand* This lack of smoothness means that high-order Gauss-type quadrature rules, or Simpson's rule and Boole's rule,[5] are not appropriate because, if our aim is to produce an algorithm for a general payoff function G, we must accept that the zones in which the integrand are rough/smooth are unknown. Thus we do not consider the use of higher-order methods beyond the simple trapezoidal or mid-point rule.

❑ *The dimensionality of the integral* This is the dominant feature of our problem because the number of integrand evaluations required by standard quadrature rules increases exponentially.

Combining these considerations, we can determine the dominant term in the work required to approximate our integral. Applying the product mid-point or trapezoidal rule in n dimensions with q quadrature points in each direction, we find that the number of "flops" required is

$$\frac{(2q)^n n^2}{2}.$$

Table 1. Computation times, q = 20

	n = 4	n = 5	n = 6
IBM AT clone	17 mins	18 hours	1,000 hours
Sun workstation	24 secs	24 mins	20 hours
IBM 6000	2 secs	2 mins	2 hours

Table 2

Exercise price	Q = 400	Q = 1,600	Q = 6,400	Q = 25,600	Boyle
35	9.3637	9.4158	9.4197	9.4198	9.420
40	5.4556	5.4891	5.4878	5.4879	5.488
45	2.7921	2.7964	2.7949	2.7950	2.795
Time (secs)	–	0.13	0.53	2.2	

PANEL 1

AVERAGED HASELGROVE

Calculates the model example of a five-dimensional integral over the unit hypercube.

```
IMPLICIT DOUBLE PRECISION (A-H,O-Z)

T1 = 0.95734608D0
T2 = 0.86730270D0
T3 = 0.09724025D0
T4 = 0.31301950D0
T5 = 0.48476582D0

WRITE (6, 1000)
FORMAT ('IQ = ?')
READ (5,*) IQ

SUM1 = G(0.0D0, 0.0D0, 0.0D0,
0.0D0, 0.0D0)
SUM2 = SUM1

DO 2 J = 1, IQ
X = DFLOAT(J)/2.0D0
X1 = 2.0D0*DABS(X*T1 - DNINT(X*T1)
X2 = 2.0D0*DABS(X*T2 - DNINT(X*T2)
X3 = 2.0D0*DABS(X*T3 - DNINT(X*T3)
X4 = 2.0D0*DABS(X*T4 - DNINT(X*T4)
X5 = 2.0D0*DABS(X*T5 - DNINT(X*T5))
SUM1 = SUM1 + 2.0D0*G
(X1, X2, X3, X4, X5)
SUM2 = SUM2 + SUM1
CONTINUE

QAPP = SUM2/((IQ + 1)**2)
QTRUE = 0.97065719D0
ERROR = DABS(QTRUE - QAPP)
WRITE (6, 2000) IQ, QAPP, ERROR
FORMAT ('IQ =', I6, 'QAPP =',
F10.8, 'ERROR =', F10.8)
STOP
END

FUNCTION G(X1, X2, X3, X4, X5)
IMPLICIT DOUBLE PRECISION (A-H, O-
Z)
G = DEXP(-X1*X2*X3*X4*X5)
RETURN
END
```

Of course, this number has no significance until we know how fast a flop is on a particular machine. For purposes of comparison we use the estimates IBM AT clone, 2×10^4, Sun workstation, 10^6, and IBM 6000, 10^7 flops a second in Table 1 to obtain representative timings for $q = 20$ and increasing dimension.

This table shows the results of rapidly advancing technology. Computations that would have taken a significant time even one year ago have suddenly become almost trivial.

To avoid the "curse of dimensionality" associated with tensor product quadrature rules, we recommend approximating the general integral

$$\int_0^1 \cdots \int_0^1 g(x_1,\ldots,x_n)\, dx_1 \ldots dx_n$$

by the formula

$$\frac{1}{(Q+1)^2} \sum_{m=1}^{Q} \left\{ g(0,\ldots,0) + 2\sum_{i=1}^{m} g \right.$$
$$\left. \times \left(2|\langle \tfrac{1}{2} i\eta_1 \rangle|,\ldots,2|\langle \tfrac{1}{2} i\eta_n \rangle|\right) \right\}$$

which performed best in our numerical tests and is far superior to the classical Monte Carlo method. Here $\langle\Theta\rangle$ denotes the fractional part, lying in the range $(-\frac{1}{2}, \frac{1}{2}]$ of a real number. Q is chosen by the user; increasing it improves the accuracy of the formula. This method is based on the Fourier-series coefficients of g and the η_i are chosen so that

$$\sum_{j=1}^{n} m_j \eta_j \text{ (with } m_1,\ldots,m_n \text{ arbitrary integers)}$$

is never an integer and should, in fact, stay as far away from integers as possible. The actual computation of η_i must be performed by an optimisation routine, but once obtained these can be stored and used for any n-dimensional integration. Suggested values for $\eta_1,\ldots,\eta_n$ are given in Haselgrove[6] for up to eight dimensions. One of the key advantages of Haselgrove's method is that increasing the number of quadrature points just means adding on a relatively small number of new function evaluations.

To illustrate the simplicity of this quadrature rule, a Fortran program for the model problem

$$g \equiv -\exp(-x_1 \ldots x_5)$$

is shown in Panel 1.

We conclude with a more practical two-dimensional example. Table 2 shows the results of applying (2) with $\eta_1 = 0.62055505$ and $\eta_2 = 0.22610245$ (as suggested by Haselgrove) to the example in Boyle.[7]

This example is of a European call option to purchase the maximum of two assets. The current value of each asset is 40 and $\sigma_1 = 0.2$, $\sigma_2 = 0.3$, ρ (the correlation coefficient) $= 0.5$, $r = 0.04879$ and $T = 0.58333$ years. In the table we show values of this option for three different exercise prices (35, 40 and 45) and for different values of Q together with the time taken on an IBM RS 6000 Unix workstation. In the final column, for comparison, are the values derived by Boyle. (The numbers in the $Q = 25{,}600$ column are accurate to four decimal places and therefore more accurate than Boyle's.)

This example is in two dimensions only. However, the method really comes into its own in higher dimensions: the simplicity of programming carries over and the sampling points become more effective compared with other methods.

Take a look at the Fortran program and write a similar code for an option problem. We hope that this will convince you that numerical integration need not be painful!

1 *The authors acknowledge the support of the directors of Adam, Harding & Lueck and Dr Simon Babbs of Midland Montagu, without whom this work would not have been possible.*

2 *See, for example, Johnson, H., 1987, "Options on the Maximum or Minimum of Several Assets",* Journal of Financial and Quantitative Analysis *22, pp. 277–83.*

3 *Black, F. and M. Scholes, 1973, "The Pricing of Options and Corporate Liabilities",* Journal of Political Economics *81, pp. 637–59.*

4 *We refer to European-style options. Recall, however, that there are analytical solutions to American option valuation problems which rely on the computation of high-order multivariate integrals, most notably that of Geske and Johnson (Geske, R. and H. Johnson, 1984, "The American Put Valued Analytically",* Journal of Finance *39, pp. 1511–24). It is therefore possible that the method in this article could find employment there.*

5 *As in Schervish, M.J., 1985, "Multivariate Normal Probabilities with Error Bound",* Applied Statistics *34, pp. 81–94.*

6 *Haselgrove, C.B., 1961, "A Method for Numerical Integration",* Mathematics of Computation *15, pp. 323–37.*

7 *Boyle, P.P., 1988, "A Lattice Framework for Option Pricing with Two State Variables",* Journal of Financial and Quantitative Analysis *23, pp. 1–12.*

GLOSSARY

This brief introduction to Monte Carlo lingo should help you survive the simulation jungle. Conventions may not be universal, but we have tried to follow the most common usage.

American Monte Carlo Twist of the standard Monte Carlo method to price American-style options. To account for early exercise, paths are bundled at each time step instead of being generated one by one to compute the final payout

Antithetic A simple way to ensure that the empirical mean of Gaussian samples is 0 to pair each sample with its' symmetric (antithetic) point with respect to 0. However, as those Gaussian samples are usually transformed in a non-linear fashion, the estimation is not perfect.

Balance sampling Structured way to generate random samples. When applied to path generation, ensures that at each time step the increments sample the Gaussian distribution correctly.

Brownian bridge Brownian motion conditioned by its terminal value.

Brownian motion The basic process of stochastic calculus. Continuous time stationary process with independent Gaussian increments. Usually (but not always, *see* "spectral truncation") simulated by first discretising time and then generating Gaussian increments over each time step.

Calibration Process by which the parameters of a model are adjusted to match a set of market prices.

Central Limit Theorem (CLT) The core mathematical principle of Monte Carlo. It states that the average of iid samples (Monte Carlo draws) converges to the expectation and follows approximately a Gaussian with vanishing standard deviation (order square root of the number of samples).

Control variate Easily computable quantity closely related to the target we try to estimate. Running Monte Carlo simulation on the target minus the control variate sharply reduces the variance if the match is close. Typical example: geometric average as control variate of arithmetic average.

Discrepancy Measure of non-uniformity of a set of points, or maximum error on a set of digital options prices.

Importance sampling Generates samples according to another probability measure. It allows for over-representing some regions, which requires payout renormalisation. A typical example is adding a drift to price out-of-the-money calls.

Latin hypercube Multi-dimensional sampling which ensures that each projection is properly sampled. Corresponds to balance sampling when applied to path generation.

Low-discrepancy sequence Deterministically generated sequence of points designed to avoid gaps and clusters.

Monte Carlo Termed after Monaco casinos. Method to compute an expectation (or an integral) by drawing random samples. Most common method to price path-dependent options.

Principal component analysis (PCA) Identifies the main directions of a set of correlated random variables by diagonalising their covariance matrix. Important tool to reduce the dimension of a problem. Usually applied to study yield curve evolution.

Risk-neutral valuation Basic valuation principle of modern finance. Thanks to a hedging argument, states that contingent claims can be priced as a discounted expectation under a proper "risk-neutral" measure. This expectation can be estimated by Monte Carlo simulations.

Stochastic differential equation (SDE) Models the evolution of financial quantities through time as a deterministic drift altered by a random noise whose amplitude is the volatility. To generate a sample path, one first generates a Brownian motion path and then transforms it through the SDE dynamics.

Smile (Also termed "skew") Describes the strike dependency of implied volatility and is a departure from the Black–Scholes model. May affect exotic option prices substantially.

Sobol' The most popular low discrepancy sequence; it proceeds by bit permutation.

Spectral truncation Alternative way to generate Brownian paths by adding harmonics coming from a principal component analysis (PCA), as opposed to the classical step-by-step construction.

Stratified sampling Partitions the event space into subsets and samples each subset to ensure well-disseminated draws.

Value-at-risk (VAR) Measure of the potential loss of a portfolio. Important for capital allocation and regulatory requirements.

Variation The variation of a function describes its ability to be approximated by a combination of characteristic functions of intervals. It determines the quality of the approximation of its' integral when using low discrepancy sequences.

INDEX